Mr. M.P. Kabra's

A COMPLETE DICTIONARY OF PHARMACY

Edition-I

Pharmamedix India Publication Pvt. Ltd.

India, Bangladesh, China, USA, UK.

Published by:

Pharmamedix India™

Pharmamedix India Publication Pvt. Ltd.

Registered Office:

220, Bhamasi, Churu (Rajasthan), 331001.

Website: www.pharmamedix.in

Online Available on: www.pmip.pharmamedix.in

Edition-I

Published on: Nov, 2013.

ISBN: 978-81-928038-6-9.

Trademark Notice

Product or corporate names may be trademarks or registered trademarks, and are used only for identification and explanation without intent to infringe.

Any brand names and product names mentioned in this book are subjected to trademark, brand or patent protection and are trademark or registered trademark of their respective holders. The publisher is not associated with any product or vendor mentioned in this book. The use of brand names, product name common names, trade name, product description etc. Even without a particular marking in this works is in no way to be construed to mean that such names may be regarded as unrestricted in respect of trademark and brand protection legislation and could thus be used by anyone. Designations used by companies to distinguish their products are often claimed as trademarks.

This publication is designed to provide accurate and authoritative information in regard to the subject matter covered. It is sold on the understanding that the publisher is not engaged in rendering professional services. If professional advice or other expert assistance is required, the services of a competent professional should be sought.

The use of general descriptive names, registered names, trademarks, etc. in this publication does not imply, even in the absence of a specific statement, that such names are exempt from the relevant protective laws and regulations and therefore free for general use.

Pharmamedix India also publishes its books in a variety of electronic formats. Some content that appears in print may not be available in electronic books.

Product liability: *This book has been published in good faith that the material provided by author is original Every effort is made to ensure accuracy of material, but the publisher, printer and author will not be held responsible for any inadvertent error's. In case of any dispute, all legal matters to be settled under Churu (Rajasthan) jurisdiction only.*

A COMPLETE DICTIONARY OF PHARMACY

Mr. M.P. KABRA

Assistant Professor & HOD
Department of Pharmacology
Kota College of Pharmacy, Kota.
Rajasthan, India.

PREFACE

A COMPLETE DICTIONARY OF PHARMACY is an ideal, reliable, authentic and comprehensive dictionary of pharmacy, which present pharmacy profession language, terms, definitions, short forms etc. The main goal of this A COMPLETE DICTIONARY OF PHARMACY is to provide correct, clear, concise meaning and definition of all terms used in pharmacy profession.

In this dictionary author's provide simplest meanings of scientific terms used in pharmacy and also provide best knowledge of all basic terms used in clinical research with latest update.

-Mr. M.P. KABRA

ABOUT THE AUTHOR

Mr. M.P. Kabra (M. Pharm.- Pharmacology) is a well renowned lecturer of Pharmacy, He is working as assistant professor & H.O.D. Dept. of pharmacology Kota college of Pharmacy, Kota (Raj.). He has presented his research work in various national conferences. He has attended many national seminars, workshops. He has published more than 50 Research/ Review articles in national and international journal. He is a managering editor of Asian journal of pharmaceutical research and development and member of editorial board of many national and international journals. He has published a book of "DRUG CONTROL OFFICER (D.I.) RECRUITMENT EXAM.

A

AAC (Actual Acquisition Cost): Pharma care pays the actual drug cost as indicated by the dispensing pharmacy. AAC is adjusted to reflect the true cost to the pharmacy and is net of any cash discounts, volume discounts, rebates or performance allowances.

Abiotic Transformation: Process in which a substance in the environment is modified by non-biological mechanisms.

Abiotic: Not associated with living organisms.

Ablate: Destroy.

ABPI: Association of the British Pharmaceutical Industry.

Abrasion: Abrasion means the rubbing-off of the surface of the skin or of a mucous membrane due to some mechanical injury. Such injuries, though slight in themselves, are apt to allow the entrance of dirt containing organisms, and so to lead to an abscess or some more severe form of inflammation.

Abscess: Absolute lethal concentration (LC100): Lowest concentration of a substance in an environmental medium which kills 100 % of test organism or species under defined conditions. This value is dependent on the number of organisms used in its assessment.

Absolute Lethal Dose (LD100): Lowest amount of a substance which kills 100% of test animals under defined conditions. This value is dependent on the number of organisms used in its assessment.

Absorbed Dose: Energy imparted to matter in a suitably small element of volume by ionizing radiations divided by the mass of that element of volume. The SI unit for absorbed dose is joule per kilogram (j kg-1) and its special name is gray (GY).

Absorption Coefficient: Ratio of the absorbed amount (uptake) of a substance to the administered amount (intake): for exposure by way of the respiratory tract, the coefficient is the ratio of the absorbed amount to the amount of the substance (usually particles) deposited (adsorbed) in the lungs.

Absorption: Process of active or passive transport of a substance into an organism: in the case of mammal or human being, this is usually through the lungs, gastrointestinal tract, or skin.

Absorption: The process by which medications reach the blood stream when administered other than intravenously, for example, through nasal membranes. See also ADME (pharmacokinetics).

Abuse: Improper use of drugs or other substances. RT "glue sniffing", solvent abuse, "solvent sniffing".

Acaricide: Substance intended to kill mites, ticks or other acaridae.

Accelerated Approval: Intended to speed the clinical development and regulatory review of important new medicines through the use of surrogate endpoints to predict the clinical benefit of the treatment.

Acceptable Daily Intake (ADI): Estimate of the amounts of a substance in food or drinking water, expressed on a body mass basis (usually mg/kg body weight), which can be ingested daily over a lifetime by humans without appreciable health risk. For calculation of the daily intake per person, a standard body mass of 60 kg is used. ADI is normally used for food additives (tolerable daily intake is used for contaminants).

Acceptable Residue Level of an Antibiotic: Acceptable concentration of a residue which has been established for an antibiotic found in human or animal foods.

Acceptable Risk: Probability of suffering disease or injury which is considered to be sufficiently small to be "negligible".

Access Record: A record showing who has accessed a patient's record in the previous fourteen months for purposes other than filling a prescription, as well as the date and time of each access.

Access: A patient's ability to obtain medical care. The ease of access is determined by components such as the availability to the patient's, availability of insurance, the location of health care facilities, transportation, hours of operation, affordability and cost of care.

Accidental Exposure: Unintended contact with a substance or change in the physical environment (including for example radiation) resulting from an accident.

Acclimatization: Processes, including selection and adaptation, by which a population of micro-organisms develops the ability to degrade a substance, or develops a tolerance to it.

Accountable Health Plans (AHP): Under the managed care act, providers and insurance companies would be encouraged (through tax incentives) to form AHPs, similar to HMOs, PPOs, and other group practice. Accountable health plans would compete on the basis of offering high-quality, low cost care and would offer insurance and health care as a single product.

Accreditation: Approval by an authorizing agency for institutions and programs that meet or exceed a set of predetermined standards.

Accredited: To meet the standards set by a non-governmental, state or national peer group.

Accrete: The addition of new enrollee to a health plan usually used in reference to medicare.

Accumulation: Successive additions of a substance to a target organism, or organ, or to part of the environment, resulting in an increasing amount or concentration of the substance in the organism, organ, or environment.

Accuracy: The use of the word "accurate" –free of error in referring to a scientific observation or scientific method sometimes obscures the fact than even the best methods and observations are only relatively free form error.

ACE Inhibitors: Are drugs inhibiting ACE (angiotensin converting enzyme). An enzyme needed for the production of the peptide angiotensin II, which causes arteries to constrict and therefore raises the blood pressure.

Acetylcholine (Ach): A neurotransmitter which plays an important role in memory. It is used for control of sensory input signals and muscular control.

Acidosis: Pathological condition in which the hydrogen ion substance concentration of body fluids is above normal and hence the pH of blood falls below the reference interval.

Acitretin: A retinoid taken in pill form that is used in the treatment of psoriasis.

Acquisition: The acquiring by contract with appropriated funds for supplies or services (including construction) by and for the use of the federal government through purchase or lease, whether the supplies or services are already in existence or must be created, developed, demonstrated and evaluated.

Action Letter: An official communication from FDA to an NDA sponsors announcing an agency decision. See also approval letter, approvable letter, and not-approvable letter.

Action Level: Concentration of a substance in air, soil, water or other defined medium at which specified emergency counter-measures, such as the seizure and destruction of contaminated materials, evacuation of the local population or closing down the sources of population, are to be taken.

Active Ingredient: Ingredient responsible for a medicine's therapeutic action.

Activities of Daily Living (ADLs): Activities performed as part of a person's daily routine of self-care such as bathing, dressing, toileting and eating.

Actual Acquisition Cost (AAC): The net cost of a drug paid by a pharmacy.

Actual Charge: The amount a physician or other provider actually bills a patient for a particular medical services, procedure or supply in a specific instance.

Actuarial Analysis: The statistical calculations used to determine the managed care company's rates and premiums charged their customers based on projections of utilization and cost for a defined population.

Actuarial Cost of Coverage: The expected dollar value of a health plan's benefits.

Actuarial Soundness: The requirement that the development of capitation rates meet common actuarial principles and rules.

Actuary: A person in the insurance field who decides policy rates and conducts various other statistical studies.

Acute Care Bed Methodology: A formula used to determine hospital bed needs.

Acute Care: Hospital care given to patients who generally require a stay of up to seven days and that focuses on a physical or mental condition requiring immediate intervention and constant medical attention, equipment and personnel.

Acute Effect: Effect of short duration and occurring rapidly (usually in the first 24h or up to 14d) following a single does or exposure to a substance or radiation.

Acute Rehabilitation: Early rehabilitation phase as soon as medically stable. Primary emphasis is to provide intensive physical and cognitive restorative services in the early months following injury.

Acute Rejection: Tissue death usually occurring within a few months after transplant. See "rejection"

Acute Services: Medical and surgical treatment and care provided typically in hospitals.

Acute Toxicity Test: Experimental animal study to determine what adverse effects occur in a short time (usually up to 14d) after a single dose of a substance or after multiple doses given in up to 24h.

Acute Toxicity: Adverse effects occurring within a short time (usually up to 14d) after administration of a single dose (or exposure to a given concentration) of a test substance or after multiple doses (exposures), usually within 24h.

Acute: Short-term, in relation to exposure or effect.

Adaptation: Change in an organism, in response to changing conditions of the environment (specifically chemical), which takes place without any irreversible disruptions of the given biological system and without exceeding normal (homeostatic) capacities of its response.

Added Risk: Difference between the incidence of an adverse effect in a treated group (of organisms or a group of exposed humans) and a control group (of the same organisms or the spontaneous incidence in humans).

Addiction: Surrender and devotion to the regular use of a medicinal or pleasurable substance for the sake of relief, comfort, stimulation, or exhilaration which it affords; often with craving when the drug is absent.

Additive Effect: Consequence which follows exposure to two or more physico-chemical agents which acts jointly but do not interact: commonly, the total effect is the simple sum of the effects of separate exposure to the agents under the same conditions.

Adduct: New chemical species AB, each molecular entity of which is formed by direct combination of two separate molecular entities A and B in such a way that there is no change in connectivity of atoms within their moieties A and B.

ADE: Adverse drug event; adverse drug effect

Adenocarcinoma: Malignant tumor originating in glandular epithelium or forming recognizable glandular structures.

Adenoma: Benign tumor occurring in glandular epithelium or forming recognizable glandular structure.

Adipose Tissue: Adipose tissue, or fat, is a loose variety of fibrous tissue in the meshes of which lie cells, each of which is distended by several small drops, or one large drop, of fat. This tissue replaces fibrous tissue when the amount of food taken is in excess of the bodily requirements. Adipose tissue occurs as a layer beneath the skin and also around several internal organs.

Adjudicate: Decide about a claim based on a set of rules or polices. Refer to claims adjudication.

Adjudication Date: The date, as determined by the process, on which the claim, transaction or request being responded to has been adjudicated.

Adjudication Results: A delayed response by the TAP to the provider (and/ or pay or to TPA) identifying invoices that have been paid in full, partially paid and not paid and reasons for non-payment.

Adjusted Payment Rate (APR): The medicare capitated payment to risk-contract HMOs. For given plan, the APR is determined by adjusted county-level AAPCCs to reflect the relative risks of the plan's entrollees.

Adjustment: A modified or changed claim that was originally paid or allowed, at least in a part, by Wisconsin Medicaid or Senior Care.

Adjuvant Therapy: Therapy given after primary therapy to kill cancer cells that may have begun to spread. Adjuvant therapy may be radiation, chemotherapy, or hormone therapy and is often used when the primary therapy is surgery to remove the cancer.

Adjuvant: In pharmacology, a substance added to a drug to speed or increase the action of the main component.

ADME: Absorption, distribution, metabolism, and excretion (used to describe pharmacokinetic process)

Administration (of a Substance): Application of a known amount of a substance to an organism in a reproducible manner and by a defined route.

Administrative Costs Savings: Reductions in prospective actuarial cost of the health care services (pure premium) for administrative, marketing expenses and profit.

Administrative Costs: Costs related to activities such as utilization review, marketing, medical underwriting, commissions, premium collection, claims processing, insurer profit, quality assurance, and risk management for purposes of insurance.

Administrative Reform: Reducing paperwork though simplified universal forms or electronic filing and processing of claims.

Administrative Services Only (ASO): An arrangement in which a licensed insurer provides administrative services to an employer's health benefits plan (such as processing claims), but doesn't insure the risk of paying benefits to enrollees.

Administrative Services Organization: An entity which only provides administrative services (including claims adjudication, member services, and management information reporting).

Administrator: The person appointed by the court to manage one's estate when he or she dies without leaving a will. Administrators have the same duties as executors.

Admission Criteria: Basis for selecting target population for a clinical trial. Subjects must be screened to ensure that their characteristics match a list of admission criteria and that none of their characteristics match any single one of the exclusion criteria set up for the study. See also inclusion criteria.

Admissions/1000 (APT): The number of hospital admissions per1.000 health plan members during a period.

Adsorption: Enrichment (positive adsorption, or briefly adsorption) of one or more components in an interfacial layer.

Adult Day Care: A program of social and health-related services provided during the day in a community group setting.

Adult Foster Care (AFC): An elderly person's placement with another family when independent living is no longer possible, but nursing care is not necessary.

Adverse Drug Reaction (ADR): In the preapproval clinical experience with a new medicinal product or its new usages particularly as the therapeutic dose(s) may not be established: all noxious and unintended responses to a medicinal product related to any dose should be considered adverse drug reactions. The phrase "responses to a medicinal product" means that a causal relationship between a medicinal product and an adverse event is at least a reasonable possibility, i.e., the relationship cannot be ruled out. Regarding marketed medicinal products: a response to a drug which is noxious and unintended and which occurs at doses normally used in man for prophylaxis, diagnosis, or therapy of diseases or for modification of physiological function.

Adverse Effect: Change in morphology, physiology, growth, development or lifespan of an organism which results in impairment of functional capacity or impairment of capacity to compensate for additional stress or increase in susceptibility to the harmful effects of other environment influences.

Adverse Event (AE): Any untoward medical occurrence in a patient or clinical investigation subject administered a pharmaceutical product and which does not necessarily have a causal relationship with this treatment. An adverse event (AE) can therefore be any unintended sign (including an abnormal laboratory finding), symptom, or disease temporally associated with the use of a medicinal (investigational) product, whether or not related to the medicinal (investigational) product.

Aerobe: Organism which needs molecular oxygen for respiration and hence for growth and life.

Aerodynamic Diameter: Diameter of a spherical particle of unit density which has the same settling velocity in air as the particle in question.

Aerosol: Dispersion of liquid or solid material in a gas.

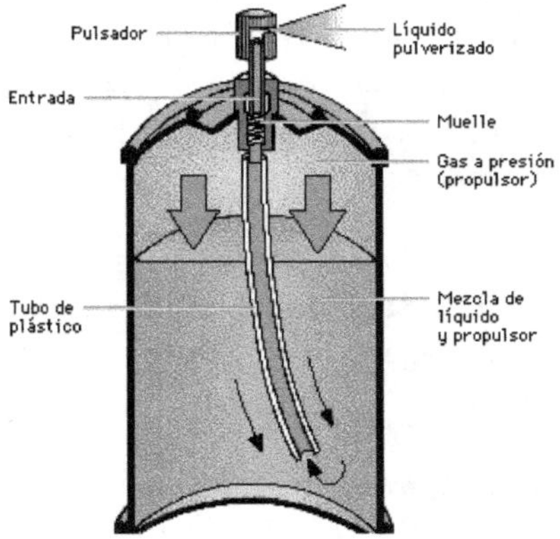

AERS: Adverse Event Reporting System (FDA).

Afferent Stage: First stage of acute rejection that begins when T cells are activated by antigen from the allograft.

Affinity: The equilibrium constant of the reversible reaction of a drug with a receptor to form a drug-receptor complex; the reciprocal of the dissociation constant of a drug-receptor complex.

Affirmative Procurement: Affirmative procurement is the purchase of supplies and services post-consumer recycled products. It also involves the purchase of supplies and services in a way that minimizes the amount of waste generated or supplies listed.

After-effect of a Poison: Ability of a poison to produce a change in an organism after cessation of contact.

Age Sensitivity: Quantitative and qualitative age dependence of an effect.

Age-related Macular Degeneration (AMD): AMD is a disease that progressively destroys the central portion of the retina, which is called the macula.

Agonist, Full: A full agonist is produces the largest maximal response of any known agonist that acts on the same receptor.

Agonist, Inverse: An inverse agonist is a ligand that by binding to a receptor reduces the fraction of receptor in an active conformational, thereby reducing basal activity form in the absence of a conventional agonist.

Agonist, Partial: A partial agonist is an agonist that produces a maximal response that is less than the maximal response produced by another agonist acting at the same receptors on the same tissue, as a result of lower intrinsic activity. See also Agonist, Full.

Agonist: A ligand that binds to a receptor and alters the receptor state resulting in a biological response.

AHCPR: Agency for Health Care Policy and Research.

AHFS Codes: American Hospital Formulary Services Codes American National Standards Institute.**AHRQ:** Agency for Healthcare Research and Quality.

Air Pollution Control System: Network of organization which monitor air pollution.

Air Pollution: Presence of substances in the atmosphere resulting either from human activity or natural processes.

Albuminuria: Presence of albumin, derived from plasma, in the urine.

Alcohol Dependence: Alcohol dependence, or alcoholism, is described under alcohol but a summary of the symptoms may be helpful in spotting the disorder. Behavioral symptoms vary but include furtiveness, aggression inappropriately generous gestures personality changes (selfishness, jealousy, irritability and outbursts of anger); empty promises to stop drinking; poor appetite; scruffy appearance; and long periods of drunkenness.

Alefacept (Amevive): An injectable biological drug used to treat moderate to severe plaque psoriasis.

Alerts: Screen pharmacy claims for clinically important potential drug therapy problems before the prescription is dispensed to the recipient.

Algicide: Substance intended to kill algae.

Algorithm: Step-by-step procedure for solving a mathematical problem; also used to describe step-by-step procedures for making a series of choices among alternative decisions to reach an outcome.

Aliquot: A part that is a definite fraction of a whole, as in aliquot samples for laboratory testing or analysis.

Alkalosis: Pathological condition in which the hydrogen ion substance concentration of body fluids is below normal and hence the PH of blood rises above the reference interval.

Alkylating Agent: Substance which introduces an alkyl substituent into a compound.

Allele: One of several alternate forms of a gene which occur at the same relative position (locus) on homologous chromosomes and which become separated during meiosis and can be recombined following fusion of gametes.

Allergen: Antigenic substance capable of producing immediate hypersensitivity.

Allergic Response: Some drugs may act as haptens or allergens in susceptible individuals; re-administration of the hapten to such an allergic response that may be sufficiently intense to call itself to the attention of the patient or the physician. The response may be so severe as to endanger the patient's life.

Allergy: Symptoms or signs occurring in sensitized individuals following exposure to a previously encountered substance (allergen) which would otherwise not cause such symptoms or signs in non-sensitized individuals.

Allograft: Tissue transplantation in which the donor and recipient are genetically similar but not genetically identical.

Allowable Charge: The maximum fee that a health plan will reimburse a provider for a given service.

Allowed Claim: A claim that has at least one service that is reimbursable.

Allowed Status: A Medicaid or Medicare claim that has at least one service that is reimbursable. Allows them to pass through the stomach and enter the intestine unaltered. Drugs coated in this way are those whose action is reduced or stopped by acid in the stomach.

Alopecia Areata: Alopecia areata is a common form of reversible hair loss which may be patchy, total on the scalp, eyebrows or eyelashes, or universal on the body. The onset is sudden at any age and the affected scalp-skin looks normal.

Alopecia: Baldness; absence or thinning of hair from areas of skin where it is usually present.

Alpha Error: Size of the likelihood acceptable to the investigators that a relationship observed between 2 variables is due to chance the probability of a Type I error.

Alports Syndrome: Hereditary disorder characterized by progressive nerve deafness, pyelonephritis or glomerulonephritis and occasionally ocular defects.

Alternative Birthing Center: A facility offering a "non-traditinol" (not like a hospital") setting for giving birth. While alternative birthing centers can range from free-standing centers to special areas within hospitals, birthing centers are generally known for a more comfortable, home-like atmosphere, allow more participation by the father and have more procedural flexibility than commonly found in hospital birth.

Alternative: Way of reducing adverse effects of hazardous materials (HAZMAT). Alternatives, as applied to HAZMAT decision-making, include, but are not limited to, substituting less hazardous material; redesigning a component to exclude use of HAZMAT in manufacture, use, or maintenance; modifying processes or procedures; restricting users; consumptive use; on-demand supply; direct ordering; extending shelf life; regenerating spent material; downgrading and reuse of spent material; use of waste as raw material in other manufacturing, and combinations of those factors.

Alveol/us (Pulmonary): Terminal air sac of the lung where gas exchange occurs.

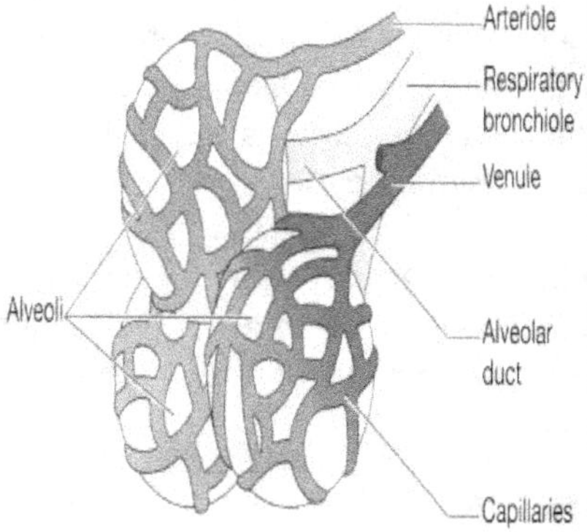

Alzheimer's Disease: Also called SDAT (senile dementia alzheimer's type). This disease is characterized by a general loss of intellectual ability and impairment of memory, judgment and abstract thinking, as well as changes in personality.

Ambient Monitoring: Continuous or repeated measurement of agents in the environment to evaluate ambient exposure and health risk by comparison with appropriate reference values based on knowledge of the probable relationship between exposure and resultant adverse health effects.

Ambient: Surrounding (applied to environmental media such as air, water, sediment or soil).

Ambulatory Care: Health care services provided without the patient being admitted to a facility. Also called outpatient care. No overnight stay in a hospital is required.

Ambulatory Surgery: Surgical procedures performed that do not require an overnight hospital stay.

AMCP: Academy of Managed care Pharmacy

American Medical Association (ADA): A professional association of dentists dedicated to serving the public and profession of dentistry.

American National Standards Institute (ANSI): Founded in 1918, ANSI itself does not develop standards. ANSI's roles include serving as the coordinator for U.S. voluntary standards efforts, acting as the approval body to recognize documents developed by other national organizations as American National Standards, acting as the U.S. representative in international and regional standards efforts, and serving as a clearinghouse for national and international standards development information.

Ames Test: In vitro test for mutagenicity using mutant strains of the bacterium salmonella typhimurium which cannot grow in a given histidine-deficient medium: mutagens can cause reverse mutations which enable the bacterium to grow on the medium.

Amino Acid: An organic acid containing an amine (ammonia like) chemical group. Amino acids are put together by your body in highly specific ways to manufacture proteins.

Amnion: Amnion is the tough fibrous membrane which lines the cavity of the womb during pregnancy, and contains between 0·5 and 1 litre (1–2 pints) of fluid in which the embryo floats. It is formed from the ovum (egg) along with the embryo, and in labour the part of it at the mouth of the womb forms the 'bag of waters'.

Amplification: The amount of change in measured output per unit change in input. The slope of the input –output, or dose-effect, curve.

Amyloid Plaque: A buildup of beta-amyloid protein. Amyloid plaques are one of the characteristic structural abnormalities found in the brain of Alzheimer structural abnormalities found in the brains of Alzheimer patients.

Amyloid: Similar to starch.

Amyloidosis: Accumulation of amyloid in various body tissues.

An Adjective Denoting: The kind of breathing heard over a cavity in the lung. The sound is like that made by blowing over the mouth of a narrow-necked vase.

An Antidote: it is a therapeutic substance used to counteract the toxic action(s) of a specific substance. Very few substances have an antidote.

Anabolism: Biochemical processes by which smaller molecules are joined to make larger molecules.

Anaemia: Condition in which there is a reduction in the number of red blood cells or amount of haemoglobin per unit volume of blood below the reference interval for a similar individual of the species under consideration, often causing pallor and fatigue.

Anaerobic: Not requiring molecular oxygen.

Anaesthetic: Substance which produces loss of feeling or sensation: general anaesthetic produces loss of consciousness; local or regional anaesthetic renders specific areas insensible to pain.

Analgesic: A drug that dulls the sense of pain .It differs from an anesthetic agent in that it relives pain without loss of consciousness.

Analogue Metabolism: Process by which a normally non-biodegradable compound is biodegraded in the presence of a structurally similar compound which can induce the necessary enzymes.

Analyte: A substance being analyzed; in chromatography, a single component (compound) of a mixture.

Analytic Study: Hypothesis testing method of investigating the association between a given disease or health state or other dependent variable and possible causative factors. In an analytic study, individuals in the study population are classified according to absence or presence of specific disease and according to attributes which may influence disease occurrence.

Anaphylaxis: Severe allergic reaction occurring in a person or animal exposed to an antigen or hapten to which they have previously been sensitized.

Anaplasia: Loss of normal cell differentiation, a feature characteristic of most malignancies.

Ancillary Care: Diagnostic and supportive services such as radiology, physical therapy, pharmacy or laboratory work.

ANDA: Abbreviated New Drug Application (FDA)

Anerobe: Organism which does not need molecular oxygen for life. Obligate anaerobes grow only in the absence of oxygen.

Anesthetic: A drug that causes loss of sensation. General anesthetics cause not only loss of sensation, but also loss of consciousness.

Aneuploid: Cell or organism with missing or extra chromosomes or parts of chromosomes.

Angitis: Angitis (or angiitis) means inflammation of a vessel such as a blood vessel, lymph vessel, or bile duct.

Anniversary Date: The day after a coverage period ends under a health benefits plan . Usually, the month and day that a health benefits plan first goes into effect become its anniversary date each year.

Annual Family Maximum: To protect

prescription coasts, an annual family maximum has been set. Once your contributions towards eligible coasts reach this maximum, Pharma Care will pay 100% of your family's further eligible prescription coasts for the remainder of the calendar year.

Annuity: A sum of money payable yearly or at other regular intervals.

Anodic Protection: A process that provides an appreciable reduction in corrosion by making a metal an anode and maintaining this highly polarized condition with very little current flow.

Anoxia: Strictly total absence of oxygen but sometimes used to mean decreased oxygen supply in tissues.

Antagonism: Combined effect of two or more factors which is smaller than the solitary effect of any one of those factors.

Antagonist: Substance that reverses or reduces the effect induced by an agonist.

Anthelmintic: Substance intended to kill parasitic intestinal worms, such as helminthes.

Anthracosis (Coal Miner's Pneumoconiosis): Form of pneumoconiosis caused by accumulation of carbon deposits in the lungs due to inhalation of smoke or coal dust.

Anthralin: A medication that reduces the increase in skin cells and inflammation.

Anthropogenic: Caused by or influenced by human activities.

Antibiotic: Substance produced by, and obtained from, certain living cells (especially bacteria, yeasts and moulds), or an equivalent synthetic substance, which is biostatic or biocidal at low concentrations to some other form of life, especially pathogenic or noxious organisms.

Antibody Medicated Immunity: Immune response mediated by antibodies.

Antibody: Protein molecule produced by the immune system (an immunoglobulin molecule) which can bind specifically to the molecule (antigen or hapten) which induced its synthesis.

Anticoagulant: Substance which prevents clotting. Anticoagulants are drugs which inhibit coagulation of the blood. They are used to prevent and treat abnormal clotting of the blood, to treat thrombosis, and sometimes to prevent or treat stroke or transient ischaemic attacks or episodes (TIA, TIE). Anticoagulant drugs are also prescribed preventively in major

Antidote: A substance, or a remedy, that counteracts the effect of a homeopathic remedy. High temperatures are thought to antidote remedies. Mint is said to be an antidote of natrum muriaticum. Many feel that coffee antidotes homeopathic remedies.

Antigen Presenting Cells: Cells with the ability to engulf antigens and present them to the cells of the immune system in a form that can be recognized by those immunecompetent cells.

Antigen: that stimulates production of an agglutinin – an antibody that causes agglutination or clumping of bacteria, blood cells or other antigenic particles. In the case of blood cells, this should not be confused with the clumping that happens in blood coagulation, which is a different process.

Antimetabolite: Substance structurally similar to a metabolite, which competes with it or replaces it, and so prevents or reduces its normal utilization.

Antimycotics: Substance used to kill a fungus or to inhabit its growth.

Antioxidant: A nutrient or chemical that reacts with and neutralizes free radicals or chemical that release free radicals. Antioxidants are also called free radical scavengers. Vitamins A, C, E and some of the B vitamins, beta-carotene, selenium and some key enzymes in your body are all antioxidents.

Antipyretic: Substance which relieves or reduces fever.

Antiresistant: Substance used as an additive to a pesticide formulation in order to reduce the resistance of insects to the pesticide.

Antiseptics: Antiseptics prevent the growth of diseasecausing micro-organisms without damaging living tissues. Among chemicals used are boric acid, carbolic acid, hydrogen peroxide and products based on coal tar, such as cresol. Chlorhexidines, iodine, formaldehyde, flavines, alcohol and hexachlorophane are also used.

Antiserum: Serum containing antibiodies to a particular antigen either because of immunization or after an infectious disease.

Antrostomy: Antrostomy is the operation in which an opening is made through the nose into the maxillary antrum.

Antrum: Antrum means a natural hollow or cavity. The maxillary antrum is now known as the maxillary sinus. The mastoid antrum is situated in the mastoid process, the mass of bone felt behind the ear. It may become the seat of an abscess in cases of suppuration of the middle ear.

Anuria: No urine excretion.

Aortography: It is the technique of rendering the aorta visible in an X-ray film by injecting a radio-opaque substance into it. The procedure is used to detect the presence of an aneurysm.

APC (Ambulatory Payment Classifications): Centers for Medicare and Medicaid Service (CMS) was mandated through the omnibus Budget Reconciliation Act of 1986 to develop a prospective payment system (PPS) for outpatient services. All services paid under the new PPS are classified into groups called Ambulatory Payment Classifications or APCS. Service in each APC is similar clinically in term of the resources they require.

Aphasia: Loss or impairment of the power of speech or writing, or of the ability to understand written or spoken language or signs, due to a brain injury or disease.

Aphicide: Substance intended to kill aphids.

Apicectomy: Apicectomy is the minor operation carried out to try to save a tooth which has an abscess on it or which does not respond to root treatment. In this, the abscess and the apex of the tooth are removed.

Aplasia: Lack of development of an organ or tissue, or of the cellular products from an organ or tissue.

Apolipoprotein E (APOE): A gene situated on chromosome 19 that codes for a protein in lipoproteins that are normal constituents of blood plasma, for example HDL (high density lipoprotine), LDL (low density lipoproptein), and VLDL (very low density lipoprotine). There are a number of common variations (alleles) of the APOE gene, the most common of which are known as: e2, e3, and e4.Research has shown that people who inherit one or more copies of the APOEe4 gene are at increased risk of developing Alzheimer's disease.

Apoptosis: Apoptosis or programmed cell death is a form of cell death in which a programmed sequence of events leads to the destruction of cells without releasing harmful substances into the surrounding area. Apoptosis plays an important role in health by eliminating aged cells, unnecessary cells, and unhealthy cells. A protein called bcl-2 prevents apoptosis in normal healthy cells. However, many cancer cells, which would normally be destroyed by apoptosis because they proliferate too quickly, produce high levels of bcl-2 in order to evade destruction.

Appeals: A process used by a patient or provider to request re-consideration of a previously denied services.

APPEs: Advance pharmacy practice Experiences, the name of the clinical rotations taken at the end of year 3 and during year 4 of the Pharm D curriculum.

Applet: A small application typically downloaded from a server.

Application (Computer): Software designed to fill specific needs of a user; for example, software for navigation, project management, or process control.

Application (Regulatory): Application made to a health authority to market or licenses a new product.

Application Programming Interface (API): In order to invoke a function within a computer, the function may need some specific data requirements, sizes, number, and order form the API for that function.

Application Programming Interface ID: An Application ID identifies a particular application or component within a Network Facility ID. The combination of a Network Facility ID and Application ID is unique across the health net BC network.

Application: Computer programs written to perform specific business tasks. They have also been described as the use of an information system or software package for a particular purpose or in a special way.

Applied Research: Applied research studies use advances in basic knowledge to discover and test potential new medicines that could prevent, diagnose, cure, or treat disease.

Appointment: A scheduled or non-scheduled interaction between a Person and Provider for the purposes of providing services or assessing health.

Appreciated Property: Property, such as real estate or stock, which has increased in value.

Approvable Letter: An official communication from FDA to an NDA sponsor that lists minor issues to be resolved before an approval can be issued.

Approval (In Relation To Institutional Review Boards): The affirmative decision of the IRB that the clinical trial has been reviewed and may be conducted at the institution site within the constraints set forth by the IRB, the institution, good clinical practice (GCP), and the applicable regulatory requirements.

Approval Letter: An official communication from FDA to inform an NDA sponsor of an agency decision that allows commercial marketing of a product.

Approval Phase Time: Time form date of submission of an NDA or BLA to date of U.S. FDA approval.

Approval Success Rate: The percentage of NCEs in clinical development that eventually obtain FDA marketing approval.

Approved Standard: For the purpose of this web site : A term, definition , or code recommended by the standards Council for adoption as a Date Standard, and approved by the Deputy Minister of Health , as per standards Council procedure.

Aqueous Cleaning: Cleaning parts with water to which may be added suitable detergents, saponifiers or other additives.

Arboricide: Substance intended to kill trees and shrubs.

Architecture: The selection, design and interconnection of the physical components of a computer system.

Archiving: The processes of saving data for later reference or use ASC X3 Accredited Standards Committee X3 for information processing systems.

Area Source: Widespread origin of emissions.

Areola: Areola literally means a small space, and is the term applied to the red or dusky ring around the nipple, or around an inflamed part. Increase in the duskiness of the areola on the breast is an important early sign of pregnancy.

Argyria: Pathological condition characterized by grey-bluish or black pigmentation of tissue caused by the accumulation of metallic silver, due to reduction of a silver compound which has entered the organism during administration or exposure.

Arm: Any of the treatment groups in a randomized trial. Most randomized trials have two "arms," but some have three or even more. A clinical trial comparing two doses of a new medication with a placebo would have three arms (dose 1, dose 2, and placebo).

Arrhythmia: Any variation form the normal rhythm of the heartbeat.

Artefact: Finding or product of experimental or observational techniques that is not properly associated with the system being studied.

Arteriosclerosis: Hardening and thickening of the walls of the arteries.

Arthralgia Saturnia: Pain in a joint resulting from lead poisoning.

Arthralgia: Pain in a joint.

Arthritis: Inflammation of a joint, usually accompanied by pain and often by changes in structure.

Articular: Articular means anything connected with a joint: for example, articular rheumatism.

Articulation: Articulation is a term employed in two senses in medicine, meaning either the enunciation of words and sentences, or the type of contact between the surfaces of joints – these surfaces are called articular surfaces.

Asbestosis: Form of pneumoconiosis caused by inhalation of asbestos fibers. BT pneumoconiosis.

ASC X9: Accredited standards Committee x9 specification for transferring clinical observations between independent systems.

Ascaricide: Substance intended to kill roundworms .

ASCII: American standard code for Information Interchange a common method of computer- encoding English-language characters.

Aseptic: Sterile; free from infection or foreign material.

ASHP: American society of Health-system Pharmacists.

ASP: Academy of students of pharmacy, the name of the student chapter of the American Pharmaceutical Association

Asphyxia: Condition resulting from insufficient intake of oxygen: symptoms include breathing difficulty, impairment of sense, and, in extreme, convulsions, unconsciousness and death.

Asphyxiant: Substance that blocks the transport or use of oxygen by living organisms.

Assay: process of quantitative or qualitative analysis of a component of a sample.

Assignment of Benefits: When a covered person authorizes his or her health benefits plan to directly pay a health care provider for covered services.

ASSP: Associated students of the school of pharmacy, the name of the school of pharmacy student government.

Asthenia: Weakness; lack or loss of strength.

Asthma: Chronic respiratory disease characterized by bronchonstriction, excessive mucus secretion and oedema of the pulmonary alveoli, resulting in difficulty in breathing out, wheezing, and cough.

Astringent: Causing contraction, usually locally after topical application.

ASUC: Associated students of the University of California, the name of the campus wide student government organization.

ATA: American Telemedicine Association.

Ataxia: Unsteady or irregular manner of walking or movement caused by loss or failure of muscular coordination.

Atherosclerosis: Pathological condition in which there is thickening, hardening, and loss of elasticity of the walls of blood vessels, characterized by a variable combination of changes of the innermost layer consisting of local.

ATP (Adenosine Triphosphate): The universal energy molecule, created in the mitochondria of your cells using energy derived from the food you eat. All the cellular activities in your body use the energy released by splitting ATP.

Atrophy: Means the dying or death, normally referring to a gland or organ.

Attractant: Substance used to attract animals with the aim of killing or sterilizing them BT Pheromone.

Attributable Risk: Difference between the risk of exhibiting a certain adverse effect in the presence of a substance and the same risk in the absence of the substance.

Attrition Rate: The percentage of NCEs that drop out of testing in a chemical phase.

AUC: The area under the plot of plasma concentration of drug (not logarithm of the concentration) against time after drug administration. The area is conveniently determined by the "trapezoidal rule": the data points are connected by straight line segments, perpendiculars are erected from the abscissa to each data point, and the sum of the areas of the triangles and trapezoids so constructed is computed. The ratio of the AUC after oral administration of a drug formulation to that after the intravenous injection of the same dose to the same subject is used during drug development to assess a drug's oral bioavailability.

Audit (of A Clinical Trial): A systematic and independent examination of trial-related activities and documents to determine whether the evaluated trial-related activities were conducted, and the data were recorded, analyzed, and accurately reported according to the protocol, sponsor's standard operating procedures (SOPs), good clinical practice (GCP), and the applicable regulatory requirement(s).

Audit Certificate: Document that certifies that an audit has taken place (at an investigative site, CRO, or clinical research department of a pharmaceutical company).

Audit Report: A written evaluation by the sponsor's auditor of the results of the audit.

Audit Trail: Documentation that allows reconstruction of the course of events [ICH]. A secure, time stamped record that allows reconstruction of the course of events relating to the creation, modification, and deletion of an electronic study record.

Autoimmune Disease: Autoimmune disease is a type of illness that occurs when the body tissues are attacked by its own immune system. People suffering from autoimmune diseases tend to have unusual antibodies circulating in their blood that target their own body tissues. Autoimmune diseases are more common in women than in men.

Autoimmune Process: A process where the body's immune system attacks and destroys body tissue that it mistakes for foreign matter.

Autophagosome: Membrane – bound body (secondary lysosome) in which parts of the cell are digested.

Autopsy: Post mortem examination of the organs and body tissue to determine cause of death or pathological condition.

Auxotroph: Organism unable to synthesize an organic molecule which is required for its growth: when the compound is given to the organisms with the other nutrients it requires, growth of the organism may occur.

Average Manufacture Price (AMP): The average price paid to a manufacture by wholesalers for drugs distributed to retail pharmacies, AMP was a benchmark created by Congress in 1990 in calculating Medicaid rebates and is not publicly available. FSS and 340B prices ' as well as prices associated with direct sales to HMO,s and hospitals , are excluded from AMP under the rebate program . The Congressional Budget Office estimates AMP to be about 20 percents less than AWP for more than 200 drug products frequently purchased by Medical recipients.

Average Sales Price (ASP): A new system created by Federal and State Government prosecutors in settlements with pharmaceutical manufacturers TAP and Bayer to ensure more accurate price reporting. ASP is the weighted average of all non-Federal sales to wholesalers and is net of chargebacks, discounts, rebates, and other benefits tied to the purchase of the drug product, whether it is paid to the wholesaler or the retailer.

Average Wholesale Price (AWP): A national average of list prices charged by wholesalers to pharmacies. AWP is sometimes reffered to as a "sticker price" because it is not the actual price that a larger purchases normally pay. For example, in a study of prices paid by retail pharmacies in eleven States, the average acquisition price was 18.3 percent below AWP. Discounts for HMOs and other large purchases can be even greater .AWP information is publicly available

Avicide: Substance intended to kill birds.

Azotarmia: Excess of urea or other nitrogenous bodies in the blood

B

B Lymphocytes: Type of lymphocyte which synthesizes and secretes antibodies in response to the response of foreign substance or one identified by its as foreign. The protective effect can be meditated to a certain extent by the antibody alone.

Back Mutation: Process which reverses the effect of a mutation which had inactivated a gene; thus it restores the wild phenotype.

Bactericide: Substance intended to kill bacteria.

Bacteriostatic: Slowing or stopping the proliferation of bacteria.

Badger Care: Badger Care extends Medicated coverage through a Medicated expansion under Titles XIX and XXI to uninsured children and parents with incomes at or below 185% of the federal poverty level and who meet other program requirements. The goal of Badger Care is to fill the gap between Medicaid and private insurance without supplanting or "crowding out" private insurance. Badger care benefits are identical to the benefits.

Badger Rx Gold Program: A program designed to address the increasing number of Wisconsin residents who either are uninsured , or whose insurance has no pharmacy benefit coverage and are affordable prescription drugs. The program is not state-funded. Instead, the program makes the purchasing makes the purchasing power of ETF available, via Navitus (TM) Health Solutions, to public and private – sector entities throughout Wisconsin.

Badger Rx: The purchasing group established by the Employee Trust Fund (ETF) that negotiates with drug manufactures to minimize the cost for the most effective and safe prescription drugs available to public . Access to this purchasing group is available to public and private sector groups so that they may also benefit from the same types of savings.

Bagassosis: Lung disease caused by the inhalation of dust from sugar-cane residues.

Balanced Study: Trial in which a particular type of subject is equally represented in each study group.

Balsamic: Soft or hard plant or tree resins composed of aromatic acids and oils. These are typically used as stimulating dressings and aromatic expectorants and diuretics. This term is also applied loosely to many plants that may not exude resins but which have a soothing, pitchy scent. Examples: Balsam Poplar, Eriodicyon.

Bandwidth: An indicator of the throughput (speed) of data flow on a transmission path; the width of the range of frequencies on which a transmission medium carries electronic signals. All digital and analog signals have a bandwidth.

Barrier: Aircraft arresting system.

Bart: Bay Area Rapid Transit, the name of a public transportation rail system for the San Francisco Bay Area Region.

Basal Metabolism: The basic rate of combustion by a person, usually measured after sleep and while resting.

Basal: At or near the base, such as leaves sprouting directly from root or crown.

Base Environmental Manager: The organization, usually the CE Environmental Management Office, possessing the responsibility for environmental management and planning activities, and for establishing base-specific policy on all aspects of environmental management.

Base Pairing: Linking of the complementary pair of polynucleotide chains of nucleic acids by means of hydrogen bonds between complementary purine and pyrimidine bases, adenine with thymine or uracil; cytosine with guanine.

Baseline Assessment: Assessment of subjects as they enter a trial and before they receive any treatment.

Baseline: Quantified starting points from which progress is measured. This could be a listing of materials, a measure of waste streams, or a set of processes are compared.

Basic Coverage: Basic coverage is the minimum prescription drug coverage that all Quebecers must have, and includes all drugs listed on the Drug Formulatory (French Only), which is published by the Regiede l' assurance maladide and lists over 4500 drugs available on prescription.

Basic Research: Basic research studies are typically designed to expand scientific knowledge of human biology, disease mechanisms and processes of life, such as how cells work.

Basic Science Research: Studies that pursue knowledge about the most fundamental processes of life, such as how cells work.

Batch: Multiple transactions sent as a single group on pharmanet for processing at once , transmitted when pharmanet is a back on-line after a network interruption.

Bayesian Approaches: Approaches to data analysis that provide a posterior probability distribution for some parameter (e.g. treatment effect), derived from the observed data and a prior probability distribution for the parameter. The posterior distribution is then used as the basis for statistical inference.

Bayesian Statistics: Statistical approach named for Thomas Bayes (1701–1761) that has among its features giving a subjective interpretation to probability, accepting the idea that it is possible to talk about the probability of hypotheses being true and of parameters having particular values.

BDC: The building letter code for the buchanan dental clinic 100 buchanan street, san Francisco.

Beacon Service: Schemes identifying examples.

Behaviour Care Service: Assessment and therapeutic services used in treatment of mental health and substance abuse problems.

Bells Palsy: An inflammatory condition of the facial, nerve, with paralysis, distortion and diminished tears.

Beneficiary Identification and Records Locator System: Birls is a set of databases indexed by social security number. With birls , a researcher can find information on veterans who obtained burial benefits from VA who receive a monetary benefit for Aid and Attendance or who had accounts for VA educational benefits (to determine the extent of contact with VA for other services besides healthcare). Birl contains other data as well, such as the type.

Beneficiary: A person who is eligible to receive benefits under a health benefits plan. Sometimes "beneficiary" is used for eligible dependents enrolled under a benefits plan, "beneficiary" can also be used to mean any person eligible for benefits , including both employees and eligible dependents.

Benefit Maximum: The total amount the insurer will contribute toward the pharmacy benefit, usually within a given timeframe. Example: insurer may pay up to $10,000 per member per year for prescription drugs.

Benefit: A drug or product defined by a Drug Identification Number (DIN) or Product Identification Number (PIN) which is eligible for full or partial payment by Pharma Care.

Benefits Package: A term informally used to refer to the employer's benefits plan or to the benefits plan options from which the employee can choose.

Benign: Benign prostatic hypertrophy, or hyperplasia (BPH), The benign buildup in the prostate of "warts" or epithelial neoplasias that can block or interrupt urination, and which are usually concurrent with moderate prostate enlargement. They cause a dull ache on urination, ejaculation, and/or defecation. The diagnosis is medical, since the same subjective conditions can result from cancer of the prostate. BPH is common in men over fifty and can be the result either of diminished production of complete testosterone or poor pelvic circulation. Alcohol, coffee, speed, and antihistamines can all aggravate the problem.

Bequest: To give or leave something by

well, typically personal property or assets.

Beryllium Disease: Serious and permanent lung damage resulting from chronic inhalation of beryllium.

Beta Blockers: Drugs used to slow the response to epinephrine (released by the adrenal medulla), usually to attempt controlling high blood pressure.

Beta Cells: Cells that make insulin. Beta cells are found in areas of the pancreas called the Islets of langerhans.

Beta Error: Probability of showing no significant difference when a true difference exists; a false acceptance of the null hypothesis.

Between-Subject Variation: In a parallel trial design, differences between subjects are used to assess treatment differences.

BIAS: When a point of view or the conditions of a study prevents impartial judgment on issues relating to the subject of that point of view. For example, bias may occur when a study is not "blinded" or it may derive from a sampling error, e.g., when the population enrolled in the study is not representative of the overall population with the disease. In clinical studies, bias is controlled by such measures as blinding and randomization.

Biased Sample: Any sample which is not a random sample.

Biliousness: A symptom-picture

resulting from a short -term disordered liver, with constipation, frontal headache, spots in front of the eyes, poor appetite, and nausea or vomiting. The usual causes are heavy alcohol consumption, poor ventilation when working with solvents, heavy binging with fatty foods, or moderate consumption of rancid fats. The term is genially archaic in medicine; people who are bilious are seldom genial, however.

Bilirubin: Orange-yellow pigment (C33H36O6N4), a breakdown product of haem-containing proteins haemoglobin, myoglobin, cytochromes, which circulates in the blood plasma bound to albumin or as water soluble glucuronides, and is secreted in the bile by the liver.

Bilirubinemia: The presence of abnormally high bilirubin in the blood, usually signifying hepatitis, with jaundice due next week.

Billiary: Related to bile and the bile ducts.

BIO: Biotechnology Industry Organization.

Bioaccumulation Potential: Ability of living organisms to concentrate a substance obtained either directly from the environment or indirectly through its food.

Bioaccumulation: Progressive increase in the amount of a substance in an organism or a part of an organism which occurs because the rate of intake exceeds the organisms ability to remove the substance from the body.

Bioactivation: Any metabolic conversion of a xenobiotic to a more toxic derivative.

Bioanalytical Assays: Methods for quantitative measurement of a drug, drug metabolites, or chemicals in biological fluids.

Bioassay: Procedure for estimating the concentration or biological activity of a substance by measuring its effect on an organism compared to an organism compared to an appropriate standard preparation.

Bioavailability: Extent to which a substance to which the body is exposed reaches the systemic circulation, and the rate at which this occurs. SN biological availability, physiological exposure from extra vascular exposure (ev) to that following intravenous exposure.

Biochemical (Biological) Oxygen Demand (BOD): Substance concentration of oxygen taken up through the respiratory activity of micro-organisms growing on organic compounds present when incubated at a specified temperature (usually 20°) for a fixed period. It is regarded as a measure of that organic pollution of water which can be degraded biologically but includes the oxidation of inorganic material such as sulphide and iron.

Biochemical Mechanism: Reaction or series of reactions, usually enzyme, associated with a specific physiological event in living organism.

Biocid: Substance intended to kill living organisms.

Bioconcentration Factor (BCF): Measure of the tendancy for a substance in water to accumulate in fish tissue or in tissues of other organisms. The equilibrium concentration of a substance in fish can be estimated by multiplying the concentration of a substance in the surrounding water by the fish bioconcentration factor for that chemical. This parameter is an important determinant for human intake but the aquatic food ingestion route.

Bioconcentration: Process leading to a higher concentration of a substance in an organism than in environmental media to which it is exposed.

Biodegradation: Breakdown of a substance catalysed by enzymes in vitro or in vitro. This may be characterised for purposes of hazard assessment as: primary. Alteration of the chemical structure of a substance resulting in a loss of a specific property of that substance.

Biodiverse: The state of life inter-dependency that is possible when large and small plants, soil organisms, insects, and fuzzy beasts exist in the ebb and flow created by the natural environment. Cut down the trees once and you lessen the biodiversity drastically. Wait fifty years and cut again and you have a small fraction of the life-form variety that you started with; the old diversity will never return.

Bioequivalence: Scientific basis on which generic and brand-name drugs are compared. To be considered bioequivalent, the bioavailability of two products must not differ significantly when the two products are given in studies at the same dosage under similar conditions.

Bioinformatics: Information about human and other animal genes and related biological structures and processes.

Biologic: A drug prepared from animal tissue or some other living source.

Biological Assessment of Exposure: Assessment of exposure to a substance by the analysis of specimens taken in the environment such as foodstuffs, plants, animals, biological material in air or water samples, or biological material from exposed subjects. when human samples are analyzed, they are usually urine and blood ; other possible samples include expired air, faces, saliva, bile, hair and biopsy or autopsy material.

Biological Cycle: Complete circulatory process through which a substance passes in the biosphere. It may involve transport through the various media.

Biological Effect Monitoring (BEM): Continuous or repeated measurement of early biological effects of exposure to a substance to evaluate ambient exposure and health risk by comparison with appropriate reference values based on knowledge of the probable relationship between ambient exposure and biological effects.

Biological Half-life or Half-time (t/2): Time required for the amount of a substance in a biological system to be reduced to one half, predominantly by biological processes, when the rate of removal is approximately exponential.

Biological Monitoring: Continuous or repeated measurement of potentially toxic substances or their metabolites or biochemical effects in tissues, secreta, excreta, expired air or any combination of these in order to evaluate occupational or environmental exposure and health risk by comparison with reference values based on knowledge of the probable relationship between ambient exposure and resultant adverse health effects.

Biological Preparation: Compound derived from living organisms and their products for use in medicine or as a pesticide etc.

Biological Specimen: Organ, tissue (including blood), secretion or excretion product taken from an organism as a sample reflecting the state of whole organism.

Biologics: New class of medications that specifically target parts of the immune system such as a tumour necrosis factor or interleukin-1. Examples include Amevive, Enbrel, and Raptiva.

Biology: Scientific study of living organisms.

Biomagnification: Sequence of processes in an ecosystem by which higher concentration are attained in organisms at higher tropic levels; at its simplest, a process leading to a higher concentration of a substance in an organism than in its food.

Biomarker: Parameter that can be used to identify a toxic effect in an individual organism and can be used in extrapolation between species. Indicator signalling an event or condition in a biological system or a sample and giving a measure of exposure, effect, or a susceptibility.

Biomass: The actual amount of existing material within a species or genus.

Biomineralization: Complete conversion of organic substances to inorganic derivatives by living organisms, especially micro-organisms.

Bioorganic Chemistry: Use if chemical synthesis to study biological systems.

Biopharmaceutical Sciences: Scientific field focused on studying the molecular, cellular and genetic principles underlying drug development and therapy.

Biopharmaceutical: Any therapeutic biological compound, including recombinant proteins, monoclonal and polyclonal antibodies, antisense oligonucleotides, therapeutic genes and recombinant and DNA vaccines.

Biophysical Chemistry: Use of chemical theory and instruments to study biological materials.

Biopsy: The removal of a tissue sample for examination under the microscope. The tissue sample can be called a biopsy. Excision of a small piece of living tissue for microscopic or biochemical examination; usually performed to establish a diagnosis.

Biosphere: Literally, the part of the earth that supports life; more broadly, a large community of life-forms sharing a similar environment, such as a rain forest or prairie grassland.

Biostatic: Arresting the growth or multiplication of living organisms.

Biostatistics: Branch of statistics applied to the analysis of biological phenomena.

Biota: All living organisms as a totality.

Biotechnology: Industrial use of biological techniques or products developed through basic research.

Biotransformation: Any chemical conversion of substances that is medicated by living organisms or enzyme preparations derived there from.

Bipinnate: A pinnate compound leaf whose leaflets, in turn, are stems that have pinnate leaflets.

Biternate: A compound leaf divided in threes, whose leaflets are in turn divided in pairs.

Bitter Tonic: A bitter-tasting substance or formula used to increase a deficient appetite, improve the acidity of stomach secretions and protein digestion, and slightly speed up the orderly emptying of the stomach. A good bitter tonic should possess little, if any, drug effect, only acting on oral and stomach functions and secretions. Dry mouth, bad gums, teeth problems with bad breath in the morning, and weak digestion, often with constipation, are the main deficiency symptoms. A bitter tonic has little effect in normal digestion. Example: Gentiana

BLA: Biologics License Application, which is an application to FDA for a license to market a new biological product in the United States.

Bladder: A hollow organ that urine drains into from the kidneys. From the bladder, urine leaves the body.

Blind or Blinding: When the researchers, the participants, or both, do not know which people are getting which potential treatment being studied in a clinical trial. The process of blinding improves the reliability of a study's results because no one is unconsciously influenced by the knowledge. If only researchers or only participants are blinded, the study is often called single blind.

Blind Review: Checking and assessing data, during the period of time between trial completion (the last observation on the last subject) and breaking the blind, for the purpose of finalizing the planned analysis.

Blind Study: One in which the subject, the investigator, or anyone assessing the outcome is unaware of the treatment assignment(s). Blinding is used to reduce the potential for bias. See also blinding\masking, double-blind study, single- blind study, triple-blind study.

Blinded (Masked) Medication: Products that appear identical in size, shape, color, flavor, and other attributes to make it very difficult for subjects and investigators (or anyone assessing the outcome) to determine which medication is being administered.

Blinding/Masking: A procedure in which one or more parties to the trial are kept unaware of the treatment assignment(s). Single-blinding usually refers to the subject(s) being unaware, and double- blinding usually refers to the subject(s), investigator(s), monitor, and, in some cases, data analyst(s) being unaware of the treatment assignment(s).

Blood Glucose: The main sugar that the body makes from the food we eat. Glucose is carried through the bloodstream to provide energy to all of the body's living cells. The cells cannot use glucose without the help of insulin.

Blood Pressure: Blood pressure is that pressure which must be applied to an artery in order to stop the pulse beyond the point of pressure. It may be roughly estimated by feeling the pulse at the wrist, or accurately measured using a sphygmo-manometer. It is dependent on the pumping force of the heart, together with the volume of blood, and on the elasticity of the blood vessels. The blood pressure is biphasic, being greatest (systolic pressure) at each heartbeat and falling (diastolic pressure) between beats. The force of the blood against the artery walls, two levels of blood pressure are measured the highest, or systolic , occurs when the heart pumps blood into the blood vessels , and the lowest, or a diastolic occurs when the heart tests.

Blood: Blood consists of cellular components suspended in plasma. It circulates through the blood vessels, carrying oxygen and nutrients to the organs and removing carbon dioxide and other waste products for excretion. In addition, it is the vehicle by which hormones and other humoral transmitters reach their sites of action.

BO: The slope of a linear plot of C (not the logarithm of C) against t; the slope of the linear plot of a zero-order reaction , in which , in equal time intervals , equal amounts of chemical undergo reaction.

Board-Certified: Any physician who has completed medical school, internship and residency in his or her chosen speciality and has successfully completed an examination conducted by a group of peers.

Body Burden: Total amount of a substance of a chemical present in an organism at a given time.

Body Weight: Sometimes, as a subscript, to indicate "of, or in, the body"; thus, a b is the amount of the drug in the body.

Bolus: single dose of a substance, originally a large pill. Dose of a substance administered by a single rapid intravenous injection. Concentrated mass of food ready to be swallowed.

Bone: Bony epiphyses have fused, the result is gigantism; if it occurs in adult life the skeletal over growthis confined to the hands, feet, cranial sinuses and jaw.

Booked Admission Schemes: An airline-style booking system for arranging NHS appointments which are easy and convenient to the patient

Borborygmus: The bubbling, gurgling

passage of gas across the transverse colon.

BPA (Blanket Purchase Agreement):

BPAs are contracts between VA and pharmacy supply manufacturers. They discounts for particular madications or supply based on the volume prescribed or used. The BPA may include additionals programs and services as well. Many are negotiated for specific customer groups under the VA federal supply schedule (fss) program.

BPS: Biopharmaceutical Sciences

Bracts: Reduced or modified leaflets that are usually parts of flowers or an inflorescence, generally subtending or beneath the floral parts.

Brady: Prefix meaning slow as in bradycardia or bradypnoea.

Bradycardia: A distinctly slow heartbeat, which may be a normal idiosyncrasy or with causes ranging from regular strenuous exercise to abnormally slow heart stimulus to the side-effects of medication. Bradycardia is usually defined as a pulse below sixty beats a minute, or seventy in children.

Bradykinesia: Bradykinesia refers to the slow, writhing movements of the body and limbs that may occur in various brain disorders

Bradykinin: Bradykinin is a substance derived from plasma proteins; it plays an important role in many of the reactions of the body, including inflammation. Its prime action is in producing dilatation of arteries and veins. It has also been described as 'the most powerful pain-producing agent known'.

Bradypnoea: Abnoramally slow breathing.

Brand Drug Name: A patented medication produced by a brand name drug company.

Brand Medically Necessary: Phrase that must appear in the prescriber's own handwriting on the face of each new prescription order for Wisconsin senior care to reimburse an innovator drug at the amount greater than the medicated maximum allowable cost (MAC).

Breakdown Through Medicine: The solution of a condition that has an important effect an all future research and development.

Breathing Zone: Space within a radius of 0.5m from person's face.

Breathlessness: Breathlessness, or dyspnoea, may be due to any condition which renders the blood deficient in oxygen, and which therefore produces excessive involuntary efforts to gain more air. Exercise is a natural cause, and acute anxiety may provoke breathlessness in otherwise healthy people.

Brick Dust: The presence of reddish brown sediment in the urine, indicating uric acid, hippuric acid and creatinine excess in the blood an anabolic grease ball who needs more liquids and alkali and who has over-acidic urine. It can be symptomatic of more serious problems as well.

Bromides: A binary salt of bromine, formerly used as a simple sedative. Given so freely and with no intent of affecting a healing, it became synonymous with a useless treatment only meant to shut up the patient. Excessive bromide use can cause some pronounced neurologic disturbances, they disappear with cessation of the drug.

Bronchitis: Inflammation of the mucus membranes on the bronchi, usually caused by an infection, sometimes by allergies or chemical irritations.

Bronchoconstriction: Narrowing of the air passages through the bronchi of the lungs.

Bronchodialation: Expansion of the air passages through the bronchi of the lungs.

Bronchorrhea: Excess mucus secretions by the bronchi; a runny nose of the lungs.

Bronchospasm: Intermittent violent contraction of the air passages of the lungs.

Browser: Computer program that runs on the user's desktop computer and is used to navigate the World Wide Web.

Bruxism: Bruxism, or teeth-grinding, refers to a habit of grinding the teeth, usually while asleep and without being aware of it. The teeth may feel uncomfortable on wakening. It is common in children and is usually of no significance.

Buffering System: The several blood factors that enable the acid waste products of metabolism to be carried in the alkaline blood without disrupting its chemistry. These include carbolic acid, carbonates, phosphates, electrolytes, blood proteins, and erythrocyte membranes.

Builder: Material which enhances or maintains the cleaning efficiency of a surfactant, in a detergent, principally by inactivating water hardness; complex phosphates, sodium carbonate, and sodium silicate are the builders most commonly used.

Bulimia: Bulimia means insatiable appetite of psycho- logical origin. This eating-disorder symptom may be of psychological origin or the result of neurological disease – for example, a lesion of the hypothalamus. Bulimia nervosa is linked to anorexia nervosa and is sometimes called the binge and purge syndrome.

Bundled Sales: The packaging of drugs of different types, where the total price for the package is less than the purchase price of the drugs if purchased separately.

Bursitis: Inflammation of a bursa, the lubricating sac that reduces friction between tendons and ligaments or tendons and bones. The more common localities for bursitis are the shoulders, the elbows, the knees, and the big toe (a bunion).

Byssinosis: Pneumoconiosis caused by inhalation of dust and associated microbial contaminants and observed in cotton, flax and hemp workers.

Bystander Exposure: Liability of members of the general public to come in contact with substance arising from operations or processes carried out by other individuals in their vicinity.

C

C, Cx: The concentration (in units of mass/ volume) of a chemical in a body fluid such a s blood, plasma, serum, urine, etc.; the specific fluid may be indicated by a subscript,i.e. CU, the concentration of drug in the urine when no subcript is used, C is commonly taken to be the concentration in the plasma.

CAC (Cost Adjusted Charges): Charges obtained from hospital bills or discharge data are frequently higher than actual costs. One can adjust the charges by multiplying reported charges by a department-level cost-to-charge ratio. We refer to the adjusted charges as cost-adjusted charges (CAC).

Cachet: An oval capsule that encloses a dose of unpleasant medicine.

Cachexia: Severe weight loss and listlessness produced by serious disease such as cancer or tuberculosis, or by prolonged starvation.

Cadaver Donor: Kidney transplanted from someone who has died.

Cadaver: A dead body.

Cadaveric: Relating to a dead body, as in the source for a donor organ

Caecum: The dilated first part of the large intestine lying in the right lower corner of the abdomen.The small intestine and the appendix open into it, and it is continued upwards through the right flank as the ascending colon.

Caesarean Section: The operation used to deliver a baby through its mother's abdominal wall. It is performed when the risks to mother or child of vaginal delivery are thought to outweigh the problems associ- ated with operative delivery.

Calcaneus: The heel-bone or os calcis, and the largest bone in the foot.

Calcicosis: Calcicosis is a traditional term applied to disease of the lung caused by the inhalation of marble dust by marble-cutters.

Calciferol: This is an outdated term for ergocaliciferol or vitamin D_2.

Calcification: Process in which organic tissue becomes hardened by deposition of calcium salts within its substance. The deposition of calcium salts in body tissues, normally bone and teeth, though abnormal deposits can occur in damaged muscles or the walls of arteries.

Calcinosis: Abnormal deposition of calcium in the body tissue.

Calcipotriene (Dovonex): A synthetic of vitamin D_3 used in the treatment of psoriasis.

Calcitonin: A hormone, produced by the thyroid gland, which is involved in the metabolism of bone. Acting to lower concentrations of calcium and phosphates in the blood,

Calculi: The general name given to concretions in, for example, the urinary bladder, kidneys or gall-bladder.

Calculus Donor: An abnormal inorganic mass occurring within the body, (a) Renal 1° Occurs in healthy urinary tract usually made up of oxalates and urates, (b) Renal 2° Occurs in association with infection and obstruction

Calibre: A talking-book service which is available to all blind and handicapped people who can supply a doctor's certificate confirming that they are unable to read printed books in the normal way. Its catalogue contains more than 370 books for adults and more than 250 for children, and additions are being made at the rate of around three a week. Full details can be obtained from Calibre.

Calliper Splint: This is applied to a broken leg in such a way that in walking, the weight of the body is taken by the hip-bone and not by the foot.

Calliper: A two-pronged instrument with pointed ends, for the measurement of diameters, such as that of the pelvis in obstetrics.

Callosities: Areas of gross thickening of the epidermis in response to trauma. They usually occur on a foot due to bony deformity or ill-fitting foot- wear.

Callus: The new tissue formed around the ends of a broken bone.

Calluses: Thick, hardened areas of the skin, generally on the foot, caused by friction or pressure. Calluses can lead to other problems, including serious infection and even gangrene.

Caloric Test: A test for vestibular function. It is performed by irrigating the external auditory meatus of the ear with alternate cold and hot water. This usually stimulates the vestibular apparatus, causing nystagmus. If the vestibular apparatus is affected by disease, the response may be absent or reduced.

Calorie: A unit of energy. Two units are called by this name. The small calorie, or gram calorie, is the amount of heat required to raise one gram of water one degree centigrade in temperature. The large Calorie or kilocalorie, which is used in the study of dietetics and physiological processes, is the amount of heat required to raise one kilogram of water one degree centigrade in temperature.

Calyx: Calyx means a cup-shaped cavity, the term being especially applied to the recesses of the pelvis of the kidney.

Canaliculus: Canaliculus means a small channel, and is applied to (a) the minute passage leading from the lacrimal pore on each eyelid to the lacrimal sac on the side of the nose; (b) any one of the minute canals in bone.

Cancellous: A term applied to loose bony tissues as found in the ends of the long bones.

Cancer: The general term used to refer to a malignant tumour, irrespective of the tissue of origin. 'Malignancy' indicates that (i) the tumour is capable of progressive growth, unrestrained by the capsule of the parent organ, and/or (ii) that it is capable of distant spread via lymphatics or the bloodstream, resulting in development of secondary deposits of tumour known as 'metas- tases'. Microscopically, cancer cells appear different from the equivalent normal cells in the affected tissue.

Cancrum Oris: Cancrum oris, also called noma, is a gangreous ulcer about the mouth which affects sickly children, especially after some severe disease such as measles. It is due to the growth of bacteria in the tissues.

Candida: Candidosis (moniliasis) is an infection with the yeast, Candida albicans. It is encouraged by pregnancy, diabetes mellitus, prolonged wide-spectrum antibiotics or cortico- steroids therapy, and is also seen in debilitated infants, the elderly and immunocompromised patients, e.g. those with AIDS/HIV.

Candidiasis: Generally, a disorder caused by Candida (Monilia) albicans. This is a common yeast-like fungus found in the mouth, vagina and rectum, as well as on the outside skin. It is a common cause of thrush in infants and vaginal yeast infections. In recent years much attention has been given to the increased numbers of people with candidiasis in the upper and lower intestinal tract. This condition is now known to occur as a result of extended antibiotic therapy and anti-inflammatory treatment. Most anti- inflammatory drugs are really immunosuppressants, and the normal, stable competition between fungus and bacteria is altered by the antibiotic use; this rather benign and common skin and mucosal fungus can then move deeply into the body.

Cannula: A tube for insertion into the body, designed to fit tightly round a trocar – a sharp, pointed instrument which is withdrawn from the cannula after insertion, so that fluid may run out through the latter.

Canthus: The name applied to the angle at either end of the aperture between the eyelids.

CAPD: Continuous ambulatory peritoneal dialysis. Technique which can continue as a patient goes day-to-day activities. Uses the peritoneum permeable membrane.

Capillaries: The minute vessels which join the ends of the arteries to venules, the tiny commencement of veins. Their walls consist of a single layer of fine, flat, transparent cells, bound together at the edges, and the vessels form a meshwork all through the tissues of the body, bathing the latter in blood with only the thin capillary wall interposed, through which gases and fluids readily pass.

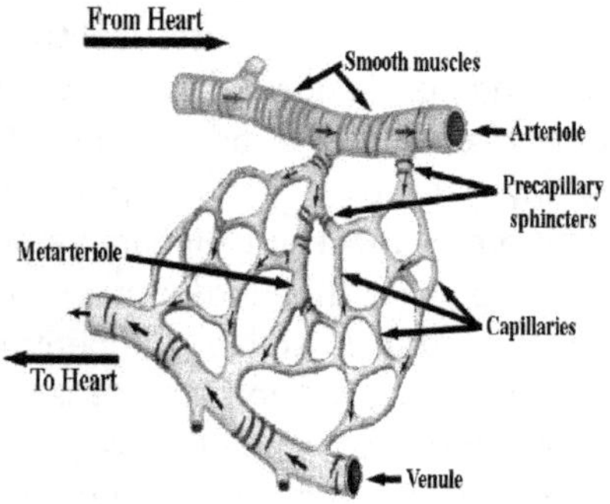

Capillary Return: A test for the adequacy of blood circulation by pressing on the skin and seeing how long it takes for the colour to return.

Capitation Rate: A monthly, per-patient rate that pays to pharmacists in lieu of dispensing fees.

Capillary: The smallest blood or lymph vessel, formed of single layers of interconnected endothelial cells, sometimes with loosely attached connective tissue basement cells for added support. Capillaries allow the transport across their membranes and between their crevices of diffusible nutrients and waste products. Blood capillaries expand and contract, depending upon how much blood is needed in a given tissue and how much is piped into them by the small feeder arteries upstream.

Capsule: A term used in several senses in medicine. It is applied to a soluble case, usually of gelatine, for enclosing small doses of unpleasant medicine. Enteric-coated capsules, which have been largely superseded by enteric-coated tablets, are capsules treated in such a manner that the ingredients do not come in contact with the acid stomach contents but are only released when the capsule disintegrates in the alkaline contents of the intestine.

Caput Medusae (Medusa's Head): The term describing the abnormally dilated veins that form around the umbilicus in cirrhosis of the liver.

Caput Succedaneum: Usually shortened by obstetricians to 'caput', this is the temporary swelling which is some- times found on the head of the newborn infant. It is due to oedema in and around the scalp, caused by pressure on the head as the child is born. It is of no significance and quickly disappears spontaneously.

Carbohydrate: One of three main groups of foods in that provides calories and energy. Carbohydrates are mainly sugars carbohydrates) and starches (complex carbc found in bread, pasta, beans) that the body into glucose.

Carbos: Carbohydrates, like starch or sugar.

Carboxyhaemoglobin: Compound which is formed between carbon monoxide and haemoglobin in the blood of animals and which is incapable of transporting oxygen.

Carboxy Haemoglobinaemia: The term applied to the state of the blood in carbon monoxide poisoning, in which this gas combines with the haemoglobin, displacing oxygen from it.

Carcinogen: Agent (chemical, physical or biological) which is reflecting the strength of the evidence derived from studies in humans and in experimental animals and from other relevant data.

Carcinogenesis: Carcinogenesis is the means or method whereby the changes responsible for the induction of cancer are brought about.

Carcinogenicity Test: Long term (chronic) test designed to detect any possible carcinogenic effect of a test substance. Sufficient evidence, Causal relationship has been established between exposure to the agent and human cancer: a positive relationship has been observed between exposure to the agent and cancer in studies in which chance, bias and confounding could be ruled out with reasonable confidence. Limited evidence. Positive association has been observed between exposure to the agent and cancer for which a causal interpretation is considered to be credible, but chance, bias or confounding could not be ruled out with reasonable confidence. Evidence suggesting lack of carcinogenicity. There are several adequate studies covering the full range of doses to which human beings are known to be exposed, which are mutually consistent in not showing a positive association between exposure to the agent and any studied cancer at any observed level of exposure

Carcinogenicity: Process of induction of malignant neoplasms by chemical, physical or biological agent

Carcinogens: Agents, such as tobacco smoke, certain chemicals, asbestos fibres and high-dose radiation, that have the property of causing cancer.

Carcinoma in Situ: The first stage of carcinoma in which the malignant tumour is present only in the epithelium, and when surgical excision of the local growth, with its pathological status confirmed in the laboratory, should ensure a cure.

Carcinoma: A type of cancer developing from cells found in the surface layer of an organ in the body.

Carcinomatosis: The spread of cancer cells from their original site of growth to other tissues in the body. Such a spread of cancer, which takes place mainly via blood and lymph vessels, is usually fatal. Chemotherapy and radiotherapy may, however, check the spread or sometimes destroy the cancerous growth.

Cardholder: A patient holding a valid care card.

Cardia: Cardia is a term applied to the upper opening of the stomach into which the oesophagus empties. The cardia lies immediately behind the heart.

Cardiac Arrest: Cardiac arrest occurs when the pumping action of the heart stops. This may be because the heart stops beating or because the heart muscle starts contracting too fast to pump effectively (ventricular systole, the period when the heart contracts). Coronary throm- bosis is the most frequent cause of arrest. Irreversible brain damage and death result without prompt treatment.

Cardiac Arrhythmia: Abnormal rhythm of the heart beat. Most commonly seen after someone has had a myocardial infarction, but also present in some normal individuals – especially if they have taken a lot of coffee or other stimulant – and in those with a congenital abnormality of the heart-muscle conducting system.

Cardiac Catheterisation: A diagnostic procedure in which a tube is inserted into a blood vessel under local anaes- thetic and threaded through to the chambers of the heart to monitor blood flow, blood pressure, blood chemistry and the output of the heart, and to take a sample of heart tissue. The tech- nique is used to diagnose congenital heart dis- ease and coronary artery disease. Another appli- cation is in the diagnosis and treatment of valvular disease in the heart.

Cardiac Cycle: The various sequential movements of the heart that comprise the rhythmic relaxation and expansion of the heart muscles as first the atria contract and force the blood into the ventricles (diastole), which then contract (systole) to pump the blood round the body.

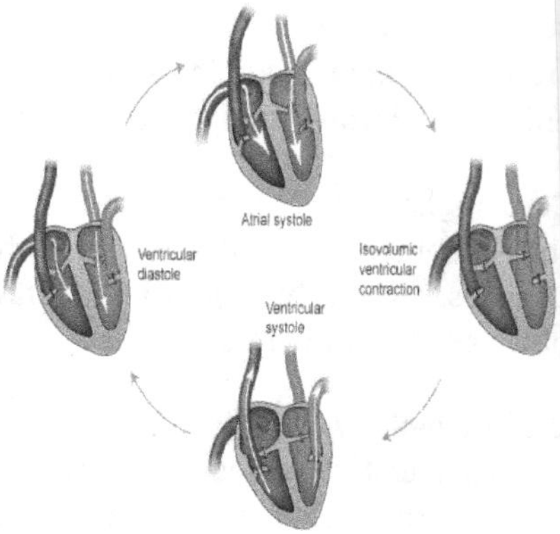

Cardiac Massage: The procedure used to restart the action of the heart if it is suddenly arrested. In many cases the arrested heart can be made to start beating again by rhythmic compression of the chest wall. This is done by placing the patient on a hard surface – a table or the floor – and then placing the heel of the hand over the lower part of the sternum and compressing the chest wall firmly, but not too forcibly, at the rate of 60–80 times a minute. At the same time artificial res- piration must be started by the mouth-to- mouth method.

Cardiac Muscle: The muscle, unique to the heart, which comprises the walls of the atria and ventricles. It consists of long broadening cells (fibres) with special physiological characteristics which enable them to keep contracting and expanding indefinitely.

Cardiac Neurosis: Obsessional fear about the state of the heart. It tends to occur after a heart attack and may result in the patient's experiencing the symptoms of another attack.

Cardiac Output: The volume of blood pumped out per minute by the ventricles of the heart. It is one measure of the heart's efficiency. At rest, the heart of a healthy adult will pump between 2.5 and 4.5 liters of blood every minute.

Cardiac Pacemaker: The natural pacemaker is the sinuatrial node, found at the base of the heart. The heart normally controls its rate and rhythm; heart block occurs when impulses cannot reach all parts of the heart. This may lead to arrhythmia, or even cause the heart to stop. Artificial pacemakers may then be used; in the United Kingdom these are required for around one person in every 2,000 of the population. Usually powered by mercury or lithium batteries, and lasting up to 15 years, they are either fixed to the outside of the chest or implanted in the armpit, and connected by a wire passing through a vein in the neck to the heart.

Cardiac Tamponade: Compression of

the heart due to abnormal accumulation of fluid within the fibrous covering of the heart (pericardium). The result is irregular rhythm and death if the fluid is not removed.

Cardiac-Assist Devices (CADs): Technological devices that assist the heart in performing its functions

Cardioangiography: Rendering the outline of the heart visible on an X-ray film by injecting a radio-opaque sub-stance into it.

Cardioglycosides: Sugar-containing plant substances that, in proper doses. Act as heart stimulants. Examples; digitoxin, strophanthin.

Cardiology: That branch of medical science devoted to the study of the diseases of the heart.

Cardiomegaly: Enlargement of the heart

Cardiomyopathy: A general term covering primary disease of the heart muscle.

Cardiopathies: Heart diseases, usually needing medical intervention.

Cardioplegia: A procedure whereby the heart is stopped by reducing its temperature (hypothermia), by injecting the muscle with a solution of salts or by electrostimulation. This enables surgeons to operate safely on the heart.

Cardiopulmonary Bypass: A procedure in which the body's circulation of blood is kept going when the heart is intention- ally stopped to enable heart surgery to be carried out. A heart-lung machine substitutes for the heart's pumping action and the blood is oxygenated at the same time.

Cardiopulmonary Resuscitation (CPR): The use of life-saving measures of mouth-to- mouth resuscitation and external cardiac compression massage in a person who has collapsed with cardiac arrest. Speedy restoration of the circulation of oxygenated blood to the brain is essential to prevent damage to brain tissues from oxygen starvation. The brain is irreversibly damaged if it is starved of oxygen for more than 4–5 minutes.

Cardiospasm: The spasmodic contraction of the muscle surrounding the opening of the oesophagus into the stomach: also termed achalasia of the cardia.

Cardiotonic: A substance that strengthens or regulates heart metabolism

Cardiotoxic: Chemically harmful to the cells of the heart.

Cardiovascular System: This refers to the whole circulatory system: the heart, the systemic circulation (the arteries and veins of the body) and the pulmonary circulation (the arteries and veins of the lungs). Blood circulates throughout the cardiovascular system bringing oxygen and nutrients to the tissues and removing carbon dioxide and other waste products.

Cardioversion: Cardioversion, or defibrillation, is indicated in patients with ventricular fibrillation or tachycardia, fast or irregular heartbeat, if other treatments have failed. A general anaes- thetic is given if the patient is conscious, following which a carefully timed direct-current shock is applied to the patient's chest wall using a defibrillator. The patient's ECG rhythm should then be monitored and anti- coagulants considered, as the risk of embolism is increased.

Care Management: A generic term which has been used in many different ways. Used by CIGNA, "Care Management" refers to a CIGNA HealthCare initiative that takes a global approach to medical care from prevention through treatment and recovery.

Care Card: A card issued by the Ministry of Health Services to each British Columbia resident registered with the Medical Services Plan. The Care card shows the cardholder's unique Personal Health Number.

Carminatives: Preparations to relieve flatulence, and any resulting griping, by the bringing up of wind, or eructation. Their essential constituent is an aromatic volatile oil, usually of vegetable extraction.

Carneous Mole: An ovum which has died in the early months of pregnancy. It usually requires no treatment and evacuates itself.

Carotene: A colouring matter of carrots, other plants, butter and yolk of egg, carotene is the precursor of vitamin A, which is formed from carotene in the liver

Carpal Tunnel Syndrome: A condition characterised by attacks of pain and tingling in the first three or four fingers of one or both hands. The attacks usually occur at night. Carpal tunnel syndrome is caused by pressure on the median nerve as it passes under the strong ligament that lies across the front of the wrist. The condition may respond to use of a night splint on the hand; otherwise a corticosteroid injection under the ligament may help. If not, pressure is relieved by surgical division of the compressing ligament.

Carpel: A simple pistil or modified leaflets forming a compound pistil.

Carpus: The Latin term for the wrist, composed of eight small bones firmly joined together with ligaments, but capable of a certain amount of sliding movement over one another.

Carrier: A term historically used for licensed insurance companies, although now is sometimes used to include both licensed insurers and HMOs.

Carry-over: Transfer in farming and agricultural processing of a component from one system such as soil or feed to another system such as a plant, animal or human being: carry-over is expressed as the concentration of a component in the second system divided by the concentration in the first. Process in analytical studies by which materials are carried into a reaction mixture ir which they do not belong.

Cartilage: A hard but pliant substance forming parts of the skeleton – for example, the cartilages of the ribs, of the larynx and of the ears. Microscopic- ally, cartilage is found to consist of cells arranged in twos or in rows, and embedded in a ground-glass-like material devoid of blood vessels and nerves. The end of every long bone has

Caruncle: Any small fleshy eminence, whether normal or abnormal.

Case Control Study: A study which starts with the identification of persons with the disease (or other outcome variable) of interest and a suitable control (comparison, reference) group of persons without the disease. The relationship of an attribute to the disease is examined by comparing the diseased and non-diseased with regard to how frequently the attribute is present or, if quantitative, the levels of the attribute, in the two groups.

Case Management: Coordination of services to help meet a patient's health care needs, usually when the patient has a condition which requires multiple services from multiple providers. This term is also used to refer to coordination of care during and after a hospital stay.

Case Mix: The mixture of clinical conditions and severity of condition met in patients in a particular healthcare setting

Caseation: A process which takes place in the tissues in tuberculosis and some other chronic diseases. The central part of a diseased area, instead of changing into pus and so forming an abscess, changes to a firm cheese-like mass which may next be absorbed or may be converted into a calcareous deposit and fibrous tissue, and so healing results in the formation of a scar.

Case-Control Study: Research in which 'cases' – that is, persons with a particular condition or receiving a particular medication are compared with a group similar in age, sex distribution and social class, etc. who do not have the condition or who are not receiving the drug.

Casein: That part of milk which forms cheese or curds. It is produced by the union of a substance, caseinogen, dissolved in the milk, with lime salts also dissolved in the milk – the union being produced by the action of rennin, a ferment from the stomach of the calf. The same change occurs in the human stomach as the first step in the digestion of milk, and therefore when milk is vomited curdled it merely shows that digestion has begun.

Cash Limit: The limit improved by the Government on the amount of cash which a public body may spend during a given financial year

Castration: This is literally defined as 'deprivation of the power of generation'. In practical terms this involves surgical removal of both ovaries, or both testicles. Such an operation is most commonly associated with the treatment of malignant lesions. In women who have reached the menopause, bilateral oophorectomy

Casts: Casts of hollow organs are found in various diseases. Membranous casts of the air passages are found in diphtheria and in one form of bronchitis, and are sometimes coughed up entire. Casts of the interior of the bowels are passed in cases of mucous colitis associated with constipation, and casts of the microscopic tubules in the kidneys passed in the urine form one of the surest signs of glomerulonephritis.

Catabolic: The part of metabolism that deals with destruction or simplification of more complex compounds. Catabolism mostly results in the release of energy. Examples: the release of glucose by the liver, the combustion of glucose by cells.

Catabolism: The breakdown by the body of complex substances to form simpler ones, a process that is accompanied by the release of energy. Among the substances catabolised are nutrients, such as carbohydrate and protein in food, and in storage in the body – for example, glycogen.

Catalepsy: A physical condition in which part or all of the body becomes rigid. It is characterised by the adoption of strange – often statue-like

– poses (catatonia), which may pass off within a few minutes or may last for several hours (rarely, days). Typically brought on by a sudden mental trauma, catalepsy may occur with prolonged depression or some other serious mental illness, and occasionally with epilepsy. Successful treatment must depend upon due recognition of all precipitating factors and circumstances.

Cataplexy: Cataplexy is a condition in which the patient has a sudden attack of muscular weakness affecting the whole body.

Cataract: An opacity of the lens sufficient to cause visual impairment.

Catarrh: Inflammation of the mucous membranes, particularly those of the air passages, associated with a copious secretion of mucus.

Catatonia: A condition in which an individual takes up odd postures, often accompanied by muteness or semi-coma. The arms and legs may be moved passively by someone else into positions that the sufferer then holds for many hours. Catatonia occurs in schizophrenia.

Catecholamines: The class of neurotra-

nsmitters that includes nor epinephrine and dopamine.

Cathartics: Substances which produce an evacuation of the bowels. The term 'cathartic' also means possessing the power to cleanse.

Catherisation: Use of a catheter.

Catheter: A tube usually made of silicone used to run fluids in and out of a person for CAPD.

Catheters: Hollow tubes, usually made of rubber or plastic, used for passing into various organs of the body, either for investigational purposes or to give some form of treatment. They are used under strict sterile conditions.

Catoplexy: Catoplexy is a condition of sudden muscular weakness or fatigue.

Cat-Scratch Fever: An infection in humans caused by a small gram-negative bacillus (Bartonella henselae). The domestic cat is a reservoir for the bacteria, and up to 50 per cent of the cat population may be affected. The disorder manifests itself as a skin lesion 3–10 days after a minor scratch; within two weeks the victim's lymph glands enlarge and may produce pus. Fever, headache and malaise occur in some patients. Antibiotics do not seem to be effective. The skin lesion and lymph-gland enlargement subside spontaneously within 2–5 months.

Cauda: A tail or a tail-like structure. For example, the cauda equina ('horse's tail') is a collection of nerve roots arising from the lumbar, sacral and coccygeal spinal nerves.

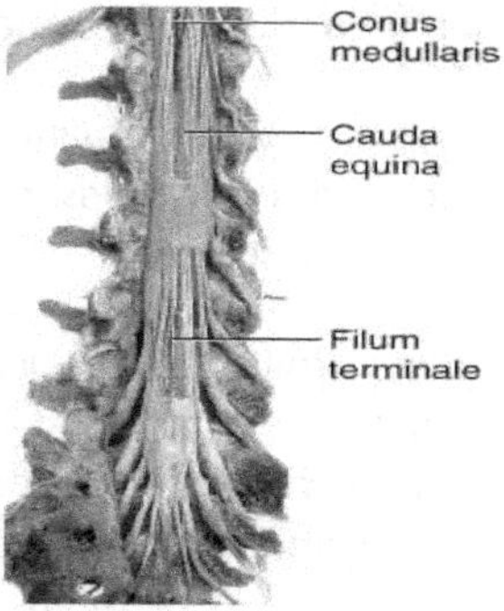

Cauliflower Ear: The term applied to the distortion of the external ear produced by repeated injury in sport. Initially it is due to a haematoma in the auricle. To prevent deformity the blood should be drawn off from this haematoma as soon as possible, and a firm pressure bandage then applied. Subsequent protection can be given to the ear by covering it with a few layers of two-way-stretch strapping wound around the head.

Cauline: Belonging to the stem, as in cauline leaves emerging from the stem

Causalgia: A severe burning pain in a limb in which the sympathetic and somatic nerves have been damaged.

Caustics and Cauteries: Caustics and cauteries are used to destroy tissues – the former by chemical action, the latter by their high temperature.

Cavernous Breathing: A peculiar quality of the respiratory sounds heard on auscultation over a cavity in the lung.

Cavernous Sinus: A channel for venous blood placed either side of the sphenoid bone at the base of the skull behind the eye sockets. Blood drains into it from the eye, the nose, the brain and part of the cheek, and leaves via the internal jugular and facial veins. Sometimes the sinus becomes blocked by a blood clot (thrombus), usually a complication of a nearby bacterial infection. A potentially serious condition, it should be treated with thrombolysis and antibiotics.

CBER: FDA's Center for Biologies Evaluation and Research

CCC: The building letter code for the Marilyn Reed Lucia Child Care Study Center, 610 Parnassus Avenue, San Francisco

CCPD: Continuous Cycle Peritoneal Dialysis. Uses the same principles as CAPD except a semi-automatic machine is used which runs for about 12 hours while the patient sleeps, or can be used 24-hours a day.

CCS (Clinical Classifications Software): CCS is a tool for clustering patient diagnoses and procedures into clinically meaningful categories. This software was developed by the Agency for Healthcare Quality and Research.

CCSC: Center for Consumer Self Care.

CD4/CD8 Count: An immunological assessment used to monitor for signs of organ rejection after transplantation; it is also used to check the progress of treatment in patients with

HIV. The count measures the ratio of helper-induced T-lymphocytes to cytotoxic- suppressor lymphocytes.

CDAs (Cost Distribution Accounts): The VA Cost Distribution Report includes patient care departments such as medicine, admitting screen, or ambulatory surgery. It also includes indirect departments such as building management. Expenditures for these departments are listed in cost distribution accounts (CDAs).

CDER: FDA's Center for Drug Evaluation and Research

CDISC: Clinical Data Interchange Standards Consortium (formerly a DIA special interest group called the Clinical Data Interchange Standards Committee)

CDL: California Digital Library, the name of a collection of journals, databases, and other reference materials available via computer

CDR (Cost Distribution Report): The CDR contains estimates of the costs expended by each patient care department in a VA medical center. Costs are distributed into Cost Distribution Accounts.

CDRH: FDA's Center for Devisees and Radiological Health

Ceiling Value (CV): U.S. term in occupational exposure indicating the airborne concentration of a potentially toxic substance which should never be exceeded in a worker's breathing zone.

Ceiling: The maximum biological effect that can be induced in a tissue by a given drug, regardless of how large a dose is administered. The maximum effect produced by a given drug may be less than the maximum response of which the reacting tissue is capable, and less than the maximum response that can be induced by another drug of greater intrinsic activity. "Ceiling" is analogous to the maximum reaction velocity of an enzymatic reaction when the enzyme is saturated with substrate.

Celecoxib (Celebrex): A nonsteroidal anti-inflammatory drug (NSAID) used to treat osteoarthritis and rheumatoid arthritis. Some research suggests that it might reduce Alzheimer's risk in persons with a family history of dementia.

Celiac: Pertaining to the abdomen.

Cell Line: Defined unique population of cells obtained by culture from a primary implant through numerous generations.

Cell Mediated Immunity: Immune response mediated by T lymphocytes, phagocytes and NK cells

Cell Salts (Biochemic Remedies, Tissue Salts): A homeopathic physician, W.H. Schuessler, developed the Biochemic system using 12 different 'cell salts'. Schuessier felt these were fundamental to the proper function of the human body. Prepared in low potency (3x or 6x) and used based on homeopathic indications.

Cell Strain: Cells having specific properties or markers derived from a primary culture or cell line.

Cell Mediated Hypersensitivity: State in which an individual reacts with allergic effects caused by the reaction of antigen-specific T-lymphocytes following exposure to a certain substance (allergen) after having been exposed previously to the same substance or chemical group.

Cell-Mediated Immunity: Immune response mediated by antigen-specific T-lymphocytes.

Cells: The basic structural unit of body tissues. There are around 10 billion cells in the human body and they are structurally and functionally linked to carry out the body's many complex activities. Every cell consists essentially of a cell-body of soft albuminous material called cytoplasm, in which lies a kernel or nucleus which seems to direct all the activities of the cell. Within the nucleus may be seen a minute body, the nucleolus; and there may or may not be a cell- envelope around all.

Cellular Biology: Cells, their components, and how they function.

Cellulitis: Inflammation taking place in cellular tissue, and usually referring to infection in the sub- cutaneous tissue. A related word, cellulite, which has no medical meaning, is used in the slimming business to refer to excess fatty tissue in the arms, buttocks and thighs.

Censored Data: Sample observations for which the complete distribution is not known : for example, a cohort study in which some persons cannot be followed to the predetermined end of the study ("right-censored data") or environmental assay data in which some results are less than the sample detection limit ("left-censored data").

Center for Chemical Diversity: Emerging collaborative research center created by the UCSF School of Pharmacy's department of pharmaceutical chemistry that promotes work at the frontier of chemistry and biology using innovative and developing techniques and technologies and focuses to date on drug discovery and combinatorial libraries

Center for Consumer Self-care: Emerging evidence-based center created by the School of Pharmacy's department of clinical pharmacy to foster smart and effective consumer decision making relative to prescription and nonprescription drugs and dietary supplements through programs in targeted research, education, patient care, and public service

Center for Pharmacogenomics: Emerging collaborative research center created by the UCSF School of Pharmacy's department of biopharmaceutical sciences with the School of Medicine's Program in Human Genetics to study the genetic variability of drug response in individuals and to use that knowledge to ultimately improve drug development and research, and to design more effective drugs with fewer side effects

Central Nervous System: A collective term for the brain, spinal cord, their nerves, and the sensory end organs. More broadly, this can even include the neurotransmitting hormones instigated by the CNS that control the chemical nervous system, the endocrine glands. This comprises the nervous tissue of the brain and spinal cord, but does not include the cranial and spinal nerves and the autonomic nervous system. This latter group makes up the peripheral nervous system.

Central Stage: Second stage of acute rejection during which activated CD4+T cells produce different cytokines which, in turn, activate CD8+ T cells, B cells, and macrophages

Central Venous Pressure: The pressure of blood within the right atrium of the heart as measured by a catheter and manometer.

Cerebellar Ataxia: Uncoordinated movements, including an unsteady gait, caused by damage to or disease of the cerebellum. Brain tumours, multiple sclerosis (MS) and stroke can result children with diplegia or athetoid symptoms may have normal or even high intelligence. Associated problems may include hearing or visual disability, behavioural problems and epilepsy.

Cerebral Palsy: The term used to describe a group of conditions characterised by varying degrees of paralysis and originating in infancy or early childhood. In some 80 per cent of cases this takes the form of spastic paralysis (muscle stiffness), hence the now obsolete lay description of sufferers.

Cerero Spinal Fluid: The fluid within the ventricles of the brain and bathing its surface and that of the spinal cord. Normally a clear, colourless fluid, its pressure when an individual is lying on one side is 50 to 150 mm water. A lumbar puncture should not be done if the intracranial pressure is raised

Cerebro Vascular Insufficiency: An inadequate supply of blood to the brain because of a narrowing of the blood vessels which lead to, or are in various areas of the brain.

Certified Reference Material: Reference material provided by a certifying body such as a National Standards Organization or Metrological Laboratory or by an international body which confirms its purity and analytical values by technically valid procedures and" provides a certificate detailing the relevant information.

Cerumen: The name for the wax-like secretion found in the external ear.

Cervical Cancer: Cancer of the cervix – the neck of the womb is one of the most common cancers affecting women throughout the world. In some areas its incidence is increasing. This cancer has clearly identifiable precancerous stages with abnormal changes occurring in the cells on the surface of the cervix: these changes can be detected by a cervical smear test. Early cancer can be cured by diathermy, laser treatment, electro coagulation or cryosurgery.

Cervical Smear: This screening test detects abnormal changes in the cells of the enabling an affected woman to have early treatment.

Cervical Vertebrae: The seven bones of the top end of the backbone that form the neck. The first cervical vertebra is the atlas and this articulates with the base of the skull. The axis is the second vertebra, which contains a shaft of bone that allows the atlas to rotate on it, thus permitting the head to turn.

Cervical: Cervical means anything pertaining to the neck, or to the neck of the womb.

Cervicitis: Inflammation of the cervix uteri or neck of the womb.

Cervico-occipital Headache: A headache of the neck and side of the head a tension headache.

Cervix Uteri: The neck of the womb or uterus, placed partly above and partly within the vagina.

CFR: Code of Federal Regulations.

CGL: Computer Graphics Laboratory.

Chagas' Disease: Chagas' disease, or American trypanosomiasis, is a disease widespread in Central and South America, and caused by the Trypanosoma cruzi. The disease is transmitted by the biting bugs, Panstrongylus megistus and Triatoma infestans. It occurs in an acute and a chronic form. The former, which is most common in children, practically always affects the heart, and the prognosis is poor. The chronic form is commonest in adolescents and young adults and the outcome depends upon the extent to which the heart is involved. There is no effective drug treatment.

Chain of Custody: Sequence of responsibility for a substance from the manufacturer to the distributor, to the user, or to the person(s) ultimately responsible for waste disposal. This term is also used in controlled transmission of samples from collection to analysis, especially of samples of materials used for medico-legal or forensic purposes.

Chalicosis: A disorder of the lungs found among stonecutters, and due to the inhalation of fine particles of stone.

Chancre: The primary lesion of syphilis.

Chancroid: A soft or non-syphilitic venereal sore, caused by a micro-organism known as Haemophilus ducreyi. It is usually acquired by sexual contact, and responds well to treatment with antibiotics. The disease is rare in the UK.

Change Window: A scheduled Pharma Net outage: usually from 12:01 a.m. to 8:00 a.m. every Thursday - during which time technical staff perform routine and scheduled maintenance on PharmaNet.

Chapped Hands: Chapped hands occur in cold weather, when reduced sweat and sebaceous activity leads to decreased natural protection of the skin. Prolonged immersion in soapy water, followed by exposure to cold air, results in cracking of the skin.

Characteristic Symptom: A symptom that is 'striking, strange, unusual, peculiar' in the case. Close attention is paid to characteristic symptoms as they must correspond to symptoms of the remedy if it is to cure. For example, burning pain better heat, or better lying on the painful side.

Characteristic Waste: A waste that exhibits any of the characteristics of toxicity, corrosiveness, ignitability, or reactivity.

Charcot's Joint: Named after a 19th-century French physician, this condition presents as a painless swelling and disorganisation of the joints resulting from damage to the pain fibres that occurs in dia- betic neuropathy.Leprosy, syringomyelia and syphylitic infection of the spinal cord.

Charge Amount: The amount billed by a provider for services rendered to a participant.

Charitable Gift Annuity: Typically an agreement in which you transfer cash or other assets to a charitable organization in exchange for its promise to pay you an annuity for life or for a term of years.

Charitable Trust: A trust having a

charitable organization as a beneficiary.

CHCE (Center for Healthcare Evaluation): CMS Centers for Medicare and Medicaid Services. CMS is a federal agency within the U.S. Department of Plealth and Human Services. CMS is responsible for the Medicare program and works with the States to run the Medicaid.

Cheilosis: Inflammation of the lips. It is common at the angles of the lips (angular stomatitis), usually as a consequence of sagging facial muscles or ill- fitting dentures producing folds at the corners of the mouth which retain food debris and allow proliferation of bacteria and candida. Atopic eczema is a common cause of cheilosis of the whole of the lips.

Cheiropompholyx: Pompholyx is an old name for vesicular eczema on the palms and fingers (cheiropompholyx) or soles of the feet (podopompholyx).

Chelating Agents: Chelating agents are compounds that will render an ion (usually a metal) biologically inactive by incorporating it into an inner ring structure in the molecule. (Hence the name, from the Greek chele = claw.) When the complex formed in this way is harmless to the body and is excreted in the urine, such an agent is an effective way of ridding the body of toxic metals such

Chelation Therapy: Treatment with a chelating agent to enhance the elimination or reduce the toxicity of a metal ion.

Chemical Biology: Composition, structure, properties, and interactions of chemicals within living organisms

Chemical Conversion: Change from one state or chemical structure to another.

Chemical Oxygen Demand (COD): Substance concentration of available oxygen (derived from a chemical oxidizing agent) required to oxidize the organic (and inorganic) matter in waste water.

Chemical Safety: Practical certainty that there will be no exposure of organisms to toxic amounts of any substance or group of substances: this implies attaining an acceptably low risk of exposure to potentially toxic substances.

Chemical Species: Set of chemically identical atomic or molecular structural units in a solid array or of chemically identical molecular entities that can explore the same set of molecular energy levels on the time scale of the experiment. For example, two conformational isomers may interconvert sufficiently slowly to be detectable by separate nuclear magnetic resonance spectra and hence be considered to be separate chemical species on a time scale governed by the radiofrequency of the spectrometer used. On the other hand, in a slow chemical reaction the same as a single chemical species, i.e., there is a virtually complete equilibrium set- of molecular energy levels belonging to the two conformers. Except where the context requires otherwise, the term is taken to refer to a set of molecular, isotopes in their natural abundance. The wording of the definition given is intended to embrace both cases such as. Graphite, sodium chloride, or a surface oxide where the basic structural units are not capable of a separate existence as well as those cases where they are.

Chemical: Caused by combination of agonist with antagonist, with resulting inactivation of the agonist, e.g., dimercaprol and mercuric ion.

Chemistry: Scientific study of the composition, structure, and properties of substances and the changes.

Chemo Sterilizer: Substance used to sterilize mites insects, rodents or other

animals.

Chemophobia: Irrational fear of chemicals.

Chemosis: Swelling of the conjunctiva of the eye, usually caused by inflammation from injury or infection.

Chemotaxis: The ability of certain cells to attract or repel others.

Chemotherapy: Drug treatment of parasitic or neoplastic disease in which the drug has a selective effect on the invading cells or organisms. The prevention or treatment of disease by chemical substances. The term is generally used in two senses: the use of antibacterial and other drugs to treat infections; and the administration of anti metabolites and other drugs to treat cancer. The discovery by Paul Ehrlich in 1910 of the action of Salvarsan in treating syphilis led to the introduction of sulphonamides in 1935, followed by penicillin during World War II, which revolutionised the treatment of common infections.

Chest: The chest, or thorax, is the upper part of the trunk. It is enclosed by the breastbone (ster- num) and the 12 ribs which join the sternum by way of cartilages and are attached to the spine behind. At the top of the thorax, the opening in between the first ribs admits the windpipe (trachea), the gullet (oesophagus) and the large blood vessels. The bottom of the thorax is separated from the abdomen below by the muscular diaphragm which is the main muscle of breathing.

Cheyne-Stokes Breathing: A type of breathing which gets very faint for a short time, then gradually deepens until full inspirations are taken for a few seconds, and then gradually dies away to another quiet period, again increasing in depth after a few seconds and so on in cycles. It is seen in some serious neurological disorders, such as brain tumours and stroke, and also in the case of per- sons with advanced disease of the heart or kidneys. When well marked it is a sign that death is impending, though milder degrees of it do not carry such a serious implication in elderly patients.

Chiasma: This is an X-shaped crossing. The optic chiasma is where the nerve fibres from the nasal half of each retina cross over the mid line to join the optic tract from the other side.

Chickenpox: Also known as varicella. An acute, contagious disease predominantly of children – although it may occur at any age characterized by fever and an eruption on the skin. The name, chickenpox, is said to be derived from the resemblance of the eruption to boiled chick- peas. Causes The disease occurs in epidemics affecting especially children under the age of ten years. It is due to the varicella zoster virus, and the condition is an extremely infectious one from child to child.

......Continued

Chigger: Another name for Trombicula autumnalis, popularly known as the harvest mite.

Chilblain: Chilblain, or erythema pernio is an inflamed condition of the hands or feet, or occasionally of the ears, and should not be confused with cracked or chapped hands. Most commonly found in child- hood and old age, it may be associated with generally poor health, though there may also be a genetic predisposition. Prevention with good food, warm clothing, a warm environment, and

Child Abuse: This traditional term covers the neglect, physical injury, emotional trauma and sexual abuse of a child. Professional staff responsible for the care and well-being of children now refer to physical injury as 'non-accidental injury'.

Children Act: The Children Act 1989 (Children Act) introduced major reforms of child-care law. It encourages negotiation and cooperation between parents, children and professionals to resolve problems affecting children.

Chimera: Chimera is an organism, whether plant, animal or human being, in which there are at least two kinds of tissue differing in their genetic constitution.

Chimeric: Hybrid produced by combining components from different sources

Chinese Avian Influenza: A variety of influenza in chickens occurring in southern China that in 1997 appeared to jump the species barrier and infect humans. Some cases of the human version of the infection occurred in Hong Kong. There were fears of a serious epidemic which, because of a lack of natural resistance among humans, might have led to its worldwide spread. This has not so far occurred.

Chiropody: Chiropody (also termed podiatry) is that part of medical science which is concerned with the health of the feet. Its practitioners are specialists capable of providing a fully comprehensive foot-health service. This includes the palliation ofestablished deformities and dysfunction, both as short-term treatment for immediate relief of painful symptoms and as long-term management to secure optimum results.

Chiropractor: A person who practises chiropractic – mainly a system of physical manipulations of minor displacements of the spinal column. These minor displacements of the spine are believed to affect the associated or neighbouring nerves and so cause malfunctions of the muscles throughout the body. By manipulating the affected part of the spinal column the patient's complaint, whatever it may be for example, backache – is relieved.

Chlamydia: A genus of micro-organisms which include those responsible for non-specific urethritis (nsu), ornithosis, psittacosis and trachoma. Chlamydia trachomitis can be sexually transmitted by both men and women and in developed countries is the most significant cause of NSU.

Chloasma: This is an increase in the melanin pigment of the skin as a result of hormonal stimulation. It is commonly seen in pregnancy and sometimes in women on the contraceptive pill.

Chloracne: Acne-like eruption caused

by exposure to certain chlorinated organic substances such as polychlorinated biphenyls or 2, 3, 7, 8-tetrachlorodibenzo-p-dioxin.

Chlorofluorocarbon: An organic chemical composed of chlorine, fluorine and carbon atoms, usually characterized by high stability contributing to a high ozone-depleting potential (ODP).

Chloroma: Chloroma, or green cancer, is the name of a disease in which greenish growths appear under the skin, and in which a change takes place in the blood resembling that in leukaemia.

Chlorophyll: The green colouring matter of plants. Its main use is as a colouring agent, principally for soaps, oils and fats. It is also being found of value as a deodorant dressing to remove, or diminish, the unpleasant odour of heavily infected sores and wounds.

Choking: Choking is the process which results from an obstruction to breathing situated in the larynx. It may occur as the result of disease causing swelling round the glottis (the entrance to the larynx), or of some nervous dis- orders that interfere with the regulation of the muscles which open and shut the larynx. Generally, however, it is due to the irritation of a piece of food or other substance introduced by the mouth, which provokes coughing but only

Cholagogues: Substances which increase the flow of bile by stimulating evacuation of the gall-bladder. The great majority of theseact only by increasing the activity of the digestive organs, and so producing a flow of bile already stored up in the gall-bladder. Substances which stimu- late the liver to secrete more bile are known as choleretic.

Cholangiography: The process whereby the bile ducts and the gall-bladder are rendered radio-opaque and therefore visible on an X-ray film.

Cholangitis: Inflammation of of only bile ducts. This word and the next three describe conditions that may be, subjectively, all the same.

Cholecystalgia: Cramps or tenesmus of the gall bladder or bile ducts.

.....Continued

Cholecystectomy: Removal of the gall-bladder by operation.

Cholecystitis: Inflammation of the gall bladder and ducts, sometimes from the presence of passing stones, sometimes following fasting or anorexia, sometimes because of a spreading intestinal tract infection, sometimes just because you eat three avocado sandwiches before going to bed.

Cholecystitis: Inflammation ofthe gall-bladder.

Cholecystography: The process whereby the gall-bladder is rendered radio-opaque and therefore visisble on an X-ray film.

Cholecystokinin: The hormone released from the lining membrane of the duodenum when food is taken, and which initiates emptying of the gall-bladder.

Cholelithiasis: The presence of gall-stones in the bile ducts and/or in the gall-bladder.

Cholelithotomy: The removal of gall-stones from the gallbladder or bile ducts, when cholecystectomy or lithotripsy are inappropriate or not possible. It involves a cholecystomy, an operation to open the gall-bladder.

Cholera: Bacterial infection caused by Vibrio cholerae. The patient suffers profuse watery diarrhoea, and resultant dehydration and electrolyte imbalance. Formerly known as the Asiatic cholera, the disease has occurred in epidemics and pandemics for many centuries. When it entered Europe in 1853, Dr John Snow, a London anaesthetist, carried out seminal epidemiological work in Soho, London, which established that the source of infection was con- taminated drinking water derived from the Broad Street pump. Several smaller epidemics involved Europe in the latter years of the 19th century, but none has arisen in Britain or the United States for many years.

Choleretic: The term applied to a drug that stimulates the flow of bile.

Cholestasis: reduction or stoppage in the flow of bile into the intestine caused either by a blockage such as a stone in the bile duct or by liver disease disturbing the production of bile. The first type is called extrahepatic biliary obstruction and the second, intrahepatic cholestasis. The patient develops jaundice and itching and passes dark urine and pale faeces. Cholestasis may occasionally occur during pregnancy.

Cholesterol: A fatty substance produced predominantly by the liver, and necessary for building cell membranes, insulating the CNS, covering fats for blood transport, forming bile acids, oiling the skin and making steroid hormones. Blood cholesterols are not derived from food (digestion breaks them down) but are intentionally synthesized by the liver, in response to seeming need.

Cholesterol: A lipid that is an important constituent of body cells and so widely distributed throughout the body. It is especially abundant in the brain, nervous tissue, adrenal glands and skin. It is also found in egg yolk and gall-stones. Cholesterol plays an portant role in the body, being essential for the production of the sex hormones as well as for the repair of membranes. It is also the source from which BILE acids are manufactured. The total amount in the body of a man weighing 70 kilograms (10 stones) is around 140 grams, and the amount present in the blood is 3.6–7.8 mmol per litre or 150–250 milligrams per 100 millilitres.

Cholestyramine: A drug of value in the treatment of the pruritus, or itching, which occurs in association with jaundice. It does this by 'binding' the bile salts in the gut and so preventing their being reabsorbed into the bloodstream, where their excess in jaundice is responsible for the itching. It reduces the level of cholesterol and

Choline: One of the many constituents of the vitamin B complex. Lack of it in the experimental animal produces a fatty liver. It is found in egg-yolk, liver, and meat. The probable daily human requirement is 500 mg, an amount amply covered by the ordinary diet. Choline can be synthesised by the body.

Cholinergic: A description of nerve fibres that release acetylcholine as a neurotransmitter.

Cholinergic: The parts of the nervous system that use acetylcholine as a neurotransmitter.

Cholinesterase Inhibitor: Substance which inhibits the action of acetylcholinesterase and related enzymes which catalyse the hydrolysis of choline esters: such a substance causes hyperactivity in parasympathetic nerves.

Cholinesterase: An enzyme that helps to break down the neurotransmitter compound acetylcholine.

Chondroma: A tumour composed in part of cartilage.

Chorda: A nerve fibre, tendon or cord.

Chorea, Sydenham's: A disease or syndrome of children, usually following or companion to rheumatic fever, and having involuntary movements, anxiety and impaired memory. It usually clears up in two or three months.

Chorea: Chorea, or St Vitus's dance, is the occurrence of short, purposeless involuntary movements of the face, head, hands and feet. Movements are sudden, but the affected person may hold the new posture for several seconds. Chorea is often accompanied by athetosis, when it is termed choreoathetosis. Choreic symptoms are often due to disease of the basal ganglion in the brain. The withdrawal of phenothiazines may cause

Choriocarcinoma: A form of cancer affecting the chorion, in the treatment of which particularly impressive results are being obtained from the use of methotrexate.

Chorion: This is the more external of the two fetal membranes.

Chorionic Gonadotrophic: Hormone A hormone produced by the placenta during pregnancy. It is similar to the pituitary gonadotrophins, which are blocked during pregnancy. Large amounts appear in a woman's urine when she is pregnant and are used as the basis for pregnancy tests. Human gonadotrophins are used to treat delayed puberty and premenstrual tension.

Choroid Plexus: An extensive web of blood vessels occurring in the ventricles of the brain and producing the cerebrospinal fluid.

Christmas Disease: A hereditary disorder of blood coagulation which can only be distinguished from haemophilia by laboratory tests. It is so-called after the surname of the first case reported in this country. About one in every ten patients clinically diagnosed as haemophiliac has in fact Christmas disease. It is due to lack in the blood of Factor IX

Chromaffin: A term applied to certain cells and organs in the body, such as part of the adrenal glands, which have a peculiar affinity for chrome salts. These cells and tissues generally are supposed to secrete substances which have an important action in maintaining the tone and elasticity of the blood vessels and muscles.

Chromatid: Either of two filaments joined at the centromere which make up a chromosome.

Chromatin: The genetic material found in the nucleus of a cell. It consists of protein and deoxyribonucleic acid (DNA). During mitotic division of the cell, chromatincondensesinto chromo-somes.

Chromic Acid: Chromic acid is used in several industries, particularly in chromium plating. Unless precau- tions are taken it may lead to dermatitis of the hands, arms, chest and face. It may also cause deep ulcers, especially of the nasal septum and knuckles.

Chromosomal Aberration: Abnorma-lity of chromosome number or structure.

Chromosome: Self-replicating struc-ture consisting of DNA complexed with various proteins and involved in the storage and transmission of genetic information; the physical structure that contains the genes.

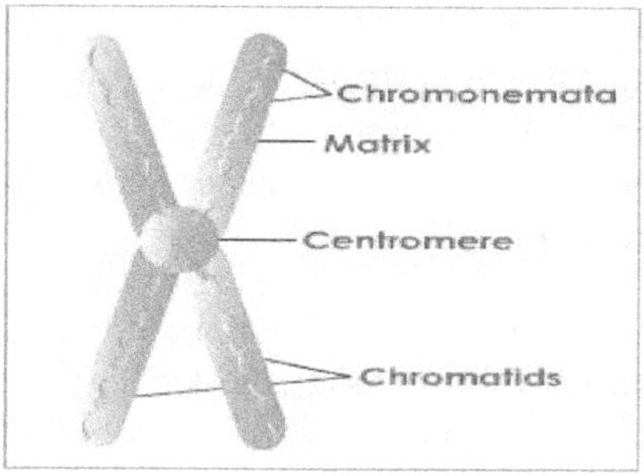

Chromosomes: The rod-shaped bodies to be found in the nucleus of every cell in the body. They contain the genes or hereditary elements, which establish the characteristics of an individual. Com- posed of a long double-coiled filament of DNA, they occur in pairs – one from the maternal, the other from the paternal – and human beings possess 46, made up of 23 pairs. The number of chromosomes is specific for each species of animal. Each chromosone can duplicate an exact copy of itself between each cell division.

Chronic: Long duration, possibly years, with frequent recurrence over a long time, and often by slowly progressing seriousness.

Chronic Disorder: A persistent or recurring condition or group of symptoms. Chronic disorders are customarily contrasted with acute diseases which start sud- denly and last a short time. The symptoms of acute disease often include breathlessness, fever, severe pain and malaise, with the patient's condition changing from day to day or even hour to hour.

Chronic Effect: Consequence which develops slowly and has a long-lasting course (often but not always irreversible).

Chronic Exposure: Continued exposures occurring over an extended period of time, or a significant fraction of the test species' or of the group of individuals', or of the population's life-time. Adverse effects following chronic exposure. Effects which persist over a long period of time whether or not they occur immediately upon exposure or are delayed.

Chronic Fatigue Syndrome: CFS is a recently designated semi-disease, often attributed to EBV (the Epstein-Barr virus) or CMV (Cytomegalovirus) infections, characterized by FUOs (Fevers of Unknown Origin) and resulting in the patient suffering FLS (Feels Like Shit). In most of us, the microorganisms involved in CFS usually provoke nothing more than a head cold; in some individuals, however, they induce a long, grinding, and debilitating disorder, characterized by exhaustion, depression, periodic fevers...a crazy-quilt of symptoms that frustrates both the sufferer and the sometimes skeptical physician. MCS (Multiple Chemical Sensitivities) are another syndrome that is often lumped with CFS, and they may often be two faces of the same condition.

Chronic Obstructive Pulmonary Disease (COPD): This is a term encompassing chronic bronchitis, emphysema, and chronic asthma where the airflow into the lungs is obstructed. Chronic bronchitis is typified by chronic productive cough for at least three months in two successive years (provided other causes such as tuberculosis, lung cancer and chronic heart failure have been excluded). The characteristics of emphysema are abnormal and permanent enlargement of the airspaces (alveoli) at the furthermost parts of the lung tissue.

Chronic Toxicity Test: Study in which organisms are observed during the greater part of the life span and in which exposure to the test agent takes place over the whole observation time or a substantial part thereof.

Chronic: A disease or imbalance of long, slow duration, showing little overall change and characterized by periods of remission interspersed with acute episodes.

Chronotoxicology: Study of the influence of biological rhythms on the

toxicity of substances.

Chyle: The milky fluid which is absorbed by the lymphatic vessels of the intestine. The absorbed portion consists of fats in very fine emulsion, like milk, so that these vessels receive the name of lacteals (L. lac, milk). This absorbed chyle mixes with the lymph and is discharged into the thoracic duct, a vessel, which passes up through the chest to open into the jugular vein on the left side of the neck, where the chyle mixes with the blood.

Chylomicrons: These are organized blobs of fats, synthesized in the submucosa of the small intestine out of dietary fats, phospholipids, specialized proteins and cholesterol, carried out of the intestinal tract by the lymph, and slowly released into the bloodstream. In the capillaries, the triglycerides inside the chylomicrons, recognized by their protein markers, are absorbed into the tissues for fuel or storage, and the outside cholesterol and phospholipid transport-cover continues through the blood to be absorbed by the liver for its use. This sideways approach takes (ideally) a large part of dietary fats into the lymph back alleys, spreading their release into the bloodstream out over many hours, thereby avoiding short-term blood fat and liver fat overload.

Chyluria: The passage of chyle in the urine. This results in the passing of a milky-looking urine. It is one of the manifestations of filariasis, where it is due to obstruction of the lymphatics by the causative parasite.

Chyme: Partly digested food as it issues from the stomach into the intestine. It is very acid and grey in colour, containing salts and sugars in solution, and the animal food softened into a semi-liquid

Chymopapain: An enzyme obtained from the paw-paw, which is being used in the treatment of prolapsed intervertebral disc. When injected into the disc it dissolves the cartilage tissue.

Chymotrypsin: An Enzyme produced by the pancreas which digests protein. It is used as an aid in operations for removal of a cataract, and also by inhalation to loosen and liquefy secretions in the windpipe and bronchi.

Cl, Clx: Clearance - in volume/unit time - of a drug or • chemical from a body fluid, usually plasma or blood, by specified route(s) and mechanism(s) of elimination, as indicated by a subscript, e.g., C1R, urinary clearance; C1H, hepatic clearance, etc

Cicatrix: Another word for scar.

Ciclosporin A: A drug used to prevent the rejection of transplanted organs such as the heart and kidneys.

Cilia: A term applied to minute, lash-like processes which are seen with the aid of the microscope upon the cells covering certain mucous membranes – for example, the trachea (or windpipe) and nose – and which maintain movement in the fluid passing over these membranes. They are also found on certain bacteria which have the power of rapid movement.

Ciliary Body: That part of the eye that connects the iris and the choroid. The ciliary ring is next to the choroid; the ciliary processes comprise many ridges behind the iris, to which the lens's suspensory ligament is attached; and the ciliary muscle contracts to change the curvature of the lens and so adjust the accommodation of the eye.

Cinchona: The general name for several trees in the bark of which quinine is found. This bark is also known as Jesuit's bark, having been first brought to notice by Spanish priests in South America, and first brought to Europe by the Countess of Cinchon, wife of the Viceroy of Peru, in 1640. The red cinchona bark contains the most quinine; quinine is usually prepared from this. Various extracts and tinctures are made direct from cinchona bark, and used in place of quinine.

Circle of Willis: A circle of arteries at the base of the brain, formed by the junction of the basilar, posterior cerebral, internal carotid and anterior cerebral arteries. Congenital defects may occur in these arteries and lead to the formation of aneurysm.

Circulation of Substances in the Environment: Movement of xenobiotic substances in the environment with air flow, river current, soil, water, etc.

Circumboreal: Plants that are found worldwide, encircling the lands around the north pole.

Circumcision: A surgical procedure to remove the prepuce of the penis in males and a part or all of the external genitalia in females. Circumcision is mainly done for religious or ethnic reasons; there is virtually no medical or surgical reason for the procedure. (The prepuce is not normally retractable in infancy, so this is not an indication for the operation – by the age of four the prepuce is retractable in most boys.)

Cirrhosis, Laennecs: The most common type of cirrhosis, caused by chronic alcoholism and a lousy diet (or malabsorption).

Cirrhosis: Cirrhosis, or fibrosis, is a diseased condition, in which the proper tissue is replaced by fibrous tissue similar to scar tissue. The name cirrhosis was originally given by Laennec to the disease as occurring in the liver, because of its yellow colour. Liver disease defined by histological examination and characterized by increased fibrous tissue, abnormal physiological changes such as loss of functional liver cells, and increased resistance to blood flow through the liver portal hypertension).

Cirsoid Aneurysm: The condition in which a group of arteries become abnormally dilated and tortuous.

Cisterna Chyli: A sac in the back of the pelvic region that drains the lymph from the intestinal tract, pelvis and legs, and acts as the beginning of the thoracic duct.

CIT: Center for Instructional Technology, the name of a promoter of the use of computers in delivering course materials

CL: The building letter code for the campus Library, 530 Parnassus Avenue, San Francisco

Claim Slip (Carnet De Reclamation): A claim slip is a document that enables a person to obtain: certain drugs prescribed by a doctor (in some cases the person must pay a deductible and a co-insurance amount); certain services, such as eye examinations and dental care.

Claim: A request to Pharma Care, submitted manually or on PharmaNet, for payment of the cost of providing a product or service.

Claims History: A record of claims and expenditure information for each patient. The claims history is contained in the PharmaCare Centralized Information System (PCIS) component of PharmaNet and contains the most recent six months.

Class I MHC: HLAs that are expressed on the membranes of most cells and tissues

Class II MHC: HLAs that are usually only expressed on the membranes of B cells, macrophages, and dendritic cells.

Clastogen: Agent causing chromosome breakage and/or consequent gain, loss or rearrangement of pieces of chromosomes.

Clastogenesis: Occurrence of chromosomal breaks and/or consequent gain, loss or rearrangement of pieces of chromosomes.

Claudication: A cramp-like pain that occurs in the legs on walking. It may cause the sufferer to limp or, if severe, stop him or her from walking. The usual cause is narrowing or blockage of the arteries in the legs due to atherosclerosis: smoking is a contributary factor. Intermittent claudication occurs when a person has to stop every so often to let the pain – caused by the build-up of waste products in the muscles – to subside.

Claustrophobia: Morbid fear of being in a confined space, or the fear experienced while in it. Claustrophobia may develop because of a previous unpleasant experience in a confined space. Cognitive behaviour therapy may help C patients whose daily lives are seriously affectedby this disorder.

Clavicle: The bone which runs from the upper end of the breastbone towards the tip of the shoulder across the root of the neck. It supports the upper limb, keeps it out from the side, and gives breadth to the shoulders. The bone is shaped like an 'f' with two curves, which give it increased strength. It is, however, liable to be broken by falls on the hand or on the shoulder, and is the most frequently fractured bone in the body.

Claw-Foot: Claw-foot, or pes cavus, is a familial deformity of the foot characterised by an abnormally high arch of the foot accompanied by shortening of the foot, clawing of the toes, and inversion, or turning inwards, of the foot and heel. Its main effect is to impair the resilience of the foot resulting in a stiff gait and aching pain. Milder cases are treated with special shoes fitted with a sponge rubber insole. More severe cases may require surgical treatment.

Claw-Hand: A (contraction) deformity of the hand and fingers, especially of the ring and little fingers. The condition is generally due to paralysis of the ulnar nerve. A somewhat similar condi- tion is produced by contraction of the fibrous tissues in the palm of the hand, partly due to rheumatic changes and partly to injury caused by the constant pressure of a tool against the palm of the hand.

Clearance: Volume of blood or plasma or mass of an organ effectively cleared of a substance by elimination (metabolism and'excretion) in a given time interval: clearance is expressed in units of volume or mass per unit of time. Total clearance for a component is the sum of the clearances of each eliminating organ or tissue for that component. In pulmonary toxicology, clearance refers specifically to removal of any inhaled substance which deposits on the lining surface of the lung, lung clearance is expressed in volume or mass of lung cleared per unit time. In renal toxicology, clearance refers to the quantification of the removal of a substance by the kidneys by the processes of filtration and secretion: clearance is calculated by relating the rate of renal excretion to the plasma concentration.

Cleft Foot: A rare congenital abnormality characterised by the absence of one or more toes and a deep central cleft that divides the foot into two. It is sometimes known as lobster foot, or lobster claw. It may be accompanied by other con- genital defects, such as cleft hand, absent permanent teeth, cleft palate (and/or lip), absence of the nails, and defects of the eye.

Cleft Hand: A rare congenital abnormality characterised by the absence of one or more fingers and a deep central cleft that divides the hand into two. It is sometimes known as lobster hand. It may be accompanied by other congenital defects, such as cleft foot, absent permanent teeth, cleft palate (and/or lip), absence of the nails and defects of the eye.

Cleft Palate: A fissure in the roof of the mouth (palate) and/ or the lip which is present at birth. It is found in varying degrees of severity in about one in 700 children. Modern plastic surgery can greatly improve the functioning of lips and palate and the appearance of the baby. Further cosmetic surgery later may not be necessary.

Clicking Finger: A condition usually occurring in middle-aged people in which the victim finds on wakening in the morning that he or she cannot straighten the ring or middle finger spontaneously, but only by a special effort, when it suddenly straightens with a painful click. Hence the name. In due course the finger remains bent at all times unless a special effort is made to straighten it with the other hand.

Client Registry: Information maintained by the Ministry of Health Services, including databases and access routines, intended to keep track of persons and organizations served by, or providing service to, the Ministry. The client registry is the control point for issuing of new personal health numbers.

Climacteric: This was a word originally applied to the end of certain epochs or stages in the life of an individual, at which some great change was supposed to take place.

Clinical Audit: A medical audit carried out by health professionals.

Clinical Governance: A concept, introduced in a UK government White Paper in 1997, aimed at 'ensuring that all NHS organisations have in place proper processes for continuously monitoring and improv- ing clinical audit'. All clinicians and managers are now expected to understand their individual and collective responsibilities for assuring accountability for the quality of patient care. Clinical governance is now included in the NHS legislation based on the 1997 white paper.

Clinical Guidelines: Systematically developed statements which assist clinicians and patients to decide on appropriate treatments for specific conditions. The guidelines are attractive to health managers and patients because they are potentially able to reduce variation in clinical practice. This helps to ensure that patients receive the right treatment of an acceptable standard.

Clinical Network: A network of health professionals from different NHS organisations working together across institutional and local boundaries to provide optimum care for a particular clinical condition

Clinical Pharmacy Care: Direct health care and consultation provided by clinically trained pharmacists to patients or consumers about prescription and nonprescription drugs, and related products.

Clinical Pharmacy Residents: Post-Pharm D professionals training in general pharmacy practice and/or specialty areas, such as oncology and pediatrics.

Clinical Pharmacy: Pharmacy practice philosophy that has shifted the focus of pharmacists from drug products themselves to the safe and optimally effective use of drug products in patients; this philosophy positions pharmacists as active members of the health care team, working side by side with physicians and nurses, to provide direct care to patients and consultation to patients' families

Clinical Phase Time: Time from filing of an IND, which is necessary to begin testing of a new compound in human subjects, to the submission of an NDA or BLA

Clinical Practice Guidelines: General procedures and suggestions about what constitutes an acceptable range of practices for particular diseases or conditions. These guidelines are usually developed by a consensus of doctors in a given field, such as radiology or cardiology.

Clinical Psychology: Psychology is the scientific study of behaviour. It may be applied in various settings including education, industry and health care. Clinical psychology is concerned with the practical application of research findings in the fields of physical and mental health. Training in clinical Clinical Trials psychology involves a degree in psychology followed by postgraduate training. Clinical psychologists are specifically skilled in applying theoretical models and objective methods of observation and measurement, and in therapeutic interventions aimed at changing patients' dysfunctional behaviour, including thoughts and feelings as well as actions.

Clinical Research (Trials): Systematic measures used to assess safety & efficacy of new medicines. They compare the new treatment under evaluation with a placebo or the previous standard treatment.

Clinical Risk Management: Initially driven by anxiety about the possibility of medical negligence cases, clinical risk management has evolved into the study of iatrogenic disease. The first priority of risk managers is to ensure that all therapies in medicine are as safe as possible. Allied to this is a recognition that errors may occur even when error-prevention strategies are in place. Lastly, any accidents that occur are analysed, allowing

Clinical Science Research: Studies involving humans, usually carried out in hospital or clinical settings, aimed at understanding the diagnosis and treatment of diseases and disorders

Clinical Scientists: Clinicians who conduct research on the effects of drugs and other interventions in humans.

Clinical Signs: The physical manifestations of an illness elicited by a doctor when examining a patient for example, a rash, lump, swelling, fever or altered physical function such as reflexes.

Clinical Study: Testing of a new compound that may be conducted in patients or healthy volunteer subjects

Clinical Symptoms: The experiences of a patient as communicated to a doctor, for example, pain, weakness, cough. They may or may not be accompanied by confirmatory clinical signs.

Clinical Therapeutic Index: Some indices of relative safety or relative effectiveness cannot be defined explicitly and uniquely, although it is presumed that the same quantifiable and precise criteria of efficacy and safety will be used in comparing drugs of similar kinds.

Clinical Trials: Clinical trials aim to evaluate the relative effects of different health-care interventions. They are based on the idea that there must a fair comparison of the alternatives in order to know which is better. Threats to a fair comparison include the play of chance and bias, both of which can cause people to draw the wrong conclusions about how effective a treatment or procedure is. Prospective studies in humans that compare the effectiveness and value of a potential new drug or therapy in one group against a control group before the drug or therapy is made available to the general population.

Clinical: Clinical means literally 'belonging to a bed', but the word is used to denote anything associated with the practical study or observation of sick people – as in clinical medicine, clinical thermometers.

Clitoris: A small, sensitive organ comprising erectile tissue at the top of the female genitalia where the labial folds meet below the pubic bone. Clitoral tissue extends into the anterior roof of the vagina. During sexual excitement the clitoris enlarges and hardens and may be the focus of orgasm.

Clone: A group of cells genetically identical to each other that have arisen from one cell by asexual reproduction.

Clonic: A word applied to short spasmodic movements.

Cloning: Cloning – from the Greek klon, meaning a cut- ting such as is used to propagate plants – is essentially a form of asexual reproduction. The initial stages were first successfully achieved in rabbits. In essence the technique consists of destroying the nucleus of the egg and replacing it with the nucleus from a body cell of the same species – either a male or a female. This pro- vides the egg with a full complement of chromosomes and it starts to divide and grow just as it would if it had retained its nucleus and been fertilised with a spermatozoon.

Clonus: A succession of intermittent muscular relaxations and contractions usually resulting from a sustained stretching stimulus. An example is the clonus stimulated in the calf muscle by maintaining sustained upward pressure on the sole of the foot. The condition is often a sign of disease in the brain or spinal cord.

Closed Biomass: The production of synthetic fuels, such as methane or hydrogen, from terrestrial and sea plants, and from animal wastes.

Closed-loop Recycling: Use or reuse of a waste as an ingredient or feedstock in the production process on-site. Recycling (in which a waste is recovered and reused in the same production process onsite) is a form of pollution prevention.

Clot: The term applied to any semi-solid mass of blood, lymph or other body fluid. Clotting in the blood is due to the formation of strings of fibrin produced by the action of ferment. Milk clots in a similar manner in the stomach when exposed to the action of the enzyme rennin. Clotting occurs naturally when blood is shed and comes into contact with tissues outside the blood vessels.

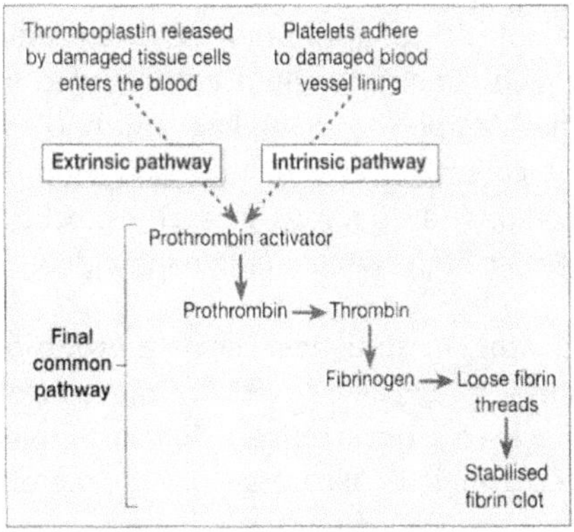

Clubbing: The term applied to the thickening and broadening of the fingertips – and, less commonly, the tips of the toes – that occurs in certain chronic diseases of the lungs and heart. It is due to interstitial oedema especially at the nail bed, leading to a loss of the acute angle between the nail and the skin of the finger. Clubbing is associated with lung cancer, empyema, bronchiectasis and congenital cyanotic heart disease.

Cluster Headaches: A type of migraine occurring in clusters – that is, a patient may have an attack daily for several days and then none for weeks or months. The pain is on one side of the head, often centred over the eye. The pain is excruciatingly severe and often associated with tearing, nasal discharge and production of thick saliva from the same side of the mouth. It is treated either with

Cluster Sampling: A method of sampling in which the population is divided into aggregates (or clusters) of items bound together in a certain manner. A sample of these clusters is taken at random and all the items which constitute them are included in the sample. A sampling method in which each unit selected is a group of persons (all persons in a city block, a family, etc.) rather than an individual.

Cluster: In statistical terms a group of subjects, closely linked in time and/or place of occurrence. For example, geographical clusters of leukaemia have been found – that is, an unexpectedly large number of persons with the disease who live in close proximity. Much research goes into trying to discover the cause of clusters but sometimes they appear to have occurred randomly.

C_{max}, C_{min}: The maximum or "peak" concentration (Cmax) of a drug observed after its administration; the minimum or "trough" concentration (Cmin) of a drug observed after its administration and just prior to the administration of a subsequent dose.

CMC: Chemistry, manufacturing and controls

CMS: Centers for Medicare and Medicaid Services. An agency housed within the U.S. Department of Health and Human Services (DHHS), CMS administers Medicare, Medicaid, related quality assurance programs, and other programs. Formerly known as the Health Care Financing Administration (HCFA).

CMV (Cytomegalovirus): This subtle, worldwide microorganism is a member of the herpes virus group. It is large for a virus, contains DNA, and has a complex protein capsid. It forms latent, lifelong infections, and, except for occasional serious infections in infants and malnourished youngsters, seldom produced a disease state. With increased use of immunosuppression therapies for conditions ranging from arthritis to cancer to organ transplants, the incidence of adults with major infections of CMV increases yearly.

CNS: Central nervous system.

Coagulation: Coagulation of the blood is the process whereby bleeding (or haemorrhage) is normally arrested in the body. Blood starts to clot as soon as the skin (or other tissue) has been cut. Coagulation is part of the process of haemostasis which is the arrest of bleeding from an injured or dis- eased blood vessel. Haemostasis depends on the combined activities of vascular, platelet and plasma elements which are off- set by processes to restrict the accumulation of platelets and fibrin to the damaged area.

Coagulopathy: Any disorder affecting the coagulability of blood.Among acute conditions precipitating the disorder are abruptio placenta, haemolysis following blood trans- fusions, infection with gram-negative bacteria, heat stroke, shock and snakebite. Chronic disorders linked with coagulopathy are septic abortion, toxaemia of pregnancy, certain cancers and leukaemia.

Coal tar: Tar distilled from coal that is applied to the skin to treat psoriasis.

Coarctation of the Aorta: A narrowing of the aorta in the vicinity of the insertion of the ductus arteriosus. It is a con- genital abnormality but may not be discovered until well into childhood or adolescence. The diagnosis is easily made by discovering a major difference between the blood pressure in the arms and that of the legs. If untreated it leads to hypertension and heart failure, but satisfactory results are now obtained from surgical treat- ment, preferably in infancy. Paediatricians screen for coarctation by feeling for femoral pulses, which are absent or weak in this condition.

Cobalamins: A group of substances which have an enzyme action and are essential for normal growth and nutrition.

Cocarcinogen: Chemical, physical or biological factor which intensifies the effect of a carcinogen.

Cocci: Spherical bacteria that cause a variety of infections. Staphylococci, streptococci and meningococci are examples.

Coccydynia: The sensation of severe pain in the coccyx.

Coccyx: The lower end of the spinal column, resembling a bird's beak and consisting of four fused nodules of bone; these represent vertebrae and correspond to the tail in lower animals. Above the coccyx lies a much larger bone, the sacrum, and together they form the back wall of the pelvis, which protects the organs in the lower abdomen.

Cochlea: That part of the inner ear concerned with hearing.

Cochlear Implants: A cochlear implant is an electronic device, inserted under a general anaesthetic, which stimulates the auditory system, restoring partial hearing in profound sensory deafness.

Cochrane Collaboration: A non-profit-making international organisation which systematically finds, appraises and reviews available evidence, mainly from ran- domised clinical trials, about the con- sequences of health care. The aim is to help people make well-informed decisions about health care. The main work is done by around

Codex alimentarius: Collection of internationally adopted food standards drawn up by the Codex Alimentarius Commission, the principal body implementing the joint FAO/WHO Food Standards Programme.

Codicil: A legal instrument made to modify an earlier will.

Coeliac Disease: Around one in 100 people suffers from coeliac disease, a condition in which the small intestine fails to digest and absorb food, but many have no or few symptoms and remain undiagnosed. The intestinal lining is permanently sensitive to the protein gliadin (an insoluble and potentially toxic peptide protein) which is contained in gluten, a constituent of the germ of wheat, barley and rye. As bread or other grain-based foods are a regular part of most people's diet, the constant presence of gluten in the intestine of sufferers of coeliac disease causes atrophy of the digestive and absorptive cells of the intestine.

Coelioscopy: A method of viewing the interior of the abdomen in patients in whom a tumour or some other condition requiring operation may be present but cannot with certainty be diagnosed. The examination is carried out by making a minute opening under local anaesthesia, and inserting an endoscope – a long flexible instrument bearing an electric lamp and telescopic lenses like that for examining the bladder (cystoscope) – into the abdominal cavity. Certain of the abdominal organs can then be directly inspected in turn.

Cognition: The mental processes by which a person acquires knowledge. Among these are reasoning, creative actions and solving problems.

Cognitive Behavior Therapy: A talking therapy that re-trains the mind to question and banish negative thoughts, change emotional responses and change behaviour. It is based on the theory that some people develop unduly negative and pessimistic thoughts (cog- nitions) about themselves, their future and the world around them, putting them at risk of depression and other mental-health problems. Put simply, the treatment involves several sessions with a trained therapist who helps to identify the negative patterns of thinking and show that they are not usually realistic.

Cognitive Service: Diagnostic services a doctor provides during delivery of medical services, consultations or care.

Cohort Study: A systemised follow-up study of people for a specific period of time, or until the occurrence of a defined event such as a particular illness or death. The aim is to follow the disease course and/or the reasons for the participants' deaths.

Cohort: Component of the population born during a particular period and identified by period of birth so that its characteristics (such as causes of death and numbers still living) can be ascertained as it enters successive time and age periods. The term "cohort" has broadened to describe any designated group of persons followed or traced over ' a period of time, as in the term cohort study (prospective study).

Coinsurance: The co-insurance is the percentage (or portion) of the drug costs that insured persons must pay once they have paid the deductible. In other words, when a person's drug costs exceed the deductible, the person pays only a portion of the remainder. The percentage of the co-insurance payable by persons covered by the public plan is adjusted on July 1 of each year.

Coitus: Coitus is sexual intercourse.

Cold, Common: An infection by any one of around 200 viruses, with about half the common-cold infections being caused by rhinoviruses. Certain corona viruses, echoviruses and coxsackie viruses are also culprits. The common cold – traditionally also called a chill – is one of several viral infections that cause respiratory symptoms and systemic illness. Others include pneumonia and gastroenteritis.

Cold-Weather Itch: Cold-weather itch is a common form of itchiness that occurs in cold weather. It is characterised by slight dryness of the skin, and is particularly troublesome in the legs of old people. The dryness may be accompanied by some mild inflammation of the skin. Treatment is by the application of emollients such as acueous cream or zinc ointment.

Colectomy: The operation for removing the colon.

Colic: Cramping or spasms of a smooth muscle tube, such as the uterus (menstrual cramps) the ureters (passing kidney stones) or the stomach (stomachache). Also called tenesmus.

Coliform Bacteria: Intestinal bacilli that are gram-negative, sugar-digesting, and both aerobic and anaerobic. They are usually from the family Enterobacteriaceae; Escherichia coli is the best known of the group.

Coliform: Description of a gram-negative bacterium found in the faeces. It covers the bacterial groups Enterobacter, Escherichia, and Klebsiella.

Colitis: Colitis: Inflammation of the colon, the first part of thelarge intestine. The subject suffers from diarrhoea, usually passing blood and mucus, abdominal pain or discomfort, and fever. Colitis can be caused by various micro-organisms: for example, campylobacter, clostridium and shigella bacteria, viruses or amoeba. Anxiety and antibiotic drugs may also cause colitis, the latter by directly irritating the lining of the gut.

Collaboratives: Shared work programmes involving different NHS organisations designed to help local clinical teams to improve the outcomes and experiences of care for patients

Collagen Diseases: A group of diseases affecting connective tissue. The term is really outdated since there is no evidence that collagen is primarily involved.

Collagen: The most abundant protein in the body. It is the major structural component of many parts of the body and occurs in many different forms. Thus it exists as thick fibres in skin and ten- dons. It is also an important constituent of the heart and blood vessels. With calcium salts it provides the rigid structure of bone. It also occurs as a delicate structure in the cornea of the eye, and in what is known as the basement membrane of many tissues including the glomeruli of the kidneys and the capsule of the lens of the eye.

College of Pharmacists of British Columbia (CPBC): The licensing and regulatory body for the profession of pharmacy and it's practice environment, obtaining authority from the Pharmacists, Pharmacy Operations and Drug Scheduling Act. Membership of the College comprises of all pharmacists registered in British Columbia.

Colles' Fracture: Colles' fracture is a fracture of the lower end of the radius close to the wrist, caused usually by a fall forwards on the palm of the hand, in which the lower fragment is displaced backwards.

Collodions: Collodions consist basically of a thick, colourless, syrupy liquid, made by dissolving gun cotton (pyroxylin) in a mixture of ether and alcohol or with acetone. When painted on the skin the solvent evaporates, leaving behind a tough protective film that is useful for covering wounds. Flexible collodion – or collodion as it is often known – contains 1•6 per cent of pyroxylin, with colophony, castor oil and alcohol (90 per cent) in solvent ether. It should be kept in a well-sealed container. Being relatively elastic, it does not crack through the movements of the skin.

Colloid: Gooey substances, usually proteins and starches, whose molecules can hold large amounts of a solvent (usually water) without dissolving. In lifeforms, virtually all fluids are held suspended in protein or starch colloids (hydrogels). (Examples: cell protoplasm, lime Jell-O.)

Coloboma: Coloboma simply means a defect, but its use is usually restricted to congenital defects of the eye. These may involve the lens, the iris, the retina or the eyelid.

Colon: The first part of the large intestine.

Colonic Irrigation: Washing out the large bowel with an enema of water or other medication.

Colonoscope: An endoscope for viewing the interior of the colon. It is made of fibreglass which ensures flexibility, and incorporates a system of lenses for magnification and a lighting system.

Colostomy: The operation for the establishment of an artificial opening into the colon. This acts as an artificial anus. The operation is carried out when there is an obstruction in the colon or rectum that cannot be overcome, or in cases such as cancer of the rectum in which the rectum and part of the colon have to be removed. A colostomy opening can be trained to function in such a way that the patient can carry on a normal life, eating a more or less n

Colostrum: The first fluid secreted by the mammary glands for two or three days after childbirth. It contains less casein and more albumin than ordinary milk.

Colporrhaphy: An operation designed to strengthen the pelvic floor in cases of prolapse of the uterus. The surgeon excises redundant tissue from the front vaginal wall (anterior colporrhaphy) or from the rear wall (posterior colporrhaphy), thus narrowing the vagina and tightening the muscles.

Colposcopy: The method of examining the vagina and cervixuteri by means of the binocular instrument known as the colposcope. It is used to screen for cancer of the cervix and in investi- gation of child sexual abuse.

Coma: A state of profound unconsciousness in which the patient cannot be roused and reflex movements are absent. Signs include long, deep, sighing respirations, a rapid, weak pulse, and low blood pressure. Usually the result of a stroke, coma may also be due to high fever, diabetes mellitus, glomerulonephritis alcohol, epilepsy, cerebral tumour, meningitis, injury to the head, overdose of insulin, carbon monoxide (CO) poisoning, or poisoning from opium and other narcotics.

Combined Effect of Poisons: Simultaneous or successive effect of two or more poisons on the organism by the same route of exposure.

Cometabolism: Process by which a normally non biodegradable substance is biodegraded only in the presence of an additional carbon source.

Commensal: Micro-organisms which live in or on the body (e.g. in the gut or respiratory tract, or on the skin) without doing any harm to the individual.

Commingled Material: The mix of several recyclables collected in one container.

Commissioning: The process of defining the health needs of a population, determining priorities, purchasing and monitoring appropriate services.

Commissure: Commissure means a joining, and is a term applied to strands of nerve fibres which join one side of the brain to the other; to the band joining one optic nerve to the other; to the junctions of the lips at the corners of the mouth, etc.

Common Symptoms: Symptoms that are common to a specific disease, for example, stiff joints in arthritis, or yellow skin in jaundice.

Communicable Disease: This is an infectious or contagious disease which can be passed from one person to another. Direct physical contact, the handling of an infected object, or the transfer by droplets coughed or breathed out are all ways in which micro-organisms can be transmitted. The government produces a list of notifiable diseases, which includes all the dangerous communicable diseases from anthrax, cholera and diphtheria through meningitis, rabies and smallpox to typhoid fever and whooping-cough (see respective entries).

Comparability: The ability of a system to deliver data that can be compared in standard units of measurement and by standard statistical techniques with the data delivered by other systems. While not a critical component of accuracy, comparability of data generated by a system is critical to evaluating its accuracy and usefulness.

Compartment(s): The space or spaces in the body, which a drug appears to occupy after it has been absorbed. Pharmacokinetic compartments are mathematical constructs and need not correspond to the fluid volumes of the body which are defined physiologically and anatomically, i.e., the intravascular, extracellular and intracellular volumes!")Some drugs make the body "behave" as if it consisted of only a single pharmacokinetic compartment. Tissue and plasma concentrations of the drug rapidly and simultaneously reach equilibrium in all the tissues to which the drug is distributed. A plot of plasma concentration against time after intravenous administration can be rectified into only a single straight line of negative slope, which can intersect the ordinate at only one point; only one volume of distribution can be calculated. Hence, the existence of only one compartment or volume of distribution can be inferred.

Compatibility: The extent to which a person's defence systems will accept invading foreign substances – for example, an injection of a drug, a blood transfusion or an organ transplant. When incompatibility occurs there is usually a rapid antibody attack on the invading antigen with a severe local or system reaction in the individual receiving the antigenic substance.

Compensation: Compensation: In medical parlance, a term applied to the counterbalancing of some defect of structure or function by some other special bodily development. The body possesses a remarkable power of adapting itself even to serious defects, so that disability due to these passes off after a time. The term is most often applied to the ability possessed by the heart to increase in size, and therefore in power, when the need for greater pumping action arises in consequence of a defective valve or some other abnormality in the circulation

Competent Authority: In the context of European Communities Directive 79/831/EEC, the Sixth Amendment to the European Community's Directive 67/548/EEC.

Competent Bacteria: Culture of bacteria (or yeast) treated in such a way that their ability to take up DNA molecules without transduction or conjugation has been enhanced.

Complement Proteins: A group of proteins involved in the control of inflammation, activation of phagocytosis, and the lysis of cell membranes cytokines which, in turn, activate CD8+ T cells, B cells, and macrophages

Complement System: This is part of the body's defence mechanism that comprises a series of 20 serum peptides. These are sequentially activated to produce three significant effects: firstly, the release of small peptides which provoke inflammation and attract phagocytes, secondly, the deposition of a substance (component C3b) on the membranes of invading bacteria or viruses, attracting phagocytes to destroy the microbes; thirdly, the activation of substances that damage cell

Complement: A large body of blood proteins (over 20), initiated in the liver, and intimately involved in nearly all aspects of immunity and nonspecific resistance. They form two types of self-mediated cascade reactions to antigens, antibody-antigen complexes, dead tissue and the like, and are almost solely able to initiate the rupture and killing of bacteria. The protein strings they form around foreign substances are the main "hooks" used for absorption by macrophages as they digest and clean up.

Complementary and Alternative Medicine (CAM): This is the title used for a diverse group of health-related therapies and disciplines which are not considered to be a part of mainstream medical care. Other terms sometimes used to describe them include 'natural medicine', 'non- conventional medicine' and 'holistic medicine'. CAM embraces those therapies which may either be provided alongside conventional medicine (complementary) or which may, in the view of their practitioners, act as a substitute for it.

Complete Mineralization: Complete breakdown of a complex organic compound to carbon dioxide, water, oxides and oxidative inorganic products such as nitrate or sulfate.

Complete Symptom: Etiology, location (including radiation or extension of sensation), sensation, modalities, and concomitants all together give a complete symptom.

Complex: The term applied to a combination of various actions or symptoms. It is particularly applied to a set of symptoms occurring together in mental disease with such regularity as to receive a special name.

Compliance: The extent to which a patient follows the advice of a doctor or other health professional, especially in respect of drug or other treatments.

This is generally increased if the patient understands both the condition and the basis for the proposed treatment. Assessment of a patient's compliance should be a routine part of treatment review.

Compliant Software: Software on a local system which has been evaluated by Pharma Care and found to comply with all related requirements for connection to Pharma Net.

Composting: The controlled biological decomposition of organic solid waste.

Compound Drug: A prescription drug prepared by a pharmacist using at least two ingredients containing at least one of which Wisconsin senior care covers.

Compound Leaves: Leaves that are made up of leaflets, such as pinnate and palmate leaves.

Compound: A substance formed by the chemical combination of two or more elements in fixed proportion. The elements lose their individual chemical properties and the compound has new properties.

Comprehensive Effect of Poisons: Simultaneous or successive effect made on an organism by poisons entering from different media, from air, from water, from food or through the skin.

Compress: Compress is the name given to a pad of linen or flannel wrung out of water and bound to the body. It is generally wrung out of cold water, and may be covered with a piece of waterproof material. It is used to subdue pain or inflammation. A hot compress is generally called a fomentation.

Computational Chemistry: Use of computers to study the general principles of chemical structures and reactions

Computed Tomography: Tomography is an X-ray examination technique in which only structures in a particular plane produce clearly focused images. Whole-body computed tomography was introduced in 1977 and has already made a major impact in the investigation and management of medical and Concentration (amount-of-substance concentration) c = n/v: Derived kind-of-quantity defined as the amount of substance (n) of a component specified by an elementary entity divided by the volume (V) of the system containing the component.

Concentration-effect Curve: Graph of the relation between exposure concentration and the magnitude of the resultant biological change.

Concentration-effect Relationship: Association between exposure concentration and the magnitude of the resultant continuously graded change, either in an individual or in a population.

Concentration-response Curve: Graph of the relation between exposure concentration and the proportion of individuals in a population responding with a quantal effect.

Concentration-response Relationship: Association between exposure concentration and the incidence of a defined biological effect in an exposed population

Conception: Conception signifies the complex set of changes which occur in the ovum and in the body of the mother at the beginning of pregnancy. The precise moment of conception is that at which the male element, or spermatozoon, and the female element, or ovum, fuse together. Only one-third of these conceptions survive to birth, whilst 15 per cent are cut short by spontaneous abortion or stillbirth. The remainder over one-half – are lost very early during pregnancy without trace.

Concomitant: Occurring simultaneously. Refers to symptoms that happen at the same time as the chief complaint. One of the parts of a complete symptom.

Concord: Pairs or groups of individuals of identical phenotype: in twin studies, a condition in which both twins exhibit or fail to exhibit a trait under investigation.

Concurrent Validity: Measurement and its criterion refer to the same point in time: an example would be a visual inspection of a wound for evidence of infection validated against bacteriological examination of a specimen taken at the same time.

Conditioned Reflex: The development of a specific response by an individual to a specific stimulus. The best-known conditioned reflex is the one described by Ivan Pavlov, in which dogs that became accustomed to being fed when a bell was sounded salivated on hearing the bell, even if no food was given. The conditioned reflex is an important part of behavioural theory.

Condom: A thin rubber or plastic

sheath placed over the erect penis before sexual intercourse. It is the most effective type of barrier contraception and is also valuable in preventing the transfer between sexual partners of pathogenic organisms such as gonococci, which cause gonorrhoea, and human immuno-deficiency virus, which may lead to AIDS. Sheaths are most effective when properly used and with spermicides.

Condyle: A rounded prominence at the end of a bone: for example, the prominences at the outer and inner sides of the knee on the thigh-bone. The projecting part of a condyle is sometimes known as an epicondyle, as in the case of the condyle at the lower end of the humerus where the epicondyles form the prominences on the outer and inner side of the elbow.

Condyloma: A localised, rounded swelling of mucous membrane around the opening of the bowel, and the genital organs, sometimes known as 'genital warts' or 'anogenital warts'. There are two main forms: condyloma latum, which is syphil- itic in origin; and condyloma accuminatum, which often occurs in association with sexually transmitted disease, but is only indirectly due to it, being primarily a virus infection.

Cone: (1) A light-sensitive cell in the retina of the eye that can also distinguish colours. The other type of light-sensitive cell is called a rod. There are around six million cones in the human retina and these are thought to comprise three types that are sensitive to the three primary colours of red, blue, and green. (2) A cone biopsy is a surgical technique in which a conical or cylindrical section of the lower part of the neck of the womb is excised.

Confidence Interval: A statistical analysis which describes the percentage of trials in which a tested value will fall between two pre-specified parameters

Confidentiality Undertakings: Agreements signed by pharmacists as part of the pharmacist registration process with the CPBC. Also, forms specific to each pharmacy which are signed by the pharmacy staff members involved in the dispensing process to ensure that they understand and comply with confidentiality procedures.

Confidentiality: The ethical principle that doctors do not reveal information to other people (or to organisations) that their patients have given to them in confidence. Normally the doctor must get per- mission to release confidential information to an employer (or other authoritative body), insurance company or lawyer. The doctor does have to provide such information if required by a court of law.

Confounding Variable: Changing factor that can cause or prevent the outcome of interest, is not an intermediate variable, and is not associated with the factor under investigation such a variable must be controlled in order to obtain an undistorted estimate of the effect of the study factor on risk.

Confounding: Situation in which the effects of two processes are not distinguishable from one another: The distortion . of the apparent effect of an exposure on risk brought about by the association of other factors which can influence the outcome. Relationship between the effects of two or more causal factors as observed in a set of data, such that it is not logically possible to separate the contribution which any single causal factor has made to an effect.

Congener: Substance which by structure, function or origin is similar -

to another.

Congenital: Congenital deformities, diseases, etc. are those which are either present at birth, or which, being transmitted direct from the parents, show themselves some time after birth.

Congestion: The accumulation of blood or other body fluid in a particular part of the body. The condition may be due to some failure in the circulation, but as a rule is one of the early signs of inflammation. Thick and boggy tissues, usually resulting from excess inflammation, or irritation that is unremitting. It is characterized by the accumulation of an excess volume of fluid, with impairment of venous and lymphatic drainage, and the buildup of unremoved cellular waste products.

Conjoined Twins: Identical twins who are united bodily but are possessed of separate personalities. Their frequency is not known, but it has been estimated that throughout the world, six or more conjoined twins are born every year who are capable of separation. The earliest case on record is that of the 'Biddendon Maids' who were born

Conjuctivitis: An inflammation of the conjunctiva, either from environmental irritation, allergies, viral or bacterial infections.

Conjugate Deviation: The term for describing the persistent and involuntary turning of both eyes in any one direction, and is a sign of a lesion in the brain.

Conjugate: Derivative of a substance formed by its combination with compounds such as acetic acid, glucuronic acid, glutathione, glycine, sulfuric acid etc. Material produced by attaching two or more substances together, for example - conjugates of antibody with fluorochromes, radio-isotopes or enzymes.

Conjunctiva: Mucous membrane which covers the eyeball and lines the under-surface of the eyelid.

Connective Tissue Disorders: A group of generalised inflammatory diseases that affect connective tissue in almost any system in the body. The term does not include those disorders of genetic origin. Rheumatic fever and rheumatoid arthritis were traditionally classified in this group, as were those diseases classed under the outdated heading collagen diseases.

Connective Tissue: Sometimes called fibrous tissue, this is one of the most abundant tissues in the body, holding together the body's many different structures. Connective tissue comprises a matrix of substances called muco polysaccharides in which are embedded various specialist tissues and cells. These include elastic (yellow), collagenous (white) and reticular fibres as well as macro-phages and mast cells.

Consanguinous: A relationship by blood: siblings are closely consanguinous; cousins, and grandparents and grandchildren, less so.

Consciousness: The state of being aware of physical events or mental concepts. A conscious person is awake and responsive to his or her surroundings.

Conservative Assessment of Risk: Assessment of risk which assumes the worst possible case scenario and therefore gives the highest possible value for risk: risk management decisions based on this value will maximize safety.

Conservative Treatment: Medical treatment which involves the minimum of active interference by the practitioner. For example, a disc lesion in the back might be treated by bed rest in contrast to surgical intervention to remove the damaged disc.

Consolidation: A term applied to solidification of an organ, especially of a lung. The consolidation may be of a permanent nature due to formation of fibrous tissue, or may be temporary, as in acute pneumonia.

Constipation: A condition in which a person infrequently passes hard faeces (stools). Patients sometimes complain of straining, a feeling of incomplete evacuation of faeces, and abdominal or perianal discomfort. A healthy individual usually opens his or her bowels once daily but the frequency may vary, perhaps twice daily or once only every two or three days. Constipation is generally.

Constitutional: Deriving from basic hereditary strengths and weaknesses, and including early environmental factors.

Construct Validity: Extent to which the measurement corresponds to theoretical concepts (constructs) concerning the phenomenon under study; for example, if on theoretical grounds, the phenomenon should change with age, a measurement with construct validity would reflect such a change.

Consultation: A discussion with another health care professional when additional feedback is needed during diagnosis or treatment. Usually, a consultation is by referral from a primary care physician.

Contact Dermatitis: Inflammatory condition of the skin resulting from dermal exposure to an allergen (sensitizer) or an irritating (corrosive, defatting) substance.

Contact Lenses: Contact lenses are lenses worn in contact with the EYE, behind the eyelids and in front of the cornea. They may be worn for cosmetic, optical or therapeutic reasons. The commonest reason for wear is cosmetic, many short-sighted people preferring to wear contact lenses instead of glasses. Optical reasons for contact-lens wear include cataract surgery (usually unilateral extraction) and the considerable improvement in overall standard of vision experienced by very short-sighted people.

Contagion: Contagion means the principle of spread of disease by direct contact with the body of an affected person.

Containment: Process by which possible release, discharge or spill of a toxic substance during normal use or after an accident is prevented by appropriate action.

Contaminant: Minor impurity present in a substance. Extraneous material inadvertently added to a sample prior to or during chemical or biological analysis. In some contexts, as in relation to gas cleaning equipment, used as a synonym for "pollutant", especially on a small scale. Unintended component in food that may pose a hazard to the consumer.

Content Validity: Extent to which the measurement incorporates the domain of the phenomenon under study; for example, a measurement of functional health status should embrace activities of daily living, occupational, family, and social functioning, etc.

Continued Fevers: Continued fevers-are typhus, typhoid and relapsing fevers, so-called because of their continuing over a more or less definite space of time.

Continuity of Care: A term describing a system of medical care in which individuals requiring advice on their health consult a named primary care physician (general practitioner (GP)) or partnership of practitioners. The availability of an individual's

Contraception: A means of avoiding pregnancy despite sexual activity. There is no ideal contraceptive, and the choice of method depends on balancing considerations of safety, effectiveness and acceptability. The best choice for any couple will depend on their ages and personal circum- stances and may well vary with time. Contraceptive techniques can be classified in various ways, but one of the most useful is into 'barrier' and 'non-barrier' methods.

Contract: A legal agreement between an individual subscriber or an employer group and a health plan that describes the benefits and limitations of the coverage.

Contracted Pharmacy: An arrangement through which a covered entity may contract with an outside pharmacy to provide comprehensive pharmacy services utilizing medications purchased under 340B.

Contracture: The permanent shortening of a muscle or of fibrous tissue. Contraction is the name given to the temporary shortening of a muscle.

Contraindication: A condition that renders a particular type of medication as undesirable or improper for use.

Contrast Medium: A material that is used to increase the visibility of the body's tissues and organs during radiography. A common example is the use of barium which is given by mouth or as an enema to show up the alimentary tract.

Contre-Coup: An injury in which a bone, generally the skull, is fractured – not at the spot where the violence is applied, but at the exactly opposite point.

Control Group: Selected group, identified as a rule before a study is done, which comprises humans, animals, or other species who do not have the disease, intervention, procedure or whatever is being studied, but in all other respects is as nearly identical to the test group as possible.

Control, Matched: Control (individual or group or case) selected to be similar to a study individual or group, or case, in specific characteristics: some commonly used matching variables are age, sex, race and socio-economic status.

Controlled Acts: Dispensing, selling or compounding a drug as defined in subsection 117 (1) of the Drug and Pharmacies Regulation Act, or supervising the part of a pharmacy where such drugs are kept.

Controlled Drugs: In the United Kingdom, controlled drugs are those preparations referred to under the Misuse of Drugs Act 1971. The Act prohibits activities related to the manufacture, supply and posses of these drugs, and they are classified into three groups which determine the penalties for offences involving their misuse. For example, class A includes cocaine, diamorphine, morphine, LSD. Class B includes amphetamines, barbiturates and codeine.

Contusions: A bruise, characterized by a trauma in which the skin is not broken but underlying blood vessels are busted, causing a deep or lateral hematoma, with disorganized blood and interstitial fluid buildup.

Convalescence: The condition through which a person passes after having suffered from some acute disease, and before complete health and strength are regained.

Convergence: (1) Inward turning of the eyes to focus on a near point, with the result that a single image is registered by both retinas. (2) The coming together of various nerve fibres to form a nerve tract that provides a single pathway from different parts of the brain.

Conversion Disorder: A psychological disorder, also called hysterical conversion, in which the affected individual presents with striking neurological symptoms – such as weakness, paralysis, sensory disturbances or memory loss – for which no organic cause can be identified. Up to 4 per cent of patients attending neurological outpatient clinics have been estimated as having con- version disorders. The disorder remains controversial, with theories about its cause unsupported by controlled research results. In clinical practice the physician's experience and intuition are major factors in diagnosis. It has been suggested that the physical symptoms represent guilt about a physical or emotional.

Conversion Option: The option to purchase individual coverage by a person who will no longer have access to group health insurance.

Convulsions: Rapidly alternating contractions and relaxations of the muscles, causing irregular movements of the limbs or body generally, usually accompanied by unconsciousness. Causes the most common reason for convulsions is epilepsy, and the underlying cause of the latter often remains uncertain. In newborns, convulsions may be due to hypoxia following a difficult labour, or to low levels of sugar or calcium in the blood.

Coomb's Test: A sensitive test that detects antibodies to the body's red cells. There are two methods: one – the direct method – identifies those antibodies that are bound to the cells; the other, indirect, method identifies those circulating unattached in the serum.

Coordination: The governing power exercised by the brain as a whole, or by certain centres in the nervous sys- to make various muscles contract in harmony and so produce definite actions (instead of meaningless movements). Coordination is intimately bound up with the complex sense of localisation, which enables a person with their eyes shut to tell, by sensations received from the bones, joints and muscles, the position of the various parts of their body.

Co-pay or Co-payment: Once any family reaches the deductible, Pharma Care will pay a percentage of further eligible prescription costs for the remainder of the year. The co-pay is the amount your family will be required to contribute towards the eligible costs of a prescription after Pharma Care has covered its portion.

Copay: A cost-sharing arrangement in which the Health Plan subscriber pays a specified flat amount for a specific service (such as $10 for an office visit or $5 for each prescription drug). The amount paid must be nominal to avoid becoming a barrier to care.

Copayment: In some Medicare health plans, this is the amount a Medicare beneficiary must pay for each medical service they receive. In the Medicare program, a copayment is usually a set amount a beneficiary pays for a service, for example, $5 or $10 for a doctor visit.

COPD (Chronic Obstructive Pulmonary Disease): COPD is comprised primarily of two related diseases: chronic bronchitis and emphysema. In both diseases, the flow of air through the airways and out of the lungs is obstructed. The condition is permanent and worserys over time.

Coprolalia: An inherited condition, usually beginning in childhood. It presents with motor tics and with irrepressible, explosive, occasionally obscene, verbal ejaculations.

Copulation: The act of coitus or sexual intercourse, when the man inserts his erect penis into the woman's vagina and after a succession of thrusting movements ejaculates his semen.

Cor Pulmonale: Another name for pulmonary heart disease, which is characterised by hypertrophy and failure of the right ventricle of the heart as a result of disease of the lungs or disorder of the pulmonary circulation.

Cordillera: The mountain ridge that spans North America, from Mexico through the Rocky Mountains into Alaska.

Cordotomy: The surgical operation of cutting the anterolateral tracts of the spinal cord to relieve otherwise intractable pain. It is also sometimes known as tractotomy.

Core grade: Quality rating, based on standard evaluation criteria established by the US Office of Pesticide Programs regulatory agencies, given to toxicological studies after submission by registrants.

Corm: The fleshy, bulblike, solid base of a stem, often rising out of a tuber or bulb.

Corneal Graft: Also known as keratoplasty. If the cornea becomes damaged or diseased and vision is impaired, it can be removed and replaced by a corneal graft. The graft is taken from the cornea of a human donor. Some of the indications for corneal grafting include keratoconus (conical- shaped cornea), corneal dystrophies, severe corneal scarring following, and alkali burns or other injury.

Corneal Reflex: Instinctive closing of the eyelids when the surface of the cornea is lightly touched with a fine hair.

Corns and Bunions: A corn is a localised thickening of the cuticle or epidermis affecting the foot. The thickening is of a conical shape; the point of the cone is directed inwards and is known as the 'eye' of the corn. A general thickening over a wider area is called a callosity. Bunion is a condition found over the joint at the base of the big toe, in which not only is there thickening of the skin, but the head of the metatarsal bone also becomes prominent. Hammer-toe is a condition of the second toe, often caused by short boots, in which the toe becomes bent at its two joints in such a way as to resemble a hammer.

Corns: A thickening of the skin of the feet or hands, usually caused by pressure against the skin.

Coronary Angioplasty: A technique of dilating atheromatous obstructions in coronary arteries by inserting a catheter with a balloon on the end into the affected artery. It is passed through the blockage (guided by X- ray fluoroscopy) and inflated. The procedure can be carried out through a percutaneous route.

Coronary Arteries: The right coronary artery arises from the right aortic sinus and passes into the right atrio-ventricular groove to supply the right ventricle, part of the intraventricular septum and the inferior part of the left ventricle.

Coronary Artery Vein Bypass Grafting (CAVBG): When coronary arteries, narrowed by disease, cannot supply the heart muscle with sufficient blood, the cardiac circulation may be improved by grafting a section of vein from the leg to bypass the obstruction. Around 10,000 people in the United Kingdom have this operation annually and the results are usually good. It is a major procedure that lasts several hours and requires the heart to be stopped temporarily, with blood circulation and oxygenation taken over by a heart-lung machine.

Coronary Care Unit (CCU): A specialised hospital unit equipped and staffed to provide intensive care for patients who have had severe heart attacks or undergone surgery on the heart.

Coronary: A term applied to several structures in the body encircling an organ in the manner of a crown. The coronary arteries are the arteries of supply to the heart which arise from the aorta, just beyond the aortic valve, and through which the blood is delivered to the muscle of the heart. Disease of the coronary arteries is a very serious condition producing various abnormal forms of heart action and the disorder, angina pectoris.

Coroner: An independent legal officer of the Crown who is responsible for deciding whether to hold a post-mortem examination and an inquest in cases of sudden or unexpected or unnatural death. He or she presides over an inquest, if held – sometimes with the help of a jury. Cor- oners are usually lawyers or doctors (some are double-qualified) who have been qualified for at least five years. In Scotland the coroner is known as the procurator fiscal.

Corporate Fiduciary: An institution that acts for the benefit of another. One example is a bank acting as trustee.

Corporate Integrity Agreement (CIA): An agreement between the Office of the Inspector General of the Department of Health and Human Services and a health care provider or other entity as part of a settlement for alleged civil wrongdoing relating to Federal health laws.

Corpus Luteum: The mass of cells formed in the ruptured Graafian follicle in the ovary from which the ovum is discharged about 15 days before the onset of the next menstrual period. When the ovum escapes, the follicle fills up with blood; this is soon replaced by cells which contain a yellow fatty acid. A temporary endocrine gland formed at ovulation from part of the former egg follicle, and the source of progesterone.

Corpuscle: Corpuscle means a small body.

Corrective action: An enforcement action that is imposed through permits or enforcement orders issued by EPA to address releases to any media from any unit at a TSD facility.

Corrigan's Pulse: The name applied to the collapsing pulse found with incompetence of the heart's aortic valve. It is so-called after Sir Dominic John Corrigan (1802–80), the famous Dublin physician, who first described it.

Corrosives: Corrosives are poisonous substances which corrode or eat away the skin or the mucous surfaces of mouth, gullet and stomach with which they come into contact. Examples are strong mineral acids like sulphuric, nitric and hydrochloric acids, caustic alkalis, and some salts like chlorides of mercury and zinc.

Corset: A support device worn around the trunk to help in the treatment of backache and spinal injuries or disorders.

Cortex: The tissues that form the outer part of an organ and which are positioned just below the capsule or outer membrane. Examples are the cerebal cortex of the brain and the renal cortex of the kidneys.

Corticosteroid: A synthetic hormone similar to that naturally produced by the adrenal gland and suppresses the immune system and reduces inflammation; available in pill, cream, and injectable forms.

Corticosteroids: Natural steroid hormones or synthetic analogues, usually taken for suppressing inflammation (and immunity) and therefore having cortisone-like functions, or taken as analogues to adrenocortical androgen or even testosterone, in order to impress the other gym members, make varsity by your junior year or to join the WWF and get newbie-mangled for two years by The Hangman or even the Hulkster Himself. Then, if your gonads don't fall off and your back holds up you get promoted to Good Guy, have your chance to Take A Name and finally wear your chosen costume a spandex violet nurse's uniform.

Corticosteroids: The generic term for the group of hormones produced by the adrenal glands, with a profound effect on mineral and glucose metabolism.

Corticotropin: Corticotropin is the British Pharmacopoeia name for the adrenocorticotrophic hormone of the pituitary gland, also known as ACTH. It is so-called because it stimulates the functions of the cortex of the suprarenal glands. This results, among other things, in an increased output of cortisone.

Coryza: Coryza is the technical name for a 'cold in the head'.

Cost Basis: The original value of an asset, such as stock, before its appreciation or depreciation.

Cost Factor: The expense and cost avoidance associated with hazmat that may be reduced to monetary terms, which includes future liability. Cost factors refer to direct and indirect costs attributable to hazmat that are encountered in operations such as acquisition, manufacture, supply, use, storage, inventory, control, treatment, recycling, emission control, training, work place safety, labeling, hazard assessment, engineering controls, personal protective equipment, medical monitoring, regulatory overhead, spill contingency, disposal, remedial action, and liability.

Cost of Illness: Traditionally, doctors have been trained to treat a patient on the basis of his or her personal clinical needs. Increasingly, however, the practice of medicine has been by patients' social circumstances and more recently by community and government-driven national priorities.

Costal: Anything pertaining to the ribs.

Costalgia: Pain in the ribs.

Cost-effective: Bringing good results or profit for little expenditure.

Cough Syncope: Temporary loss of consciousness that may be induced by a severe spasm of coughing. This is the result of the high pressure that may be induced in the chest – over 200 millimetres of mercury – by such a spasm, which prevents the return of blood to the heart. The veins in the neck begin to bulge and the blood pressure falls; this may so reduce the blood flow to the brain that the individual feels giddy and may then lose consciousness.

Cough, Paroxysmal: Attacks of uncontrollable coughing or "whooping", often relating to whooping cough or bronchiectasis, but they can also be caused by the smoke from burning plastics and (memories of yesteryear) hash oil.

Cough, Hectic: The dry and unproductive coughing in early bronchitis, when the mucosa is irritated but still too infected to secrete mucus.

Cough, Reflex: A cough induced by intestinal, gastric or uterine irritation -

and not from respiratory causes.

Cough: A natural reflex reaction to irritation of the air passages and lungs. Air is drawn into the air passages with the glottis wide open. The inhaled air is blown out against the closed glottis, which, as the pressure builds up, suddenly opens, expelling the air – at an estimated speed 960 kilometres (600 miles) an hour. This Cexplosive exhalation expels harmful substances from the respiratory tract.

Counselling: Psychological support and advice provided by a trained therapist or health professional. The aim is to help an individual manage a particular personal or family problem: this may be a diagnosis of cancer, mental and physical trauma following an accident or assault, or a bereavement. Cowper's Glands: Also known as the bulbourethral glands, these are a pair of glands whose ducts open into the urethra at the base of the penis. They secrete a fluid that is one of the constituents of the semen which carries the spermatozoa and is ejaculated into the vagina during coitus (sexual intercourse).

Count Median Diameter: Calculated diameter in a population of particles in a gas or liquid phase above which there are as many particles with larger diameters as there are particles below it with smaller diameters.

Counterirritant: A substance applied to the skin to produce an irritating, heating, or vasodilating effect, in order to speed local healing by increasing circulation of blood, radiating the heat inward to inflamed tissues deep below the skin. It can also be used to induce reflex stimulation to seemingly unrelated internal organs.

Coverage End Date: Coverage end date displays the date that coverage ends for a participant. This field is blank, if the participant is considered covered as of the date of the inquiry.

Coverage: Services provided within a given health or pharmacy care plan. Health care and drug benefit services provided or authorized by the payer's medical staff or payment for health care services.

Covered Entities: The statutory name for facilities and programs eligible to purchase discounted drugs through the 340B Drug Pricing Program. Covered entities include certain disproportionate share hospitals owned by, or under contract with, State or local governments and 11 categories of facilities or programs funded by Federal grant dollars, including federally qualified health centers, AIDS drug assistance programs, hemophilia treatment centers, and family planning clinics.

Covered Services: Hospital, medical, and other health care services incurred by the enrollee that are entitled to a payment of benefits under a health benefit contract. The term defines the type and amount of expense that will be considered in the calculation of benefits.

Cowpox: Cowpox is a disease affecting the udders of cows, on which it produces vesicles. It is communicable to humans, and there has for centuries been a tradition that per- sons who have caught this disease from cows do not suffer afterwards from smallpox. This formed the basis for Jenner's experiments on vaccination.

Coxa Vara: A condition in which the neck of the thighbone is bent so that the lower limbs are turned outwards and lameness results.

Coxalgia: Pain in the hip-joint; also referred to as coxodynia.

CP: Clinical Pharmacy

CPBC: College of Pharmacists of British Columbia

CPhA (Canadian Pharmacists Association): National voluntary organization of pharmacists self-mandated to promote the profession of pharmacy

CPI (Consumer Price Index: The CPI measures the price paid by consumers for a fixed group of goods and services. Changes in the CPI over time constitute a common measure of inflation.

CPMP: Committee for Proprietary Medicinal Products (EU)

CPSBC: College of Physicians and Surgeons of British Columbia

Crab-Louse: Another name for-

Pediculus pubis, a louse that infests the pubic region.

Cracked-Pot Sound: A peculiar resonance heard sometimes on percussion of the chest over a cavity in the lung, resembling the jarring sound heard on striking a cracked pot or bell. It is also heard on percussion over the skull in patients with diseases of the brain such as haemorrhages and tumours, and in certain cases of fracture of the skull.

CRADA: Cooperative research and development agreement (with NIH)

Cradle Cap: Crusta lactea, or cradle cap as it is technically known, is a form of seborrhoea of the scalp which is not uncommon in nursing infants.

Cradle: A cage which is placed over the legs of a patient in bed, in order to take the weight of the bedclothes the legs.

Craniotomy: The removal of part of the skull to provide surgical access for an operation on the brain. This may be to obtain a biopsy, to remove a tumour or to drain an infection or a blood clot. Following the operation the bone is replaced, along with the membranes, muscle and skin.

Cranium: The part of the skull enclosing the brain as distinguished from the face.

Cream: The oily or fatty part of milk-from which butter is prepared. Various medicinal preparations are known also as cream – for example, cold cream, which is a simple ointment containing rose- water, beeswax, borax, and almond oil scented with oil of rose.

Creatine Kinase: An enzyme which is proving to be of value in the investigation and diagnosis of muscular dystrophy, in which it is found in the blood in greatly increased amounts.

Creatine: A nitrogenous substance, methyl-guanidine-acetic acid. The adult human body contains about 120 grams – 98 per cent of which is in the muscles. Much of the creatine in muscles is combined with phosphoric acid as phosphocreatine, which plays an important part in the chemistry of muscular contraction.

Creatinine Clearance: A method of assessing the function of the kidney by comparing the amount of creatinine – a product of body metabolism which is normally excreted by the kidneys – in the blood with the amount appearing in the urine.

Creatinine: It is the waste product of creatine, an enzyme found in large amounts throughout the tissues, and mainly excreted in the urine. The parent compound creatine enables the body to use the "blue flame" of anaerobic combustion (as opposed to the yellow flame of oxidation). Elevated creatinine in the blood may be an early symptom of kidney disease.

Creeping Eruption: Creeping eruption is a skin condition caused by the invasion of the skin by the larvae of various species of nematode worms. It owes its name to the fact that as the larva moves through and along the skin it leaves behind it a long creeping thin red line.

Crenation: Abnormal microscopic appearance of blood cells in which their usually smooth margins appear irregular. It usually occurs after a blood specimen has been stored for a long time, but may occasionally indicate a blood disorder.

Crenelated: Leaves having rounded, scalloped teeth along the edges.

Creosote: A clear, yellow liquid, of aromatic smell and burning taste, prepared by distillation from pine-wood or beech-wood. It mixes readily with alcohol, ether, chloroform, glycerin, and oils. Creosote is a powerful antiseptic and disinfectant; it is also an ingredient of some disinfectant fluids.

Crepitations: Abnormal respiratory sounds heard on auscultation of the chest, produced by passage of air through passages which contain secretion or exudate or which are constricted by spasm or a thickening of their walls; more usually referred to as crepitations or rhonchi (auscultation is the process of listening for sounds within the body by ear unassisted or using a stethoscope).

Crepitations: Certain sounds which occur along with the breath sounds, as heard by auscultation, in various diseases of the lungs. They are signs of the presence of moist exudations in the lungs or in the bronchial tubes, are classified as fine, medium, and coarse crepitations, and resemble the sound made by bursting bubbles of various sizes.

Crepitus: Crepitus means a grating sound. It is found in cases of fractured bones when the ends rub together; also, in cases of severe chronic arthritis, by the rubbing together of the dried internal surfaces of the joints.

Cretinism: An out-of-date name for congenital hypothyroidism, a disease caused by defective thyroid function in fetal life or early in infancy.

CRF: Chronic Renal Failure. Recurrent or long standing renal failure-not reversible.

Crisis: Crisis is a word used with several distinct meanings. (1) The traditional meaning is that of a rapid loss of fever and return to comparative health in certain acute diseases. For example, pneumonia, if allowed to run its natural course, ends by a crisis usually on the eighth day, the temperature falling in 24 hours to normal, the pulse and breathing becoming slow and regular and the patient passing from a partly delirious state into natural sleep.

Criterion Validity: Extent to which the measurement correlates with an external criterion of the phenomenon under study.

Criterion: Validated set of data used as a basis for judgement.

Critical Concentration (for a Cell or Organ): Concentration of a potentially toxic substance at which undesirable (or adverse) functional changes, reversible or irreversible, occur in the cell or organ.

Critical Effect: For deterministic effects, the first adverse effect which appears when the threshold (critical) concentration or dose is reached in the critical organ. Adverse effects, such as cancer, with no defined threshold concentration are often regarded as critical. Decision on whether an effect is critical is a matter of expert judgment.

Critical End-point: Toxic effect used by the USEPA as the basis for a reference dose.

Critical Group: Part of a target population most in need of protection because it is most susceptible to a given toxicant.

Critical Organ Concentration: Mean concentration in the critical organ at the time the most sensitive type of cell reaches the critical concentration.

Critical Organ: In toxicology Organ which first attains the critical concentration (of a potentially toxic substance) under specified circumstances of exposure and for a given population.

Critical Period (of Development): Stage of development of an organism that is of particular importance in the life cycle if the normal full development of some anatomical, physiological, metabolic, or psychological structure or function is to be attained: such a period may be associated with very high susceptibility to specific potentially toxic substances.

Critical Study: Investigation yielding the no-observed adverse effect level that is used by the USEPA as the basis of the reference dose.

CRO: Contract research organization

Crohn's Disease: A chronic inflammatory bowel disease which has a protracted, relapsing and remitting course. An autoimmune condition, it may last for several years. There are many similarities with ulcerative colitis; sometimes it can be hard to differentiate between the two conditions. A crucial difference is that ulcerative colitis is confined to the colon, whereas Crohn's disease can affect any part of the gastrointestinal tract, including the mouth and anus. The sites most commonly affected in Crohn's disease (in order of frequency) are terminal ileum and right side of colon, just the colon, just the ileum and finally the ileum and jejunum.

Cross linking: An oxidation reaction in which undesirable bonds form between nucleic acids (RNA and DNA) or between proteins (bit into an apple and watch it do yellow over time, this is protein cross linking).

Cross: Media Transfer - The transfer of hazardous materials and wastes from one environmental medium to another, (e.g. from water to air).

Crossover Claim: A Medicare-allowed claim for a dual entitle sent to Wisconsin Medicaid for possible additional payment of the Medicare coinsurance and deductible.

Cross-over Experiment: A form of experiment in which each subject receives the test preparation at least once, and every test preparation is administered to every subject. At successive experimental sessions each preparation is "crossed-over" from one subject to another. The purpose of the cross-over experiment is to permit the effects of every preparation to be studied in every subject, and to permit the data for each preparation to be similarly and equally affected by the peculiarities of each subject. In a well-designed cross-over experiment, if it is at all possible, the sequence in which the test preparations are administered is not the same for all subjects, in order to avoid bias in the experiment as a result of changes in the behavior of the subjects that are a function of time rather than of drug administration, or a function of drug interactions.

Cross-tolerance: Tolerance to a drug-that generalizes to drugs that are chemically related of that produce similar affects. For example, a patient who is tolerant to heroin will also exhibit cross-tolerance to morphine.

Crosswalk: The systematic process of changing a provider submitted value for a specific field on a claim to a value required by the system when they are not the same.

Croup: Also known as laryngo-tracheo-bronchitis, croup is a household term for a group of diseases characterised by swelling and partial blockage of the entrance to the larynx, occur- ring in children and characterised by crowing inspiration. There are various causes but by far the commonest is acute laryngo-tracheo- bronchitis.

CRS: The building letter code for Central Receiving South, 616 Forbes Boulevard, South San Francisco

Crude Botanical: on the other hand, is one of our herbs that has no official standing. Examples: Digitalis leaves (crude drug), White Sage (crude botanical).

Crude Drug: A dried, unprocessed plant, and referring to one that was or is an official drug plant or the source of a refined drug substance.

Crural: Crural means connected with the leg.

Crush Syndrome: A condition in which kidney failure occurs in patients who have been the victims of severe crushing accidents. The fundamental injury is damage to muscle. The limb swells. The blood volume falls. Blood urea rises; there is also a rise in the potassium con- tent of the blood.

Crutch Palsy: Crutch palsy is weakness or paralysis of muscles in the wrist and hand, due to pressure exerted by the crutch head on the nerves that control the affected muscles.

Crutch: (1) An aid to support the weight of the body for a person unable to bear weight on one of his or her legs. Made of wood or metal, usually long enough to reach from the person's armpit to the ground, it has a concave surface that fits under the arm and a cross-bar for the hand. An elbow crutch provides weight-bearing support using the forearm and elbow and is usually recommended when the leg can take some weight.

Cryoanalgesia: The induction of analgesia by the use of cold that is produced by means of a special probe. The use of cold for the relief of pain dates back to the early days of mankind: two millennia ago, Hippocrates was recommending snow and ice packs as a preoperative analgesic.

Cryoprecipitate: When frozen plasma is allowed to thaw slowly at 4 ^{0}C, a proportion of the plasma protein remains undissolved in the cold thawed plasma and stays in this state until the plasma is warmed. It is this cold, insoluble precipitate that is known as cryoprecipitate. It can be recovered quite easily by centrifuging. Its value is that it is a rich source of factor VIII, which is used in the treatment of haemophilia.

Cryopreservation: Maintenance at very low temperatures of the viability of tissues or organs that have been excised from the body.

Cryoscopy: The method of finding the concentration of blood, urine, etc., by observing their freezing- point.

Cryosurgery: The use of cold in surgery. Its advantages include little associated pain, little or no bleeding, and excellent healing with little or no scar formation. Hence it's relatively wide use in eye surgery, some abdominal surgery, skin cancers and treatment of haemorrhoids. The coolants used include liquid nitrogen with which temperatures as low as 196 ^{0}C can be obtained, carbon dioxide (78 ^{0}C) and nitrous oxide (88 ^{0}C).

Cryotherapy: The treatment of disease by refrigeration. The two main forms in which it is now used are hypothermia and refrigeration anaesthesia.

Cryptorchidism: An undescended testis. The testes normally descend into the scrotum during the seventh month of gestation; until then, the testis is an abdominal organ. If the testes do not descend before the first year of life, they usually remain undescended until puberty – and even then, descent is not achieved in some instances.

CSHP: California Society of Health-System Pharmacists

CSPCC: Cooperative Studies Program Coordinating Center

CSR: Client Service Representative at the PharmaNet Help Desk.

Css: The concentration of a drug or chemical in a body fluid - usually plasma - at the time a "steady state" has been achieved, and rates of drug administration and drug elimination are equal. Css is a value approached as a limit and is achieved, theoretically, following the last of an infinite number of equal doses given at equal intervals.

CT Scanne: The machine which combines the use of a computer and X-rays to produce cross-sectional images of the body

CTD: Common technical document

Culdoscopy: Culdoscopy is a method of examining the pelvic organs in women by means of an instrument comparable to a cystoscope, inserted into the pelvic cavity through the vagina. The instrument used for this purpose is known as a culdoscope.

Cumulative Effect: Overall adverse change which occurs when repeated doses of a harmful substance or radiation have biological consequences which are mutually enhancing.

Cumulative Incidence, Cumulative Incidence Rate: Number and Cumulative median lethal dose: Estimate of the total administered amount of a substance which is associated with the death of half a population of animals when the substance is administered repeatedly in doses which are generally fractions of the median lethal dose.

Cushing's Syndrome: Described in 1932 by Harvey Cushing, the American neurosurgeon, Cushing's syndrome is due to an excess production of cortisol. It can thus result from a tumour of the adrenal glands secreting cortisol, or from a pituitary gland tumour secreting ACTH and stimulating both adrenal cortexes to hypertrophy and secrete excess cortisol. It is sometimes the result of ectopic production of ACTH from non-endocrine tumours in the lungs and pancreas. The patient gains weight and the obesity tends to have a characteristic distribution over the face, neck, and shoulder and pelvic girdles.

Custodial Care: Care that is provided primarily to meet the personal needs of a patient. The care is not meant to be curative or providing medical treatment.

Cutaneous: Cutaneous means belonging to the skin.

Cyanogenic: Compounds able to produce cyanide; examples are the cyanogenic glycosides such as amygdalin in peach and apricot stones.

Cyanosis: A condition in which the skin – usually of the face and extremities – takes on a bluish tinge. It accompanies states in which the blood is not properly oxygenated in the lungs, and appears earliest through the nails, on the lips, on the tips of the ears, and over the cheeks. It may be due to blockage of the air passages, or to disease in the lungs, or to a feeble circulation, as in heart disease.

Cyanosis: Bluish coloration, especially of the skin and mucous membranes and fingernail beds, caused by abnormally large amounts of reduced haemoglobin in the blood vessels as a result of deficient oxygenation.

Cybernetics: The science of communication and control in the animal and in the machine.

Cyclical Oedema: This is a syndrome in women, characterised by irregular intermittent bouts of generalised swelling. Sometimes the fluid retention is more pronounced before the menstrual period.

Cycloplegia: Paralysis of the ciliary muscle of the eye, which results in the loss of the power of accommodation in the eye.

Cyclosporine: A medication developed to prevent organ transplant rejection that also is helpful in treating psoriasis.

Cyclothymia: The state characterised by extreme swings of mood from elation to depression, and vice versa.

Cyclotron: A machine in which positively charged atomic particles are so accelerated that they acquire energies equivalent to those produced by millions of volts. From the medical point of view, its interest is that it is a source of neutrons.

Cyesis: Another term for pregnancy.

Cystectomy: The surgical excision of the bladder. When this is done – usually to treat cancer of the bladder – an alternative means of collecting urine from the kidneys must be arranged.

Cystic Duct: The tube that runs from the gall-bladder and joins up with the hepatic duct (formed from the bile ducts) to form the common bile duct. The bile produced by the liver cells is drained through this system and enters the small intestine to help in the digestion of food.

Cystic Fibrosis: This is the most common serious genetic disease in Caucasian children, with an incidence of about one per 2,500 births, and more than 6,000 patients in the UK (30,000 in the USA). It is an autosomal recessive disorder of the mucus-secreting glands of the lungs, the pancreas, the mouth, and the gastrointestinal tract, as well as the sweat glands of the skin.

Cysticercosis: This disease rarely occurs except in Central Europe, Ethiopia, South Africa, and part of Asia. It results from ova (eggs) being swallowed or regurgitated into the stomach from an adult pork tapeworm in the intestine. In the stomach the larvae escape from the eggs and are absorbed.

Cystitis: An inflammation, often infectious, of the urinary bladder. It usually arises from a distal infection of the urethra or prostate.

Cystitis: Inflammation of the urinary bladder. The presenting symptom is usually dysuria – that is, a feeling of discomfort when urine is passed and frequently a stinging or burning pain in the urethra. There is also a feeling of wanting to pass water much more often than usual, even though there is very little urine present when the act is performed. The condition may be associated with a dragging ache in the lower abdomen, and the urine usually looks dark or stronger than normal. It is frequently associated with haematuria, which means blood in the urine and is the result of the inflammation.

Cystocoele: A prolapse of the base of the urinary bladder in a woman. The pelvic floor muscles may be weakened after childbirth and, when the woman strains, the front wall of the vagina bulges. Stress incontinence often accompanies a cystocoele and surgical repair is then advisable.

Cystogram: An X-ray picture of the urinary bladder.

Cystometer: An instrument for measuring the pressure in the urinary bladder.

Cystometry: A technique for measuri-

ng the pressure in the urinary bladder as part of a urodynamic investigation to assess the functioning of the bladder.

Cystorrhea: Mucus in the urine, usually following infection or from chronic congestion of the bladder mucosa.

Cystoscope: An instrument for viewing the interior of the urinary bladder. It consists of a narrow tube carrying a small electric lamp at its end; a small mirror set obliquely opposite an opening near the end of the tube; and a telescope which is passed down the tube and by which the reflec- tion of the brightly illuminated bladder wall in the mirror is examined.

Cysts: Hollow tumours containing fluid or soft material. They are almost always simple in nature.

Cyto: A prefix meaning something connected with a cell or cells.

Cytochrome P-420: Inactive derivative of cytochrome P-450 found in microsomal preparations.

Cytochrome P-448: Obsolete term for cytochrome P-450 I, Al, and A2, one of the major families of the cytochromes P- 450 haemoproteins. During the mono-oxygenation of certain substances, often a detoxification process, these isoenzymes may produce intermediates which initiate mutations, chemical carcinogenesis, immunotoxic reactions and other forms of chemical toxicity.

Cytochrome P-450: laemoproteins which form the major part of the enzymes concerned with the mono-oxygenation of many endogenous and exogenous substrates. .The term includes a large number of iso-enzymes which are coded for by a superfamily of genes.

Cytochrome: Haemoprotein whose characteristic mode of action involves transfer of reducing equivalents associated with a reversible change in oxidation state of the haem prosthetic group: strictly, the cytochrome P450 family are not cytochromes but haem-thiolate proteins.

Cytogenetics: Branch of genetics which correlates the structure and number of chromosomes as seen in isolated cells with variation in genotype and phenotype.

Cytogenetics: The study of the structure and functions of the cells of the body, with particular reference to the chromosomes.

Cytokine: Also lymphokine, a broad term for a variety of proteins and neuropeptides that lymphocytes and macrophages use to communicate between themselves, often from long distances. They stimulate organization and antibody responses, seem to induce the bone marrow to proliferate the type of white blood cells needed for immediate resistance, and generate sophistication and fine tuning for an overall strategy of resistance.

Cytokines: A family of protein molecules that carry signals locally between cells. Cytokines are released by cells when activated by antigens, behaving as enhancing mediators for immune response. These proteins include interleukins (produced by leucocytes), lymphokines (produced by lymphocytes), interferon, and tumour necrosis factor, one of whose many functions is killing tumour cells.

Cytology: The study of cells.

Cytometer: An instrument for counting and measuring cells.

Cytoplasm: Fundamental substance or matrix of the cell (within the plasma membrane) which surrounds the nucleus, endoplasmic reticulum, mitochondria and other organelles.

Cytoprotectant: A substance or reaction that acts against chemical or biological damage to cell membranes. The most common cytoprotectant actions are on the skin and the liver (hepatoprotectant), although there has been recent research involving lymphocyte T-cell cytoprotectants.

Cytotoxic: Causing damage to cell structure or function. proportion of a group of people who experience the onset of a health-related event during a specified time interval; this interval is generally for all members of the group, but, as in lifetime incidence, it may vary from person to person without reference to age. The classification, packaging and labelling of dangerous substances, official government organization or group receiving and evaluating notifications of new substances.

Cytotoxic: Cytotoxic means destructive to living cells.

D

D: The building letter code for the Dentistry Building, 707 Parnassus Avenue, San Francisco

Dactylitis: Inflammation of a finger or toe.

Daily Nursing Facility Rate: The amount that a nursing facility is reimbursed for providing each day of routine health care services to a recipient who is a patient in the home.

Dandruff: Also known as scurf. The white scales shed from the scalp, due to increased production of epidermal surface cells. Treatment is regular washing with an antidandruff shampoo.

Dartos: The thin muscle just under the skin of the scrotum which enables the scrotum to alter its shape.

Date of Service: The date the service was provided to the participant as specified on the claim.

Day Blindness: A condition in which the patient sees better in a Dead, Disposal of the dim light or by night than in daylight. It is only found in conditions in which the light is very glaring, as in the desert and on snow, and is relieved by resting the retina for example, by wearing coloured glasses for a time.

Day Surgery: Surgery done in a clinic or a hospital without an overnight stay either before or after the operation. Improvements in surgery – especially the introduction of minimally invasive surgery (MIS) – as well as more effective methods of anaesthesia have simplified many procedures and reduced the physical and mental stress on patients. Patients undergoing day surgery should be accompanied home by a friend or relative. Occasionally a patient may develop complications that require a post-operative stay in hospital.

Day Treatment Center: An outpatient facility that is licensed to provide outpatient care and treatment, usually for mental or nervous disorders or substance abuse.

Daydreams: Daydreams occur when an individual during waking hours imagines enjoyable or exciting events or images. Most people day dream at some stage during their lives, but it tends to occur when someone is stressed or unhappy. Children and teenagers in particular may sometimes daydream a lot. This should not usually worry their parents or teachers unless their work suffers or it affects the individual's personal relationships.

Days' Supply: The estimated days' supply of tablets, capsules, fluids cc's, etc., that has been prescribed for the recipient. Days' supply is not the duration of treatment, but the expected number of days the drug will be used.

De Morgan's Spots: De Morgan's spots are a type of small haemangioma occuring in the skin of middle-aged people. No more than 3 mm in diameter, they are rarely widespread and are not malignant.

Dead Space: Gas exchange only occurs in the terminal parts of the pulmonary airways. That portion of each breath that is taken into the lungs but does not take part in gas exchange is known as dead space. Anatomical dead space describes air in the airways up to the terminal bronchioles. Physiological dead space also includes gas in alveoli (air sacs) which are unable to take part in gas exchange because of structural abnormalities or disease.

Dead, Disposal of the: Practically, only three methods have been used from the earliest times: burial, embalming and cremation. Burial is perhaps the earliest and most primitive method. It was customary to bury the bodies of the dead in consecrated ground around churches up until the earlier half of the 19th century, when the utterly insanitary state of churchyards led to legislation for their better control. Burials in Britain take place usually upon production of a certificate from a registrar of deaths, to whom notice of the death, accompanied by a medical certificate, must be given without delay by the nearest relatives.

Deadly Nightshade: The popular name of *Atropa belladonna*, from which atropine is procured. Its poisonous black berries are sometimes eaten by-children.

Deafness: Impairment of hearing, which affects about million adults in the UK. In infants, permanent deafness is much less common: about 1–2 percent.

Deamination: The process of removal of the amino group, NH_2, from amino acids not required for building up body protein. This is carried out mainly in the liver by means of an enzyme, deaminase. The fatty acid residue is either burnt up to yield energy, or is converted into glucose.

Death Rate: Estimate of the proportion of a population which dies during a specified period. The numerator is the number of persons dying during the period; the denominator is the size of the population, usually estimated as the mid-year population

Death, Causes of: The final cause of death is usually the failure of the vital centres in the brain that control the beating of the heart and the act of breathing. The important practical question, however, is what disease, injury or other agent has led to this failure. Sometimes the cause may be obvious – for example, pneumonia, coronary thrombosis, or brain damage in a road accident. Often, however, the cause can be uncertain, in which case a postmortem examination is necessary.

Death, Signs of: There are some minor signs, such as: relaxation of the facial muscles (which produces the staring eye and gaping mouth of the

'Hippocratic countenance'), as well as a loss of the curves of the back, which becomes flat by contact with the bed or table; discoloration of the skin, which takes on a wax-yellow hue and loses its pink transparency at the finger-webs; absence of blistering and redness if the skin is burned (Christison's sign); and failure of a ligature tied round the finger to produce, after its removal, the usual change of a white ring, which, after a few seconds, becomes redder than the sur- rounding skin in a living person.

Death, Sudden: If deaths from accidents are excluded, this term means the unexpected death of an apparently healthy person. Cardiac arrest is the most common cause of sudden death. Older people (35 years or above) who suffer cardiac arrest commonly have coronary artery disease with restriction or stop- page of blood supply to part of the heart which causes infarction (heart attack).

Debility: A state of weakness.

Debridement: The surgical removal of foreign material and damaged tissue from a wound.

Decibel: The unit of hearing. One decibel is the least intensity of sound at which a given note can be heard. The usual abbreviation for decibel is dB.

Decidua: The soft coat which lines the interior of the womb during pregnancy and which is cast off at birth.

Deciduous: A plant that drops its leaves in the fall or, in some cases, during drought.

Decimal: The first experiments with the decimal scale were performed by Constantine Bering in 1833. 1 part medicinal substance (dry or tincture), mixed with 9 parts diluent (lactose or alcohol), and then successed (shaken), yields the 1X (D) potency.

Decoction: A preparation made by boiling various plants in water and straining the fluid.

Decompensation: A failing condition of the heart in a case of valvular disease. The failure of the heart to maintain full and adequate circulation.

Decongestants: Drugs which relieve nasal congestion and stuffiness. They may be given orally or by nasal spray, and most are sympathomimetic drugs which cause vasoconstriction in the nasal mucosa. Too frequent use reduces their effectiveness, and there is a danger of 'rebound' worsening if they are used for more than 10–14 days. A safer option for babies is simple sodium chloride drops. Warm moist air is also a traditional effective decongestant.

Decontamination: Process of rendering harmless (by neutralization, elimination, removal etc.) a potentially toxic substance in the natural environment, laboratory areas, the workplace, other indoor areas, clothes, food, water, sewage etc.

Decubitus: Decubitus refers to the positions taken up in Dehydration bed by patients suffering from various conditions such as pneumonia, peritonitis, or severe exhaustion. Such patients are liable to develop bed sores, or decubitus ulcer.

Decussation: Any point in the nervous system at which nerve fibres cross from one side to the other: for example, the decussation of the pyramidal tracts in the medulla, where the motor fibres from one side of the brain cross to the other side of the spinal cord.

Deductible: The deductible is a fixed amount that constitutes the first portion of the costs that insured persons must pay when obtaining insured drugs.

Defaecation: Opening the bowels.

Defibrillation: If a heart is fibrillating, the application of a large electric shock via paddles applied to the chest wall causes simultaneous electrical depolarisation of all the cardiac cells, and may allow the heart's natural pacemaker to re-establish sinus rhythm. One paddle is placed below the right clavicle and the other over the cardiac apex. Care must be taken that no one is in contact with the patient or the bed when the shock is given, to avoid electrocution.

Defibrillator: Apparatus that delivers a controlled electric shock to restore normal heart rhythm in patients whose hearts have developed ventricular fibrillation or have stopped beat- ing. The shock is delivered by electrodes placed on the chest wall or directly to the heart after the chest has been surgically opened. Defibrillators are a standard item of equipment for para- medical staff in ambulances, and aeroplanes of some airlines now routinely carry the apparatus.

Deficiency Disease: Any disease resulting from the absence from the diet of any substance essential to good health: for example, one of the vitamins.

Defoliant: Substance used for removal of leaves by its toxic action on living plants.

Deformities: Malformations or distortions of part of the body. They may be present at birth, or they may be the result of injuries, or disease, or simply produced by bad posture, like the curved spine occasionally found in children.

Degeneration: A change in structure or in chemical composition of a tissue or organ, by which its vitality is lowered or its function interfered with. Degeneration is of various kinds, the chief being fatty, where cells become invaded by fat globules; calcareous, where calcium is deposited in tissue so that it becomes chalky in consistency; and mucoid, where it becomes semi-liquefied.

Degenerative Disorders: An umbrella description for a wide variety of conditions in which there is increased deterior- ation of the structure or function (or both) of the body. Ageing causes a steady degeneration of many tissues and organs – for example, wrinkling of the skin, cataract and poor neuromuscular coordination. In degenerative disorders the changes occur earlier in life. The nervous system, muscles, arteries, joints and eyes are all susceptible. Specialised tissues are replaced by connective tissue. The commonest example in the nervous system is alzheimer's disease, which causes dementia; while in huntington's chorea, a genetic disorder, dementia is accompanied by in coordination of movements.

Deglutition: Deglutition means the act of swallowing.

Dehiscence: The breaking open of a wound that is partly healed, usually after surgery.

Dehydration: A fall in the water content of the body. Sixty per

Dehydrogenase: Enzyme which catalyses oxidation of compounds by removing hydrogen.

Djea Vu: A feeling of having already experienced an event which the person is doing or seeing at the moment. French for 'already seen', déjà vu is quite common but no satisfactory explanation for the phenomenon has yet been discovered.

Delayed Discharge: Where a patient who is fit for discharge remains in an acute hospital bed when other more suitable forms of care cannot be provided

Delayed Effect: Consequence occuring after a latent period following the end of exposure to a toxic substance or other harmful environmental factor.

Delegation: Delegation is a formal process by which CIGNA HealthCare gives another entity the authority to perform the claims payment administration on behalf of CIGNA Health Care.

Delhi Boil: Delhi boil is a form of chronic body sore occurring in Eastern countries, caused by a protozoan parasite, Leishmania tropica.

Delinquency: Behaviour by a young person that would be judged a crime if carried out by an adult. Delinquency may also include non-criminal activities – for example, running away from home, missing school lessons, drug or alcohol abuse, and unruly behaviour in public places. Delinquency is now a serious social problem in the UK, especially in deprived areas, and it is increasingly accompanied by alcohol and drug abuse.

Delirium Tremens (DTS): A distinct neurologic disorder suffered by late-in-the-game alcoholics, characterized by sensory confusion (is it red or sour, hot or loud, smelly or wet, am I thinking or screaming); part of the problem is the result of diminished myelination of nerves and decreased brain antioxidant insulation (cholesterol), with nerve impulses "shorting out" across temporary synapses.

Delirium: A condition of altered consciousness in which there is disorientation (as in a confusional state), incoherent talk and restlessness but with hallucination, illusions or delusions also present. Delirium (confusion) in some old people, acute confusion is a common effect of physical illness. Elderly people are often referred to as being 'confused'; unfortunately this term is often inappropriately applied to a wide range of eccentricities of speech and behaviour as if it were a diagnosis.

Delivery: The final expulsion of the child in the act of birth.

Delta Waves: Abnormal electrical waves observed in the electroencephalogram. The frequency of the normal alpha waves is 10 per second; that of the delta waves is 7 or fewer per second. They occur in the region of tumours of the brain, and in the brains of patients with epilepsy.

Deltoid: The powerful triangular muscle attached above to the collar-bone and shoulder-blade, and below, by its point, to the humerus, nearly half- way down the outer side of the upper arm. Its action is to raise the arm from the side, and it covers and gives roundness to the shoulder.

Delusions: An irrational and usually unshakeable belief (idee fixe) peculiar to some individuals. They fail to respond to reasonable argument and the delusion is often paranoid in character with a belief that a person or persons is/are persecuting them. The existence of a delusion, of such a nature as to influence conduct seriously, is one of the most important signs in reaching a decision to arrange for the compulsory admission of the patient to hospital for observation.

Dementia: An acquired and irreversible deterioration in intellectual function. Around 10 per cent of people aged over 65 and 20 per cent of those aged 75 or over are affected to some extent. The disorder is due to progressive brain disease. It appears gradually as a disturbance in problem-solving and agility of thought which may be considered to be due to tiredness, boredom or depression. As memory failure develops, the affected person becomes bewildered, anxious and emotional when dealing with new surroundings and complex conversations. In professional skilled workers this is frequently first recognised by family and friends.

Demography: The study of populations and factors affecting their health.

Demulcent: An agent that soothes internal membranes, traditionally separated from exernal soothing agents, emollients.

Demyelination: Destruction of the fatty myelin sheath around nerve fibres which interferes with the nerve function. It can occur after injury to the nerve, but is particularly associated with multiple sclerosis (MS).

Denaturation: Addition of methanol or acetone to alcohol to make it unfit for drinking. Change in molecular structure of proteins so that they cannot function normally, often caused by splitting of hydrogen bonds following exposure to reactive substances or heat.

Dendrites: The fine network of branches that extend from the body of a nerve, receiving impulses and carrying them into the center of the cell.

Dendritic Cells: Antigen presenting cells found in T cell areas of lymph nodes

Dendritic Ulcer: A branching ulcer on the surface of the cornea of the eye, caused by herpes simplex infection.

Denervation: Interruption of the nerve supply to an organ or other structure.

Dengue: Also known as dengue fever, breakbone fever, and dandy fever, dengue is endemic and epidemic in tropical and subtropical regions. It is an acute infection caused by a flavivirus (family togaviridae) transmitted by mosquitoes – especially Aedes aegypti. Incubation period is 5–8 days, and is followed by abrupt onset of symptoms: fever, facial erythema with intense itching (which spreads throughout the body), sore throat, running eyes, and painful muscles and joints are common accompaniments. The symptoms subside within a few days and are frequently succeeded by a relapse similar to the

Denied Claim: Claims that are not issued a bank draft/ remittance due to a specific reason code.

Denitrification: Reduction of nitrates to nitrites, nitrous oxides or dinitrogen (N2) catalysed by facultative aerobic soil bacteria under anaerobic conditions.

Dental Fluorosis: Variety of tooth enamel malformations due to excessive fluoride exposure during dental development.

Dental Hygienist: A person qualified to carry out the scaling (removal of calculus [deposits]) from the teeth and to advise patients on how to keep their teeth and gums healthy. Hygienists usually work in a qualified dentist's surgery.

Dental Surgeon: A dental surgeon, or dentist, is an individual trained to diagnose and treat disorders of the teeth and gums, as well as to advise on preventive measures to ensure that these areas remain healthy. Dentists qualify after a four-year course at dental school and then register with the general dental council, which is responsible for maintaining education-nal and professional standards. Around 25,000 dentists practise in the NHS and private sector.

Denture: A plate or frame bearing false teeth. It may be complete (replacing all the teeth in one jaw) or partial.

Deodorants: Substances which remove or lessen objectionable odours. Some, which have a powerful odour, simply cover other smells, but the most effective act by giving off oxygen, so as to convert the objectionable substances into simple and harmless ones. Varieties Volatile oils of plants, such as eucalyptus and turpentine, chlorine water and chlorinated lime, peroxide of hydrogen and charcoal have been used as deodorants. There are now many commercial products available.

Deoxyribonucleic Acid (DNA): Constituent of chromosomes which stores the hereditary information of an organism in the form of a sequence of purine and pyrimidine bases.

Dependence: A somatic state which develops after chronic administration of certain drugs; this state is characterized by the necessity to continue administration of the drug in order to avoid the appearance of uncomfortable or dangerous (withdrawal) symptoms. Withdrawal symptoms, when they occur, may be relieved by the administration of the drug upon which the body was "dependent."

Dependence: Physical or psychological reliance on a substance or an individual. A baby is naturally dependent on its parents, but as the child develops, this dependence lessens. Some adults, however, remain partly dependent, making abnormal demands for admiration, love and help from parents, relatives and others.

Dependent Child for Pharmacare Purposes: A child is a resident who is the legal ward or child of the registrant or spouse, is supported by the registrant or spouse, is neither married nor living and cohabiting in a marriage-like relationship, and is either age 18 or younger or .age 19 to 24 and attending school or university full-time.

Dependent: A person eligible for coverage under an employee benefits plan because of that person's relationship to an employee. Spouses, children and adopted children are often eligible for dependent coverage. Depends on which part of the lungs is affected. It may need to be carried out for up to three hours daily in divided periods. It is of particular value in bronchiectasis and lung abscess.

Depigmentation: Also called hypo-pigmentation, this congenital or acquired disorder is one in which the skin loses its pigmentation because of reduced melanin production.

Depilation: The process of destroying hair – substances and processes used for this purpose being known as depilatories. The purpose may be effected in three ways: by removing the hairs at the level of the skin surface; by pulling the hairs out (epilation); and by destroying the roots and so pre- venting the growth of new hairs.

Depilatory: Substance causing loss of hair.

Deposition: Process by which a substance arrives at a particular organ or tissue site, for example the deposition of particles on the ciliated epithelium of the bronchial airways.

Depression: Depression is a word that is regularly misused. Most people experience days or weeks when they feel low and fed up (feelings that may recur), but generally they get over it without needing to seek medical help. This is not clinical depression, best defined as a collection of psychological symptoms including sadness; unhappy thoughts characterised by worry, poor self-image, self-blame, guilt and low self- confidence; downbeat views on the future; and a feeling of hopelessness. Sufferers may consider suicide, and in severe depression may soon develop hallucinations and delusions. Doctors make the diagnosis of depression when they believe a patient to be ill with the latter condition, which may affect physical health and in some instances be life- threatening. This form of depression is common, with up to 15 per cent of the population suffering from it.

Depressor: (1) A muscle that lowers or flattens a part of the body. (2) The name given to a nerve by whose stimulation motion, secretion, or some other function is restrained or prevented: for example, the depressor nerve of the heart slows the beating of this organ.

Deprivation Score: A measure of an individual's or group's lack of normal social amenities such as proper housing, diet and warmth. It was devised in the 1980s to help assess the medical services needed by a socially deprived population.

Dermabrasion: Dermabrasion, or 'skin planing', is a method of removing the superficial layers of the skin, useful in the treatment of tattoos and acne scars.

Dermal Irritation: Skin reaction resulting from a single or multiple exposure to a physical or chemical entity at the same site, characterised by the presence of inflammation; it may result in cell death.

Dermal: Pertaining to the skin.

Dermatitis: Synonymous with eczema in all respects. Although the lay term 'eczema' usually refers to atopic or endogenous eczema, there are many other causes. Susceptibility to dermatitis is genetically determined in some cases; in others, environmental irritants and allergens are implicated. Symptoms typically include itching, dryness or cracking and, occasionally, sore- ness of the skin. Physical signs include redness (erythema), scaling, and vesiculation (tiny blisters just beneath the surface of the skin).

Dermatofibroma: Also known as histiocytoma. A firm, painless nodule in the skin, typically on a leg, due to excessive formation of collagen. A common disorder, it is often a slow response to an insect bite and persists indefinitely.

Dermatoglyphics: Dermatoglyphics is the study of the patterns made by the ridges and crevices of the hands and the soles of the feet.

Dermatologist: A medically qualified specialist who diagnoses and treats disorders of the skin.

Dermatology: In essence, this is the study of the skin. As well as being an organ in its own right, the skin is a stage on which other organs as well as the emotions most visibly play out their roles. Changes in its blood vessels – and hence blood flow through the skin – may indicate a major immunological response to a range of potential factors

Dermatome: (1) Embryological tissue which has developed from the somites to become the dermis and subcutaneous tissue. The cutaneous area that is derived from each dermatome is supplied by a single dorsal spinal nerve root.

Dermatomes: As spinal chord nerves branch out into the body, some segments fan out across the skin; these are the nerves that monitor the surface and are the source of senses of touch, pain, hot, cold and distension.

Dermatomyositis: A rare disease, possibly caused by an autoimmune reaction, in which muscle inflammation and weakness is associated with a characteristic heliotrope erythema of the face and backs of the hands. In adults it may be associated with underlying malignancy. Tissue changes are similar to those in polymyositis.

Dermatophytes: Fungi which can infect skin, hair and nails. About 30 species in three genera are pathogenic to humans.

Dermographism: Dermographism, or factitious urticaria, refers to transient erythema and wealing caused by trauma to the skin.

DES1: Drug Efficacy Study Implementation notice (FDA, to evaluate drugs in use before 1962)

Descriptive Epidemiology: Study of the occurrence of disease or other health-related characteristics in populations, including general observations concerning the relationship of disease to basic characteristics such as age, sex, race, occupation, and social class; it may also be concerned with geographic location. The major characteristics in descriptive epidemiology can be classified under the headings: individuals, time and place.

Desensitisation: In psychiatry, a method for treating phobias used in behaviour therapy. The affected individual is slowly acclimatised to the cause of his or her fear.

Designer Drugs: A group of chemical substances produced illegally whose properties and effects are similar to those of drugs of abuse. They may be derived from narcotic analgesics, amphetamines or hallucinogens. Ecstasy is a widely used designer drug and has caused deaths among teenagers. Designer drugs are potentially dangerous, especially if taken with alcohol.

Desorption: Opposite of adsorption; a decrease in the amount of adsorbed substance.

Desquamation: Shedding of an outer layer of skin in scales or shreds.

Detached Retina: Separation of the retina from the choroid in the eye. It may be due to trauma or be secondary to tumour or inflammation of the choroid, and causes blindness in the affected part of the retina. It can be treated surgically using photocoagulation.

Detergents: Substances which clean the skin surface. This means that, strictly speaking, any soap, or soap-like substance used in washing, is a detergent. At the present day, however, the term is largely used for the synthetic detergents which are now used on such a large scale. These are prepared

Detoxication: Also called detoxification, this is a process whereby toxic agents are removed from the body and toxic effects neutralised.

Detoxification: Process, or processes, of chemical modification which make a toxic molecule less toxic. Treatment of patients suffering from poisoning in such a way as to promote physiological processes which reduce the probability or severity of harmful effects. Detoxification by haemosorption perfusion: Passage of a patient's blood through a set of columns filled with a haemosorbent (activated charcoal, ion-exchange resin, etc.): the purpose of the operation is to remove a toxic substance from the organism, particularly in an emergency.

Detriment: Estimated measure of the expected harm or loss associated with an adverse event, usually in a manner chosen to facilitate meaningful addition over different events. It is generally the integrated product of arbitrary values of risk and hazard and is often expressed in terms such as costs in US dollars, loss in expected years of life or loss in productivity, and is needed for numerical exercises such as cost-benefit analysis.

Developmental Toxicity: Adverse effects on the developing organism (including structural abnormality, altered growth, or functional deficiency or death) resulting from exposure prior to conception (in either parent), during prenatal development, or postnatally up to the time of sexual maturation.

Deviance: Variation from normal. Often used to describe unusual sexual behaviour.

Dextran: The name given to a group of polysaccharides first discovered in sugar-beet preparations which had become infected with certain bacteria. A homogenous preparation of dextran, with a consistent molecular weight and free from protein, is in appropriate clinical circumstances used as a substitute for plasma for transfusion purposes.

Dextrocardia: A condition in which a person's heart is situated on the right of the chest in a mirror image of its usual position. This may be associated with similar inversion of the abdominal organs situs inversus.

DHCP: Decentralized Hospital Computer Program.

DHFS: Department of Health and Family Services. The DHFS administers Wisconsin Medicaid and Senior Care. Its primary mission is to foster healthy, self-reliant individuals and families by promoting independence and community

DHHS: Department of Health and Human Services. The United States government's principal agency for protecting the health of all Americans and providing essential human services, especially for those who are least able to help themselves.

DHT (Di Hydro Testosterone): A conversion of testosterone that is considered to be an aging-bio-marker. Among its affects are the appearance of body-hair, the loss of scalp hair and the onset of prostate gland problems.

Dia-: Dia is a prefix meaning through or thoroughly.

Diabetes Insipidus: Diabetes insipidus is a relatively rare condition and must be differentiated from diabetes mellitus which is an entirely different disease. It is characterised by excessive thirst and the passing of large volumes of urine which have low specific gravity and contain no abnormal constituents.

Diabetes Mellitus: Diabetes mellitus is a condition characterized by a raised concentration of glucose in the blood due to a deficiency in the production and/or action of insulin, a pancreatic hormone made in special cells called the islet cells of langerhans. Insulin-dependent and non-insulin- dependent diabetes have a varied pathological pattern and are caused by the interaction of several genetic and environmental factors. Insulin-dependent diabetes mellitus (IDDM) (juvenile-onset diabetes, type 1 diabetes) describes subjects with a severe deficiency or absence of insulin production. Insulin therapy is essential to prevent ketosis – a disturbance of the body's acid/base balance and an accumulation of ketones in the tissues. The onset is most commonly during childhood, but can occur at any age. Symptoms are acute and weight loss is common.

Diabetes Pills: Pills or capsules that are taken by mouth to help lower the blood glucose level. These pills may work for people whose bodies are still making insulin.

Diabetes: Properly diabetes mellitus, it is a disease characterized by high blood sugar levels and sugar in the urine. Diabetes is really several disorders, generally broken down into juvenile onset and adult onset. The first, currently called insulin-dependent diabetes mellitus (IDDM or Type I), is somewhat hereditary, and results from inadequate synthesis of native insulin or sometimes from auto-immunity or a virus, and occurs most frequently in tissue-types HLA, DR3, and DR4. These folks tend to be lean. The other main group is known as non-insulin-dependent diabetes mellitus (NIDDM or Type II). It is caused by a combination of heredity, constitution, and lifestyle, where high blood sugar and high blood fats often occur at the same time, and where hyperglycemic episodes have continued for so many years that fuel-engorged cells start to refuse glucose, and the person is termed insulin resistant. These folks are usually overweight, tend to have fatty plaques in their arteries, and usually have chunky parents.

Diabetic Eye Disease: A disease of the small blood vessels of the retina of the eye in people with diabetes. In this disease, the vessels swell and leak liquid into the retina, blurring the vision and sometimes leading to blindness.

Diabetic Ketoacidosis: High blood glucose with the presence of ketones in the urine and bloodstream, often caused by taking too little insulin or during illness.

Diabetic Kidney Disease: Damage to the cells or blood vessels of the kidney.

Diabetic Nephropathy: Kidney disease due to damage and destruction of blood vessels in the kidney which leads to chronic renal failure in some patients with diabetes.

Diabetic Nerve Damage: Damage to the nerves of a person with diabetes. Nerve damage may affect the feet and hands, as well as major organs.

Diagnosis: The skill of distinguishing one disease from another; it is essential to scientific and successful treatment. The name is also given to the opinion arrived at as to the nature of a disease. It is in diagnosis more than in treatment that the highest medical skill is required, and, for a diagnosis, the past and hereditary history of a case, the symptoms complained of, and the signs of disease found upon examination are all weighed. Many methods of laboratory examination are also used at the present day in aiding diagnosis. Computers are also being used to help clinical and laboratory diagnostic procedures.

Diagnostic and Treatment Centres:

Centres providing routine hospital operations but not emergency treatment intended to ensure that non-emergency patients are more quickly provided with diagnostic procedures and operations

Diagnostic Tests: Tests and procedures ordered by a physician to help diagnose or monitor a patient's condition or disease. Diagnostic tools include radiology, ultrasound, nuclear medicine, laboratory and pathology services or tests.

Dialysate: A salt and water mixture used for haemodialysis. Usually produced in the form of a concentrate, which must be diluted before use with water.

Dialyser: The apparatus through which dialysis occurs. It acts as the semi-permeable membrane. Also known as the "filter", artificial kidney.

Dialysis: A procedure used to filter off waste products from the blood and remove surplus fluid from the body in someone who has kidney failure. The scientific process involves separating crystalloid and colloid substances from a solution by interposing a semi-permeable membrane between the solution and pure water. The crystalloid substances pass through the membrane into the water until a state of equilibrium, so far as the crystalloid substances are concerned is established between the two sides of the membrane. The colloid substances do not pass through the membrane. Dialysis is available as either haemodialysis or peritoneal dialysis.

Diaphoresis: Profuse perspiration.

Diaphoretic: A substance that increases perspiration, either by (1) dilating the peripheral blood vessels (Capsicum), (2) directly stimulating by drug action the nerves that affect the sweat glands (Asclepias tuberosa), or by (3) introducing a volatile oil into the bloodstream that performs both tasks (Asarum canadensis).

Diaphragm: The diaphragm is the thin, dome-shaped muscular partition which separates the cavity of the abdomen from that of the chest. It is of great importance in respiration, playing the chief part in filling the lungs. During deep respiration its movements are responsible for 60 per cent of the total amount of air breathed, and in the horizontal posture, or in sleep, an even greater percentage.

Diaphysectomy: The operation whereby a part of the shaft of a long bone (e.g. humerus, femur) is excised.

Diaphysis: The shaft of a long bone.

Diarrhea: Diarrhoea or looseness of the bowels is increased frequency, fluidity or volume of bowel movements compared to usual. Most people have occasional attacks of acute diarrhoea, usually caused by contaminated food or water or excessive alcohol consumption. Such attacks normally clear up within a day or two, whether or not they are treated. Chronic diarrhoea, on the other hand, may be the result of a serious intestinal disorder or of more general disease.

Diastole: The relaxation of a hollow organ. The term is applied in particular to the heart, to indicate the resting period between the beats (systole), while blood is flowing into the organ.

Diastolic Pressure: The pressure exerted by the blood against the arterial wall during diastole. This is the lowest blood pressure in the cardiac cycle. A normal reading of diastolic pressure in a healthy adult at rest is 70 mm Hg.

Diastolic: The lower number of a blood pressure reading signifying the myocardial and arterial relaxation between pump strokes. Too close to the higher number (systolic) usually signifies inadequate relaxation of the heart and arteries between heartbeats.

Diathermy: A process by which electric currents can be passed into the deeper parts of the body so as to produce internal warmth and relieve pain; or, by using powerful currents, to destroy tumours and diseased parts bloodlessly. The form of electricity used consists of high-frequency oscillations, the frequency of oscillation ranging from

Diathesis: An archaic term meaning constitutional or inherited state giving an individual a predisposition towards a disease, a group of diseases or a structural or metabolic abnormality. An example is haemophilia, a bleeding disorder.

Dicephalus: The term applied to symmetrical conjoined twins with two separate heads.

Dicrotism: A condition in which the pulse occurs as a beat each time the heart contracts. A dicrotic wave is naturally present in a tracing of any pulse as recorded by an instrument for the purpose, but in health it is imperceptible to the finger. In fevers, a dicrotic pulse is a serious sign in which the heart continues to beat violently while the small blood vessels have lost their tone.

Die-off: The phenomenon of killing so many infectious organisms so quickly that the amount of dead biomass itself causes liver overload, allergic reactions, or a mild foreign-body response. It can occur with antibiotic therapy, treatment of candidiasis, and even with use of some herbal antivirals. Outside of prescription antifungals, it is seldom acknowledged as a medical problem. If you use a liver stimulant, diaphoretic, and diuretic, you will increase the efficiency of transport, catabolism, and excretion, and lessen the effects of die-off.

Diet: The mixture of food and drink consumed by an individual. Variations in morbidity and mortality between population groups are believed to be due, in part, to differences in diet. A balanced diet was traditionally viewed as one which provided at least the minimum requirement of energy, protein, vitamins and minerals needed by the body.

Dietetics: Dietitians apply dietetics, the science of nutrition, to the feeding of groups and individuals in health and disease. Their training requires a degree course in the nutritional and biological sciences. The role of the dietitian can be divided as follows.

Differential Diagnosis: A list of the possible diagnoses that might explain a patient's symptoms and signs, and from which the correct diagnosis will be extracted after further investigations.

Differentiation: The gradual diversification of the stem cells of the early embryo into the specialised cells, tissues and organs that go to make up the fully developed organism.

Digestion: The three processes by which the body incorporates food are digestion, absorption, and assimilation. In digestion, food is soft- ened and converted into a form soluble in the watery fluids of the body; or, in the case of fat, into minute globules. The substances formed are then absorbed from the bowels and carried throughout the body by the blood. In assimilation, these substances, deposited from the blood, are used by the various tissues for their growth and repair.

DIN/GP: A number assigned to each drug product marketed under the Food and Drug regulations by the Health Protection Branch of Health Canada. The DIN identifies only one product.

Dioptre: A term used in the measurement of the refractive or focusing power of lenses; one dioptre is the power of a lens with a focal distance of one metre and is the unit of refractive power. As a stronger lens has a greater refractive power, this means that the focal distance will be shorter. The strength in dioptres therefore is the reciprocal of the focal length expressed in metres.

Diphtheria: Diphtheria is an acute infectious disease of the respiratory tract. Rarely seen in the UK since the introduction of inoculation in 1940, it is still an important cause of disease in many parts of the world. The infection is caused by the Corynebacterium diphtheriae and is spread by water droplets. It usually presents with a sore throat, and there is a slightly raised membrane on the tonsils surrounded by an inflammatory zone. There may be some swelling of the neck and lymph nodes, though the patient's temperature is seldom much raised. Occasionally the disease occurs in the eye or genital tract, or it may complicate lesions of the skin.

Diplegia: Extensive paralysis on both sides of the body but affecting the legs more than the arms.

Diplo: A prefix meaning twofold.

Diploe: The layer of spongy bone which intervenes between the compact outer and inner tables of the skull.

Diploid: Chromosome state in which the chromosomes are present in homologous pairs. Normal human somatic (non- reproductive) cells are diploid (they have 46 chromosomes), whereas reproductive cells, with 23 chromosomes, are haploid. RT haploid, meiosis, mitosis.

Diplopia: Double vision. It is due to some irregularity in action of the muscles which move the eyeballs, in consequence of which the eyes are placed so that rays of light from one object do not fall upon corresponding parts of the two retinae, and two images are produced. It is a symptom of several nervous diseases, and often a temporary attack follows an injury to the eye, intoxication, or some febrile disease like diphtheria.

Diprosopus: The term applied to a fetus which has two faces instead of one.

Dipsomania: A morbid and insatiable craving for alcohol.

Dipygus: A fetus with a double pelvis.

Disability: An observable mental or physical loss or impairment which is measurable and which may be permanent or temporary. If the disability is serious enough to affect a person's normal function adversely, it is described as a handicap.

Disarticulation: The amputation of a bone by cutting through the joint of which the bone forms a part.

Disc: An anatomical term describing a rounded flattened structure. Examples are the cartilagenous disc positioned between two vertebrae and the optic disc.

Discharge (Effluent, Emission) Standard or Release Limit: Maximum amount of a pollutant released from a given source to a specified medium which is acceptable under specified circumstances.

Discharge Planning: Identifying a patient's health care needs after discharge from inpatient care.

Discharge: Abnormal emission or emissions from any part of the body. It usually applies to purulent material – for example, the septic material which comes away from an infected ear, or nose– but can be the result of excess secretions from the mucous linings of the vagina or rectum.

Discission: The term applied to an operation for destroying a structure by tearing it without removal: for example, the operation of needling the lens of the eye for cataract.

Discordance (Genetic): Any difference in a character between individuals due to genetic differences such as may occur in dizygotic twins, or between matched pairs in a case cohort study.

Disease Management: A continuous, coordinated health care process that seeks to manage and improve the health status of a carefully defined patient population over the entire course of a disease. The patient populations targeted are high-risk, high-cost patients with chronic conditions that depend on appropriate pharmaceutical care for proper maintenance. Disease management services include disease prevention efforts as well as patient management once the disease states have developed.

Disease: Literally, disease, lack of ease; pathological condition that presents a group of symptoms peculiar to it and which establishes the condition as an abnormal entity different from other normal or pathological body states.

Disease-Modifying Antirheumatic Drugs (DMARDs): A class of medication that has the potential to reduce or prevent joint damage, preserve joint function, and maintain productivity of the patient. Examples include Rheumatrex, Plaquenil, and Arava.

Disenrollment: Voluntarily terminating

one's participation in a health benefits plan.

Disinfectants: Substances that destroy micro-organisms, thus preventing them from causing infections. The name is usually applied to powerful chemicals that are also capable of destroying tissue and so are used only to sterilise inanimate surfaces.

Disinfection: Processes by which vegetative organisms, excluding spores, are killed in order to prevent the items disinfected from passing on infection. Equipment, bedlinen and hard surfaces may all be disinfected – the method chosen will depend on the material and size of the object. One of the most important procedures in pre-venting the spread of infection is the careful washing of hands before handling equipment and between treating different patients. Sterilisation is different from disinfection in that the methods used kill all living organisms and spores.

Disinfestation: The destruction of insect pests, especially lice, whether on the person or in dwelling-places.

Disintegration time: The time required for a tablet to break up into granules of specified size (or smaller), under carefully specified test conditions. The conditions of the laboratory test, in vitro, are set to simulate those that occur in vivo. Factors such as the kind and amount of tablet binders and the degree of compression used in compacting the tablet ingredients help determine disintegration time. The active ingredients in a disintegrated tablet are rtot necessarily found to be in solution and available for absorption. A long disintegration time is incompatible with rapid drug absorption; a short disintegration time, by itself, does not ensure rapid absorption.

Dislocations: Injuries to joints of such a nature that the ends of the opposed bones are forced more or less out of connection with one another. Besides displacement of the bones, there is bruising of the tissues around them, and tear- ing of the ligaments which bind the bones together.

Disorientation: Orientation in a clinical sense includes a person's awareness of time and place in relation to him- or herself and others, the recognition of personal friends and familiar places, and the ability to remember at least some past experience and to register new data. It is therefore dependent on the ability to recall all learned memories and make effective use of memory. Disorientation can be the presenting feature of both delirium (confusion) and dementia; delirium is reversible, developing dramatically and accompanied by evidence of systemic disease, while dementia is a gradually evolving, irreversible condition.

Dispensing Fee: Represents the charge for the professional services provided by the pharmacist when dispensing a prescription (including overhead expenses and profit). Medicaid and most direct pay insured prescription programs use dispensing fees to establish pharmacy payment for prescriptions. Dispensing fees do not include any payment for the drugs being dispensed.

Displacement: A term used in psychological medicine to describe the mental process of attaching to one object, painful emotions associated with another object.

Disposal: The discharge, deposit, injection, dumping, spilling, leaking, or placing of a waste into its permanent, final location.

Disposition: Natural tendency shown by an individual or group of individuals, including any tendency to acquisition of specific diseases, often due to hereditary factors.

Disproportionate Share Hospital (DSH): A hospital with a disproportionately large share of low-income patients. Under Medicaid, States augment payment to these hospitals. Medicare inpatient hospital payments are also adjusted for this added burden.

Dissection: (1) The cutting of tissue to separate the structural components for identification or removal during an operation or the study of anatomy. (2) Dissection of an artery involves tearing of the inner part of the wall, allowing blood to track through the media occluding the origins of smaller arteries and often leading to vessel rupture.

Disseminated: Spread of disease from its original site throughout an organ or the body. Often used to describe the spread of cancer.

Dissipation: Reduction in the amount of a pesticide or other compound which has been applied to plants, soil etc. (used when it is not clear whether this is by mineralization degradation, binding, or leaching).

Dissociation: A psychiatric term describing the process whereby an individual separates his or her ideas and thoughts from consciousness, thus allowing them to function independently. The result may be that the individual holds contrary views on the same subject.

Dissociative Disorder: A collection of psychological disorders in which a particular mental function becomes cut off from a person's mind. Hysterical amnesia is one example, when the person forgets his or her personal history but can still absorb and talk about new events. Other examples are fugue, depersonalisation (detachment from self and environment), and multiple personality disorder.

Dissolution Time: The time required for a given amount (or fraction) of drug to be released into solution from a solid dosage form. Dissolution time is measured in vitro, under conditions that simulate those that occur in vivo, in experiments in which the amount of drug in solution is determined as a function of time. Needless to say, the availability of a drug in solution - rather than as part of insoluble particulate matter - is a necessary preliminary to the drug's absorption.

Distal: An adjective applied to a body part that is further away from another part, with reference, for example, to the trunk.

Distention: An excess expansion of a tissue or organ, either from inflammation, injury or, as in the Bean Syndrome, gas.

Distichiasis: Distichiasis is the term applied to the condition in which there are two complete rows of eye- lashes in one eyelid (or in both).

Distoma: Distoma is a general term including various forms of trematodes, or fluke-worms, parasitic in the intestine, lung and other organs.

Distribution: Dispersal of a substance and its derivatives throughout the natural environment. Dispersal of a substance within an organism, including metabolism, storage and excretion. Final location of a substance within an organism after dispersal.

Disuse Atrophy: The wasting of muscles after prolonged immobility. This can be seen after lengthy immobilisation in a plaster cast, and is particularly severe following paralysis of a limb through nerve injury.

Diuresis: An increase in the production of urine. This may result from increased fluid intake, decreased levels of antidiuretic hormone, renal disease, or the use of drugs.

Diuretics: Substances which increase urine and solute production by the kidneys. They are used in the treatment of heart failure, hypertension, and sometimes for ascites secondary to liver failure. They may work by extra-renal or renal mechanisms.

Diverticulosis: Having congenital pouches of the type found in many organs, particularly the colon, that are benign, but, being little cul-de-sacs, are likely to become inflamed from time to time. Diverticulitis is the term for inflamed diverticula.

Diverticulum: A pouch or pocket leading off a main cavity or tube. The term is especially applied to protrusions from the intestine, which may be present either at the time of birth as a developmental peculiarity, or which develop in numbers upon the large intestine during the course of life.

Dizygotic Twins: Two people born at the same time to the same parents after fertilisation of two separate oocytes. They may be of different sexes and are no more likely to resemble each other than any other sibling pairs.

Dizziness: This means different things to different people, so it is important to establish what the indi means by dizziness. It may encompass a feeling of disequilibrium; it may be light-headedness, faintness, a sensation of swimming or floating, an inbalance or unsteadiness, or episodes of mental confusion. It may be true vertigo, which is an hallucination of movement. These symptoms may be due to diseases of the ear, eye, central nervous system, cardiovascular system, or endocrine system, or they may be a manifestation of psychiatric disease.

DMAE (De Methyl Amino Ethanol): Demethylaminoethanol is found in small amounts in the brain and is known for its brain enhancing affects.

DME: Drug metabolizing enzyme.

DMO: Disease Management Organizations.

DNA (Deoxyribo Nucleic Acid): The genetic blueprint that resides in the nucleus of every cell of every living organism ever studied. Many researchers believe that free radicals damage to the DNA is directly involved in aging and cancer.

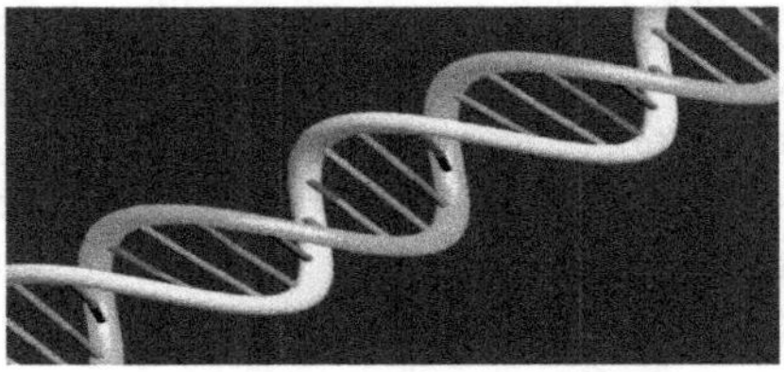

DND: Department of National Defence.

DOB: Date of birth.

Dobutamine Therapy: An intravenous form of inotropic therapy that is administered by an infusion pump to monitor the dosage of medicine.

Doctor: The academic title granted to someone who has a university degree higher than a master's degree. Some

UK universities grant a medical doctorate (MD) for a research thesis of approved standard. In Britain, 'doctor' is also the title given to a qualified medical practitioner registered by the General Medical Council, usually after he or she has obtained a bachelor's degree or a diploma in medicine and surgery. In the UK a doctor has to spend a year of supervised practice in a recognised hospital post before he or she is registered as fully qualified, but specialists have to obtain further training and higher qualifications before they can be accredited.

Dolichocephalic: Dolichocephalic means long-headed, and is a term applied to skulls the breadth of which is less than four-fifths of the length.

Donor Insemination: Use of the semen of an anonymous donor to produce fertilisation in cases of infertility where the male partner has oligospermia or impotence. The donor is chosen for ethnic and physiognomic similarity to the male partner and is screened for transmissible diseases (e.g. HIV, syphilis, hepatitis, gonorrhoea, and genetic disorders). Insemination is performed at the time of ovulation by introducing the semen into the upper vagina. Semen may be fresh or have been stored frozen in liquid nitrogen.

Donors: People who donate parts of their bodies for use in other people. Many organs and tissues can be donated – most commonly blood, but skin, corneas, kidneys, livers and hearts can all be used. Combined heart and lung transplants are being increasingly used for patients with severe lung diseases, and, if the recipients have a condition such as cysticfibrosis in which the heart is normal,

Dopamine: A neurotransmitter critical to fine motor coordination, immune function, motivation, insulin regulation, physical energy, thinking, short term memory, emotions such as sexual desire and autonomic nervous system balance.

Dopaminergic: The parts of the nervous system which use dopamine as a neurotransmitter.-

Dorsal Root Ganglia: These are swellings on the dorsal roots of spinal nerves just proximal to the union of the dorsal and ventral nerve roots. They are situated in the inter-vertebral foramina and contain the cell bodies of sensory neurones.

Dorsum: (Adjective: dorsal.) The back or posterior part of an organ or structure. The dorsum of the hand is the opposite surface to the palm.

DOS: Date of service. The calendar, date on which a specific medical service is performed.

Dosage Form: The physical state in which a drug is dispensed for use. For example: a frequent dosage form of procaine is a sterile solution of procaine. The most frequent dosage form of aspirin is a tablet.

Dosage: The quantity of drug, or dosage form, administered to a subject at a given time; for example, the usual dose of aspirin for relief of pain in an adult is 300-600 milligrams. Dose may be expressed in terms appropriate to a specific dosage form, i.e., one teaspoonful of a liquid medication, rather than the weight of drug in the teaspoonful.

Dose-duration Curve: The curve describing the relationship between dose (as the independent variable) and duration of drug effect (as the dependent variable, T). The slope of the curve is always positive, in contrast to the slope of the time-concentration curve. There has been increased interest in the dose-duration curve as a useful measure of drug action since Levy's demonstration that the constants describing the straight log dose-duration curve of a drug can be used to infer pharmacokinetic and pharmacodynamic properties of the drug, such as the elimination half-life and the threshold dose.

Dose-effect Curve: A characteristic, even the sine qua non, of a true drug effect is that a larger dose produces a greater effect than does a smaller dose, up to the limit to which the cells affected can respond. While characteristic of a drug effect, this relationship is not unique to active drugs, since increasing doses of placebos can, under certain conditions, result in increasing effects. Distinguishing between "true" and "inactive" drugs requires more than demonstration of a relationship between "dose" and effect.

Dose-effect Relationship: Association between dose and the magnitude of a continuously graded effect, either in an individual or in a population or in experimental animals.

Dose-related Effect: Situation in which the magnitude of a biological change is related to the dose.

Dose-response Curve: Graph of the relation between dose and the proportion of individuals in a population responding with an all-or-none effect.

Dose-response Relationship: Association between dose and the incidence of a defined biological effect in an exposed population.

Dosette Packaging: Single doses packaged individually, e.g., in blister packs.

Double Blind Trial: A scientific study in which different patients receive a different drug, the same drug at a different dose, or a placebo – with neither the investigators assessing the outcome nor the subjects being treated knowing which of these the latter are receiving. The aim is to remove any hint of bias due to the investigators' or patients' preferences or preconceptions.

Double Blind: A type of scientific experiment in which neither the subjects nor the researchers know who is receiving an active substance and who is receiving a placebo. Researchers who do not know which subjects received the active substance then usually evaluate the data generated from the experiment. This type of experiment helps to eliminate personal bias from research.

Double-blind Crossover: This is a study where at one point in the experiment all of the subjects switch from an active substance to a placebo or vice versa.

Douche: An application to the body of a jet of fluid via a pipe or tube. It may be used to clean any part of the body but is used most commonly with reference to the vagina (although used as a method of contraception it is ineffective).

Down's (Down) Syndrome: A genetic disorder in which the affected person usually carries an extra chromosome – 47 instead of the usual 46. The extra chromosome occurs in the no. 21 group, hence the disorder is described as trisomy 21. The condition was named after Dr J L H Down, the London doc- tor who first described it in 1866. The incidence is around one in 600 births. The disorder is characterised by a particular physical appear- ance and learning difficulties, with the affected individuals having an intelligence quotient (IQ) ranging from 30 to 80 (normal is 100).

DPC-PTR Act: Drug Price Competition and Patent Term Restoration Act of 1984 (also Waxman-Hatch or Hatch-Waxman bill)

DPT Vaccine: Often called the triple vaccine, the injections produce immunity against diphtheria, whooping cough (pertussis) and tetanus.

Draize Test: Evaluation of materials for their potential to cause dermal or ocular irritation and corrosion following local exposure; generally using the rabbit model (almost exclusively the New Zealand White) although other animal species have been used.

Draught: A draught is a small mixture intended to be taken at one dose. It consists generally of two or four tablespoonful of fluid.

Drepanocytosis: Another term for sickle-cell anaemia which is characterised by the presence in the blood of red blood corpuscles that are sickle-like in shape. The anaemia is a severe one and afflicts black people and to a lesser extent people of Mediterranean background.

DRG (Diagnosis Related Group): The DRG system is an inpatient classification system based on several factors: principal diagnosis; secondary diagnosis; surgical factors; age; sex; and discharge status. Under the Medicare prospective payment system, hospitals are paid a set fee for treating patients in a single DRG category, regardless of the actual cost of care for the individual.

Drop Attack: A brief episode affecting the nervous system that causes the person to fall suddenly. There is no loss of consciousness. The loss of tone in the muscles, responsible for the fall, may per- sist for several hours; in such cases moving the patient or applying pressure to the soles of the feet may restore muscle tone. In most cases, however, recovery is immediate. The cause is probably a temporary inter- ference with the blood supply to the brain. In others there may be some disturbance of the vestibular apparatus which controls the balance of the body.

Drop Foot: This is the inability to dorsiflex the foot at the ankle. The foot hangs down and has to be swung clear of the ground while walking. It is commonly caused by damage to the lateral popliteal nerve or the peroneal muscles.

Drop Wrist: This is the inability to extend the hand at the wrist. It is usually due to damage to the radial nerve which supplies the extensor muscles.

Drug Absorption: Drugs are usually administered distant to their site of action in the body; they must then pass across cell membranes to reach their site of action.

Drug Abuse: Use or misuse of a drug under conditions, or to an extent, considered "more destructive than constructive for society and the individual." More specifically, the use of drugs for their effects chiefly on the central nervous system, to an extent and/or at a frequency and/or for a duration of time that is inimical to the welfare of the user and/or the total of social groups in which she/he lives. Drug abusers to produce a 'high' and is referred to as 'poppers'.

Drug Benefit Plan: A plan that states which people are covered under what plan, what drugs are paid for, and how these drugs are paid for. Some drug plans are private, while others are public.

Drug Benefit: A benefit normally included with a medical plan that allows prescription medicine to be obtained using a benefit card. The member normally pays only a small co-payment or coinsurance amount for each prescription obtained.

Drug Binding: The process of attachment of a drug to a receptor or plasma protein, fat, or other tissue component. This process may be reversible or irreversible.

Drug Clearance: The volume of blood from which a drug is completely removed in one minute is known as clearance. Renal clearance of a drug is the amount of blood completely cleared of the drug by the kidney in one minute.

Drug Cost / Product Value: Total cost of ingredients in a prescription, or total value of supplies issued.

Drug Delivery Systems: Methods of-transporting drugs into the body or to specific sites of action in the body.

Drug Dependence: "Drug Dependence" has been recommended as a term to be substituted for such words as "addiction" and "habituation" since it is frequently difficult to classify specific agents as being only addictive, habituating, or non-addicting or non-habituating. It has been suggested that the general term be used and modified, appropriately, in specific instances, e.g., drug dependence of the barbiturate type.

Drug Development: Research process following drug discovery that takes a molecule with desired biological effects in animal models and prepares it as a drug that can be used in humans

Drug Discovery: Research process that identifies molecules with desired biological effects in animal models, and thus have promise as new therapeutic drugs in humans.

Drug Formulary: A list of covered or reimbursable drugs approved for use within a health care system or by clients of a PBM.

Drug Identification Number (DIN): A number assigned to each drug product marketed under the Food and Drug regulations by the Health Protection Branch of Health Canada. The DIN identifies only one product.

Drug Information System (DIS): A component of Pharma Net that contains the patient record and drug information. Access to DIS is controlled by the College of Pharmacists of British Columbia.

Drug Interactions: Many patients are on several prescribed drugs, and numerous medicines are available over the counter, so the potential for drug interaction is large.

Drug Metabolism and Transport: How drugs are processed, broken down, and moved within the body

Drug Metabolism: A process by which the body destroys and excretes drugs, so limiting their duration of action. Phase 1 metabolism consists of transformation by oxidation, reduction, or hydrolysis. In phase 2 this transformed product is conjugated (joined up) with another molecule to produce a water-soluble product which is easier to excrete.

Drug Monograph: Clinical data available on Pharma Net that provides the pharmacist with information on the proper use, side effects, etc., of the medication being dispensed. Often, a monograph contains educational or counseling information intended to be understood by and printed out for a patient.

Drug Policy: Any policy that affects the use and application of drugs, including legislative drug policies, insurance company drug policies, and hospital drug policies

Drug Therapy Management:

Management by pharmacists of the range of factors, such as drug dose, method of delivery, toxicity, and cost, that can affect the access to and effective use of drugs by patients

Drug Use Evaluation (DUE): Information on potential drug therapy and dispensing problems, which Pharma Net provides.

Drug Utilization Check: A system of drug use review that can identify potential adverse drug interactions, drug- condition conflicts, therapeutic duplication, drug-age conflicts, etc.

Drug Utilization Review (DUR): A review and assessment of the appropriateness of prescription drug use and also the prescribing patterns of participating physicians.

Drug: A chemical used in the diagnosis, treatment, or prevention of disease. More generally, a chemical, which, in a solution of sufficient concentration, will modify the behavior of cells exposed to the solution.

Dry Weight: The patients weight which they achieve after dialysis and they try to maintain.

DSH Payment: Additional payments in the Medicaid and Medicare programs that, along with local tax appropriations, help hospitals finance care to low income and uninsured patients.

DSS (Decision Support System): The VHA Decision Support System (DSS) is an automated management information system that tracks health care utilization (VHA workload) and assigns an approximate cost to it. Each VA facility has a separate implementation of DSS]The source of DSS data are the VISTA systems at each VA facility. On a quarterly basis, DSS staff at the VA Austin Automation Center extract data from the DSS systems at each facility. After creating additional data fields, they create a series of extracts that cover all VA facilities.

DTC: Direct-to-consumer (drug advertising).

Dual Choice: The opportunity to choose between health care coverage from two or more group health plans, such as indemnity insurance. This choice is usually offered during an employer's annual open-enrollment period.

Dual Eligibles: Persons who are entitled to Medicare (Part A and/or Part B) and who are also eligible for some form of medicaid benefit.

Dual Entitlee: A recipient who is eligible for Medicaid and Medicare, either Medicare Part A, Part B, or both.

Duchenne Muscular Dystrophy: An X-linked recessive disorder (that is, the abnormal gene is carried on the X chromosome). This means that the disease occurs almost exclusively in males, as its presence in a female is counteracted by the normal gene likely to be in her other X chromosome. The disorder is characterised by progressive muscular weakness and wasting.

Duct: The name applied to a passage leading from a gland into some hollow organ, or on to the surface of the body, by which the secretion of the gland is discharged: for example, the pancreatic duct and the bile duct opening into the duodenum, and the sweat ducts opening on the skin surface.

Ductless Gland: Any one of certain glands in the body the secretion of which goes directly into the bloodstream and so is carried to different parts of the body. These glands – the pituitary, thyroid, parathyroid, adrenal and reproductive – are also known as the endocrine glands.

Ductus Arteriosus: The blood vessel in the fetus through which blood passes from the pulmonary artery to the aorta, thereby bypassing the lungs, which do not function during intra-uterine life. The ductus normally ceases to function soon after birth and within a few weeks is converted into a fibrous cord. Occasionally this obliteration does not occur: a condition known as patent ductus arteriosus. This is one of the more common congenital defects of the heart, and one which responds particularly well to surgical treatment. Closure of the duct can also be achieved in some cases by the - administration of indomethacin.

Ductus Deferens: Ductus deferens, or vas deferens, is the tube which carries spermatozoa from the epidydimis to the seminal vesicles.

Dummy: "A counterfeit object;" a form of treatment - as in an experimental investigation of drug effects - which is intended to have no effects, to be biologicaily inert. The dummy treatment should mimic in every way (dosage form, route of administration, etc.) the purportedly active ingredient upon which the effectiveness of the active treatment is expected to depend.

Dumping Syndrome: A sensation of weakness and sweating after a meal in patients who have undergone gastrectomy. Rapid emptying of the stomach and the drawing of fluid from the blood into the intestine has been blamed, but the exact cause is unclear.

Duodenal Ileus: This disorder is related to gastric ulcer, both being a form of chronic peptic ulcer. Although becoming less frequent in western communities, peptic ulcers still affect around 10 per cent of the UK population at some time. Duodenal ulcers are 10–15 times more common than gastric ulcers, and occur in people aged from 20 years onwards.

Duodenum: This is the beginning of the small intestines, and it empties the stomach. It is 9 or 10 inches long, holds about the same amount of food as the digestive antrum or bottom of the stomach, and, through a papilla or sphincter, squirts a mixture of bile and pancreatic juices onto the previous stomach contents. These juices neutralize the acidic chyme; the pancreatic alkali and bile acids form soap to emulsify and aid fat digestion; and the duodenum walls secrete additional fluids and enzymes to admix with the pancreatic enzymes to initiate the final upper digestive investment. The duodenal wall secretes blood hormones to excite the brain, and gallbladder and pancreas secretions, and, if overwhelmed, can inhibit the stomach from sending anything else down for a while, until they can catch all their collective breath.

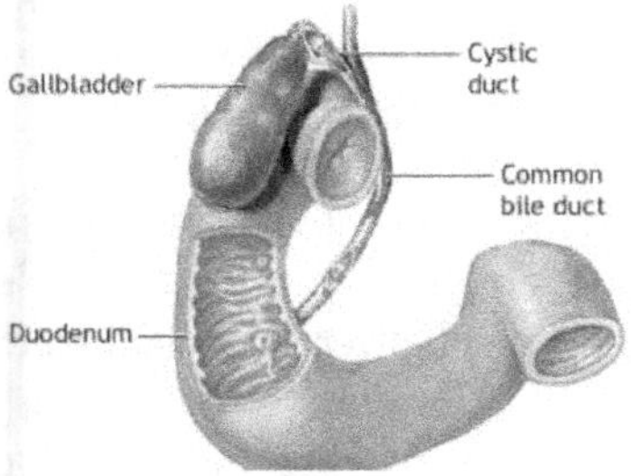

Duplicate Coverage: When a person has coverage for the same health services under more than one health benefits plan.

Duplicate Portion Sampling Method (Diet/Food): Method frequently used for the same purposes as the total diet study technique. Test persons consume their ordinary diet but for each meal, they prepare for subsequent analysis a duplicate portion of all food as prepared, served and consumed.

Dupuytren's Contracture: A condition of unknown aetiology in which there is progressive thickening and contracture of the fascia in the palm of the hand with adherence of the overlying skin. A clawing deformity of the fingers, particularly the little and ring fingers, develops. It is associated with liver disease, diabetes, epilepsy, and gout. Treatment is surgical to excise the affected fascia. Recurrence is not uncommon.

DUR: Drug Utilization Review. There are two components of DUR, prospective and retrospective. Prospective DUR is a system within the Pharmacy point-of-sale (POS) system that assists pharmacy providers in screening selected drug categories for clinically important potential drug therapy problems before the prescription is dispensed to the recipient.

Dura Mater: The outermost and strongest of the three membranes or meninges which envelop the brain and spinal cord. In it run vessels which nourish the inner surface of the skull.

Durable Medical Equipment: Equipment that can withstand repeated use and is primarily and usually used to serve a medical purpose, is generally not useful to a person in the absence of illness or injury, and is appropriate for use in the home.

Durable Power of Attorney: A written legal document that lets an individual designate another person to act on his or her behalf, even in the event the individual becomes disabled or incapacitated.

Dural Headaches: The most common type, resulting from autotoxicity or an excess of blood metabolites, such as from liver dysfunction or hangovers.

Duty of Partnership: A duty placed on NHS bodies and local government to work collaboratively General Medical Services (GMS) - services provided by family doctors and their staff, as defined in the General Medical Services Regulations 1992.

Dwarfism: Dwarfism, or short stature, refers to under - development of the body. The condition, which has various causes, is not common. All children who by the age of five years are at least what is technically known as 'three standard deviations below the mean' – well below aver- age size for children of that age – should be referred for specialist advice.

Dynamometer: An elliptical ring of steel to which is attached a dial and moving index. It is used to test the strength of the muscles of the forearm, being squeezed in the hand, and registering the pressure in pounds or kilograms.

Dys: A prefix meaning difficult or painful.

Dysarthria: A general term applied when weakness or incoordination of the speech musculature pre- vents clear pronunciation of words. The individual's speech may sound as if it is slurred or weak. It may be due to damage affecting the centres in the brain which control movements of the speech muscles, or damage to the muscles themselves.

Dysarthria: Imperfect articulation of speech due to neuromuscular damage.

Dyscalculia: A condition commonly seen when the brain's parietal lobe is diseased or injured, in which an individual finds it hard to carry out simple mathematical calculations.

Dyschezia: Constipation due to retention of faeces in the rectum. This retention is the outcome of irregular habits, which damp down the normal reflex causing defaecation.

Dyscrasia: Presently a term referring to inadequate synthesis of blood proteins by the liver, especially clotting factors. Formerly the term described an improper balance between blood and lymph in an organ or a whole person. Archaically, it referred to an imbalance between the four humors: blood, phlegm, yellow bile, and the postulated black bile.

Dysentery: A clinical state arising from invasive colorectal disease; it is accompanied by abdominal colic, diarrhoea, and passage of blood/mucus in the stool. Although the two major forms are caused by Shigella spp. (bacillary dysentery) and Entamoeba histolytica (amoebic dysentery), other organisms including entero haemorrhagic Escherichia coli.

Dysfunction : Abnormal, impaired, or incomplete functioning of an organism, organ, tissue or cell.

Dysidrosis: Disturbance of sweat secretion.

Dyskinesia: Abnormal movements of the muscles resulting from disorder of the brain. Movements are uncoordinated and involuntary and occur in facial as well as limb muscles. They include athetosis (writhing movements), chorea (jerking movements predominate), choreoathetosis (a combined type), myoclonus (spasms), tics and tremors.

Dyslexia: Dyslexia is difficulty in reading or learning to read. It is always accompanied by difficulty in writing, and particularly by difficulties in spelling. Reading difficulties might be due to various factors – for example, a general learning problem, bad teaching or understimulation, or a perceptive problem such as poor eyesight. Specific dyslexia ('word blindness').

Dysmenorrhea: Painful menstruation.

Dyspareunia: Dyspareunia means

painful or difficult coitus.

Dyspepsia: Poor digestion, with heartburn and regurgitation of stomach acids.

Dysphagia: Difficulty in swallowing. This may be caused by narrowing of the oesophagus because of physical disease such as cancer or injury. Disturbance to the nervous control of the swallowing mechanism for example, in stroke or motor neurone disease (MND) – can also cause dysphagia.

Dysphasia: Dysphasia is the term used to describe the difficulties in understanding language and in self-expression, most frequently after stroke or other brain damage. When there is a total loss in the ability to communicate through speech or writing, it is known as global aphasia. Many more individuals have a partial understanding of what is said to them; they are also able to put their own thoughts into words to some extent.

Dysphonia: Abnormal vocal sounds when speaking. It is caused by injury or disease of the larynx or of the nerves suplying the laryngeal muscles.

Dysplasia: Abnormal tissue growth. Classically midway between hyper-plasia (overgrowth) and neoplasia.

Dyspnea: Air hunger with pained breathing. It occurs normally from physical exertion, and abnormally either from impaired respiration, emotional distress, or a breakdown in nerve responses

Dyspnoea: Difficulty in breathing.

Dysrhythmia: Disturbance in the

rhythmical contractions of the heart. It is also called arrhythmia.

Dystocia: Slow or painful birth of a child. This may occur because the baby is large and/or the mother's pelvis is small or wrongly shaped for the baby to pass through easily. Abnormal presentation of the baby is another cause.

Dystonia: Dystonia refers to a type of involuntary movement characterize-ed by a sustained muscle contraction, frequently causing twisting and repetitive movements or abnormal postures, and caused by inappropriate instructions from the brain. It is sometimes called torsion spasm, and may be synonymous with athetosis when the extremities are involved. Often the condition is of unknown cause (idiopathic), but an inherited predisposition is increasingly recognised among some cases. Others may be associated with known path- ology of the brain such as cerebral palsy or wilson's disease.

Dystrophia Myotonica: A type of muscular dystrophy in which the affected person has weakness and wasting of the muscles, particularly those in the face and neck. Other effects are cataract, ptosis baldness and malfunctioning of the endocrine system.

Dystrophy: Dystrophy means defective or faulty nutrition, and is a term applied to a group of develop- mental changes occurring in the muscles, independently of the nervous system. The best-known form is progressive muscular dystrophy, a group of hereditary disorders characterised by sym- metrical wasting and weakness, with no sensory loss. There are three types: Duchenne (usually occurring in boys within the first three years of life); limb girdle (occurring in either sex in the second or third decade); and facio-scapulo-humeral (either sex, any age). The three types have different prognoses, but may lead to severe disability and premature death, often from respiratory failure. The third type progresses very slowly, however, and is compatible with a long life.

Dysuria: Painful urination. only quantitative changes in the behavior of cells; i.e., drugs increase or decrease the magnitude, frequency, of duration of the normal activities of cells Drugs used in therapy never produce qualitative changes'in cell behavior short of producing death of the cell, e.g., a nerve cell cannot be made to contract or a muscle cell cannot be made to secrete saliva by use of a drug. The degree to which this point of view will be modified by the discovery and development of agents which act on cells at a genetic level remains to be seen. responsibility; strengthening families; encouraging healthy behaviors; protecting vulnerable children, adults, and families; preventing individual and social problems; and providing services of value to taxpayers.

E

Ear: The ear is concerned with two functions. The more evident is that of the sense of hearing; the other is the sense of equilibration and of motion. The organ is divided into three parts: (1) the external ear, consisting of the auricle on the surface of the head, and the tube which leads inwards to the drum; (2) the middle ear, separated from the former by the tympanic membrane or drum, and from the internal ear by two other membranes, but communicating with the throat by the Eustachian tube; and (3) the internal ear, comprising the complicated labyrinth from which runs the vestibule cochlear nerve into the brain.

Eating Disorders: The term 'eating disorders' covers obesity, feeding problems in childhood, anorexia nervosa, and bulimia nervosa. The latter two are described here. Anorexia nervosa often called the slimmer's disease, this is a syndrome characterised by the loss of at least a quarter of a person's normal body weight; by fear of normal weight; and, in women, by amenorrhoea. An individual's body image may be distorted so that the sufferer cannot judge real weight and wants to diet even when already very thin.

Eburnation: Eburnation is a process of hardening and polishing which takes place at the ends of bones, giving them an ivory-like appearance. It is caused by the wearing away, in consequence of osteoarthritis, of the smooth plates of cartilage which in health cover the ends of the bones.

EBV: Epstein-Barr Virus, a relative of the herpes virus, is the cause of infectious mononucleosis, an African malignancy called Burkitt's lymphoma, and at least part of Chronic Fatigue Syndrome. A very common virus, most of the time it only causes a head cold.

Ecchymosis: The development of a discoloured skin patch resulting from escape of blood into the tissues just under the skin, often from bruising.

ECG or EKG: Electrocardiogram.

Echinococcus: The immature form of a small tapeworm, Taenia echinococcus, found in dogs, wolves and jackals and from which human beings become infected, so that they harbour the immature parasite in the form known as hydatid cyst.

Echocardiography: The use of ultrasonics for the purpose of examining the heart. By thus recording the echo (hence the name) from the heart of ultrasound waves, it is possible to study, for example, the movements of the heart valves as well as the state of the interior of the heart. Safe, reliable and painless, the procedure cuts the need for the physically interventionist procedure of cardiac catheterisation.

Echolalia: Echolalia is the meaningless repetition, by a person suffering from mental deterioration, of words and phrases addressed to him/her.

Eclampsia: A rare disorder in which convulsions occur during late pregnancy. This condition occurs in around 50 out of every 100,000 pregnant women, especially in the later months and at the time of delivery, but in a few cases only after delivery has taken place.

Eclectics: The name commonly applied to the American School Physicians, a distinct group of Medical Doctors who trained in their own schools, and were licensed as M.D.s. They specialized in low-tech, nonhospital rural health care, the famous country doc with a black bag. Besides standard medical procedures, they used a more wholistic approach to disease, sometimes terming themselves Vitalists. They were the most sophisticated of the many movements that arose in response to the almost maniac medical practices of the first half of the 19th century, especially in the United States, where, as always, medicine was philosophically invasive and heroic (often a wonderment to visiting physicians from Paris or London).

Ecogenetics: Study of the influence of hereditary factors on the effects of xenobiotics on individual organisms.

Ecology: Branch of biology which studies the interactions between living organisms and all factors (including other organisms) in their environment: such interactions encompass environmental factors which determine the distributions of living organisms.

Economy: The ability of a system to deliver data of high information content at a low overall cost per item of data; economy does not of course, contribute to " accuracy" but is an important determinant of the practical usefulness of a system or method.

Ecosystem: Grouping of organisms (micro-organisms, plants, animals) interacting together, with and through their physical and chemical environments, to form a functional entity.

Ecotoxicology: Study of the toxic effects of chemical and physical agents on all living organisms, especially on populations and communities within defined ecosystems; it includes transfer pathways of these agents and their interactions with the environment.

Ecstasy: Ecstasy refers to a morbid mental condition, associated with an extreme sense of well- being, with a feeling of rapture, and temporary loss of self-control. It often presents as a form of religious obsession, with a feeling of direct communication with God, saintly voices and images being perceived. In milder cases the patient may preach as though with a divine mission to help others.

Ectasia: A term that means widening, usually referring to a disorder of a duct bearing secretions from a gland or organ (e.g. mammary duct ectasia).

Ecthyma: In debilitated or immune-depressed subjects, staphylococcal impetigo or folliculitis (inflammation of hair follicles may become ulcerated. This is called ecthyma and is seen in vagrants, drug addicts, and individuals with AIDS/HIV or uncontrolled diabetes mellitus.

Ecto: Ecto is a prefix meaning on the outside.

Ectoderm: The outer of the three germ layers of the embryo during its early development. The ectoderm develops into the nervous systems, organs of sensation, teeth and lining of mouth, and the skin and its associated structures such as hair and nails.

Ectomorph: A thumbnail description of the somato type who is dominated by the ectoderm, specifically the skin, nervous system, and endocrine glands. Less arcane, a tall and thin person, with long limbs, narrow chest, and a somewhat oversensitive nervous system.

Ectoparasiticide: Substance intended to kill parasites living on the exterior of the host.

Ectopic Beat: A heart muscle contraction that is outside the normal sequence of the cardiac cycle. The impulse is generated outside the usual focus of the sinoatrial node. Also known as extrasystoles, ectopic beats are called ventricular if they arise from a focus in the ventricles and supraventricular if they arise in the atria. They may cause no symptoms and the affected subject may be unaware of them. The beat may, however, be the result of heart disease or may be caused by nicotine or caffeine. If persistent, the individual may suffer from irregular rhythm or ventricular fibrillation and need treatment with anti-arrhythmic drugs.

Ectopic Pregnancy: An ectopic pregnancy most commonly develops in one of the fallopian tubes. Occasionally it may occur in one of the ovaries, and rarely in the uterine cervix or the abdominal cavity. Around one in 200 pregnant women have an ectopic gestation. As pregnancy proceeds, surrounding tissues may be damaged and, if serious bleeding happens, the woman may present as an 'abdominal emergency'. A life-threatening condition, this needs urgent surgery.

Ectopic: Ectopic means out of the usual place. For example, the congenital displacement of the heart outside the thoracic cavity is said to be ectopic. An 'ectopic gestation' means a pregnancy outside of the womb.

Ectromelia: Ectromelia means the absence of a limb or limbs, from congenital causes.

Eczema: A chronic dermatitis, more common in those with thin skin or allergies of an atopic or IgE-mediated type, and often clearly and distinctly aggravated by emotional stress.

EDC: Electronic data capture/collection

Edema: A localized or systemic condition in which the body tissues contain an excessive amount of fluid. Systemic edema can be as mild as premenstrual water retention (I mean mild by comparison) or involve loss of blood proteins or kidney and heart failures. Local edema is the result of extensive or extended inflammation, with blood protein leakage and the loss of interstitial colloid. An accumulation of excessive amounts of water or other fluid in cells, tissues, or body cavities

Edentulous: Lacking teeth: this may be because teeth have not developed or because they have been removed or fallen out.

EDTA: Ethylenediamine tetra-acetic acid is used to treat poisoning with metals such as lead and strontium. One of the chelating agents, EDTA is used in the form of sodium or cal- cium salts. The stable chelate compounds resulting from the treatment are excreted in the urine.

Efalizumab: An injectable biologic drug used to treat moderate to severe plaque psoriasis.

Effective Concentration (EC): Concentration of a substance that causes a defined magnitude of response in a given system: EC50 is the median concentration that causes 50 % of maximal response.

Effective Date: The date on which coverage under a health benefits plan begins.

Effective Dose (ED): Dose of a substance that causes a defined magnitude of response in a given system: ED50 is the median dose that causes 50 % of maximal response.

Effector Cell: Cell capable of participating in the immune response is destruction of foreign cells or tissues

Effector Stage: Third and final stage of acute rejection in which effector cells and humoral factors enter the graft, leading to its destruction.

Efferent: The term applied to vessels which convey away blood or a secretion from a body part, or to nerves which carry nerve impulses outwards from the nerve-centres.

Efficacy: Refers to whether or not the mechanism or compound treats the disease, and whether or not it is able to differentiate over existing treatments

Effluent: Fluid, solid or gas discharged from a given source into the external environment.

Effort Syndrome: Also known as Da Costa's syndrome, this is a condition in which symptoms occur, such as palpitations and shortness of breath, which are attributed by the patient to disorder of the heart. There is no evidence, however, of heart disease, and psychological factors are thought to be of importance.

Effusion: The passage of fluid through the walls of a blood vessel into a tissue or body cavity. It commonly occurs as a result of inflammation or damage to the blood vessel. A pleural effusion may occur in heart failure (as a result of increased blood pressure in the veins forcing out fluid) or as a result of inflammation in the lung tissue (pneumonia). Effusions may also develop in damaged joints.

Ego: A psychoanalytical term to describe that part of the mind which develops as a result of the individual's interactions with the outside world. Freud describes the Ego as reconciling the demands of the Id.

EHS: The building letter code for the Environmental Health and Safety, 50 Medical Center Way, San Francisco.

EHT Essential Hypertension the early, mesomorphic stages of high blood pressure, cause mostly by thick blood and accompanying sodium retention.

Eisenmenger Syndrome: A condition in which the subject suffers from a defect in one of the dividing walls (septum) of the heart and this is accompanied by pulmonary hypertension. The defect allows blood low in oxygen to flow from the right to the left side of the heart and be pumped into the aorta, which normally carries oxygenated blood to the body. The patient has

Ejaculation: The expulsion of semen from the penis during orgasm. The stimulation of sexual intercourse (coitus) or masturbation produces a spinal reflex action that causes ejaculation. As well as containing spermatozoa (male germ cells), the semen comprises several constituents arising from cowper's glands, the prostate gland, the testicles and seminal vesicles and these are discharged in sequence.

EKG: A test that measures the heart's action. Also called an electrocardiogram.

Elastic Tissue: Connective tissue which contains a profusion of yellow elastic fibres. Long, slender and branching, these fibres (made up of elastin, an albumin-like protein) ensure that the elastic tissue is flexible and stretchable. The dermis layer of the skin, arterial walls and the alveolar walls in the lungs all contain elastic tissue.

Electrical Injuries: These are usually caused by the passage through the body of an electric current of high voltage owing to accidental contact with a live wire or to a discharge of lightning. The general effects produced are included under the term electric shock, but vary greatly in degree.

Electrocardiogram (ECG): A record of the variations in electric potential which occur in the heart as it contracts and relaxes. Any muscle in use produces an electric current, but when an individual is at rest, the main muscular current in the body is that produced by the heart. This can be recorded by connecting the outside of the body by electrodes with an instrument known as an electrocardiograph.

Electrocardiography: A method of recording the electrical activity of the heart muscles. Electrodes from a recording machine (electrocardiograph) are placed on the skin of the chest wall, arms and legs. The record of the electrical changes is called an ECG (electrocardiogram). The number of electrodes used depends on the complexity of the heart disorder being monitored. The procedure can be done in hospital, doctors' surgeries and the patient's home, and should not cause any discomfort.

Electrocautery: The use of an electrically heated needle or loop to destroy diseased or unwanted tissue. Benign growths, warts and polyps can be removed with this technique.

Electrocoagulation: A method of sealing blood vessels using heat generated by high-frequency electric current through fine needles or a surgical knife. The procedure is used during surgery to close newly cut vessels. It can also be used to stop nose- bleeds and to remove vascular deformities such as naevi.

Electrocochleography: Electrocochleography is a method of recording the activity of the cochlea, the part of the inner ear concerned with hearing.

Electroconvulsive Therapy (ECT): A controversial but sometimes rapidly effective treatment for cases of severe depression, particularly where psychotic features are present or in high-risk patients such as suicidal or post-partum patients. ECT is only indicated after antidepressants have been tried and shown to be ineffective; the full procedure of treatment should be explained to the patient, whose consent must be obtained.

Electroencephalography (EEG): In the brain there is a regular, rhythmical change of electric potential, due to the rhythmic discharge of energy by nerve cells. These changes can be recorded graphically and the 'brain waves' examined – a procedure introduced to medicine in the 1920s. These records electroencephalograms are useful in diagnosis: for example, the abnormal electro- encephalogram occurring in epilepsy is characteristic of this disease. The normal waves, known as alpha waves, occur with a frequency of 10 per second. Abnormal waves, with a frequency of 7 or fewer per second, are known as delta waves and occur in the region of cerebral tumours and in the brains of epileptics. An electroencephalogram can assess whether an individual is awake, alert or asleep. It may also be used during surgery to monitor the depth of unconsciousness in anaesthetised patients.

Electrolysis: The use of short-wave electric currents to des- troy the roots of hairs and so remove unwanted hair from the skin surface. If used by a trained operator, the procedure is safe, but care must be taken in the vicinity of the eyes and, as electrolysis of hair on the legs is such a lengthy process, it is best avoided there.

Electrolytes: Substances, for example, potassium chloride, whose molecules split into their constituent electrically charged particles, known as ions, when dissolved in fluid. In medicine the term is customarily used to describe the ion itself. The description 'serum electrolyte concentration' means the amounts of separate ions – for example, sodium and chloride in the case of salt – present in the serum of the circulating blood. Various diseases alter the amounts of electrolytes in the blood, either because more than normal are lost through vomiting or diarrhea.

Electronic Prescription: Refers to prescription information that is created, stored and transmitted via electronic means by computer or hand-held device. The term electronic prescriptions would not apply to prescriptions communicated either by facsimile (fax) or over the phone.

Electro-Oculography: A method of recording movements of the eyes, which is of value in assessing the - function of the retina.

Electrophoresis: The migration of charged particles between electrodes. A simple method of electrophoresis, known as paper electrophoresis, has been introduced to analyse protein in body fluids. This method consists in applying the protein- containing solution as a spot or a streak to a strip of filter paper which has been soaked in buffer solution and across the ends of which a potential difference is then applied for some hours. Comparison is made between filter strips of normal fluids and that of the patient under investigation.

Electroretinogram: An electroretino-gram is the record of an electrical response of visual receptors in the retina which can be measured with corneal electrodes.

Electrostatic System: Painting/coating system in a much charged paint droplets or particles are attracted tow he finish surface having opposite charge.

Elephantiasis: Chronically oedematous and thickened tissue, especially involving the lower extremities and genitalia, which arises from repeated attacks of inflammation of the skin and subcutaneous tissue, with concurrent obstruction of lymphatic vessels. In a tropical country, the usual cause is lymphatic filariasis.

Eligibility: A person entitled to receive benefits for health care services.

Eligible: Provisions contained in each health benefits plan that specify who qualifies for coverage under that plan.

Elimination Half-life or Half time: Period taken for the plasma concentration of a substance to decrease by half.

Elimination: Expulsion of a substance or other material from an organism (or a defined part thereof), usually by a process of extrusion or exclusion, sometimes after metabolic transformation.

Eliminator: Substance that contributes to the elimination of a poison from an organism.

Elixir: A liquid preparation of a potent or nauseous drug made pleasant to the taste by the addition of aromatic substances and sugar.

Emaciation: Pronounced wasting; a common symptom of many diseases, particularly those which are associated with a prolonged or repeated rise of temperature, such as tuberculosis. It is also associated with diseases of the alimentary sys- tem in which digestion is inefficient, or in which the food is not fully absorbed: for example, in long-standing diarrhoea, whatever its cause. It is also a marked feature of severe malnutrition and malignant disease.

Embolectomy: Surgical removal of a clot or embolus to clear an obstruction in an artery. The obstruction may be cleared by inserting a balloon (Fogarty) catheter into the blood vessel or by surgical incision through the arterial wall. Embolectomy may be a life-saving operation when a patient has a pulmonary embolism.

Embolism: The plugging of a small blood vessel by an embolus which has been carried through the larger vessels by the bloodstream. It is due usually to fragments of a clot which has formed in some vessel, or to small portions carried off from the edge of a heart-valve when this organ is diseased. However, the plug may also be a small mass of bacteria, or a fragment of a tumour, or even a mass of air bubbles sucked into the veins during operations on the neck. The result is usually more or less destruction of the organ or part of an organ supplied by the obstructed vessel.

Embolus: Substances – for example, air, amniotic fluid, blood clot, fat or foreign body – that are carried by the blood from a vessel (or vessels) in one part of the body to another part where the matter lodges in a blood vessel causing a blockage.

Embrocations: Embrocations are mixtures, usually of an oily nature, intended for external application in cases of rheumatism, sprains, and other painful conditions. Their action is due mainly to the massage employed in rubbing in the embrocations, in part to the counter-irritant action of the drugs which they contain.

Embryo Research: When a woman is

treated for infertility it is necessary to nurture human embryos for a few days (until the first cell divisions of the fertilised egg have occurred) in a specialised laboratory. More eggs are fertilised than are usually needed because not all fertilisations are successful. Sur- plus embryos may be frozen for use in later attempts to implant an embryo in the womb. Research has been done on very early embryos but the practice is controversial and some countries have either forbidden it or imposed tight restrictions. In the UK such research is controlled by the government Human Fertilisation

Embryo Transfer: Embryo transfer is the process whereby the initial stages of procreation are produced outside the human body and completed in the uterus or womb. The procedure is also known as 'embryo transplantation' and 'in vitro fertilisation' (IVF). It consists of extracting an ovum (or egg) from the prospective mother's body and placing this in a dish where it is mixed with the male partner's semen and special nutrient fluids. After the ovum is fertilised by the sperm it is transferred to another dish containing a special nutrient solution.

Embryo: Stage in the developing mammal at which the characteristic organs and organ systems are being formed: for humans, this involves the stages of development from the second to the eighth week (inclusive post conception). In birds, the stage of development from the fertilization of the ovum up to hatching. In plants, the stage of development within the seed.

Embryology: The study of the growth and development of an embryo and subsequently the fetus from the fertilisation of the ovum by the spermatozoon through the gestational period until birth. Embryology is valuable in the under-standing of adult anatomy, how the body works and the occurrence of congenital deformities.

Embryotoxicity: Production by a substance of toxic effects in progeny in the first period of pregnancy between conception and the fetal stage. Any toxic effect on the conceptus as a result of prenatal exposure during the embryonic stages of development: these effects may include malformations and variations, malfunctions, altered growth, prenatal death, and altered postnatal function.

Embryotropic Effect: Change in the embryo and the regulation of its development.

EMC: Electronic Media Claims. Method of claims submission through a personal computer or mainframe system. Claims can be mailed on tape or transmitted via telephone and modem.

EMEA: European Agency for the Evaluation of Medicinal Products. Similar to the FDA, the EMEA is responsible for protecting public health by mobilizing the best scientific resources within the European Union, promoting health care through the effective regulation of new pharmaceuticals, and supporting the pharmaceutical research and development industry by developing efficient, effective and responsive operating procedures

Emergency Services: Those services which are necessary to prevent death or serious impairment of the health of the individual.

Emergency: An accident or sudden illness that a person with an average knowledge of medical science believes needs to be treated right away or it could result in loss of life, serious medical complications or permanent disability. Emergencies are covered by your CIGNA HealthCare plan 24 hours a day, seven days a week, no matter where you are. Whenever there's a serious accident or sudden illness, and symptoms are severe and they occur unexpectedly, seek medical help immediately.

Emesis: Vomiting.

Emetics: An emetic is a substance which induces vomiting (emesis). Emetics were previously used for gut decontamination in the treatment of poisoning but are now considered obsolete. This is because the efficacy of emesis as a means of gut decontamination is unproved; there is a delay between administration and actual emesis, during which time continued absorption of the poison may occur; and some emetics have effects other than vomiting which may mask the clinical features of the ingested poison.

Emission and Exposure Control: Technical and administrative procedures and specifications applied for the monitoring, reduction or elimination of emissions from a source or exposure to a target.

Emission Standard: Quantitative limit on the emission or discharge of a substance from a source, usually expressed in terms of a time-weighted average concentration or a ceiling value.

Emission: A discharge. The term is commonly used to describe the orgasmic flow of semen from the erect penis that occurs during sleep. Described as a nocturnal emission or, colloquially, as a 'wet dream', it is a common event in late puberty.

Emmetropia: The normal condition of the eye as regards refraction of light rays. When the muscles in the eyeball are completely relaxed, the focusing power is accurately adjusted for parallel rays, so that vision is perfect for distant objects.

Emollients: Emollients are substances which have a softening and soothing effect upon the skin. They include dusting powders such as French chalk, oils such as olive oil and almond oil, and fats such as the various pharmacopoeial preparations of paraffin, suet, and lard. Glycerin is also an excellent emollient.

Emotion: Mental arousal that the individual may find enjoyable or unpleasant. The three components are subjective, physiological and behavioural. The instinctive fear and flee response in animals comprises physiological reaction – raised heart rate, pallor and sweating – to an unpleasant event or stimulus. The loving relationship between mother and child is another well- recognised emotional event. If this emotional bond is absent or inadequate, the child may suffer emotional deprivation, which can be the trigger for behavioural problems ranging from attention-craving to aggression.

Empathy: The facility to understand and be sympathetic to the feelings and thoughts of another individual. Empathy in the therapist is an essential component of successful psychotherapy and is a valuable characteristic in anyone who is a member of a caring profession.

Emphysema: A pulmonary condition with loss of elasticity in the alveoli and the interalveolar septa the meat-foam and their interleaving sheaths that you fill up when you breathe. If a septum gets too stretched over time, several of the little sacs will coalesce together, decreasing the surface area for oxygen and carbon dioxide exchange. If enough of these sacs lose their separateness, like small soap bubbles joining to make a few larger ones, breathing gets harder because each breath accomplishes less interchange of gases, resulting in emphysema caused by years of bad asthma, tobacco smoking, chemical damage, and other chronic lung disorders, it can be halted but not reversed. The first breath you take as a newborn defines forever the number of the alveolar bubbles, they cannot be regenerated if they coalesce together.

Empirical: Method of treatment founded simply on experience rather than on scientific evidence from, for example, clinical trials. Because a given remedy has been successful in the treatment of a certain group of symptoms, it is assumed, by those who uphold this principle, that it will be successful in the treatment of other cases presenting similar groups of symptoms, without any inquiry as to the cause of the symptoms or reason underlying the action.

Employee Assistance Program (EAP): An EAP is an assessment and referral program or a short-term counseling program that is pre-purchased by some employers and is available to their employees, their dependents and household members. Visits to the EAP are separate from your behavioral health care benefits plan with no copayment required.

Empyema: An accumulation of pus within a cavity, the term being generally reserved for collections of pus within one of the pleural cavities. Since the advent of antibiotics, the condition is relatively uncommon in developed countries. The condition is virtually an abscess, and therefore gives rise to the general symptoms accompanying that condition. However, on account of the thick, unyielding wall of the chest, it is unlikely to burst through the surface, and therefore it is of particular importance that the condition should be recognised early and treated adequately.

EMR (Electronic Medical Record): This technology, when fully developed, meets provider needs for real-time data access and evaluation in medical care. In concert with clinical workstations, point-of-care devices, and clinical data repository technologies, the EMR provides the means for longitudinal data storage and access. The result will be increased efficiency, reduced costs and improved quality of care.

Emulsions: Emulsions are oil-in-water or water-in-oil dispersions.

Encephalitis: Encephalitis means inflammation or infection of the brain, usually caused by a virus; it may also be the result of bacterial infection. It occurs throughout the world and affects all racial groups and ages. Rarely it occurs as a complication of common viral disease such as measles, mumps, glandular fever, or chickenpox. It may occur with no evidence of infection elsewhere, such as in herpes simplex encephalitis, the most common form seen in Europe and America. Rabies is another form of viral encephalitis, and the HIV virus which causes AIDS invades the brain to cause another form of encephalitis. In some countries – North and South America, Japan and east Asia and Russia

Encephaloid: A form of cancer which, to the naked eye, resembles the tissue of the brain.

Encephalomyelitis: Inflammation of the substance of both brain and spinal cord.

Encephalopathy: The term covering certain conditions in which there are signs of cerebral irritation without any localised lesion to account for them. Examples are hypertensive encephalopathy, spongiform encephalopathy, wernicke's encephalopathyand lead encephalopathy. In the first, which occurs in the later stages of chronic glomerulonephritis, or uraemia, the headache, convulsions and delirium which constitute the main symptoms are supposed to be due to a deficient blood supply to the brain.

Enchondroma: A tumour formed of cartilage.

Encryption: The process of encoding data to prevent unauthorized access to the data, especially during transmission on a network.

Encysted: Enclosed within a bladder-like wall. The term is applied to parasites, collections of pus, etc., which are shut off from surrounding tissues by a membrane or by adhesions.

End Organ: A structure at the end of a peripheral nerve that acts as receptor for a sensation. For example, the olfactory nerves have end organs that identify smells.

End Stage Renal Disease: (ESRD) Severe kidney dysfunction reached when kidney function is reduced to 10 percent or less of normal function

Endarterectomy: Surgical reopening of an artery obstructed by atheroma. If a blood clot is present, the reboring process is called thrombo endarterectomy. Restored patency allows arterial blood supply to restart. The carotid arteries and arteries to the legs are those most commonly operated on.

Endarteritis: Inflammation of the inner coat of an artery.

Endemic: A term applied to diseases which exist in particular localities or among certain races. Some diseases, which are at times epidemic over wide districts, have a restricted area where they are always endemic, and from which they spread. For example, both cholera and plague are endemic in certain parts of Asia.

Endo-: Prefix meaning situated inside.

Endocarditis: Inflammation of the lining, valves and muscle of the heart. The main causes are bacterial and virus infections and rheumatic fever, and the condition occurs most often in patients whose endocardium is already damaged by con- genital deformities or whose immune system has been suppressed by drugs. Infection may be introduced into the bloodstream during dental treatment or surgical procedures, especially on the heart or on the gastrointestinal system. The condition is potentially very serious and treat- ment is with large doses of antibiotic drugs.

Endocardium: A thin membrane consisting of flat endothelial cells; it lines the four chambers of the heart and is continuous with the lining of arteries and veins. The endocardium has a smooth surface which helps the blood to flow easily. The valves at the openings of the heart's chambers are made from folded-up membranes. Inflammation of the endocardium is called endocarditis.

Endocrine Glands: Organs whose function it is to secrete into the blood or lymph, substances known as hormones. These play an important part in general changes to or the activities of other organs at a distance. Various diseases arise as the result of defects or excess in the internal secretions of the different glands.

Endocrine System: Endocrine System is a term for a group of glands, specifically the pituitary, thyroid, thymus, pancreas, adrenal, testes and ovaries.

Endocrine: Pertaining to hormones or-to the glands that secrete hormones directly into the bloodstream.

Endocrinology: The study of the endocrine system, the substances (hormones) it secretes and its disorders

Endoderm: The inner layer of the three germ layers of the embryo during its early development. The endoderm develops into the lining of the gastrointestinal tract and associated glands, the lining of the bronchi and alveoli of the lungs and much of the urinary tract.

Endogenous: From within the body, either a native function or the product of the extended colony. Normal flora in the colon are considered endogenous.

Endolymph: The fluid that fills the membranous labyrinth of the inner ear.

Endometriosis: The presence of endometrial tissue outside of the uterus. The endometrium is the mucus membrane inner lining of the uterus, with glandular cells and structural cells, both responding to estrogen by increasing in size (the proliferative phase), the first responding to progesterone (the secretory phase); if there is endometrial tissue outside of the uterus, the tissue expands and shrinks in response to the estrus cycle, but the normal shedding of the menstrual phase can be difficult.

Endometritis: Inflammation of the - mucous membrane lining the womb.

Endometrium: The mucous membrane which lines the interior of the uterus.

Endoplasmic Reticulum: Intracellular complex of membranes in which proteins and lipids, as well as molecules for export, are synthesized and in which the biotransformation reactions of the mono-oxygenase enzyme systems occur: may be isolated as microsomes following cell fractionation procedures.

Endorphins: Peptides produced in the brain which have a pain-relieving action; hence their alternative name of opiate peptides. Their name is derived from endogenous morphine. They have been defined as endogenous opiates or any naturally occurring substances in the brain with pharmacological actions resembling opiate alkaloids such as morphine.

Endoscope: A tube-shaped instrument inserted into a cavity in the body to investigate and treat disorders. It is flexible and equipped with lenses and a light source. Examples of endoscopes are the cystoscope for use in the bladder, the gastro scope for examining the stomach and the arthroscope for looking into joints.

Endoscopy: Examination of a body cavity for example, pleural cavity, gastrointestinal tract, bile duct and urinary bladder – using an endoscope in order to diagnose or treat a disorder in the cavity. The development of endoscopy has reduced the need for major surgery, as many diagnostic procedures can be per- formed with an endoscope (as can minimally invasive surgery (MIS). The development of fibre optics (The transmission of light along bundles of glass or plastic fibres) has greatly advanced the practice of endoscopy and hospitals now routinely run endoscopy.

Endothelial: Pertaining to the layer of flat cells lining the inner surface of blood and lymphatic vessels, and the surface lining of serous and synovial membranes.

Endothelium: The membrane lining various vessels and cavities of the body, such as the pleura (lining the lung, the pericardium (lining the heart), the peritoneum (lining the abdomen and abdominal organs), the lymphatic vessels, blood vessels, and joints.

Endotoxin: A poison produced by certain gram-negative bacteria that is released after the micro- organisms die. Endotoxins can cause fever and shock, the latter by rendering the walls of blood vessels permeable so that fluid leaks into the tissues, with a consequent sharp fall in blood pressure.

Endotracheal Intubation: Insertion of a rubber or plastic tube through the nose or mouth into the trachea. The tube often has a cuff at its lower end which, when inflated, provides an airtight seal. This allows an anaesthetist to supply oxygen or anaesthetic gases to the lungs with the knowledge of exactly how much the patient is receiving. Endotracheal intubation is necessary to undertake artificial ventilation of a patient.

Enema: Introduction of fluid into the rectum via the A fibreoptic bronchoscope used to examine the larger bronchial tubes of the lungs to determine, for example, whether a tumour is present.

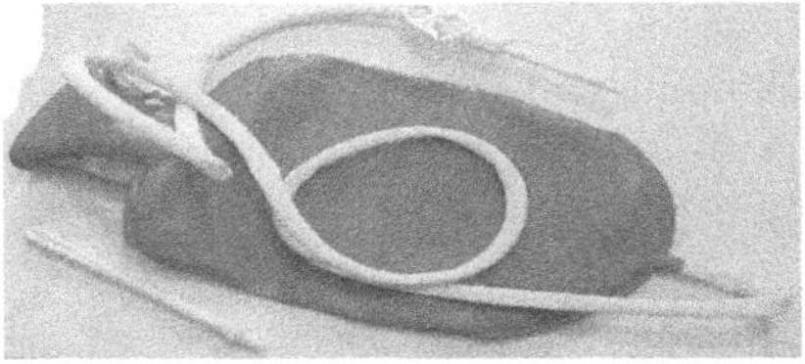

Energy Balance: A concept based on a fundamental law of physical science (conservation of energy) known as the first law of thermodynamics, which says that energy may be transformed, but not created or destroyed.

Energy Recovery Processes: Processes that recover the energy content of combustible wastes directly by burning, or indirectly by being converted to another fuel form such as gas or oil.

Energy Recovery: The retrieval of energy from waste by converting heat from incineration, or collecting methane gas from landfills. Conversion of waste energy, generally through the combustion of processed or raw refuse to produce steam or electricity.

Engagement: The event during pregnancy when the presenting part of the baby, usually the head, moves down into the mother's pelvis.

Enophthalmos: Abnormal retraction of the eye into its socket: for example, when the sympathetic nerve in the neck is paralysed.

Enrollee: An individual who is enrolled and eligible for coverage under a health plan contract. Also called Member.

ENS: Enterprise Network Services, the name of a UCSF communications service provider

Enteral Feeding: In severely ill patients, the metabolic responses to tissue damage may be sufficient to cause a reduction of muscle mass and of plasma pro- teins. This state of catabolism may also impair the immune response to infection and delay the healing of wounds. It is probable that as many as one-half of patients who have had a major operation a week previously show evidence of

Enteral Nutritional Support: A liquid nutritional supplement is administered directly to the gastrointestinal tract through a feeding tube.

Enteralgia: Another name for colic.

Enteric Fever: Enteric fever is caused by bacterial infection with either Salmonella typhi or Salmonella paratyphi A, B or C. These infections are called typhoid fever, or paratyphoid fever respectively. Transmission usually occurs by ingestion of water or food that has been contaminated with human faeces – for example, by drinking water contaminated with sewage, or eating foods pre- pared by a cook infected with or carrying the organisms. Enteric fever is endemic in many areas of the world, including Africa, Central and South America, the Indian subcontinent

Enteric: pertaining to the small intestines.

Enteritis: Intestinal inflammation.

Enterobiasis: Infection with enterobius vermicularis, the threadworm (or pinworm as it is known in the USA). It is the most common of all the intes- tinal parasites in Britain, and the least harmful. The male is about 6 mm (¼ inch) in length and the female about 12 mm (½ inch) in length. Each resembles a little piece of thread. These worms live in considerable numbers in the lower bowel, affecting children particularly. They usually cause no symptoms but can result in great irritation round the anus or within the female genitalia, especially at night when the female worm emerges from the anus to lay its eggs and then die. The most effective form of treatment is either viprynium embonate or piperazine citrate, which needs to be taken by the whole family. Bedclothes must then be laundered.

Enterocele: A hernia of the bowel.

Enterogastrone: A hormone derived from the mucosal lining of the small intestine which inhibits the movements and secretion of the stomach.

Enterohepatic Circulation: Cyclical process involving intestinal re-absorption of a substance that has been excreted through the bile followed by transfer back to the liver, making it available for biliary excretion again.

Enterokinase: The enzyme secreted in the duodenum and jejunum which converts the enzyme, trypsinogen, secreted by the pancreas, into trypsin.

Enteroptosis: A condition in which, owing to a lax condition of the mesenteries and liga- ments which support the intestines, the latter descend into the lower part of the abdominal cavity.

Enterostomy: An operation by which an artificial opening is formed into the intestine and joined to another part of the intestine or brought to the exterior via the abdominal wall.

Enterotoxin: A type of toxin that causes inflammation of the intestinal lining and results in vomiting and diarrhea.

Enteroviruses: A family of viruses which include the polio- myelitis, coxsackie and echo groups of viruses. Their importance lies in their tendency to invade the central nervous system. They receive their name from the fact that their mode of entry into the body is through the gut.

Entire: A leaf with a straight, untoothed margin.

Entonox: A proprietary analgesic drug taken by inhalation and comprising half nitrous oxide and half oxygen. It is valuable in providing relief to casualties who are in pain, as it provides analgesia without making them unconscious. Entonox is also used in obstetric practice to ease the pains of childbirth.

Enuresis: Bed-wetting, or the involuntary passage of urine at night. It can occur at all ages but is a particular problem with children and the elderly. In general, paediatricians prefer not to treat enuresis much before the age of six, as it may be a normal phenomenon and usually stops as the child grows older. However, when the condition persists, the children (and parents) need advice.

Environment: Aggregate, at a given moment, of all external conditions and influences to which a system under study is subjected.

Environmental Damage: Adverse effects to the natural environment.

Environmental Exposure Level (EEL): Level (concentration or amount or a time integral of either) of a substance to which an organism or other component of the environment is exposed in its natural surroundings.

Environmental Fate: Destiny of a chemical or biological pollutant after release into the natural environment.

Environmental Health Impact Assessment: Estimate of the adverse health effects or risks likely to follow from a proposed or expected environmental change or development.

Environmental Health: Human welfare and its influence by the environment, including technical and administrative measures for improving the human environment from a health point of view.

Environmental Hygiene: Practical control measures used to improve the basic environmental conditions affecting human health, for example clean water supply, human and animal waste disposal, protection of food from biological contamination, and housing conditions, all of which are concerned with the quality of the human environment.

Environmental Impact Statement

(EIS): Report resulting from an environmental impact assessment.

Environmental Manager: The base environmental management function supervisor or designated representative. Synonymous with the term environmental coordinator.

Environmental Medicine: The study of the consequences for people's health of the natural environment. This includes the effects of climate, geography, sun- light and natural vegetation.

Environmental Monitoring: Continuous or repeated measurement of agents in the environment to evaluate environmental exposure and possible damage by comparison with appropriate reference values based on knowledge of the probable relationship between ambient exposure and resultant adverse effects.

Environmental Protection: Actions taken to prevent or minimize adverse effects to the natural environment. Complex of measures including monitoring of environmental pollution, development and practice of environmental protection principles (legal, technical, and hygienic), including risk assessment, risk management and risk communication.

Environmental Quality Objective (EQO): Overall state to be aimed for in a particular aspect of the natural environment, for example, "water in an estuary such that shellfish populations survive in good health". Unlike an environmental quality standard, the EQO is usually expressed in qualitative and not quantitative terms.

Environmental Quality Standard (EQS): Amount concentration or mass concentration of a substance that should not be exceeded in an environmental system, often expressed as a time-weigh ted average measurement over a defined period.

Environmental Transformation: Chemical transformation of substances resulting from interactions in the environment.

Enzootic: Present in a community or among a group of animals; said of a disease prevailing continually in a region.

Enzyme: A protein that acts as a catalyst for the body's metabolic processes. The body contains thou-sands of enzymes, with each cell producing several varieties. The first enzyme was obtained in a reasonably pure state in 1926. Since then, several hundred enzymes have been obtained in pure crystalline form. They are present in the digestive fluids and in many of the tissues, and are capable of producing in small amounts the transformation on a large scale of various com- pounds.

Enzyme-Linked Immunosorbent Assay (ELISA): This is a sensitive method for measuring the quantity of a substance. An antibody to the substance is prepared along with an enzyme which binds to the antibody and which can be accurately measured using colour changes that occur as a result of the chemical reaction.

EOB: Explanation of Benefits. Appears on the provider's Remittance and Status (R/S) Report and notifies the Medicaid provider of the status of or action taken on a claim.

Eosinophil: Any cell in the body with granules in its substance that stain easily with the dye, eosin. Granulocytes which form about 2 per cent of the white cells of the blood are eosinophils.

Eosinophilia: A group of conditions having the characteristic elevation of eosinophils. These somewhat mysterious granulocytic leukocytes (white blood cells filled with cottage cheese) are definitely involved in parasite resistance, seem to initiate strong inflammation under some conditions, can facilitate clotting by inhibiting heparin, yet also are a part of the process of healing and inflammation control as an infection winds down. Eosinophilia is on one hand an inherited condition associated with atopic dermatitis (common, relatively benign, and irritating as hell), but, when acquired from chemical contact, drug reaction or spontaneously surfaced auto-immune response, it can destroy muscles, nerve, lungs, even kill. It caused the notorious string of chemical reactions that was triggered by tainted Japanese tryptophan.

EP: The building letter code for Executive Park, 250 Executive Park Boulevard, San Francisco

Ephelides: The technical term for freckles.

Epi-: A prefix meaning situated on or outside of.

Epicanthic Fold: A vertical skinfold that runs from the upper eyelid to the side of the nose. These folds are normal in oriental races but uncommon in others, although babies may have a temporary fold that disappears. Folds are present in people with down's (down) syndrome.

Epicondyle: The protuberance above a condyle at the end of a bone with an articulating joint – for example, at the bottom of the humerus, the bone of the upper arm.

Epidemic: Epidemic is a term applied to a disease which affects a large number of people in a particular locality at one time. The term is, in a sense, opposed to endemic, which means a disease always found in the locality in question. A disease may, however, be endemic as a rule – for example, malaria in swampy districts – and may become at times epidemic, when an unusually large number of people are affected. The rapid expansion of air travel has extended the scope for the spread of epidemic and endemic disease.

Epidemiology: The study of disease as it affects groups of people. Originating in the study of epidemics of diseases like cholera, plague and smallpox, epidemiology is an important discipline which contributes to the control not only of infectious diseases but also of conditions such as heart disease and cancer. Their distributions in populations can provide important pointers to possible causes. The relation between the environment and disease is an essential part of epidemiology.

Epidermis: The outer layer of the skin, which forms the protective covering of the body.

Epidermophyton: A fungus infection of the skin causing ringworm or tinea.

Epididymis: An oblong body attached to the upper part of each testicle, composed of convoluted vessels and ducts, that connects the vasa efferentia to the vas deferens. Sperm cells produced in the testis pass along the epididymis, maturing as they go, to be stored in the seminal vesicles until ejaculation occurs. The epididymis may be damaged by trauma or infection resulting sometimes in sterility. Cysts may also occur.

Epigastric: Pertaining to the upper-middle region of the abdomen.

Epigastrium: The region lying in the middle of the abdomen over the stomach.

Epigen: Changes in an organism brought about by alterations in the expression of genetic information without any change in the genome itself: the genotype is unaffected by such a change but the phenotype is altered.

Epigenetics: The science of how the -

activity of DNA (deoxyribonucleic acid, which is the fundamental genetic material of cells) can be altered semi-permanently by chemical processes rather than by natural mutation. Genes contain instructions for making proteins. The natural process of implementing these instructions – gene expression – can be altered by chemical groups attaching themselves to the chemical bases that make up a strand of DNA. This, in turn, affects the generation of proteins from the genes so tagged. Some chemical groups can even stop a gene from being expressed. Recently, research in Australia showed that such a chemically induced alteration could be inherited, at least in mice.

Epiglottis: A leaf-like piece of elastic cartilage covered with mucous membrane, which stands upright between the back of the tongue and the glottis, or entrance to the larynx. In the act of swallowing, it prevents fluids and solids from passing off the back of the tongue into the larynx.

Epignathus: Maldevelopment of the fetus in which the deformed remains of one twin are united to the upper jaw of the other.

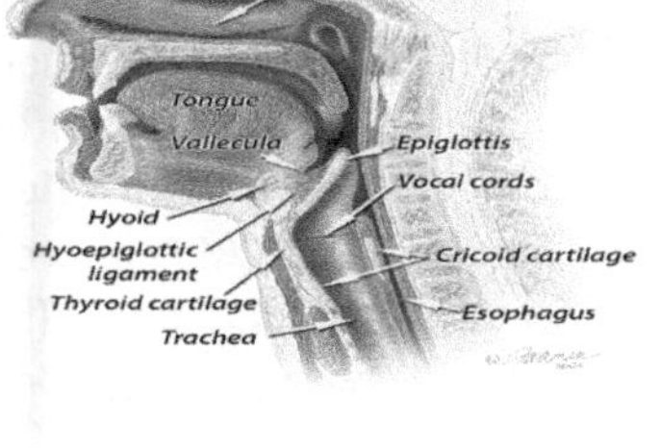

Epilation: Removal of hair by the roots.

Epilepsy: Epilepsy is the name given to any condition in which a person suffers repeated fits or seizures. It is present in one in 200 (0.5 per cent) of the population and up to 5 per cent of all children will have had a fit by the age of 12, although most of these are harm- less accompaniments of an acute feverish illness.

Epileptiform: Occurring in severe or sudden spasms, as in convulsion or epilepsy.

Epiphora: Inadequate drainage of tears in the eyes with the result that they 'overflow' down the cheeks. The condition is caused by an abnormality of the tear ducts which drain away the normal secretions that keep the eyeball moist.

Epiphysitis: Inflammation of an epiphysis.

Epiphyte: An air plant, growing on or with other plants but not in any way parasitic.

Epirubicin: A cytotoxic anthracycline antibiotic drug used in the treatment of solid tumours, acute leukaemia and lymphoma. It is related structurally to doxorubicin and is given intravenously, and by instillation into the urinary bladder to treat bladder cancer under specialist supervision.

Episclera: The most superficial layer of the sclera of the eye. It sometimes becomes inflamed (episcleritis) but the condition usually clears without treatment.

Episiotomy: A cut made in the perineumto enlarge the vaginal opening and facilitate childbirth during a difficult birth when the baby's head (or in a breech delivery, the buttocks) is making slow progress down the birth canal, or when forceps have to be applied.

Epispadias: An inherited abnormality of the penis in which the opening of the urethra is on the upper surface instead of at the end of the organ. Surgical correction carried out in infancy has a high success rate.

Epistasis: Stopping a flow or discharge of, for example, blood from a wound. (2) In genetics the term describes a type of gene action where a gene is able to block the action of another one.

Epistaxis: Nosebleeds.

Epithelioma: Epithelioma is a tumour of malignant nature arising in the epithelium covering the surface of the body.

Epithelium: Epithelium is the cellular

layer which forms the epidermis on the skin, covers the inner surface of the bowels, and forms the lining of ducts and hollow organs, like the bladder. It consists of one or more layers of cells which adhere to one another, and is one of the simplest tissues of the body. It is of several forms: for example, the epidermis is formed of scaly epithelium, the cells being in several layers and more or less flattened.

Epitope: Any part of a molecule that acts as an antigenic determinant: a macromolecule can contain many different epitopes each capable of stimulating production of a different specific antibody.

Epizootic: Any disease in animals which diffuses itself widely. The term corresponds to the word epidemic as applied to human beings. In plague, for example, an epizoötic in rats usually pre- cedes the epidemic in human beings.

Eponym: A species, structure or disorder named after a particular individual, customarily the one who first described or discovered it. The use of eponyms has been widespread in medicine, but more descriptive – and so more practical – terms are replacing them.

Epsom Salts: The popular name for magnesium sulphate, which was used as a saline purgative.

Epstein-barr Virus: A large, ubiquitous, and normally benign, herpes-like virus with both DNA and capsid. It is sometimes implicated in mononucleosis and at least two types of lymphomas. Recently it has been become connected with the symptom picture called chronic fatigue syndrome (as has been CMV) and can produce many ill-defined (but subjectively distressful) symptoms, including fatigue, fevers of an unknown origin and emotional lability.

Epulis: Epulis is a term applied to any tumour connected with the jaws.

Equivalent Diameter: Diameter of a spherical particle of the same density as a particle under investigation that, relative

Erb's Paralysis: Erb's paralysis is a form of paralysis of the arm due to stretching or tearing of the fibres of the brachial nerve plexus. Such damage to the brachial plexus may occur during birth, especially when the baby is unusually large, and it is found that the arm lies by the side of the body with elbow extended, forearm pronated, and the fingers flexed. The infant is unable to raise the arm.

Erection: The rigid state of the penis when it responds to sexual stimulus. An erection is necessary for effective penetration of the vagina. As a result of sexual arousal, the three cylinders of erectile tissue in the penis become engorged with blood, lengthening, raising and hardening the penis. Muscles surrounding the blood vessels contract and retain the blood in the penis. Erections also occur during sleep and in young boys. Inability to have or maintain an erection is one cause of impotence.

Erg: A prefix indicating activity or work.

Erogenous: A term to describe those parts of the body – for example, the mouth, breasts and genitals which, when stimulated, result in the individual's sexual arousal.

Erosion: Erosion means a process of gradual wearing down of structures in the body. The term is applied to the effect of tumours, when they cause destruction of tissue in their neighbour- hood without actually growing into the latter: for example, an aneurysm may erode bones in its neighbourhood. The term is also applied to minute ulcers – for example, erosions of the stomach, caused by extreme acidity of the gastric juice.

Eroticism: The emotional nature and characteristics of sexual arousal. This may occur as a result of visual, auditory or physical stimuli and also as a result of sexually oriented memories or imaginings.

Eructation: Eructation, or belching, is the sudden escape of gas or of portions of half-digested food from the stomach up into the mouth.

Eruption: Eruption, or rash, means an outbreak, in a scattered form, upon the surface of the skin. The skin is usually raised and red, or it may be covered with scales, or crusts, or vesicles containing fluid. Eruptions differ in appearance: for example, the eruption of measles is always distinguishable from that of chickenpox. But the same disease may also produce different eruptions in different people; or in the same person in different states of health; or even on different parts of the body at one time.

Erysipelas: A streptococcal infection of the skin characterised by an acute onset with fever, malaise and a striking, usually unilateral, rash almost always on a lower leg or the face. Shivering, local pain and tenderness are associated with a sharply defined, spreading, Bright red swollen zone of skin inflammation. On the leg, blistering and purpura may follow. The bacteria enter the skin through a fissure in a toe cleft (often associated with tinea pedis [ringworm] or via a crack in the skin behind an ear or in a nostril.

Erythema: Redness of the skin due to dilatation of dermal blood vessels. It may be transient or chronic, localised or widespread, and it can be blanched by pressure. Erythema may be caused by excessive exposure to heat or ultraviolet light, or by inflammation of the skin due to infection, dermatitis, and various allergic reactions – for example, to drugs. It may be emotional (e.g. as in flushing), mediated by the autonomic nervous system.

Erythrasma: A superficial mild infection of the skin caused by corynebacteria. It produces pink or slightly brown flaky areas of skin usually on the upper inner thighs or axillae. Toe clefts may be affected with thickened, white, macerated skin. The affected areas fluoresce coral pink under ultraviolet light. Clotrimazole or ketoconazole cream clears the rash rapidly. Very extensive erythrasma responds to oral erythromycin given for seven days.

Erythroblasts: A series of nucleated cells in the bone marrow that go through various stages of development until they form erythrocytes. They may appear in the blood in certain diseases.

Erythrocytes: The biconcave red blood cells that carry oxygen from the lungs to the tissues, and return carbon dioxide. They have an excess of membrane, some of which may be lost in various disorders, as a result of which they become progressively more spherical and rigid. Erythrocytes, which have no nuclei, are formed during erythropoeisis from erythroblasts in the bone marrow, and each mm3 of blood contains 5 million of them.

Erythroderma Psoriasis: The least common form of psoriasis; causes widespread reddening and scaling of the skin; may be a reaction to severe sunburn or to taking corticosteroids (cortisone) or other medications. It also can be caused by a prolonged period of increased activity of psoriasis that is poorly controlled.

Erythroderma: A rare inflammation of the skin which causes universal itching. The skin is red, thickened and scaly. It is also called generalised exfoliative dermatitis. It may complicate chronic eczema or psoriasis, particularly in men, in the second half of life. It may also result from hypersensitivity to a drug, such as gold injections used in rheumatoid arthritis.

Erythromelalgia: A condition in which the fingers or toes, or even larger portions of the limbs, become purple and bloated in appearance, and very painful. In people suffering from the condition – which is not a common one – the attacks come and go, being worse in summer (unlike chil- blains), and worse on exertion or when the affected parts are warmed or allowed to hang down. The condition may appear without apparent cause, but is often associated with vascular diseases, such as hypertension and polycythaemiavera. It aso occurs in association with certain diseases of the central nervous system, and in cases of metallic poisoning (e.g. arsenic, mercury and thallium). Treatment is unsatisfactory but aspirin provides sympomatic relief.

Erythropoeisis: The process by which erythrocytes or red blood cells are produced. The initiating cell is the haemopoietic stem cell from which an identifiable proerythroblast develops. This goes through several stages as a normoblast before losing its nucleus to become an erythrocyte. This process takes place in the blood-forming bone-marrow tissue.

Erythropoietin: The protein, produced mainly in the kidney, that is the major stimulus for the production of erythrocytes, or red blood corpuscles. It is used when treating anaemia dure to end-stage kidney failure and in premature newborns with anaemia.

Escalation Procedure: A procedure followed by the pharmanet help desk in responding to a problem call. The steps of the procedure for a particular problem call depends on the severity level assigned.

Eschar: Hard adherent crust caused by tissue killed by heat, chemicals or disease.

Escherichia: The generic name for the group of gram negative, usually motile, rod-shaped bacteria that can ferment carbohydrate. They occur naturally in the intestines of humans and some animals. E. coli, which ferments lactose, is not normally harmful but some varieties, particularly E. coli 0157, cause gastrointestinal infections which may be severe in old people. E. coli is also used in laboratory experiments for gen- etic and bacteriological research.

Esmarch's Bandage: A rubber bandage which is applied to a limb before surgery from below upwards, in order to drive blood from the limb. The bandage is removed after an inflated pneumatic tourniquet has been placed round the limb; the operation can then proceed.

Esophagus: The dense, muscular tube, 9 to 10 inches long, that extends from the back of the throat (pharynx) to the stomach.

ESR: The ESR or erythrocyte sedimentation rate is a test that measures the rate at which red blood cells settle out of suspension in blood plasma. In certain diseases, such as infection and malignancy, the amount of proteins in the plasma increases; the result is that red cells settle out more quickly and this test is used to show whether inflammation is present and, to some extent, its severity.

Essential Fatty Acids: Three acids arachidonic, linolenic and tinoleic – which are essential for life, but which the body cannot produce. They are found in natural vegetable and fish oils and their functions are varied. EFAs have a vital function in fat metabolism and transfer and they are also precursors of prostaglandins.

Estate Tax: A tax imposed at one's death on the transfer of most types of property.

Ester: An organic compound formed from an alcohol and an acid by the removal of water.

Estimated Daily Intake (EDI): Prediction of the daily intake of a residue of a potentially harmful agent based on the most realistic estimation of the residue levels in food and the best available food consumption data for a specific population.

Estimated Exposure Concentration (EEC): Measured or calculated amount or mass concentration of a substance to which an organism is likely to be exposed, considering exposure by all sources and routes.

Estimated Exposure Dose (EED): Measured or calculated dose of a substance to which an organism is likely to be exposed, considering exposure by all sources and routes.

Etenercept: An injectable biologic drug used to treat psoriatic arthritis.

Ethmoid: A bone in the base of the SKULL which separates the cavity of the nose from the membranes of the brain. It is a spongy bone with numerous cavities or sinuses. Suppuration in the ethmoidal sinuses is sometimes responsible for inflammation in neighbouring parts such as the eye.

Etiology: Etiology, or aetiology, means the group of conditions which form the cause of any disease.

Eu: A prefix meaning satisfactory or beneficial. Eugenics: The study and cultivation of conditions that may improve the human race, in particular the detection and elimination of genetic disease.

Eugeroic: A unique class of drugs that have stimulatory properties.

Eukaryote: A cell that has a nucleus bounded by a membrane and with chromosomes containing DNA, RNA and proteins. The cell divides by mitosis and also contains mitochondria. Animals, plants and cellular organisms made up of this type of cell are included in the biological superkingdom of Eukaryote.

Eukaryote: Cell or organism with the genetic material packed in a membrane-surrounded structurally discrete nucleus and with well-developed cell organelles. The term includes all organisms except archaebacteria, eubacteria and cyanobacteria (until recently classified as cyanophyta or blue-green algae).

Eunuch: A man whose testes have been removed or seriously damaged so that he is unable to produce male hormones and thus is sterile. A male castrated before puberty willhave a feminine appearance and under-developed secondary sexual characteristics. The term was historically used to describe boys, castrated to make them suitable for working in harems, or boy singers, castrated to retain their higher-register voices (castrati singers).

Euphoria: A feeling of well-being. This may occur normally; for instance, when someone has passed an examination. In some neurological or psychiatric conditions, however, patients may have an exaggerated and quite unjustified feeling of euphoria. This is then a symptom of the underlying condition. Euphoria may also be drug- induced – by drugs of addiction or by thera- peutic drugs such as corticosteroids.

Euphoriants: Drugs which induce a state of euphoria or well-being.

Eustachian Tubes: The passages, one on each side, leading from the throat to the middle ear. Each is about 38 mm (1½ inches) long and is large at either end, though at its narrowest part it only admits a fine probe. The tubes open widely in the act of swal- lowing or yawning. The opening into the throat is situated just behind the lower part of the nose, so that a catheter can be passed through the corresponding nostril into the tube for inflation of the middle ear.

Euthanasia: Literally meaning the procuring of an easy and painless death, euthanasia (or 'mercy killing') has come to be understood as a deliberate act or omission whose primary intention is to end another person's life.

Euthyroid: The descriptive term for a person with a normally functioning thyroidgland, or someone who has had successful treatment for an underactive (hypothyroid) or overactive (hyper- thyroid) gland.

Eutrophic: Describes a body of water with a high concentration of nutrient salts and a high or excessive rate of biological production.

Eutrophication: Adverse change in the

chemical and biological status of a body of water following depletion of the oxygen content caused by decay of organic matter resulting from high primary production as a result of enhanced input of nutrients.

Evacuant: Evacuant is a name for a purgative medicine

Evacuator: A device for extracting fluid from a cavity. In its basic form it comprises a hollow flexible bulb attached (using a valve system) to a tube inserted into the cavity. Another valve leads to the discharge tube. One use of an evacuator is to empty the urinary bladder of extraneous material during surgery for removal of calculi or for prostatectomy.

Evidence-Based Medicine: The process of systematically identifying, appraising and using the best available research findings, integrated with clinical expertise, as the basis for clinical decisions about individual patients. The aim is to encourage clinicians, health-service managers and consumers of health care to make decisions, taking account of the best available evidence, on the likely con- sequences of alternative decisions and actions. Evidence-based medicine has been developing internationally for the past 25 years, but since around 1990 its development has accelerated.

Evisceration: Extrusion of the abdominal viscera or internal organs, usually as the result of serious injury. (Usually described as disembowelment when deliberately carried out by one person on another.) In surgery the term refers to part removal of the viscera, and in ophthalmology it is an operation to remove the contents of the eyeball.

Evolution: An uninterrupted process of change from one condition, form or state to another. In bio- logical evolution, all varieties of living things are seen as having developed by inheritable, incremental changes from unicellular structures to complex organisms such as humankind. Although the likelihood of some form of evolution had been postulated by scientists.

Ewing's Sarcoma: An uncommon but very malignant cancer of the bone in children and young adults, the condition was first identified as being different from osteosarcoma by Dr J Ewing in 1921. It usually occurs in the limbs or pelvis and soon spreads to other parts of the body.

Exanthem: Rash caused by a systemic infection. Several childhood infections –for example, measles and rubellahave characteristic exanthemata.

Excess Lifetime Risk: Additional or excess risk incurred over the lifetime of an individual by exposure to a toxic substance.

Exchange Transfusion: A method of treating newborn infants with haemolytic disease. Blood is taken out of the baby through the umbilical vein and is replaced with the same quantity of blood from a donor that is compatible with the mother's blood. The procedure is repeated several times to get rid of damaged cells while maintaining the infant's blood volume and keeping its red cell count constant.

Excimer Laser: A type of laser that is used to remove thin sheets of tissue from the surface of the cornea thus changing the curvature of the eye's corneal surface. The procedure is used to excise diseased tissue or to correct myopia, when it is known as photorefractive keratectomy or lasik.

Excipient: An inactive or inert substance which is added to a drug formulation, usually to provide stability or bulk

Excipient: An inert substance added to a prescription in order to make the remedy as prescribed more suitable in bulk consistence, or form for administration.

Excision: A term applied to the removal of any structure from the body, when such removal necessitates a certain amount of separation from surrounding parts: for example, the excision of a tumour, of a gland or of a joint. When an opening is simply made into the body, the term incision is used; when a limb, or part of one, is removed, the term amputation is employed.

Excitation: When used in neurophysiology, the term means the triggering of a conducted electrical impulse in the membrane of a muscle cell or the nerve fibre controlling it.

Excluded Product: A product that is not listed on the drug formulary and, as such, is not covered nor available at a reduced copayment or coinsurance amount under normal conditions.

Exclusions: Specific conditions or services that are not covered under the benefit agreement.

Excoriation: Excoriation means the destruction of small pieces of the surface of skin or mucous membrane.

Excreta: Waste material, especially faeces.

Excretion Rate: Amount of substance (and/or its metabolites) or fraction that is excreted per unit time. It should be noted that according to this definition excretion does not include the passingof a substance through the intestine without absorption.

Excretion: Discharge or elimination of an absorbed or endogenous substance, or of a waste product, and/or their metabolites, through some tissue of the body and its appearance in urine, faeces, or other products normally leaving the body. Excretion of chemical compounds from the body occurs mainly through the kidney and the gut.

Exfoliation: The separation, in layers, of pieces of dead bone or skin.

Exhalation: Also called expiration, this is the act of breathing air from the lungs out through the bronchi, trachea, mouth and nose.

Exhibitionism: Public flaunting of a person's characteristics. Also a term describing the public exposure of genitals to another person, regarded as a form of sexual deviation, colloquially called 'flashing'.

Exocrine Gland: A gland that secretes its products through a duct to the surface of the body or of an. The sweat glands in the skin and the salivary glands in the mouth are examples. The secretion is set off by a hormone or a neurotransmitter.

Exogenous: Arising outside the body. For example, exogenous depression is an illness caused by an out- side factor such as the death of a close relative; an exogenous illness may occur because of something eaten in the diet.

Exomphalos: The term applied to a congenital hernia formed by the projection of abdominal organs through the umbilicus.

Exophthalmometer: Also known as a proptometer. An instrument used to measure the extent of protrusion of the eyeball – a development that occurs in certain disorders such as goitre, tumour, oedema, injuries, orbital inflammation or cavernous venous thrombosis (a blood clot in the cavern- ous sinus in the base of the skull behind each eye).

Exophthalmos: Exophthalmos, or proptosis, refers to forward displacement of the eyeball and must be distinguished from retraction of the eyelids, which causes an illusion of exophthalmos. Lid retraction usually results from activation of the autonomic nervous sytem. Exophthalmos is a more serious disorder caused by inflammatory and infiltrative changes in the retro-orbital tissues and is essentially a feature of Graves' disease, though it has been described in chronic thyroiditis.

Expoththalmic Goitre: Sometimes called Graves' disease, this is a disorder in which there is overactivity of the thyroid gland, protrusion of the eyes, and other symptoms.

Exostosis: An outgrowth from a bone: it may be due to chronic inflammation, constant pressure or tension on the bone, or tumour-formation.

Exotoxin: A powerful poison produced by a bacterial cell and secreted into its surrounding environment. Exotoxins are often damaging to only few tissues and they are usually inactivated by chemicals, heat and light. Bacteria causing botulism, diphtheria and tetanus all produce exotoxins.

Expectant: A form of treatment in which the cure of the patient is left mainly to nature, while the physician simply watches for any unsatisfactory developments or symptoms, and relieves them if they occur.

Expectorant: Drugs which are claimed to help the removal of secretions from the air passages although there is no convincing evidence that they do this. A simple expectorant may, however, be a useful placebo. Most preparations are available without a doctor's prescription and pharmacists will advise on which might be helpful for particular patients with dry or congestive coughs.

Expectoration: Expectoration means either material brought

Experimental Procedures: Experimental, investigational or unproven procedures and treatments.

Expiration Date: The date indicated in an insurance contract as the date coverage expires.

Expiration: (1) Breathing out air from the lungs. (2) The act of dying.

Explant: Living tissue removed from its normal environment and transferred to an artificial medium for growth.

Exploration: A surgical operation to investigate the cause of a patient's illness.

Exposed Group (Sometimes Abbreviated to Exposed) in Epidemi-ology: Group whose members have been exposed to a supposed cause of a disease or health state of interest, or possess a characteristic that is a determinant of the health outcome of interest.

Exposed: Subject to a factor that is under study in the environment, for instance an environmental hazard.

Exposure Assessment: Process of measuring or estimating concentration (or intensity), duration and frequency of exposures to an agent present in the environment or, if estimating hypothetical exposures, that might arise from the release of a substance, or radionuclide, into the environment.

Exposure Limit: General term defining an administrative substance concentration or intensity of exposure that should not be exceeded.

Exposure Ratio: In a case control study, value obtained by dividing the rate at which persons in the case group are exposed to the risk factor (or to the protective factor) by the rate at which persons in the control group are exposed to the risk factor (or to the protective factor) of interest

Exposure Test: Determination of the level, concentration or uptake of a potentially toxic compound and/or its metabolite(s) in biological samples from an organism (blood, urine, hair etc.) and the interpretation of the results to estimate the absorbed dose or degree of environmental pollution; or the measuring of biochemical effects, usually not direct adverse effects of the substance, and relating them to the quantity of substance absorbed, or to its concentration in the environment.

Exposure: (1) A term used in behaviour therapy to describe a method of treating fears and phobias. The subject is confronted by the circumstances that he or she fears, either gradually or suddenly, with the aim of defusing the fear or phobia. (2) The term is also a colloquialism for public exposure by a man of his genitals to achieve sexualgratification.

Exsanguinate: The removal of blood from the body. This may occur as the result of a serious accident in which the victim bleeds extensively. Rarely, it may happen that bleeding becomes uncontrollable during an operation.

Extended Care Facility (ECF): A medical care institution for patients who require long-term custodial or medical care, especially for chronic disease or a condition requiring prolonged rehabilitation therapy.

Extension of Benefits: When a person's coverage is extended under certain conditions, such as disability, after their group health coverage would otherwise have ended.

Extension: Extension is the process of straightening or stretching a limb. When used in the natural sense, it involves the contraction of the muscles opposing those used in flexion. In cases of fractured limbs, extension is employed during the application of splints, in order to reduce the displacement caused by the fracture, and pre- vent movement of the broken ends of bone. It is effected by gently and steadily pulling upon the part of the limb beyond the fracture. Extension of a more permanent type is used in the after- treatment of some fractures, as well as in diseases of the spine, by placing the patient upon an inclined bed and affixing weights to his or her lower limbs, or to his or her head by means of adhesive plaster or of straps.

Exteriorisation: In surgery, the procedure to transfer an organ from its normal place in the body to the skin surface. It may be temporary or permanent. A common example is when the intestine is brought to the abdominal surface as a colostomy: this may be permanent because of serious disease in the lower part of the colon, or temporary to allow a disorder in the colon to be treated.

External Validity: Generalizability of the results of a particular study, beyond the limits of the population actually studied.

Extinction: (1) In behavioural psychology, the lessening of a conditioned reflex that occurs when it is not regularly reinforced. An example is in the treatment of unsatisfactory behaviour such as violence, when by preventing the individual from enjoying the rewards of violent acts – namely, the attention he or she attracts – the conditioned reflex that is part of the cycle of violence is reduced in strength. (2) Elimination of life or of a biological species.

Extirpation: The total removal of a growth, organ or tissue by surgery.

Extra Risk: Probability that an agent produces an observed response, as distinguished from the probability that the response is caused by a spontaneous event unrelated to the agent.

Extra: Extra is the Latin prefix meaning outside of, or in addition – such as extracapsular, meaning outside the capsule of a joint, and extrasystole, meaning an additional contraction of the heart.

Extracellular: An adjective that describes an object or event outside a cell. An example is extracellular fluid, the medium surrounding a cell.

Extracts: Extracts are preparations, usually of a semi-solid consistency, containing the active parts of various plants extracted in one of several ways. In the case of some extracts, the juice of the fresh plant is simply pressed out and purified; in the case of others the active principles are dissolved out in water, which is then to a great extent driven off by evaporation. Other extracts are similarly made by the help of alcohol, and in some cases ether is the solvent.

Extradural: Outside the dura mater, the outermost of the three membranes that cover the brain and spinal cord. The extradural or epidural space is the space between the vertebral canal and the dura mater of the spinal cord.

Extrapolation: Calculation, based on quantitative observations in exposed test species or in vitro test systems, of predicted dose-effect and dose-response relationships for a substance in humans and other biota including interspecies extrapolations and extrapolation to susceptible groups of individuals: the term may also be used for qualitative information applied to species or conditions that are different from the ones in which the original investigations were carried out.

Extrapyramidal System: This is a complex part of the nervous system, extending from the cortex to the medulla in the brain, from which emerge descending spinal pathways which influence voluntary motor activity throughout the body. Although the normal functions of the system are poorly understood, there are characteristic signs of an extrapyramidal lesion. These include disturbance of voluntary movements, notably slowness and 'poverty' of movement.

Extrasensory Perception (ESP): An alleged way of perceiving current events (clairvoyance), future events (precognition) or the thoughts of other people (telepathy). ESP has never been scientifically proven and does not involve the use of any known senses.

Extrasystole: Extrasystole is a term applied to premature contraction of one or more of the chambers of the heart. A beat of the heart occurs sooner than it should do in the ordinary rhythm and is followed by a longer rest than usual before the next beat. In an extrasystole, the stimulus to contraction arises in a part of the heart other than the usual. Extrasystoles often give rise to an unpleasant sensation as of the heart stumbling over a beat, but their -

occurrence is not usually serious.

Extrasystoles: A premature contraction of the heart. It can be caused by nervousness, indigestion, a tired and enlarged heart - anything up to overt organic heart.

Extravasation: An escape of fluid from the vessels or passages which ought to contain it. Extravasation of blood due to tearing of vessel walls is found in stroke, and in the commoner condition known as a bruise. Extravasation of urine takes place when the bladder or the urethra is ruptured by a blow on the abdomen or on the crutch (perineum), or torn in a fracture of the pelvis. Intravenous infusions frequently extravasate.

Extremely Hazardous Substance: Any substance listed in Appendix A or Appendix B of 40 CFR 255. Appendix A is an alphabetical listing of extremely hazardous substances. Appendix B lists extremely hazardous substances in order of each substance's Chemical Abstracts Service registry number.

Extrinsic: (1) Originating outside the body. (2) An extrinsic muscle is one whose origin is some way from the part of the body it acts upon– for example, the muscles controlling the movement of the eyeball which are attached to the bony orbit in which the eye sits.

Extrovert: A person, who is outgoing, enjoys mixing with others and looks for fresh activities to take part in. Tends to act emotionally rather than intellectually.

Exudates: The feral and congested fluids built up in a bruise or infection. Unlike a transudate, which is merely edema from lymphatic congestion, exudates contain dead cells, erythrocytes, white blood cells and often pus. To a given phenomenon or property, would behave in the same way as the particle under investigation.

Exudation: The process in which some of the constituents of the blood pass slowly through the walls of the small vessels in the course of inflammation, and also means the accumulation resulting from this process. For example, in pleurisy the solid, rough material deposited on the surface of the lung is an exudation.

Eye: The eye is the sensory organ of sight. It is an elaborate photoreceptor detecting information, in the form of light, from the environment and transmitting this information by a series of elec- trochemical changes to the brain. The visual cortex is the part of the brain that processes this information.

F

Face Lift: An operation to reduce wrinkles and lift 'loose' skin on the face of mature persons with the aim of making them look younger. It can be per- formed under local anaesthetic as an outpatient or with general anaesthesia as an inpatient. The operation is regarded as cosmetic and is usually done in the private health-care sector in the UK.

Facial Nerve: The seventh cranial nerve (arising from the brain), supplying the muscles of expression in the face, being purely a motor nerve. It enters the face immediately below the ear after split- ting up into several branches.

Facies: Facies is a term applied to the expression or appearance of the face. Facilitation of the drainage of secretions from

Factor VIII: A coagulative blood protein that is a constituent of the coagulation cascade – an essential component in the clotting of blood. Those people with the inherited disorder, haemophilia, have abnormally low amounts of factor VIII and so bleed more when cut. They are treated with a concentrated version to reduce the tendency to bleed.

Faeces: Faeces, or stools, consist of the remainder of the food after it has passed through the alimentary canal and been subjected to the action of the digestive juices, and after the nutritious parts have been absorbed by the intestinal mucous membrane. The stools also contain various other matters, such as pigment derived from the bile, and large quantities of bacteria which are the main component of human stools. The stools are passed once daily by most people, but infants have several evacuations of the bowels in 24 hours and some adults may defaecate only two or three times weekly. Sudden changes in

Fainting: Fainting, or syncope, is a temporary loss of consciousness caused by inadequate supply of blood to the brain. It may be preceded by nausea, sweating, loss of vision, and ringing in the ears. It is most often caused by pooling of blood in the extremities, which reduces venous return and thus cardiac output: this may be due to hot weather or prolonged standing. Occasionally, fainting on standing occurs in people with low blood pressure, autonomic neuropathy (in which normal vasomotor reflexes are absent), or those taking antihypertensive drugs.

Faith Healing: The facility, claimed by some people, to cure disease by a healing force present in their

Fallen Arches: Weakness in the muscles that support the bony arches of the foot. The result is flat feet, a condition that can adversely affect a person's ability to walk and run normally.

Falling Sickness: An old name for epilepsy.

Fallopian Tubes: Tubes, one on each side, lying in the pelvic area of the abdomen, which are attached at one end to the uterus, and have the other unattached but lying close to the ovary. Each is 10–12.5 cm (4–5 inches) long, large at the end next to the ovary, but communicating with the womb by an opening which admits only a bristle. These tubes conduct the ova from the ovaries to the interior of the womb. Blockage of the Fallopian tubes by a chronic inflammatory process resulting from infection is a not uncommon cause of infertility in women.

Familial: A description of a disorder, illness or characteristic that runs in families.

Fansidar: A combination of pyrimethamine and sulfadoxine used in conjunction with other anti- malarial drugs to treat falciparum malaria.

FAQ: Frequently asked questions, a list of the most frequently asked questions and their corresponding answers

Farmer's Lung: A form of external allergic alveolitis caused by the inhalation of dust from mouldy hay or straw.

Fascia: Sheets or bands of fibrous tissue which enclose the body tissues beneath the skin and connect the muscles.

Fasciitis: Inflammation of fascia. The most common site is the sole of the foot, where it is known as plantar fasciitis. It is characterised by gnawing pain. There is no specific treatment, but the condition usually clears up spontaneously – though over a considerable time.

Fascioliasis: Disease caused by the liver fluke, Fasciola hepatica. This is found in sheep, cattle and other herbivorous animals, in which it is the cause of the condition known as liver rot. In Britain it is the most common disease found in animal slaughterhouses. The danger to humans is in eating vegetables – particularly wild water- cress – that have been infected by snails; there have been several outbreaks of fascioliasis in Britain due to eating contaminated wild water- cress, and much larger outbreaks from the same cause have been reported in France. The disease is characterised by fever, dyspepsia (indigestion), heavy sweating, loss of appetite, abdominal pain, urticaria, and a troublesome cough. In the more serious cases there may be severe damage to the liver, with or without jaundice. The diagnosis is clinched by the finding of the eggs of the fluke in the stools. The two drugs used in treatment are bithionol and chloroquine. Even though many cases are quite mild and recover spontaneously, prevention is particularly important. This consists primarily of never eating wild watercress, as this is the main cause of infestation. Lettuces have also been found to be infested.

Fascictomy: An operation to relieve pressure on muscles caused by tight surrounding fascia or fibrous connective tissue. The fascia is slit with an incision.

Fast Track: FDA designation for a therapeutic development program intended for treating a serious or life-threatening condition and which has the potential to address unmet medical needs for the condition

Fasting: Fasting is the abstention from, or deprivation of, food and drink. It may result from a genuine desire to lose weight – in an attempt to improve one's health and/or appearance – or from a mental illness such as depression, or from one of the eating disorders. Certain religious customs and practices may demand periods of fasting. Forced fasting, often extended, has been used for many years as an effective means of torture.

Fat Necrosis: In injury to, or inflammation of, the pancreas, the fat-splitting enzyme in it may escape into the abdominal cavity, causing death of fat- containing cells.

Fat: A food that has more energy-producing power, weight for weight, than any other. Animal fat is a mixture in varying proportions of stearic, palmitic, and oleic acids combined with glycerin. Butter contains about 80 per cent of fat; ordinary cream, 20 per cent; and rich cream 40 per cent – whilst olive oil is practically a pure form of fat. When taken to a large extent in the diet, fat requires to be combined with a certain proportion of either carbohydrate or protein in order that it may be completely consumed, otherwise harmful products, known as ketones, may form in the blood.

Fatigue: Tiredness: a physiological state in which muscles become fatigued by the lactic acid accumulating in them as the result of their activity. For the removal of lactic acid in the recovery phase of muscular contraction, oxygen is needed. If the supply of oxygen is not plenti- ful enough, or cannot keep pace with the work the muscle is doing, then lactic acid accumulates and fatigue results. There is also a nervous element in muscular fatigue: it is diminished by stimulation of the sympathetic nervous system.

Fatty Degeneration: As a result of anaemia, interference with blood or nerve supply, or because of the action of various poisons, body cells may undergo abnormal changes accompanied by the appearance in their substance of fat droplets.

Fauces: The somewhat narrowed opening between the mouth and throat. It is bounded above by the soft palate, below by the tongue, and on either side by the tonsil. In front of, and behind, the tonsil are two ridges of mucous membrane, the anterior and posterior pillars of the fauces.

Favism: A type of haemolytic anaemia, attacks of which occur within an hour or two of eating broad beans (Vicia fava). It is a hereditary disease due to lack of an essential enzyme called glucose-6- phosphate dehydrogenase, which is necessary for the continued integrity of the red cell. This defect is inherited as a sex-linked dominant trait, and the red cells of patients with this abnormality have a normal life-span until chal- lenged by certain drugs or fava beans when the older cells are rapidly destroyed, resulting in haemolytic anaemia.

Favus: Another name for honeycomb ringworm.

FDA: Food and Drug Administration. The U.S. federal agency charged with promoting and protecting the public health by helping safe and effective products reach the market in a timely way and monitoring products for continued safety after they are in use.

Fear: An emotional condition provoked by danger and usually characterised by unpleasant subjective feelings accompanied by physiological and behavioural changes. The heart rate increases, sweating occurs and the blood pressure rises. Sometimes fear of certain events or places may develop into a phobia: for example, agora- phobia, a fear of open spaces.

Feasibility Analysis Phase: The point in a pollution prevention program at which screened waste reduction options are evaluated technically, economically, and environmentally. The results are used to select options to be recommended for implementation.

Febrile Convulsion: Convulsion occurring in a child aged six months to six years with a high temperature in which the limbs twitch; he or she may lose consciousness. The condition is common, with one child in 20 suffering from it. It is a result of immature homeostatic control and is not usually serious, occurring generally during an infection such as measles or tonsillitis. The brain and nervous system are normal in most cases.

Febrile: Feverish.

Febrile: Having a fever. Describes a patient whose body temperature is greater than normal.

Fecundity: Ability to produce offspring frequently and in large numbers. In demography, the physiological ability to reproduce.

Federal Ceiling Price (FCP): The maximum price manufacturers can charge for FSS-listed brand name drugs to the Big 4 (VA, DOD, PHS, and the Coast Guard) even if the FSS price is higher. FCP must be at least 24 percent below the non-Federal average manufacturer price. FCP prices are not publicly available.

Federal Supply Schedule (FSS): The collection of multiple award contracts used by Federal agencies, U.S. territories, Indian tribes and other specified entities to purchase supplies and services from outside vendors. FSS prices for the pharmaceutical schedule are negotiated by the VA and are based on the prices that manufacturers charge their "most-favored" non-Federal customers under comparable terms and conditions.

Feedback Mechanism: Many glands which produce hormones are influenced by other hormones, particularly those secreted by the hypothalamus (a con- trolling centre in the brain) and the pituitary gland. If the amount of hormone produced by a gland rises, negative feedback mechanisms operate by instructing the pituitary gland, via the hypothalamus, to produce less of the stimulating hormones. This cuts activity in the target gland. Should the amount of hormone produced fall, the feedback mechanism weakens with the result that the output of stimulating hormones increases.

Fee-for-Service: The traditional health care payment system under which physicians and other providers receive a payment for each unit of service provided rather than a capitation payment for each recipient.

Fellow: A scientist or health care professional who has already earned a doctoral degree and has received an award (fellowship) in support of gaining additional experience and training in a focused area of study at UCSF.

Fellowship: An award given to a doctoral-degreed scientist or health care professional in support of gaining additional experience and training in a focused area of study at UCSF

Feminisation: The development of a feminine appearance in a man, often the result of an imbalance in the sex hormones. Castration, especially before puberty, causes feminisation, as may the use of hormones to treat an enlarged prostate gland.

Femoral: Adjective relating to the femur or the region of the thigh. For example, the femoral nerve, artery, vein and canal.

Femur: The thigh bone, which is the longest and strongest bone in the body. As the upper end is set at an angle of about 120 degrees to the rest of the bone, and since the weight of the body is entirely borne by the two femora, fracture of fetus one of these bones close to its upper end is a common accident in old people, whose bones are often weakened by osteoporosis. The femur fits, at its upper end, into the acetabulum of the pelvis, forming the hip-joint, and, at its lower end, meets the tibia and patella in the knee-joint.

Fenestration: A largely obsolete surgical operation to form a new opening in the bony labyrinth of the inner ear in the treatment of deafness caused by otosclerosis . Nowadays the disorder is usually surgically treated by stapedectomy.

Ferritin: An iron storage protein used as an indication of iron depletion.

Fertilisation: The process by which male and female gametes (spermatozoa and oocytes respectively) fuse to form a zygote which develops, by a complex process of cell division and differentiation, into a new individual of the species. In humans, fertilisation occurs in the fallopian tubes. Sperm deposited in the upper vagina traverse the cervix and uterus to enter the Fallopian tube. Many sperm attempt to penetrate the zona pellucida surrounding the oöcyte, but only one is able to penetrate the oöcyte proper and this prevents any other sperm from entering. Once the sperm has entered the oöcyte, the two nuclei fuse before the zygote begins to divide.

Fertility Rate: The number of live births that occur in a year for every 1,000 women of childbearing age (this is usually taken as 15–44 years of age). The fertility rate in the UK (all ages) was 54.9 in 2002 (UK Health Statistics, 2001 edition, The Stationery Office).

Fertility Toxicant: Produces abnormalities of male or female reproductive functions or impairs reproductive capacity.

Fertility: Ability to conceive and to produce offspring: for litter- bearing species the number of offspring per litter is used as a measure of fertility. Reduced fertility is sometimes referred to as subfertility.

Fertilizer: Substance applied to soil or hydroponic systems for improving the root nutrition of plants with the aim of increasing crop yields and/or controlling production.

Fester: A popular, not a medical, term used to mean any collection or formation of pus. It is applied to both abscesses and ulcers.

Festination: The involuntary quickening of gait seen in some nervous diseases, especially in parkinsonism.

Fetal Alcohol Syndrome: A disorder of newborn infants that is caused by the toxic effects on the growing fetus of excessive amounts of alcohol taken by the mother. Low birth-weight and retarded growth are the main consequences, but affected babies may have hand and facial deformities and are some-times mentally retarded.

Fetal Blood Sampling: A procedure performed during a mother's labour in which a blood sample is taken from a vein in the scalp of the fetus. This enables tests to be performed that indicate whether the fetus is, for example, suffering from a shortage of oxygen (hypoxia). If so, the obstetrician will usually accelerate the baby's birth.

Fetal Transplant: A procedure in which cells – for example, from the pancreas – are taken from an aborted fetus and then transplanted into the malfunctioning organ (pancreas) of an individual with a dis- order of that organ (in this case, diabetes). The cells from the fetus are intended to take over the function of the host's diseased or damaged cells. Fetal brain cells have also been transplanted into brains of people suffering from parkinsonism. These treatments are at an experimental stage.

Fetishism: This is a form of sexual deviation in which the person becomes sexually stimulated by parts of the body, such as the feet, which are not usually regarded as erotogenic. Some people are sexually aroused by items of clothing or shoes.

Fetoscopy: Inspection of a fetus by passing a fibreoptic instrument called a fetoscope through the abdominal wall of a pregnant woman into her uterus. The procedure is usually conducted in the 18th to 20th week of pregnancy to assess the fetus for abnormalities and to take blood samples to preclude diseases such as haemophilia, duchenne muscular dystrophy and sickle-cell anaemia. The procedure should be used only if there is a serious possibility of abnormality, the presence of which will usually have been indicated by other screening tests such as ultrasound and tests of blood obtained by (intrauterine) cordocentesis (withdrawal of blood by syringe inserted into the umbilical cord).

Fetotoxicity: Toxicity to the fetus.

Fetus: The name given to the unborn child after the eighth week of development. Humans, like all animals, begin as a single cell, the ovum, in the ovary. After fertilisation with a spermatozoon, the ovum becomes embedded in the mucous membrane of the uterus, its covering being known as the decidua. Increase in size is rapid, and development of complexity is still more marked. The original cell divides repeatedly to form new cells, and these become arranged in three layers known as the ectoderm, mesoderm and endoderm.

Fever: Fever, or pyrexia, is the abnormal rise in body temperature that frequently accompanies disease in general. Fibreoptic Endoscopy Causes the cause of fever is the release of fever-producing proteins (pyrogens) by phagocytic cells called monocytes and macrophages, in response to a variety of infectious, immuno- logical and neoplastic stimuli. The lymphocytes play a part in fever production because they recognise the antigen and release substances called lymphokines which promote the production of endogenous pyrogen. The pyrogen then acts on the thermoregulatory centre in the hypothalamus and this results in an increase in heat generation and

Fibreoptic Endoscopy: A visualising technique enabling the operator to examine the internal organs with the minimum of disturbance or damage to the tissues. The procedure has transformed the management of, for example, gastrointestinal disease. In chest disease, fibreoptic bronchoscopy has now replaced the rigid wide-bore metal tube which was previously used for examination of the tracheo-bronchial tree.

Fibrillation: A term applied to rapid contraction or tremor of muscles, and especially to a form of abnormal action of the heart muscle in which individual bundles of fibres take up independ- ent action. It is believed to be due to a state of excessive excitability in the muscle associated with the stretching which occurs in dilatation of the heart. The main causes are athero- sclerosis, chronic rheumatic heart disease and hypertensive heart disease. Fibrillation is distinguished as atrial or ventricular, depending on whether the muscle of the atria or of the ventricles is affected. In atrial fibrillation, the heartbeats and the pulse become extremely irregular, both as regards time and force; when the atrium is fibrillating there is no significant contraction of the atrial muscle but the cardiac output is maintained by ventricular contraction. In ventricular fibrillation there is no significant contractile force, so that there is no cardiac output.

Fibrin: A substance formed in the blood as it clots: indeed, its formation causes clotting. The sub- stance is produced in threads; after the threads have formed a close meshwork through the blood, they contract, and produce a dense, felted mass. The substance is formed not only from shed blood but also from lymph which exudes from the lymph vessels. Thus fibrin is found in all inflammatory conditions within

Fibrinogen: The soluble protein in the blood which is the precursor of fibrin, the substance in blood clot.

Fibrinolysis: The way in which blood clots are removed from the circulation. The insoluble protein fibrin is broken down by the enzyme plasmin which is activated at the same time as the coagulation process of blood. There is normally a balance between coagulation and fibrinolysis; an abnormal increase in the latter causes excessive bleeding.

Fibrinolytic Drugs: A group of drugs, also known as thrombolytics, with the ability to break down the protein fibrin, the prime constituent of blood clots. They are used to disperse blood clots that have formed in the vessels of the circulatory system. The group includes streptokinase, alteplase and reteplase. The drugs work by activating plasminogen to form plasmin which degrades fibrin and breaks up the blood clot .

Fibroadenoma: A benign tumour of glandular epithelium containing fibrous elements. The commonest benign tumour is of the breast, often occurring in young women.

Fibroblasts: Cells distributed widely throughout connective tissue that produce the precursor sub- stances of collagen, elastic fibres and reticular fibres.

Fibroid: Fibroid, or fibromyoma, is the commonest form of tumour of the uterus and one of the most common tumours of the human body. It is composed of a mixture of muscular and fibrous tissue. The tumour may be small or as large as a grapefruit. Fibroids may cause pain and heavy menstrual

Fibroids: Also called a leiomyoma or fibromyoma (or myofibroma, for that matter), it is an encapsulated tumor made up of disorganized and irregular connective tissue. A uterine fibroid is benign, there may be one or many, they grow slowly, have unknown causes, and may or may not cause painful menses or mid-cycle bleeding. Much depends on where they are in the uterus and whether or not they extend far enough into the cavity to impair and thin out the endometrium. If they do, they cause distress.

Fibroma: A benign tumour comprising mainly connective tissue – the substance that surrounds body structures, binding them together. Examples are neurofibroma, affecting connective tissue around nerves, and ovarian fibroma which develops around the follicles from which eggs (ova) develop in the ovaries. Unless the fibroma is causing symptoms (as a result of pressure on surrounding tissues) it does not require treatment. If symptoms occur, the tumour is removed surgically.

Fibromyalgia Syndrome: Symptoms These vary, with pain and fatigue generally prominent, sometimes causing considerable disability. Patients can usually dress and wash independently but cannot cope with a job or household activities. Pain is mainly axial, but may affect any region. Analgesics, non- steroidal anti-inflammatorydrugs (NSAIDS) and local physical treatments are generally ineffective.

Fibrosarcoma: A cancer of the connective tissue arising in the fibroblasts, stem cells that produce connective tissue cells. The tumours can develop in bone or in soft tissue and occur most commonly in the limbs. Treatment is by surgery or radiotherapy.

Fibrosis: Formation of fibrous tissue.

Fibrositis: Pain, muscular stiffness and inflammation affecting the soft tissues of the arm, legs and trunk. The cause is unknown but may include immunological factors, muscular strain and psychological stress. Treatment is usually palliative.

Fibrous Dysplasia: A rare disease in which areas of bone are replaced by fibrous tissue. This renders the bone fragile and liable to fracture. It may involve only one bone – usually the thigh bone or femuror several bones. This latter form of the disease may be accompanied by pigmentation of the skin and the early onset of puberty.

Fibula: The slender outer bone of the leg. The head of this bone articualtes with the tibia just below the knee, and at the ankle it articulates with the talus bone.

Fiducial Limit: Form of confidence limit given as a stated probability, for example P 0.95 : in toxicology the terms fiducial limits and confidence limits are generally considered to be synonymous.

Fiduciary: A person or institution legally responsible for the management, investment and distributions of funds. Examples include trustees, executors and administrators.

Field: A space on a PharmaNet screen where specific information is entered.

Filariasis: The term used to describe several clinical entities caused by one or other of the nematode filariae; these include Wuchereria bancrofti/Bru- gia malayi, Onchocerca volvulus, Loa loa, Dra- cunculus medinensis (dracontiasis or guinea-worm disease), Mansonella perstans, etc. These organisms have widely differing geographical distributions. Whereas lymphatic filariasis is present throughout much of the tropics and subtropics, onchocerciasis (river-blindness) is largely confined to west and central Africa and southern America. Loaiasis is an infection of west and central Africa, and dracontiasis involves west and central Africa and western India only.

Filaricide: A generic term for drugs used to treat filarial infections.

Fill: The only filling of a particular prescription by a pharmacist. For multi-fill prescriptions, the terms first-fill and refill are used.

Filler: Informal term for "excipient"

Filling: The insertion, in denistry, of a specially prepared material into a cavity drilled into a tooth, usually for the treatment of dental caries.

Finasteride: Finasteride is a drug which inhibits the enzyme that metabolises testosterone into the more potent androgen, dihydrotestosterone. This action results in a reduction of prostate tissue. The drug is used to treat an enlarged prostategland, thus improving urinary flow. Its side- effects include reduced libido, and impotence. Finasteride offers an alternative to prostatectomy for some men but a significant minority do not improve. Women of childbearing age should not handle broken or crushed tablets.

Finger Print: The unique pattern of fine ridges in the outer horny layer of the skin at the front of the tip of each finger and thumb. The ridges are of three types: loops (70 per cent), whorls (25 per cent) and arches (5 per cent). Fingerprint patterns are used as a routine forensic test by police forces to identify individuals. Some patterns can indicate that the subject has an inherited disorder.

First Aid: Emergency procedures to help an ill or injured person before he or she receives expert medical attention or is admitted to hospital. Courses of instruction in first aid comprise 6–12 sessions, each of about 2 hours' duration. Syllabuses of instruction are published by various organisations, the principal ones being the British Red Cross, the St John Ambulance Association, and the St Andrew's Ambulance Association.

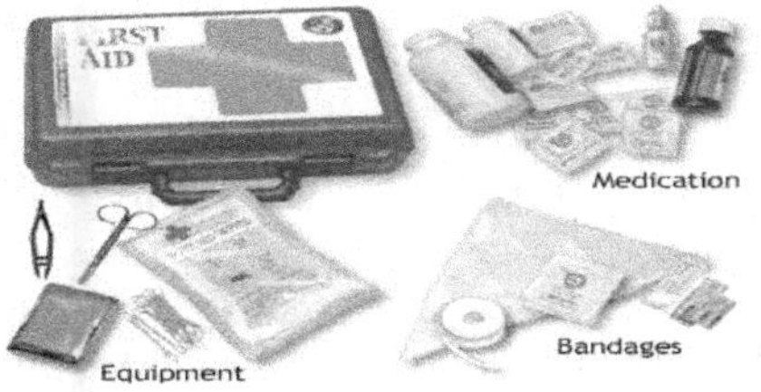

First Data Bank (FDB): An independent agency that supplies drug monograph information to PharmaNet.

First-pass Effect: Biotransformation of a substance in the liver after absorption from the intestine and before it reaches the systemic circulation.

Fiscal Year: July 1 through June 30 of any given year

Fissure: A term applied both to clefts of normal anatomical structure and also to small narrow ulcers occurring in skin and mucous membrane. The latter type of fissure occurs especially at the corners of the mouth and at the anus.

Fistula: An unnatural, narrow channel leading from some natural cavity – such as the duct of a gland, or the interior of the bowels – to the surface. Alternatively a fistula may be a communication between two such cavities where none should exist – as, for example, a direct communication between the bladder and bowel.

Fit: A popular name for a sudden convulsive seizure, although the term is also extended to include sudden seizures of every sort. During the occurrence of a fit of any sort, the chief object should be to prevent the patient from doing any harm to him or herself as a result of the convulsive movements. The person should therefore be laid flat, and the head supported on a pillow or other soft material.

Fitness: An ability to perform daily activities without becoming overtired. Fitness is dependent on strength, flexibility and endurance, and the level of an individual's fitness will often depend on their type of employment and the extent to which they indulge in physical exercise, whether training in the local health club or at home or regularly participating in sports. Regular fitness improves one's health and well-being. Fitness exercises should be matched to a person's age and abilities and there is a health danger if someone regularly exercises beyond their capabilities.

Fixed Dose Procedure: Acute toxicity test in which a substance is tested initially at a small number (3 or 4) predefined doses to identify which produces evident toxicity without lethality: the test may be repeated at one or more higher or lower defined discriminating doses to satisfy the criteria.

Flaccid: Relaxed or lacking in stiffness. Used to describe muscles that are not contracting (or following denervation), and organs – for example, the penis – that are lying loose, empty, or with wrinkles.

Flap: A section of tissue (usually skin) separated from underlying structures but still attached at its distal end by a pedicle through which it receives its blood supply. The free end may then be sutured into a new position to cover a defect caused by trauma or excision of diseased tissue. A free flap involves detachment of a section of tissue, often including bone and muscle, to a distant site where the artery and vein supplying it are anastomosed to adjacent vessels and the tissue is sutured into place.

Flare-up: A worsening of a disease.

Flat-Foot: Flat-foot, or pes planus, is a deformity of the foot in which its arch sinks down so that the inner edge of the foot comes to rest upon the ground. Causes The disorder may develop in infancy or occur in adult life, usually resulting from a combination of obesity and/or an occupation involving long periods of standing. Symptoms Often none, but there may be pain along the instep and beneath the outer ankle. The foot is stiff and broad, walking is tiresome, and the toes turn far out.

Flatulence: A collection of gas in the stomach or bowels. In the former case the gas is expelled from time to time in noisy eructations by the mouth; in the latter it may produce rum- blings in the bowels, or be expelled from the anus. Causes when gas is found in large amount in the bowels, its production is usually due to fermentation set up by bacteria.

Flatus: Intestinal or stomach gas. If it rises upwards, it is an eructation (burp or belch).

Flavonoids: From flavus, Latin for yellow. A 2-benzene ring, 15-carbon molecule, it is formed by many plants (in many forms) for a variety of oxidative-redox enzyme reactions. Brightly pigmented compounds that make many fruits and berries yellow, red, and purple, and that are considered in European medicine to strengthen and aid capillary and blood vessel integrity, they are sometimes (redundantly) called bioflavonoids.

Flexible Benefits Plan: A type of benefits program that offers employees a menu of benefit options, allowing them to create a benefits package which best suits their individual needs.

Flexible Spending Account (FSA): An account that reimburses the participant for qualified health costs, dependent care expenses and commuter costs through one pre-tax savings account. Employees or employers or both fund the account. At the end of each year, unused dollars are forfeited by the account holder.

Flexible Training: A term applied to the system of postgraduate medical training that allows young doctors to integrate their domestic commitments with the training requirements necessary to become a fully qualified specialist, usually by working part-time.

Flexion: The bending of a joint in the sagittal plane. Usually an anterior movement, it is occasionally posterior, as in the case of the knee-joint. Lateral flexion refers to the bending of the spine in the coronal plane – that is, from side to side.

Flexor: A muscle that causes bending of a limb or other body part.

Flexure: A bend in an organ or body part. The term is used, for example, to describe the skin on the inner aspect of the elbow or knee, as in the 'hepatic flexure' of the colon.

Floaters: Particles that appear to be floating in a per- son's field of vision. They move quickly as the eye moves, but when the eye is still they seem to drift. Vision is not usually affected. Most floaters are shadows on the retina from minute particles in the vitreous humour which lies in the main part of the eyeball behind the lens. As a person ages, the jelly-like vitreous humour usually shrinks a little and becomes detached from the retina; this pro- duces floaters which vanish over time. If a per- son notices a sudden cloud of floaters, some- times accompanied by flashes of light, it is likely that a tear in or detachment of the retina has occurred. This requires prompt medical attention.

Flooding: A popular name for an excessive blood-stained discharge from the womb (uterus) In the majority of cases, flooding is the sign of a miscarriage.

Flu: An infection caused by the "flu" (short for "influenza") virus. The flu is a contagious viral illness that strikes quickly and severely. Signs include high fever, chills, body aches, runny nose, sore throat, and headache.

Fluctuation: A sign obtained from collections of fluid by laying the fingers of one hand upon one side of the swelling, and, with those of the other, tapping or pressing suddenly on a distant point of the swelling. The 'thrill' communicated from one hand to the other through the fluid is one of the most important signs of the presence of an abscess, or of effusion of fluid into joints or into the peritoneal cavity.

Fluid Balance: The appropriate balance of fluid input and output (along with dissolved salts essential for life) over 24 hours. During this period, about 2,500 millilitres (ml) of fluid should be taken in by a 70-kg man and the same amount excreted; of this, 1,500 ml will be drunk, 800 ml will be in the food eaten, and 200 ml produced by food Metabolism.

Fluid Extract: An extract of an herb that is made according to official (and unofficial) pharmaceutical practice, with a strength of 1:1. That means each ounce of the fluidextract has the solutes found in an ounce of the dried herb. dvantageous for some herbs (such as Arctium or Taraxacum), where the active constituents retain the same proportions as in the plant, even though reduced to a very small volume of menstruum, it is deadly for others (such as Hydrastis or Lobelia), whose constituents may have wildly varying solubility, and whose fluidextract will contain only the most soluble constituents and lack others completely.

Flukes: Flukes are a variety of parasitic worms.

Fluoroscope: An apparatus for rendering X-rays visible after they have passed through the body, by projecting them on a screen of calcium tungstate. The technique is known as fluoroscopy. It provides a method of being able to watch, for instance, the beating of the heart, or the movements of the intestine after the administration of a barium meal.

Fluorosis: Adverse effects of fluoride, as in dental or skeletal fluorosis.

Flutter: The term applied to a form of abnormal cardiac arhythm, in which the atria contract at a rate of 200–400 beats a minute, and the ventricles more slowly. The abnormal rhythm is the result

FM: The building letter code for Facilities Management, 3130 20th Street, San Francisco

FMS (Financial Management System): FMS is the electronic general ledger for VA. Its purpose is to track obligations and expenditures by facility and time period (month, quarter and fiscal year). Cost centers and subaccounts (budget object codes) are used to organize the data by purpose, such as labor costs, medical supplies, and overhead.

Foci (Singular Focus): Small groups of cells distinguishable, in appearance or histochemically, from the surrounding tissue: Indicative of an early stage of a lesion that may lead to the formation of a neoplastic nodule.

Folium: (Plural: Folia) Latin term for leaf: for example, digitalis folium is digitalis leaf.

Follicle: Follicle is the term applied to a very small sac or gland: for example, small collections of adenoid tissue in the throat, and the small digestive glands on the mucous membrane of the intestine.Follow-on drug or me-too drug: Result from a development race in which only one drug can be the first approval in a new therapeutic class.

Follow-up Study: Investigation in which

individuals or populations, selected on the basis of whether they have been exposed to risk, have received a specified preventive or therapeutic procedure, or possess a certain characteristic, are followed to assess the outcome of exposure, the procedure, or effect of the characteristic, for example, occurrence of disease.

Fomentation: Any warm application to the surface of the body in the form of a cloth. Usually, the fomentation cloth is heated by being wrung out of hot water. A hot, wet poultice used on painful, inflamed areas. The usual form is a towel dipped in tea and applied hot or warm to the swollen tissue, being changed when it cools.

Fomites: A traditional term used to include all articles which have been brought into sufficiently close contact with a person sick of some infectious disease to retain the infective material and spread the disease. For example, clothes, bed- ding, carpets, toys and books may all be fomites until they are disinfected.

Fontanelle: Areas on the head on which bone has not yet formed. The chief of these is the anterior fontanelle, situated on the top of the head between the frontal and two parietal bones. In shape it is four-sided, about 25 mm (1 inch) square at the time of birth, gradually diminishing until it is completely covered by bone, which should happen by the age of 18 months.

Food Additive: Any substance not normally consumed as a food by itself and not normally used as a typical ingredient of the food, whether or not it has nutritive value, the intentional addition of which to food for a technological (including organoleptic) purpose in the manufacture, processing, preparation, treatment, packing, packaging, transport or holding of such food results, or may be reasonably expected to result (directly or indirectly) in it or its byproducts becoming a component of or otherwise affecting the characteristics of such foods.

Food Allergy: Hypersensitivity reaction to substances in the diet to which an individual has previously been sensitised.

Food and Drug Administration (FDA): U.S. government agency responsible for the regulation of food and medical materials

Food Chain: Sequence of transfer of matter and energy in the form of food from organism to organism in ascending or descending trophic levels.

Food Exchanges: A way to help people stay on special food plans by letting them replace items from one food group with items from another group.

Food Intolerance: This is divided into food aversion, where a person simply avoids a food they dislike; food intolerance, where taking the food causes symptoms; and food allergy, where the symptoms are due to an immunological reaction. Some cases of food intolerance are due to idiosyncrasy – that is, a genetic defect in the patient, such as alactasia, where the intestine lacks the enzyme that digests milk sugar, with the result that individuals so affected develop diarrhoea when they drink milk. Intolerance to specific foods, as distinct from allergy, is probably quite common and may be an important factor in the aetiology of the irritable bowel syndrome (IBS).

Food Poisoning: This illness is characterised by vomiting, diarrhoea and abdominal pain, and results from eating food contaminated with metallic or chemical poisons, certain micro-organisms or microbial products. Alternatively, the foods – such as undercooked red kidney beans or fish of the scombroid family (mackerel and tuna) – may contain natural posions. Food poisoning caused by chemical or metallic substances usually occurs rapidly, within minutes or a few hours of eating. Among micro-organisms, bac- teria are the leading cause of food poisoning, particularly

Food Web: Network of food chains.

Food: Mixture of substances containing Carbohydrate, Fat, Protein, Vitamins, Trace Elements and water consumed by animals, including humans, to provide the necessary nutrients to maintain the body's metabolism.

Foramen: The Latin term for a hole. It is especially applied to natural openings in bones, such as the foramen magnum, the large opening in the base of the skull through which the brain and spinal cord are continuous.

Forced Diuresis: A means of encouraging excretion via the kidneys of a compound by altering the pH and increasing the volume of the urine. Forced diuresis is occasionally used after drug over- doses, but is potentially dangerous and so only suitable where proper intensive monitoring of the patient is possible. Excretion of acid com- pounds, such as salicylates, can be encouraged by raising the pH of the urine to 7.5–8.5 by the administration of an alkali such as bicarbonate (forced alkali diuresis) and that of bases, such as Amphetamines, by lowering the pH of the urine to 5.5–6.5 by giving an acid such as ammonium chloride (forced acid diuresis).

Forceps: Surgical instruments with a pincer-like action which are used, for example, during operations, for grasping tissues and other materials. There are many different designs for different uses. Obstetric forceps are designed to fit around the infant's head and allow traction to be applied to aid its delivery or to protect the soft skull of a very premature baby.

Foreign: Tissue, cells or other material not normally present in a particular body, recognized as being "non-self" by the immune system.

Forensic Medicine: That branch of medicine concerned with mat- ters of law and the solving of crimes, for example, by determining the cause of a death in suspicious circumstances or identifying a crim- inal by examining tissue found at the scene of a crime. The use of DNA identification to establish who was present at the 'scene of the crime' is now a widely used procedure in forensic medicine.

Formulary: A list of formulae used as drugs and other medical preparations. The British National Formulary is an authoritative six-monthly publication containing information and advice on medicines and drugs. Published jointly by the British Medical Association and the Royal Pharmaceutical Society of Great Britain, with input from the Department of Health, it is distributed to all NHS doctors by the government.

Fossa: A term applied to various depressions or holes, both on the surface of the body and in internal parts, such as the iliac fossa in each lower corner of the abdomen, and the fossae within the skull which lodge the different parts of the brain.

Fovea: A small depression. In the eye this is an area near the fundus which contains predominantly cones and is the area with greatest visual acuity.

Frame-shift Mutation: Point mutation involving either the deletion or insertion of one or two nucleotides in a gene: by the frame shift mutation, the normal reading frame used when decoding nucleotide triplets in the gene is altered.

Freckles: Also known as ephelides, these are small, brown, flat spots on the skin. They occur mostly in blonde or red-haired subjects in exposed areas, and darken on exposure to the sun. Melanocytes are not increased in the basal layer of the epidermis.

Free Association: A psychoanalytic technique in which the therapist encourages the patient to follow up a specific line of thought and ideas as they enter his or her consciousness.

Free Radical: A highly chemically reactive atom, molecule or molecular fragment with a free or unpaired electron. Free radicals are produced in many different ways such as, normal metabolic processes, ultraviolet radiation from the sun, nuclear radiation and the breakdown in the body of spoiled fats. Free radicals have been implicated in aging, cancer, cardiovascular disease and other kinds of damage to the body.

Free-radical Scavenger: The cascade of chemical reactions that occurs when a free radical reacts with another molecule in order to gain an electron. The molecule that loses an electron to the free radical then becomes a free radical, repeating the process until the energy of the free radical is spent, or the reaction is stopped by an antioxidant. In biological systems, this cascade can damage important molecules like DNA.

Freeze Drying: A technique for fixating specimens of tissue, involving a minimum of chemical and physical alteration. The histological specimen is immersed in a chemical, isopentane, which has been cooled in liquid air to a temperature just below 200 ^{0}C. This preserves the tissue instantly without large ice crystals forming – these would result in structural damage. The specimen is then dehydrated in a vacuum for three days, after which it can be examined using a microscope.

Fremitus: Tremors or vibrations in an area of the body, detected by palpating (feeling) with the fingers or the hand or by auscultation (listening). The procedure is most commonly used when examining the chest and assessing what happens when the patient breathes, coughs or speaks. This helps the doctor to diagnose whether dis- orders such as fluid in the pleural cavity or solidification of a section of the lung have occurred. Friction fremitus is a grating feeling communicated to the hand by the movements of lungs or heart when the membrane covering them is roughened, as in pleurisy or pericarditis. Vocal fremitus means the sensation felt by the hand when a person speaks; it is increased when the lung is more solid than usual. The 'thrills' felt over a heart affected by valvular disease are also varieties of fremitus.

Frenum: Also known as the fraenum or frenulum, this comprises the folds of mucous membrane that anchor the bottom of the tongue to the floor of the mouth.

Frequency: (1) The number of regular recurrences of an event during a given period of time. Examples in medicine are the heartbeat, and sound vibrations in the ear or vocal cords.

Frequency: The number of times each day that a medication is administered.

Freudian Theory: A theory that emotional and allied diseases are due to a psychic injury or trauma, generally of a sexual nature, which did not produce an adequate reaction when it was received and therefore remains as a subconscious or 'affect' memory to trouble the patient's mind.

Friction: The name given either to the fremitus felt, or to the grating noise heard, when two rough surfaces of the body move over one another. It is characteristically obtained over the chest in cases of dry pleurisy.

Friedreich's Ataxia: A hereditary disease resembling locomotor ataxia, and due to degenerative changes in nerve tracts and nerve cells of the spinal cord and the brain. It occurs usually in children, or at any rate before the 20th year of life, and affects often several brothers and sisters. Its chief symptoms are unsteadiness of gait, with loss of the knee jerks, followed later by difficulties of speech, tremors of the hands, head and eyes, deformity of the feet, and curvature of the spine. There is often associated heart disease. The sufferer gets gradually worse, but may live, with increasing disability, for 20–30 years.

Frigidity: A term used to describe a lack of interest in sexual intercourse (coitus) or the inability to achieve intercourse or orgasm. Though applicable to both sexes, frigidity is usually applied to women with these sexual problems.

Fröhlich's Syndrome: A condition in children characterised by obesity, physical sluggishness, and retarded sexual development. It is the result of disturbed pituitary gland function.

Frontal Sinus: One of the airspaces that form the paranasal sinuses within some of the frontal bones of the skull. These sinuses are lined with mucous membrane and open into the nasal cavity.

Frontal: Describing the anterior part of a body or organ.

Frostbite: This results from the action of extreme cold on the skin. Vasoconstriction results in a reduced blood – and hence, oxygen– supply, leading to necrosis of the skin and, in severe cases, of the underlying tissues. Chiefly affecting exposed parts of the body, such as the face and the limbs, frostbite occurs especially in people exercising at high altitudes, or in those at risk of peripheral vascular disease, such as diabetics who should take particular care of their fingers and toes when in cold environments.

Frozen Shoulder: A painful condition of the shoulder accompanied by stiffness and considerable limitation of movement. The usual age-incidence is between 50 and 70. The cause is inflammation and contracture of the ligaments and muscles of the shoulder joint, probably due to overuse. Treatment is physiotherapy and local steroid infections. There is practically always complete recovery, even though this may take 12–18 months.

Fugue: The term literally means flight, and it is used to describe the mental condition in which an individual is suddenly seized with a subconscious motivation to flee from some intolerable reality of everyday existence: this usually involves some agonising interpersonal relationship. As a rule, fugue lasts for a matter of hours or days, but may go on for weeks or even months. During the fugue the individual seldom behaves in a particularly odd manner – though he or she may be considered somewhat eccentric. When it is over there is no remembrance of events during the fugue.

Full Listing: Include virtually all Health Canada (TPP) approved drugs and physicians may write prescriptions for any medicine since they will be covered (or partially) by the drug plan. There are NO open formularies offered by the government in Canada (either at the federal or provincial level). These formularies are mostly found in private insurance plans.

Fumigant: Substance that is vaporized in order to kill or repel pests.

Fumigation: A means of disinfection by the vapour of powerful antiseptics.

Functional Area: The operation, or area of responsibility, that affects, or are affected by, the use of hazardous material. These areas include, but are not limited to, budget and fiscal planning, legal support, research and development, weapons systems acquisition and maintenance, material and performance specifications and standards, design handbooks, and technical manuals; maintenance and repair procedures, industrial processes, procurement policy, contracting provisions, new material identification, public works operations, construction, management of munitions, chemical agents, propellants, medical and other personnel support, safety and occupational health, transportation, and logistics analysis; supply; warehousing; distribution; recycling; disposal; spill prevention, control, and cleanup; contaminated site remediation; staffing, education, and training; information exchange; public affairs; general administration; and oversight.

Functional: An imbalance of response, without permanent tissue damage, and generally reversible.

Fungicide: Substance intended to kill fungi.

Fungus: A simple plant that is parasitic on other plants and animals. Included in this group are mil- dews, moulds, mushrooms, toadstools and yeasts. Unlike other plants, they do not contain the green pigment chlorophyll. Most of the world's 100,000 different species of fungus are harmless or even beneficial to humans. Yeasts are used in the preparation of food and drinks, and antibiotics are obtained from some fungi. A few, however, can cause fatal disease and illness in humans

Funiculitis: Inflammation of the spermatic cord, usually arising in men with epididymitis (inflammation of the epididymis in the testicle). The condition can be painful. Antibiotics and analgesics are effective treatment.

Funny Bone: Colloquial name for the small area at the back of the elbow where the ulnar nerve goes over a prominence at the lower end of the humerus bone. If the nerve is hit, acute pain results, accompanied by tingling in the forearm and hand.

Furuncle: Another term for a boil.

Fusidic Acid: A valuable antistaphylococcal antibiotic used both orally and topically. It is particularly useful in osteomyelitis.

G

GABA (Gamma Amino Butyric Acid): GABA, or gamma amino butyric acid, is an amino acid that occurs in the central nervous system, mainly in the brain tissue. It is a chemical substance that transmits inhibitory impulses from nerve endings across synapses to other nerves or tissues.

Gag Reflex: Assessment of victims of major trauma must include maintenance of their airways and breathing. Any false teeth, vomitus and foreign bodies should be removed, and the response to digital stimulation of the posterior pharyngeal wall – the 'gag reflex' – assessed. Even with a normal gag reflex, the airway may be seriously threatened if vomiting occurs. During the initial stages of resuscitation, careful and constant supervision of the airway is essential, with a high-volume sucker immediately available.

Gag: A device that, when placed between a person's teeth, keeps the mouth open.

Gait: The way in which an individual walks. Gait may be affected by inherited disorders; by illness especially neurological disorders; by injury; or by drug and alcohol abuse. Children, as a rule, begin to walk between the ages of 12 and 18 months, having learned to stand before the end of the first year. If a normal-sized child shows no ability to make movements by this time, the possibility of mental retardation must be borne in mind, and if the power of walking is not gained by the time the child is a year.

Galactocele: A cyst-like swelling in the breast which forms as a result of obstruction in the milk-duct draining the swollen area.

Galactorrhoea: This term may refer to unusually copious secretion of milk from the mammary glands when a mother is feeding her baby. It is also used to describe secretion of milk after the mother has stopped breast feeding.

Galactosaemia: A very rare, recessively inherited disease, with an incidence of around one in 75,000 births. Its importance lies in the disastrous consequences of it being overlooked, and results from the deficiency of an enzyme essential for the metabolism of galactose. Normal at birth, affected infants soon develop jaundice, vomiting, diarrhoea, and fail to thrive on starting milk feeds.

Gall: Another name for bile.

Gamete: A sexual or germ cell: for example, an ovum or spermatozoon.

Gamete: Reproductive cell (either sperm or egg) containing a haploid set of chromosomes.

Gametocide: Substance intended to kill gametes.

Gamgee Tissue: A surgical dressing

composed of a thick layer of cotton-wool between two layers of absorbent gauze, introduced by the Birmingham surgeon, Sampson Gamgee (1828–1886). Gamgee tissue has been a registered trademark since 1911.

Gamma Rays: Short-wavelength penetrating electromagnetic rays produced by some radioactive compounds. More powerful than X-rays, they are used in certain radiotherapy treatments and to sterilise some materials.

Gamma-Globulin: Gamma-globulin describes a group of proteins present in the blood plasma. They are characterised by their rate of movement in an electrical field, and can be separated by the process of electrophoresis. Most gamma-globulins are immunoglobulins.Gamma-globulin injection provides passive or active immunity against hepatitis A.

Gammexane: The proprietary name for a synthetic insecticide which is a formulation of benzene hexachloride. It is active against a large range of insects and pests, including mosquitoes, fleas, lice, cock- roaches, house-flies, clothes moths, bed-bugs, ants, and grain pests.

Ganglia: (singular: ganglion) Colonies of neurons outside the brain and spinal cord sometimes acting to control local functions. These are little affected by normal stress conditions. (Example: the solar plexus made of two separate ganglions.)

Ganglion: This term is used in two senses. In anatomy, it means an aggregation of nerve cells found in the course of certain nerves. In surgery, it means an enlargement of the sheath of a tendon, containing fluid. The latter occurs particularly in connection with the tendons in front of, and behind, the wrist.

Gangrene: The death and decay of body tissues caused by a deficiency or cessation of the blood supply. There are two types: dry and moist. The former is a process of mummification, with the blood supply of the affected area of tissue stopping and the tissue withering up. Moist gangrene is characterised by putrefactive tissue decay caused by bacterial infection. The dead part, when formed of soft tissues, is called a slough and, when part of a bone, is called a sequestrum.

GAP: Geriatrics Awareness Project, the name of a community service project of ASP

Garble: Rummaging through and cleaning out herbs; sorting.

Gardnerella: Formerly Hæmophilus, this is an anaerobic bacteria that is a main contributor to bacterial vaginosis. It is sometimes sexually transmitted, but can stick around for years as a passive part of the vaginal flora, only to flare up. It seems to occur in up to a quarter of relatively monogamous women and in half of women with multiple male partners. As bacterial vaginosis, Gardnerella is one of the three main causes of vaginal discharges, along with Trichomonas and Candida albicans. Antibiotic therapy for male partners seems of only marginal value, and the distinguishing characteristic of the infection is nearly no Lactobacillus vaginal presence, the main part of the flora thatretains the lactic acid and peroxide balance so important in a healthy vagina. Live culture yogurt, as both food and douches help the problem.

Gargles: Gargling is a process by which various substances in solution are brought into contact with the throat without being swallowed. The watery solutions used for the purpose are called gargles. Gargles are used in the symptomatic treatment of infections of the throat: for example,'sore throat', pharyngitis and tonsillitis.

Gargoylism: Also known as Hurler's syndrome, gargoylism is a rare condition due to lack of a specific enzyme. It is a progressive disorder usually leading to death before the age of 10 years. The affected child is usually normal during the first few months of life; mental and physical deterioration then set in. The characteristic fea- tures include coarse facial features (hence the name of the condition), retarded growth, chest deformity, stiff joints, clouding of the cornea, enlargement of the liver and spleen, deafness, and heart murmurs, with mental deterioration. It occurs in about one in 100,000 births.

Gas Chromatography (GC): A process by which the components of a mix are separated from one another by volatilizing the sample into a carrier gas stream and passing the gas through a column containing a substance that selectively retains (adsorbs) and releases the volatile constituents.

Gastralgia: A stomach ache.

Gastrectomy: A major operation to remove the whole or part of the stomach. Total gastrectomy is a rare operation, usually performed when a person has cancer of the stomach; the oesphagus is then connected to the duodenum. Sometimes cancer of the stomach can be treated by doing a partial gastrectomy: the use of partial gastrectomy to treat pepticulcer used to be common before the advent of effective drug therapy.

Gastric Ulcer: A usually chronic condition, started by irritation, with congestion in time, leading to edema, blistering, and the formation of an ulcer. Hylobacter infections seem to prolong and aggravate the condition, but the presence alone of the bacteria, without functional impairment, will not begin the disease. Possessing certain "workaholic" panache even boasted of in some business circles as if to validate one's work ethic, it nonetheless is fatal if untreated.

Gastric Lavage: A method of gastric decontamination used in the treatment of poisoning. It is not used routinely. Lavage involves the passage of a lubricated tube via the mouth and oesophagus into the stomach. Patients are positioned on their side with the head lower than the feet. A small quantity of fluid (300 ml) is passed into the stomach and the contents then drained out (by gravity) by lowering the end of the tube. This is repeated until the solution is clear of particulate matter. The procedure should be done only by an experienced health professional.

Gastric: Pertaining to the stomach.

Gastrin: A hormone produced by the mucous membrane in the pyloric part of the stomach. The arrival of food stimulates production of the hormone which in turn stimulates the production of gastric juice.

Gastritis: Inflammation of the stomach lining. This may take an acute form when excess alcohol or other irritating substances have been taken, resulting in vomiting. Chronic gastritis may be the result of regular smoking and chronic alcoholism, or the condition may be caused by the back flow of bile from the duodenum.

Gastrocnemius: The large double muscle which forms the chief bulk of the calf, and ends below in the tendo calcaneus.

Gastro Duodenostomy: A surgical operation to join the duodenum to a hole made in the stomach wall to circumvent an obstruction in the gut for example, pyloric stenosis – or to improve the passage of food from the stomach into the duodenum.

Gastroenteritis: Inflammation of the stomach and intestines, usually resulting from an acute bacterial or viral infection. The main symptoms are diarrhoea and vomiting, often accompanied by fever and – especially in infants – dehydration. Although generally a mild disease in western countries, it is the number-one killer of infants in the developing world, with more than

Gastroenteritis: Inflammation of the stomach and small intestines. It is more likely to be infectious than simple gastritis and is often accompanied by fever and general malaise.

Gastroenterostomy: An operation performed usually in order to relieve some obstruction to the outlet from the stomach. One opening is made in the lower part of the stomach; another in a neighbouring loop of the small intestine. The two are then stitched together.

Gastroesophageal Reflux: The involuntary regurgitation of stomach contents or surface acids into the throat, with heartburn; it can be simple or serious.

Gastrointestinal Tract: The passage along which the food passes, in which it is digested, and from which it is absorbed by lymphatics and blood vessels into the circulation. The tract consists of the mouth, pharynx or throat, oesophagus or gullet, stomach, small intestine, and large intestine, in this order.

Gastrointestinal: Pertaining or communicating with the stomach and intestine.

Gastro-Oesophageal Reflux: A disorder in which the contents of the stomach back up into the oesophagus because the usual neuromuscular mechanisms for preventing this are intermittently or permanently failing to work properly. If persistent, the failure may cause oesophagitis. If a person develops heartburn, regurgitation, discomfort and oesophagitis, the condition is called gastro-oesophageal reflux disease (GORD) and sometimes symptoms.

Gastroscope: An endoscopic instrument for viewing the interior of the stomach. Introduced into the stomach via the mouth and oesophagus, the long flexible instrument (also called an oesophagogastroduodenoscope) transmits an image through a fibreoptic bundle or by a small video camera. The operator can see and photograph all areas of the stomach and also take biopsy specimens when required.

Gastrostomy: An operation on the stomach by which, when the gullet is blocked by a tumour or other cause, an opening is made from the front of the abdomen into the stomach, so that fluid food can be passed into the organ.

Gatekeeper: A primary care physician who provides a broad range of routine medical services and refers patients to specialists, hospitals and other providers as necessary. This traditional primary care physician role is called a "gatekeeper" function.

Gateway: A device that connects networks so that information can be both transferred and converted to a form compatible with the receiving network.

Gaucher's Disease: A disease characterised by abnormal storage of lipid, particularly in the spleen, central nervous system, bone marrow, and liver. This results in enlargement of the spleen and the

Gavage: Administration of materials directly into the stomach by oesophageal intubation.

GCP: Good clinical practice

Gel: The term applied to a colloid substance G which is firm in consistency although it contains much water: for example, ordinary gelatin.

Gelatin: This is derived from collagen, the chief constituent of connective tissue. It is a colourless, transparent substance which dissolves in boiling water, and on cooling sets into a jelly. Such a jelly is a pleasant addition to the invalid diet, especially when suitably flavoured, but it is of relatively little nutritive value as not more than one ounce can be taken in the day (i.e. the amount required to make one pint of jelly). Although it is a protein, it is lacking in several of the vital amino acids.

Gemeprost: One of the prostaglandins administered vaginally as pessaries for the medical induction of late therapeutic abortion. Gemeprost also softens the cervix before surgical abortion, being particularly useful for women in their first pregnancy. Prostaglandins induce contractions of the uterus while keeping blood loss t

Gender Identity Disorders: Gender identity is the inner sense of masculinity or femininity, and gender role is an individual's public expression of being male, female, or a 'mix' (androgynous). Most people have no difficulty because their gender identity and role are congruous. A person with a gender identity disorder, however, has a conflict between anatomical sex and gender identity.

Gender: Subject self-identification as a male or female person . Gender signifies an individual's personal, legal, and social status without reference to

genetic sex, which is an objective biological fact.

Gene Amplification: Production of extra copies of a chromosomal sequence found either as intra- or extra- chromosomal DNA; with respect to a plasmid, it refers to the increase in the number of plasmid copies per cell induced by a specific treatment of transformed cells.

Gene Delivery: Development of methods to deliver genes to cells

Gene Therapy: Gene therapy is the transfer of normal genes into a patient to combat the effects of abnormal genes which are causing disease(s). The genetic engineering technique used is somatic cell gene therapy in which the healthy gene is put into somatic cells that produce other cells for example, stem cells that develop into bone marrow. Descendants of these altered cells will be normal and, when sufficient numbers have developed, the patient's genetic dis-order should be remedied. The abnormal gene, however, will still be present in the treated individual's germ cells (eggs or sperm) so he or she can still pass the inherited defect on to succeeding generations.

Gene Therapy: Use of genes to treat disease

Gene: A unit of heredity that contains the instructions a cell uses to make a specific product; the functional and physical unit of heredity passed from parent to offspring. Genes are pieces of DNA, and most genes contain the information for making a specific protein.

General Paralysis of The Insane: An

outdated term for the tertiary stage of syphilis.

General Practitioner (GP): A general practitioner ('family doctor'; 'family practitioner') is a doctor working in primary care, acting as the first port of professional con- tact for most patients in the NHS. There are approximately 35,000 GPs in the UK and their services are accessed by registering with a GP practice – usually called a surgery or health centre. Patients should be able to see a GP within 48 hours, and practices have systems to try to ensure that urgent problems are dealt with immediately. GPs generally have few diag- nostic or treatment facilities themselves, but can use local hospital diagnostic services (X-rays,

Generalization: The extent to which the findings of a clinical trial can be reliably extrapolated from the subjects who participated in the trial to a broader patient population and a broader range of clinical settings.

Generic Drug: A medicinal drug that is sold under its official (generic) name instead of its proprietary (patented brand) name. NHS doctors are advised to prescribe generic drugs where possible as this enables any suitable drug to be dispensed, saving delay to the patient and sometimes expense to the NHS.

Generic Substitution: The process of substituting a bio- equivalent generic for a brand name drug. In private-sector programs, generic substitution is encouraged by the use of different copayments (higher for brand name products), differential dispensing fees that provide incentives to pharmacists, and physician education. In most public- sector programs, generic substitution is mandated by provincial laws.

Genes: Humans possess around 30,000 genes which are the biological units of heredity. They are arranged along the length of the 23 pairs of chromosomes and, like the chromosomes, therefore come in pairs. Human beings have 46 chromosomes, comprising two sex chromosomes and 44 autosomes, but there is also a mitochondrial chromo- some outside the cell nucleus which is inherited from the mother.

Genetic Code: The message set out sequentially along the human chromosomes. The human gene map is being constructed through the work of the international, collaborative human genome project; so far, only part of the code has been translated and this is the part that occurs in the genes. Genes are responsible for the protein synthesis of the cell they instruct the cell how to make a particular polypeptide chain for a particular protein. Genes carry, in coded form, the detailed specifications for the thousands of kinds of protein molecules required by the cell for its existence, for its enzymes, for its repair work and for its reproduction. These proteins are synthesised from the 20 natural amino acids, which are uniform throughout nature and which exist in the cell cytoplasm as part of the metabolic pool.

Genetic Counselling: The procedure whereby advice is given about the risks of a genetic disorder and the various options that are open to the individual at risk. This may often involve establishing the diagnosis in the family, as this would be a pre- requisite before giving any detailed advice. Risks can be calculated from simple Mendelian inheritance in many genetic disorders. However, in many disorders with a genetic element, such as cleft lip or palate

Genetic Engineering: Genetic engineering, or recombinant DNA technology, has only developed in the past decade or so; it is the process of changing the genetic material of a cell. Genes from one cell – for example, a human cell – can be inserted into another cell, usually a bacterium, G and made to function. It is now possible to insert the gene responsible for the production of human insulin, human growth hormone and interferon from a human cell into

Genetic Polymorphism: Condition in which a genetic character occurs in more than one form, resulting in the coexistence of more than one morphological type in a given population.

Genetic Research: The study of the human body's genes to determine the cause behind diseases or disorders.

Genetic Screening: A screening procedure that tests whether a person has a genetic make-up that is linked with a particular disease. If so, the person may either develop the disease or pass it on to his or her offspring. When an individual has been found to carry a genetically linked disease, he or she should receive genetic counselling from an expert in inherited diseases.

Genetic Toxicology: Study of substances that can produce adverse heritable changes.

Genetics: The science which deals with the origin of the characteristics of an individual or the study of heredity.

Genitalia: The external organs of reproduction. The term is usually applied to the external parts of the reproductive system: the vulva in females and penis and scrotum in males. Rarely the sex of an individual may not be apparent from the genitalia. Genitals develop from a common embryonic structure, and disturbances in the hormone controls of the developing genitalia may produce an individual whose external genitalia are ambiguous. The condition is known as intersex. The individual may be hermaphrodite or pseudohermaphrodite.

Genito-Urinary Medicine: The branch of medicine that deals with the effects of sexually diseases (STDS) on the urinary tract, reproductive system and other systems in the body. The specialty overlaps with gynaecology (women's urinary and reproductive systems) and urology (men's urinary and reproductive system).

Genito-Urinary Tract: This consists of the kidneys, ureters, urinary bladder and urethra –and, in the male, also the genital organs.

Genome: A complete set of chromosomes derived from one parent, or the total gene complement of a set of chromosomes. An international study is well underway to produce a complete map of the human genome.

Genomics: Genomics, or genomic research, means studying the whole genome - including DNA and genes - in humans, animals, plants, or other living organisms. Genetic Research is a type of genomic research.

Genotoxicity: Ability to cause damage to genetic material. Such damage may be mutagenic and/or carcinogenic.

Genotype: All of an individual's genetic information that is encoded in his or her chromosomes. It also means the genetic information carried by a pair of alleles which controls a particular characteristic.

Genu Valgum: The medical term for knock-knee – a deformity of the lower limbs in such a direction that when the limbs are straightened, the legs diverge from one another. As a result, in walking, the knees knock against each other. The amount of knock-knee is measured by the distance between the medial malleoli of the ankles, with the inner surfaces of the knee touching and the knee-caps facing forwards. The condition is so common in children between the ages of 2–6 years that it may almost be regarded as a normal phase in childhood. When marked, or persist- ing into later childhood, it can be corrected by surgery (osteotomy).

Genu Varum: Genu varum is the medical term for bow leg.

Geriatrics: Now increasingly termed 'medicine of the elderly', this is a branch of medicine that deals with disorders and diseases associated with old age, and particularly their social consequences

Germ Cell: Those embryonic cells with the potential to develop into ova or spermatozoa

Germ Layer: Any one of the three discrete varieties of body tissue that develop in the early stages of growth of the embryo. Development of the layers can G be followed throughout the embryo's stages of growth and specialisation into the body's full range of tissues and organs.

Germ-free Animal: Animal grown under sterile conditions in the period of postnatal development: such animals are usually obtained by Caesarean operation and kept in special sterile boxes in which there are no viable microorganisms (sterile air, food and water are supplied).

Germinal Aplasia: Complete failure of

gonad development.

Gerontology: The study of alterations in the body and mind that occur as a person ages, and the problems that result.

Gestaltism: A school of psychology based on the concept that an individual's sense of wholeness is more valuable than a piecemeal approach to perception and behaviour. Founded in Germany early in the 20th century, the school's practitioners regarded the whole as more than a sum of its parts. Aimed at resolving personal problems, the therapy increased subjects' self-awareness of all aspects of themselves in their environment.

Gestation: Gestation is another name for pregnancy.

Gestational Diabetes: A type of diabetes that can occur in pregnant women who have not been known to have diabetes before. Although gestational diabetes usually subsides after pregnancy, many women who've had gestational diabetes develop type 2 diabetes later in life.

GHC (Group Health Cooperative): GHC is a not-for-profit health maintenance organization (HMO). It is a staff-model HMO, meaning that its physicians are salaried employees.

GI: Gastrointestinal

Giardiasis: A condition caused by a parasitic organism known as Giardia lamblia, which is found in the duodenum and the upper part of the small intestine. This organism is usually harmless, but is sometimes responsible for causing diarrhoea. The illness develops one or two weeks after exposure to infection, and usually starts as explosive diarrhoea, with the passage of pale fatty stools, abdominal pain and nausea. It responds well to metronidazole or mepacrine.

Gift Tax: Tax on gifts generally paid by the person making the gift, rather than the recipient.

Gift-tax Annual Exclusion: The provision in the tax law that exempts the first $11,000 (as adjusted for inflation) in present-interest gifts a person gives to each recipient during a year from federal gift taxes.

Gigantism: Excessive growth (mainly in height) caused by overproduction, during childhood or adolescence, of growth hormone by a tumour of the pituitary gland. Untreated, the affected individual may die in early adulthood. Sometimes the tumour appears after the individual has stopped growing and the result then

Gilles De La Tourette's Syndrome: Also known as Tourette's syndrome, this is a hereditary condition of severe and multiple tics of motor or vocal origin. It usually starts in childhood and becomes chronic (with remissions). With a prevalance of one in 2,000, a dominant gene with variable expression may be responsible. The disorder is associated with explosive vocal tics and grunts, occasionally obscene. The patient may also involuntarily repeat the words or imitate the actions of others. Haloperidol, pimozide (an oral antipsychotic drug similar to chlorpromazine hydrochloride) and clonidine are among drugs that may help to control this distressing, but fortunately rare, disorder.

Gingivitis: A swelling and soreness of the gums that, without treatment, can cause serious gum problems and disease.

Gland: A collection of cells or an organ with a specialised ability to make and secrete chemical substances such as enzymes and hormones essential for the normal functioning of the body. Glands are classified into two groups: endocrine and exocrine. The former secrete their products, hormones, straight into the bloodstream; the latter's secretions are dis- charged through ducts. (These functional differences are the reason why glands have been defined as ductless and ducted.) Examples of endocrine glands are the adrenals, pituitarygland and thyroid gland. Exocrine glands include sebaceous glands (in the skin) and the salivary glands in the mouth whose enzymes start the digestion of food. The breasts or mammary glands are exocrine glands that secrete milk. Though strictly speaking not a gland, lymph nodes.

Glans: The term applied to the ends of the penis and the clitoris. In the penis the glans is the distal, helmet-shaped part that is formed by the bulbous corpus spongiosum (erectile tissue). In an uncircumcised man the glans is covered by the foreskin or prepuce when the penis is flaccid.

Glasgow Coma Scale: A method developed by two doctors in Glasgow that is used to assess the depth of coma or unconsciousness suffered by an individual. The scale is split into three groups – eye opening, motor response, and verbal response – with the level of activity within each group given a score. A person's total score is the sum of the numbers scored in each group, and this provides a reasonably objective assessment of the patient's coma state – particularly useful when monitoring people who have suffered a head injury.

Glaucoma: An eye disease, usually chronic and slow, with increased pressure of fluid within the eye causing degrees of impairment to the optic nerve, and slowing circulation between the eye chambers sufficient to also contribute to lens deposits and corneal opacities. When under adrenalin stress or under the effect of most stimulants, pupils dilate, the eyeball changes shape, and pressure within the eye increases. This may not itself start glaucoma, but adrenergic stress will surely make it worse.

Gleet: Gleet means a chronic form of gonorrhoea.

Glenoid: The term applied to the shallow socket on the shoulder-blade into which the humerus fits, forming the shoulder-joint.

Glia: Also called neuroglia, this is the specialized connective tissue of the Central Nervous System. Providing support and nutrition to neurones, glia comprises various cells including oligodendrocytes, astrocytes and ependymal cells. There arearound ten times as many glial cells as neurons and they form about 40 per cent of the total volume of the brain and spinal cord, playing an essential role in the neurochemical transmission function of neurons.

Gliobastoma: A type of brain tumour arising from tissue. It grows rapidly, destroying brain cells and causing a progressive loss of brain function. The patient suffers from headache as a result of raised cranial pressure, eventually vomiting regularly and becoming increasingly drowsy. The prognosis is poor and palliative treatment is required as surgical removal, radiotherapy and chemotherapy are not effective.

Glioma: A tumour in the brain or spinal cord, composed of neuroglia, which is the special connective tis- sue that supports the nerve cells and nerve fibres. Low-grade malignant gliomas cause symptoms by putting pressure on surrounding tissues and organs. Highly malignant gliomas are usually invasive. Gliomas, like other space-occupying tumours in the brain, may present with headaches, seizures, neurological symptoms or symptoms of mental disturbance. Treatment may include surgery, radiotherapy and chemotherapy and should be done in a specialist neurological centre. Gliomas tend to spread within the brain and can be difficult to remove surgically.

Global Assessment Variable: A single variable, usually a scale of ordered categorical ratings, which integrates objective variables and the investigator's overall impression about the state or change in state of a subject.

Globus: A term applied generally to any structures of ball shape, but especially to the sensation of a ball in the throat causing choking, which forms a common symptom of acute anxiety (globus hystericus).

Glomerular Filtration Rate (GFR): Each of the two kidneys filters a large volume G of blood – 25 per cent of cardiac output, or around 1,300 ml – through its two million glomeruli every minute. The glomeruli filter out cell, protein, and fat-free fluid which, after reabsorption of certain chem- icals, is excreted as urine. The rate of this ultra- filtration process, which in health is remarkably constant, is called the glomerular filtration rate (GFR).

Glomerular: Pertaining to a tuft or cluster, as of a plexus of capillary blood vessels or nerve fibres, especially referring to the capillaries of the glomeruli of the kidney.

Glomerulonephritis: Inflammation of the capillary tubes in the glomeruli of the kidney.

Glomerulus: A small knot of blood vessels about the size of a grain of sand, of which around 1,000,000 are found in each of the two kidneys, and from which the excretion of fluid out of the blood into the tubules of the kidney takes place.

Glossary: A collection of specialized terms with their meanings.

Glossitis: Inflammation of the tongue.

Glossopharyngeal: The glossopharyngeal nerve is the ninth cranial nerve, which in the main is a sensory nerve, being the nerve of taste in the posterior third of the tongue and the nerve of general sensation for the whole upper part of the throat and middle ear. It also supplies the parotid gland and one of the muscles on the side of the throat.

Glottis: The narrow opening at the upper end of the larynx. The glottis is made up of the true vocal cords.

GLP: Good laboratory practice

Glucagon: A hormone produced by the alpha cells of the pancreas that increases the release of sugar by the liver: it is hyperglycemic. The substance produced by the beta cells, insulin, induces many tissues (muscles particularly) to absorb glucose through their membranes and out of the blood; it is hypoglycemic.

Gluconeogenesis: The formation of sugar from amino acids in the liver.

Glucose: A sugar in our blood and a source of energy for our bodies.

Glucose-6-Phosphate: Dehydrogenase An enzyme that performs an essential function in the metabolism of carbohydrate. A deficiency in this enzyme – acronym G6PD – results in the breakdown of erythrocytes (haemolysis), usually in the presence of oxidants such as infections or drugs.

Glucose-Tolerance Test: A way of assessing the body's efficiency at metabolising Glucose. The test is used in the diagnosis of Diabetes Mellitus. The patient is starved for up to 16 hours, after which he or she is fed glucose by mouth. The concentrations of glucose in the blood and urine are then meas- ured at half-hour intervals over a period of two hours.

Glucoside: A plant compound containing glucose and another substance (the bioactive part). A special-case glycoside.

Glue Ear: Another name for secretory otitis media

Glutaminase: An enzyme occurring in the kidneys which catalyses the breakdown of glutamine to ammonia – a phase in the production of the metabolic waste product, urea.

Gluten: The constituent of wheat-flour which forms an adhesive substance on addition of water, and allows the 'raising' of bread. It can be separated from the starch of flour, and being of a protein nature is used to make bread for those diabetics

Glycoside: A plant compound containing one or more alcohols or sugars and a biologically active compound. The sugar part is called a glycone, the other stuff is called an aglycone. The important things to remember about some glycosides is that they may pass through much of the intestinal tract, with the hydrolysis of the molecule only occurring in the brush borders of the small intestine. The result is that the bioactive part, the aglycone, is absorbed directly into the bloodstream, and is often not floating around the intestinal tract contents at all.

Glycosuria: The presence of sugar in the urine. By far the most common cause of glycosuria is Diabetes Mellitus, but it may also occur as a result of a lowered renal threshold for sugar when it is called renal glycosuria, and is not indicative of disease. Measurements of the amounts of sugar in the urine is a standard method used by patients (and health professionals) to assess the stability of treatment for diabetes mellitus, indicating whether adjustment is required in the hypoglycaemic (sugar-lowering) agents they are taking.

Glycosylated Haemoglobin: (HbA1c) This forms a small proportion of the total haemoglobin in the blood. It differs from the major component, HbA, in that it has a glucose group attached. The rate of synthesis of HbA1c is a function of the blood-glucose concentra- tion, and since it accumulates throughout the life span of the red blood cell – normally 120 days – the concentration of HbA1c is related to the mean blood-glucose concentration over the past 3–4 months. It is thus a useful indicator of medium-term diabetic control a good target range would be a concentration of 5–8 per cent. When interpreting the HbA1c level, however, it is important to remember that wide fluctuations in blood- glucose concentration, together with anaemia or a reduced erythrocytes life span, may give misleading results.

GMP: Good manufacturing practice

Gnotobio: Specially reared laboratory animal whose microflora and microfauna are specifically known in their entirety.

Gnotobiota: Specifically and entirely known microfauna and microflora of a specially reared laboratory animal.

Goblet Cell: A columnar secretory cell occurring in the epithelium of the respiratory and intestinal tracts. The cells produce the main constituents of mucus.

Goekerman Regimen: A psoriasis treatment using coal tar together with UVB phototherapy.

Goiter, Exophthalmic: The physical -

Symptoms often associated with Grave's disease or thyrotoxicosis, with an inflamed, sometimes enlarged thyroid gland and, most noticeably, protruding eyes.

Goitre: Goitre is a term applied to a swelling in the front of the neck caused by an enlargement of the thyroid gland. The thyroid lies between the skin and the front of the windpipe and in health is not large enough to be seen. The four main varieties of goitre are the simple goitre, the nodular, the lymphadenoid goitre and the toxic goitre.

Gold Carecard: A card issued bv the Ministry of Health Services to each British Columbia resident registered with the Medical Services Plan who is 65 years of age or older. The CareCard shows the cardholder's unique Personal Health Number.

Golfer's Elbow: A term applied to a condition comparable to tennis elbow. It is not uncommon in the left elbow of right-handed golfers who catch the head of their club in the ground when making a duff shot.

Gonad: A gland which produces a gamete – that is, an ovary or a testis.

Gonadorelin: A hormone that stimulates the pituitary gland to secrete three hormones: gonadotrophic, luteinising and follicle-stimulating. Gonadorelin can be made artificially and given by intravenous injection. It is used to stimulate the ovaries when treating infertile women, and to investigate suspected disease of the hypothalamus. Analogues of the hormone (buserelin and goserelin) are chemically similar and can be used to suppress release of gonadorelin, so cutting the production of pituitary hormones. The two analogues are given to treat

Gonadotropic: Pertaining to effects on sex glands and on the systems that regulate them.

Gonorrhoea: Gonorrhoea is an inflammatory disease caused by Neisseria gonococcous, affecting especially the mucous membrane of the urethrA in the male and that of the vagina in the female, but spreading also to other parts. It is the most common of the sexually transmitted diseases (STD).

Good Agricultural Practice in the Use of Pesticides: Nationally authorised safe uses of pesticides under actual conditions necessary for effective and reliable pest control. It encompasses a range of levels of pesticide applications up to the highest authorised use, applied in a manner that leaves a residue which is the smallest amount practicable.

Good Clinical Practice (GCP): A standard for the design, conduct, performance, monitoring, auditing, recording, analyses, and reporting of clinical trials that provides assurance that the data and reported results are credible and accurate, and that the rights, integrity, and confidentiality of trial subjects are protected.

Good Laboratory Practice Principles: Fundamental rules incorporated in national regulations concerned with the process of effective organization and the conditions under which laboratory studies are properly planned, performed, monitored, recorded, and reported.

Good Manufacturing Practice (GMP) Principles: Fundamental rules incorporated in national regulations concerned with the process of effective organization of production and ensuring standards of defined quality at all stages of production, distribution and marketing; minimization of waste and its proper disposal are part of this process.

Good Medical Practice: Guidelines for doctors on the provision of good medical care laid down by the general medical council (GMC).

Good Operating Practice: A procedural, administrative, or institutional measure that can be used to prevent pollution.

Gout: A disease that causes episodes of acute arthritis and inflammatory swelling in one or more joints. Gout usually starts in a well-used, oft traumatized joint like the right big toe or knee, and usually starts in the night, during the time that Traditional Chinese Medicine calls "liver hour," 2:00 to 4:00 A.M. (allowing for daylight saving time). The inflammation is caused by uric acid crystals that have lodged in the joint's white blood cells and results from the condition called hyperuricemia.

Graded Effect: Consequence that can be measured on a graded scale of intensity or severity and its magnitude related directly to the dose or concentration of the substance producing it.

Graft Versus Host Disease: A condition that is a common complication of bone marrow transplant. It results from certain lymphocytes in the transplanted marrow attacking the trans- plant recipient's tissues, which they identify as 'foreign'. GVHD may appear soon after a transplant or develop several months later. The condition, which is fatal in about a third of victims, may be prevented by immunosuppressant drugs such as ciclosporin.

Graft: The term applied to a piece of tissue removed from one person or animal and implanted in another, or the same, individual in order to remedy some defect. Skin grafts are commonly used, and artificial skin for grafting has recently been developed. Bone grafts are also used to replace bone which has been lost by disease: for example, a portion of rib is sometimes removed in order to furnish support for a spine weakened by disease, after removal of the damaged bone.

Gram's Stain: Bacteria can be stained-with an iodine-based chemical dye called Gram's stain (after the scientist who discovered the technique). Different bacteria react differently to exposure to the stain. Broadly, the bacterial specimens are stained first with gentian violet, then with Gram's stain, and finally counterstained with a red dye after a decolorising process. Bacteria that retain the gentian stain are called gram-positive; those that lose it but absorb the red stain are called gram-negative. Some species of staphylococcus, streptococcus and clostridium are gram-positive, whereas salmonella and Vibrio cholerae are gram-negative.

Gram-Positive/Negative: Gram's Method is a staining procedure that separates bacteria into those that stain (positive) and those that don't (negative). Gram- positive bugs cause such lovely things as scarlet fever, tetanus, and anthrax, while some of the gram negs can give you cholera, plague, and the clap. This is significant to the microbiologist and the pathologist; otherwise I wouldn't worry. Still, knowing the specifics (toss in anaerobes and aerobes as well), you can impress real medical professionals with your knowledge of the secret, arcane language of medicine.

Grand Mal: An out-of-date and now colloquial name for a tonic-clonic seizure (in contrast to petit mal).

Granisetron: One of several serotonin antagonists used to treat nausea and vomiting induced by cytotoxic chemotherapy.

Grantor: The person who transfers assets into a trust for the benefit of another.

Granulations: Small masses of formative cells containing loops of newly formed blood vessels which spring up over any raw surface, as the first step in the process of healing of wounds.

Granulocytes: These are a group of white blood cells that have many and well- pigmented granules, and derive from the bone marrow myeloblasts. The granules are sources of digestive, immunologic, and inflammatory proteins. The classic granulocytes are neutrophils, eosinophils, and basophils, but one should also include mast cells. Also, macrophages, which start out as agranulocytic monocytes but get lots of granules when they grow up.

Granuloma: A non-malignant or new growth made up of granulation tissue. This is caused by various forms of chronic inflammation, such as syphilis and tuberculosis.

Gravel: The name applied to any sediment which precipitates in the urine, but particularly to small crystal masses of uric acid. It produces dysuria and other urinary symptoms.

Gravid: Pregnant.

Greenstick Fracture: An incomplete fracture, in which the bone is not completely broken across. It occurs in the long bones of children and is usually due to indirect force.

Grey Matter: Those parts of the brain and spinal cord that comprise mainly the interconnected and tightly packed nuclei of neurons (nerve cells). The tissue is darker than that of the white matter, which is made of axons from the nerve cells. In the brain, grey matter is mainly found in the outer layers of the cerebrum, which is the zone responsible for advanced mental functions. The inner core of the spinal cord is made up of grey matter.

Grievance Process: The process by which members may appeal coverage decisions.

Groin: The region which includes the upper part of the front of the thigh and lower part of the abdomen. A deep groove runs obliquely across it, which corresponds to the inguinal ligament, and divides the thigh from the abdomen. The principal diseased conditions in this region are enlarged glands, and hernia.

Grommet: A small bobbin-shaped tube used to keep open the incision made in the ear drum in the treatment of secretory otitis media. It acts as a ventilation tube by allowing the Eustachian tube to recover its normal function. The operation is now less commonly performed than 20 years ago.

Gross Estate: The total property or assets held by an individual as defined for federal estate tax purposes.

Ground Treatment of Plants: Dusting or spraying of plants with pesticides by hand, by special machines, or by apparatus fixed to tractors or driven by them.

Group Health Coverage: A health benefits plan that covers a group of people as permitted by state and federal law.

Group Plan Sponsor: A group plan sponsor is the intermediary that represents a group of people in the context of a group insurance contract. It may be an employer, a professional order or association, a union or a group of employees.

Group Therapy: Psychotherapy in which at least two, but more commonly up to ten, patients, as well as the therapist, take part. The therapist encourages the patients to analyse their own and the others' emotional and psychological difficulties. Group therapy is also used to help patients sharing the same condition – for instance, alcoholism or compulsive gambling. They discuss their problems for perhaps an hour twice a week and explore ways of resolving them.

Growing Pains: Ill-defined discomfort and pains that occur in G the limbs of some children. They occur mainly at night between the ages of 6–12 years. The cause is unknown, but the condition is not significant and does not require treatment once other more important conditions have been ruled out.

Growth Hormone (GH): A hormone secreted by the pituitary gland. GH stimulates growth and repair of the body as well as the activities of the immune system. With age, GH release diminishes (also known as GH or human growth hormone).

Growth: A popular term applied to any new formation in any part of the body.

GSA: Graduate Students' Association, the name of the student government body for UCSF graduate students

GU: Genital-urinary tract of particular application to males.

Guaranteed Income Supplement (GIS): The Guaranteed Income Supplement (GIS) is an amount added to the Old Age Security Pension and is paid at the same time as that

Guardian: An individual legally appointed to manage the rights and/or property of a person incapable of taking care of his or her own affairs.

Guest Privileges: A CIGNA HealthCare

benefits option that, under certain circumstances, can provide an in-network level of coverage to covered persons who are temporarily outside of their "home" network service area for 90 days or longer.

Guideline Value: Quantitative measure (a concentration or a number) of a constituent of an environmental medium that ensures aesthetically pleasing air, water, or food and does not result in a significant risk to the user.

Guides to Air Quality: Sets of atmospheric concentrations and exposure times that are associated with specific effects of varying degrees of pollution on man, animals, vegetation, and the environment in general.

Guillain-Barré Syndrome: A disease of the peripheral nerves causing weakness and numbness in the limbs. It customarily occurs up to three weeks after an infection – for example, campylobacter infection of the gastrointestinal tract provoking an allergic response in the nerves. It may begin with weakness of the legs and gradually spread up the body. In the worst cases the patient may become totally paralysed and require to be artificially ventilated. Despite this, recovery is the

Guinea-pig Maximisation Test: Widely used skin test for screening possible contact allergens :considered to be a useful method to identify likely moderate and strong sensitizers in humans.

Gulf War Syndrome: A collection of varying symptoms, such as persistent tiredness, headaches, muscle pain and poor concentration, reported by members of the Coalition Armed Forces who served in the 1991 Gulf War. Whilst there is strong evidence for a health effect related to service, there is no evidence of a particular set of signs and symptoms (the definition of a 'syndrome') unique to those who served in the Gulf War. Symptoms have been blamed on multiple possible hazards, such as exposure to depleted uranium munitions, smoke from oil-well fires and use of pesticides. However, the only clearly demonstrated association is with the particular pattern of vaccinations used to protect against biological weapons. Many conflicts in the past have generated their own 'syndromes', given names such as effort syndrome and shell-shock, suggesting a link to the psychological stress of being in the midst of warfare.

Gullet: Gullet, or oesophagus, is the tube down which food passes from the throat to the stomach.

Gum: The soft tissue surrounding the teeth. Also called the gingiva, this tissue protects under- lying structures and helps to keep the teeth tightly in the jawbones. Inflammation of the gum is called gingivitis and may be due to dental plaque or to systemic disorders such as leukaemia and scurvy.

Gumboil: A painful condition of inflammation, ending sometimes as an abscess, situated in the gum about the root of a carious tooth.

Gumma: A hard swelling, or granuloma, characteristic of tertiary syphilis. It normally develops in the skin or subcutaneous tissue, mucous membranes or submucosa, and the long bones.

Guttate Psoriasis: Characterized by small, drop-shaped lesions that appear on the trunk, limbs, and scalp. Guttate psoriasis is most often triggered by upper respiratory infections. pension to certain persons age 65 and over. Persons may receive the maximum GIS, a partial GIS or no GIS. The amount of the Guaranteed Income Supplement depends on the person's income.

Gynaecology: The branch of medicine dealing with the female pelvic and urogenital organs, in both the normal and diseased states. It encompasses aspects of contraception, abortion, and in vitro or IVF.

Gynaecomastia: An abnormal increase in size of the male breast.

Gyrus: Gyrus is the term applied to a convolution of the brain.

These are caused when there are mutations or other abnormalities which disrupt the code of a gene or set of genes. These are divided into autosomal (one of the 44 chromosomes which are not sex-linked), dominant, autosomal recessive, sex-linked and polygenic disorders.

H

Habit: A behavioural response or practice that is established by the individual frequently repeating the same act. The process is called habituation, and the more a person is exposed to a particular stimulus, the less is he or she aroused by it. People may also become habituated to certain drugs, requiring more and more of a substance to produce the same effect – a process known as tolerance.

Haem: An iron-containing porphyrin compound that combines with the protein globulin to make haemoglobin, a constituent of erythrocytes (red blood cells).

Haemangioma: These can be acquired or congenital. The acquired type presents as a red papule which bleeds easily; treatment is normally by cautery. A 'strawberry naevus' is a 'capillary- cavernous' haemangioma appearing at or soon after birth, which may grow to a large size. Treatment is not usually required, as most of them fade although this may take a few years. Where a haemangioma is disfiguring or interfering with vision or breathing, treatment is necessary: this may be by laser, by using corticosteroids or interferon treatment, or by surgery.

Haemarthrosis: Haemarthrosis is the process of bleeding into, or the presence of blood in, a joint. It may occur as a result of major trauma (for example, fracture of the patella may lead to bleeding into the knee-joint), or, more commonly, following minor trauma. It may even occur spontaneously, in cases of haemophilia or other disorders of blood clotting. If repeated several times, haemarthrosis may lead to fibrosis of the joint-lining and inflammation of the cartilage, causing marked stiffness and deformity.

Haematemesis: Haematemesis means the vomiting of blood. Blood brought up from the stomach is generally dark in colour and may have been so far digested as to form small brown granules resembling coffee grounds. Vomiting of blood is one of the main symptoms of peptic ulcer, but it may occur in gastritis, from varicocoeles in the oesophagus, or, rarely, in cancer of the stomach. Gastritis caused by an irritant poison, sustained intake of alcohol, or the regular use of certain drugs such as aspirin and non-steroidal anti- inflammatory drugs (NSAIDS) may cause bleeding. Blood may also originate from the nose and throat, be swallowed and then vomited. Persistent haematemesis or a sudden severe bleed is a potentially serious medical emergency and the patient should be referred urgently to hospital.

Haematemesis: Vomiting of blood.

Haematinic: A drug that raises the quantity of haemoglobin in the blood. Ferrous sulphate is a common example of iron-containing compounds given to anaemic patients whose condition is due to iron deficiency. Trad- itionally, haematinics have been used to prevent anaemia in pregnant women, but nowadays a maternal diet containing iron-rich foods and regular antenatal checks of haemoglobin con- centrations in the blood should make the rou- tine use of haematinics unnecessary.

Haematocoele: A cavity containing blood. Generally as the result of an injury which ruptures blood vessels, blood is effused into one of the natural cavities of the body, or among loose cellular tissue, producing a haematocoele.

Haematocolpos: The condition in which menstrual blood cannot drain from the vagina because of an imperforate hymen.

Haematocrit: Also known as packed cell volume, this is an expression of the fraction of blood volume occupied by the erythrocytes. It is determined by centrifuging a sample of blood in a capillary tube and measuring the height of the resulting packed cells as a percentage of the total sample height.

Haematogenous: An adjective applied to a biological process which produces blood, or to an agent produced in or coming from blood. For example, a haematogenous infection is one resulting from contact with blood that contains a virus or bacterium responsible for the infection.

Haematologist: A doctor or scientist who specialises in the study and treatment of blood and blood disorders.

Haematologist: A doctor who specializes in blood and bone marrow disorders.

Haematology: The study of diseases of the blood.

Haematoma: Haematoma means a collection of blood forming a definite swelling. It is found often upon the head of newborn children after a protracted and difficult labour (cephalhaematoma). It may occur as the result of any injury or operation.

Haematuria: Blood in the urine. The blood may come from any part of the urinary tract. When the blood comes from the kidney or upper part of the urinary tract, it is usually mixed throughout the urine, giving the latter a brownish or smoky tinge. This condition is usually the result of glomerulonephritis, or it may be present in persons suffering from high blood pressure or pyelitis. Blood may also appear in the urine when a stone or gravel is present in the pelvis of the kidney setting up irritation, especially after exercise. The blood may also originate from a bladder that is inflamed or infected or which contains benign growths (papilloma) or malignant growths. Inflammation or injury to the urethra can also cause haematuria. Someone with haematuria should seek medical advice.

Haematuria: Presence of blood in the urine.

Haemic Murmur: Unusual sounds heard over the heart and large blood vessels in severe cases of anaemia. They disappear as the condition improves.

Haemochromatosis: A disease in which cirrhosis of the liver, enlargement of the spleen, pigmentation of the skin, and diabetes mellitus are associated with the abnormal and excessive deposit in the organs of the body of the iron-containing pigment, haemosiderin. It is caused by an increase in the amount of iron absorbed from the gastrointestinal tract.

Haemodialysis: A method of removing waste products or poisons from the circulating blood using the principle of dialysis. The procedure is used on patients with malfunctioning or non-functioning kidneys. It is done using an artificial kidney or dialyser which restores blood to its normal state. The process has to be repeated, sometimes for many months, until a donor kidney is available for transplantation to replace the patient's failing one.

Haemofiltration: A technique similar to haemodialysis. Blood

Haemoglobin: The colouring compound which produces the red colour of blood. Haemoglobin is a chromo- protein, made up of a protein called globin and the iron-containing pigment, haemin. When separated from the red blood corpuscles – each of which contains about 600 million haemoglobin molecules – it is crystalline in form. Haemoglobin exists in two forms: simple haemoglobin, found in venous blood; and oxyhaemoglobin, which is a loose compound with oxygen, found in arterial blood after the blood has come into contact with the air in the lungs. This oxyhaemoglobin is again broken down as the blood passes through the tissues, which take up the oxygen for their own use. This is the main function of haemoglobin: to act as a car- rier of oxygen from the lungs to all the tissues of the body. When the haemoglobin leaves the lungs, it is 97 per cent saturated with oxygen; when it comes back to the lungs in the venous blood, it is 70 per cent saturated. The oxygen content of 100 millilitres of blood leaving the lungs is 19.5 millilitres, and that of venous blood returning to the lungs, 14.5 millilitres. Thus, each 100 millilitres of blood delivers 5 millilitres of oxygen to the tissues of the body. Human male blood contains 13–18 grams of

Haemoglobinopathies: Abnormal haemoglobin formation occurs in the haemoglobinopathies, which are hereditary haemolytic anaemias, genetically determined and related to race. The haemoglobin may be abnormal because: (1) there is a defect in the synthesis of normal adult haemoglobin as in thalassaemia, when there may be an absence of one or both of the polypeptide chains characteristic of normal adult haemoglobin; or (2) H there is an abnormal form of haemoglobin such as haemoglobin S which results in sickle-cell

Haemoglobinuria: The presence of blood pigment in the urine caused by the destruction of blood corpuscles in the blood vessels or in the urinary passages. It turns urine a dark red or brown colour. In some people this condition, known as intermittent haemoglobinuria, occurs from time to time, especially on exposure to cold. It is also pro- duced by various poisonous substances taken in the food. It occurs in malarious districts in the form of one of the most fatal forms of malaria: blackwater fever.

Haemolysin: Substance that damages the membrane of erythrocytes causing the release of haemoglobin.

Haemolysis: The destruction of red blood corpuscles by the action of poisonous substances, usually of a protein nature, circulating in the blood, or by certain chemicals. It occurs, for example, gradually in some forms of anaemia and rapidly in poisoning by snake venom.

Haemolytic Disease: Newborn A potentially serious disease of the newborn, characterised by haemolytic anaemia (excessive destruction of red blood cells) and jaundice. If severe, it may be obvious before birth because the baby becomes very oedematous and develops heart failure – so-called hydrops fetalis. It may first present on the first day of life as jaundice and anaemia.

Haemolytic Uraemic Syndrome: A disease of children resulting in acute renal failure. A febrile illness of the gastrointestinal or respiratory tracts is followed by intravascular coagulation of blood which results in haemolysis, anaemia, thrombocytopaenia.

Haemoperfusion: Passing blood through a column of charcoal or adsorbent resin for the removal of drugs or toxins.

Haemopericardium: The presence of blood in the pericardium, the membranous sac which surrounds the heart. The condition may result from a myocardial infarction, leaking aneurysm, injury, or tumour. Because the pericardial blood compresses the heart, the latter's pumping action is impeded, reducing the blood pressure and causing cardiac failure. Urgent surgical drainage of the blood may be required.

Haemophilia: An inherited disorder of blood coagulation which results in prolonged bleeding even after minor injury. There is a deficiency of factor VIII, an essential clotting factor in the coagula- tion cascade – the complex series of bio- chemical events that leads from injury of the wall of a blood vessel to the formation of a blood clot that checks bleeding. Haemophilia is

Haemopoiesis: The formation of blood cells and platelets – a continuous process throughout life. As ageing cells are removed from the circulation, new ones, generated in the bonemarrow, replace them.

Haemopoietic Stem Cell: This is the basic cell from which all types of blood cells originate. Its appearance is believed to be similar to that of a lymphocyte.

Haemoptysis: The coughing-up of blood from the lungs. The blood is usually bright red and frothy, thus distinguishing it from blood brought up from the stomach. It is a potentially serious sign of lung disease, although in elderly people haemoptysis may be due to a varicose condition of the small veins in the throat. In young people this condition is often due to bleeding from the nose, in which, owing to the position of the head, the

Haemorrhage: The escape of blood from any of the blood vessels, normally in response to some trauma, or as a result of a clotting disorder such as haemophilia. The bleeding may be external for example, following a skin laceration; or it may be internal – for example, haematemesis (bleeding into the stomach), haemoptysis (bleeding from the lungs), or haematuria (bleeding from the kidneys or urinary tract). For more information about these conditions, see separate entries. Bleeding into or around the brain is a major

Haemorrhoids: Haemorrhoids, or piles, are varicose (swollen) veins in the lining of the anus. They are very common, affecting nearly half of the UK population at some time in their lives, with men having them more often and for a longer time.

Haemosiderin: Iron-containing pigment that is formed from haemoglobin released during the disintegration of red

Haemosiderosis: An increase in the amount of iron stored in the body. Rarely, it may be due to ingestion of too much iron, but a more likely cause is repeated blood transfusions. The extra iron may affect the function of the heart and liver.

Haemostasis: The process by which bleeding stops. It involves constriction of blood vessels, the formation of a platelet plug, and blood clotting. The term is also used for surgical interventions to stop bleeding – for example, the use of diathermy.

Haemostatics: A group of drugs used to treat bleeding disorders such as haemophilia. Factor VIII is one of the clotting factors available for treatment: preparations of it are injected after abnormal bleeding or before surgery. Vitamin K preparations are another haemostatic group used to treat an overdose of anticoagulants. Haemostatic preparations of gelatine and cel- lulose are used to stem bleeding from the skin and gums, or as a result of tooth extractions.

Haemothorax: An effusion of blood into the pleural cavity.

Half Life: The time taken for the plasma concentration of an administered drug to decline by half as a result of redistribution, metabolism and excretion.

Halfway House: A residential home for mentally ill individuals where they can live under supervision after dis- charge from hospital. They may be fit to work but cannot manage an independent life.

Halitosis: Bad breath. This may be a sign of illness – for example, lung disease or sinusitis; and a person with Diabetes Mellitus may have breath tainted with acetone if his or her sugar metabolism is poorly controlled. Usually, however, halitosis is caused by smoking, drinking alcohol,

Hallucinations: False perceptions arising without an adequate external stimulus, as opposed to illusions, which are misinterpretations of stimuli arising from an external object. Hallucinations come from 'within', although the affected individual may see them as coming from 'without'. Nevertheless, they may occur at the same time as real perceptions, and may affect any sense (vision, hearing, smell, taste, touch, etc.).

Hallucinogens: Compounds characterised by their ability to produce distortions of perception, emotional changes, depersonalisation, and a variety of effects on memory and learned behaviour. They include cannabis, lysergic acid diethyl- amide (LSD) and Mescaline.

Hallux Rigidus: Stiffness of the joint between the great toe and the foot, which induces pain on walking. It is usually due to a crush injury or stubbing of the toe. Such stubbing is liable to occur in adoles- cents with a congenitally long toe. If trouble-

Hallux Valgus: Outward displacement of the great toe – always associated with a bunion. It is due to the pressure of footwear on an unduly broad foot. In adolescents, this broad foot is inherited; in adults it is due to splaying of the foot as a result of loss of muscle tone. The bunion is produced by pressure of the footwear on the protruding base of the toe. In mild cases the wearing of comfortable shoes may be all that is needed. In more severe cases the bunion may need to be removed, while in the most severe the operation of arthroplasty may be needed.

Hallux: The anatomical name of the great toe.

Halon: Halogen-containing compounds with long atmospheric lifetimes whose breakdown in the stratosphere cause depletion of the ozone layer that protects humans from harmful rays of the sun.

Halothane Hepatitis: A very rare form of hepatitis following exposure to Halothane during anaesthesia (1:35,000 halothane anaesthetics).

Jaundice develops three to four days after exposure and will occasionally develop into a fatal massive hepatic necrosis. It is of unknown aetiology but probably has an immunological basis. It is more common following multiple exposures in a short time (less than 28 days), and in obesity, middle age and females. It is rare in children.

Hamartoma: These are benign tumours, usually in the lung, containing normal components of pulmonary tissue such as smooth muscle and connective tissue.

Hamate: One of the carpal bones of the hand that articulate between the metatarsal bones and the radius and ulnar bones of the forearm.

Hammer-Toe: The deformity in which there is permanent flexion, or bending, of the middle joint of the toe. The condition may affect all the toes, as in claw-foot; more commonly it affects one toe, usually the second. It is due to a relatively long toe and the pressure on it of the footwear. A painful bunion usually develops on it. In mild cases, relief is obtained by protecting the toe with adhesive pads. If this does not suffice then an operation is necessary.

Hamstrings: The name given to the tendons at the back of the knee – two on the inner side and one on the outer side – which bend this joint. They are attached to the tibia below. Strains, or 'pulls', and tears of the hamstring are common in people taking part in strenuous sports such as athletics, football, rugby and tennis.

Hand: In structure, the hand has a bony basis of eight small carpal bones in the wrist, five metacarpal bones in the fleshy part of the hand, and three phalanges in each finger – two only in the thumb. From the muscles of the forearm, 12 strong tendons run in front of the wrist. Of these, nine go to the fingers and thumb and are bound down by a strong band, the flexor retinaculum, in front of the wrist. They are enclosed in a complicated synovial sheath, and pass through the palm and down the fingers. Behind the wrist, 12 tendons likewise cross from forearm to hand.

Handicap: The inability to carry out a social, occupational or other activity that could normally be done by a person without such a handicap. The handicap may be partial or complete, physical or mental, and may result from disease, injury or inherited disorder. The extent of the handicap is measured against the normal function of those in a peer group.

Hanging: Hanging is a form of death due to suspension of the body from the neck – either suddenly, as in judicial hanging (although not in the United Kingdom, as the death penalty is not used in the UK justice system), so as to damage the spinal column and cord, or in such a way as to constrict the air passages and the blood vessels to the brain. Death is, in any case, speedy, resulting in

Hang-Nail: A splitting of the skin (cuticle) at the side of a fingernail. In manual workers it is usually caused by trauma, but ischaemia of the fingers may predispose to the condition. Secondary infection and inflammation may make hang-nail a very tender condition; treatment consists of reducing trauma, and the use of emollients and antibiotics if necessary.

Hantavirus: A group of viruses that infect mice, rats and voles and can also infect humans who come into contact with the excreta or secretions of these animals. Widely distributed in Asia, the USA and Europe, in Britain hantavirus usually affects rural and sewage workers, as well as people engaged in watersports. Many victims have a mild feverish illness; severe cases are characterised by headache, high temperature, nausea, vomiting and even shock, accompanied by skin petechiae. The kidneys and sometimes the lungs are affected, and in severely affected patients mortality is high.

Haploid (Monoploid): State in which a cell contains only one set of chromosomes.

Haploid: An adjective describing organisms, cells or nuclei that have a single set of unpaired chromosomes. Human beings have haploid gametes following meiosis.

Hapten: Low-molecular-weight molecule that contains an antigenic determinant (epitope) that may bind to a specific antibody but which is not itself antigenic unless complexed with an antigenic carrier such as a protein or cell; once bound it can cause the sensitization of lymphocytes, possibly leading to allergy or cell-mediated hypersensitivity.

Hard Water: The term applied to water that contains a large amount of calcium and magnesium salts (lime salts). These form an insoluble curd with soap and thus interfere with the use of the water for washing. Hard water is especially found in districts where the soil is chalky. T

Harm: Damage or adverse effect to a population, species, individual organism, organ, tissue or cell.

Harmful Occupational Factor: Component of the work environment the effect of which on a worker under certain conditions leads to ill health or reduction of working ability.

Harmful Substance: Substance that, following contact with an organism can cause ill health or adverse effects either at the time of exposure or later in the life of the present and future generations.

Harmonized Standard: A European Norm (EN) that has been accepted by all Member States and has been published in the Official Journal of the European Communities (OJEC).

Hartmann's Solution: A solution commonly used as a means of fluid replacement in dehydrated patients.

Each litre contains 3•1 grams of sodium lactate, 6 grams of sodium chloride, 0•4 grams of potassium chloride, and 0•7 grams of calcium chloride.has been circulating in many of the abdominal organs.

Hashimoto's Disease: A condition in which the whole of the thyroid gland is diffusely enlarged and firm. It is one of the diseases produced by autoimmunity. The enlargement is due to diffuse infiltration of lymphocytes and increase of fibrous tissue. This form of goitre appears in middle-aged women, does not give rise to symptoms of thyrotoxicosis, and tends to produce myxoedema.

Haversian Canals: The fine canals in bone which carry the blood vessels, lymphatics and nerves necessary for the maintenance and repair of bone.Having the qualities of both sexes. The term is used to describe people who are sexually attracted to both men and women.

Hay Fever: Also known as allergic rhinitis, this is caused by an allergy to the pollen of grasses, trees and other plants. Contact with the particular pollen to which the sufferer is allergic causes Histamine release, resulting in a blocked, runny nose and itchy, watering eyes. It affects approximately 3 million people each year in the UK.

Hazard Assessment: Determination of factors controlling the likely effects of a hazard such as the dose-effect and dose- response relationships, variations in target susceptibility, and mechanism of toxicity.

Hazard Communication Standard: US OSHA standard requiring all employers to inform employees of the hazard of substances in the workplace and the steps necessary to avoid harm.

Hazard Evaluation: Establishment of a qualitative or quantitative relationship between hazard and benefit, involving the complex process of determining the significance of the identified hazard and balancing this against identifiable benefit: this may subsequently be developed into a risk evaluation.

Hazard Identification: Determination of substances of concern, their adverse effects, target populations, and conditions of exposure, taking into account toxicity data and knowledge of effects on human health, other organisms and their environment.

Hazard Quotient (HQ): Ratio of toxicant exposure (estimated or measured) to a reference value regarded as corresponding to a threshold of toxicity : if the total hazard quotient from all toxicants to a target exceeds unity, the combination of toxicants may produce (will produce under assumptions of additivity) an adverse effect.

Hazard: Set of inherent properties of a substance, mixture of substances or a process involving substances that, under production, usage or disposal conditions, make it capable of causing adverse effects to organisms or the environment, depending on the degree of exposure; in other words, it is a source of danger.

Hazardous Material (HAZMAT): Any material that poses a threat to human health and/or the environment due to its

Hazardous Material Pharmacy: A program that places central control on hazardous materials and waste management for an installation.

Hazardous Material Pollution Prevention Plan: Typically a plan including actions and milestones, responsibilities, regulations, and other variations under an appropriate formal, issued implementing document.

Hazardous Production Factor: Production factor the effect of which on a worker under certain conditions results in injury or some impairment of health. Blood cells and that accumulates in individuals who have ingested excess iron.

Hazardous Waste: A hazardous waste is a solid waste which because of its quantity, concentration, or characteristics may cause an increase in mortality or serious irreversible illness or pose a substantial hazard to human health or the environment when improperly treated, stored, transported, disposed of, or otherwise managed.

Hazmin: Minimization of hazardous-materials and wastes through improved management and planning practices.

HCPCS: Health Care Common Procedure Coding System. A listing of services, procedures, and supplies offered by physicians and other providers. HCPCS includes Current Procedural Terminology (CPT) codes, national alphanumeric codes, and local alphanumeric codes. The national codes are developed by the CMS to supplement CPT codes.

HCUP (HealthCare Cost and Utilization Project): HCUP is a family of healthcare databases and related software. HCUP databases bring together the data collection efforts of state data organizations, hospital associations, private data organizations, and the federal government to create a national information resource of patient-level healthcare data.

HDL (High-Density Lipoprotein): A combined protein and fatlike substance. Low in cholesterol, it usually passes freely through the arteries. Sometimes called "good cholesterol."

Head Injury: Any injury to the head, whether associated with a skull fracture or not. Patients with head injuries should be assessed for signs of neurological damage, which may not develop at once. Patients who after a head injury are or have been unconscious or who are drowsy, vomiting, confused or have any focal neurological signs – for example, blurred vision or a motor or sensory malfunction – should be seen by a doctor.

Headache: A very common condition which may vary considerably in severity, type, significance and cause. At one extreme, headache may indicate the presence of a tumour or meningitis, while at the other it may merely indicate a common cold or tiredness. Even so, persistent or recurrent headaches should always be taken seriously. Although the brain itself is insensitive to pain, the surrounding membranes – meninges – are very sensitive, and changes in intracranial arteries, or spasm of the neck or scalp muscles, which may occur for various reasons, may cause considerable pain. In most cases a clinical diagnosis should be possible; further investigations should only be necessary following head injury,

Heaf Test: A skin test to find out if a person is immune to tuberculosis. Tuberculin (a preparation derived from the tubercle bacillus) is injected via punctures in the skin of the forearm, using a spring-loaded gunlike instrument with six very short needles set in circular form. A positive test is indicated by a red raised reaction of the skin: this means that the subject is immune. If the result is negative, the subject can be given BCG Vaccine.

Health Action Zones (HAZ): Strategies aiming to improve health services by forming new partnerships and - reducing poor health

Health Care Financing Administration (HCFA): The federal agency responsible for administering Medicare and federal participation in Medicaid.

Health Care Providers: Health professionals, such as physicians, pharmacists, registered nurses, nurse practitioners, and dentists, who provide patient care.

Health Centre: An imprecise term that may refer to a group of doctors and supporting staff operating from one building owned or leased by the doctors, or to a building owned or leased by a health authority that contains staff or services from one or more sections of the National Health Service.

Health Check: Program which provides Medicaid-eligible children under age 21 with regular health screenings.

Health Delivery: Direct delivery'of health care to patients.

Health Economy: Health authority and the primary care organisations, main NHS trusts, and social services department(s) that are co-terminous with it.

Health Education: The process of educating the public to adopt a healthy lifestyle and abandon dangerous or unhealthy behaviour. With the rising cost of health care and the increasing amount of illness and injury resulting from preventable causes, health professionals, governments and the World Health Organisation are strongly supporting more and better health education for everyone from schoolchildren to the elderly. Information on all aspects of health education in the United Kingdom can be obtained from

Health Hazard: Any factor or exposure that may adversely affect health.

Health Level 7 (HL7) Messaging: An application protocol for electronic data exchange in health care environments. The HL7 protocol is a collection of standard formats that specify the implementation of interfaces between computer applications from different vendors. This communication protocol allows health care institutions to exchange key sets of data between different application systems.

Health Level Seven (HL7): A data interchange transaction protocol for health care technology applications that simplifies the ability of different vendor-supplied IS systems to assure interoperability. Although not a software program in itself, HL7 requires that each healthcare software vendor program HL7 interfaces for its products.

Health Maintenance Organization (HMO): An organization that arranges a wide spectrum of health care services which commonly include hospital care, physicians' services and many other kinds of health care services with an emphasis on - preventive care.

Health Maintenance Services: Any health care service or program that helps maintain a person's good health. Health maintenance services include all standard preventive medical practices, such as immunizations and periodic examinations, as well as health education and special self-help programs.

Health Plan: A term that has different meanings depending upon the context. "Health plan" can be used to mean an HMO, a health benefits plan provided by an employer to its employees, or a health benefits plan offered to employers by an insurer or third party administrator.

Health Policy Research: Studies of the health care system and health policy-making processes

Health Policy: Plans and actions of government and other organizations designed to maintain and improve public health

Health Promotion: A surveillance plan based on a community, and aimed at maintaining an optimum state of health and quality of life for those who make up the community – both as individuals and as a group. Programmes would include immunization targets, health education and screening tests, as well as environmental procedures such as monitoring atmosphere, improving housing, water and food supplies, and encouraging safe and healthy working practices.

Health Reimbursement Arrangement (HRA): An arrangement in which the participant is reimbursed for covered health expenses by his/her employer up to a predetermined amount.

Health Resources and Services Administration (HRSA): HRSA is an agency within the Department of Health and Human Services. Its mission is to improve and expand access to quality health care for all.

Health Savings Account (HSA): A reimbursement account in which the participant pays for health costs through a fully insured, tax-exempt savings account. Employees or employers or both fund the account.

Health Services Research: Studies of the organization, delivery, and financing of health care

Health Surveillance: Periodic medico-physiological examinations of exposed workers with the objective of protecting health and preventing occupationally related disease.

Health: The state of health implies much more than freedom from disease, and good health may be defined as the attainment and maintenance of the highest state of mental and bodily vigour of which any given individual is capable. Environment, including living and working conditions, plays an important part in deter- mining a person's health. The UK government is now placing much greater emphasis on health promotion and the prevention of disease, and has published national targets for reducing the incidence of some major diseases.

Health-based Exposure Limit: Maximum concentration or intensity of exposure that can be tolerated without significant effect (based on only scientific and not economic evidence concerning exposure levels and associated health effects).

Healthcare Flexible Spending Account: Allows a person to set aside pre-tax dollars to pay for healthcare-related services and costs not covered by insurance plans. Eligible expenses include over-the-counter drugs for colds, flu, and pain, as well as eyeglasses and contact lenses.

Health-Care Priorities: As the needs and demands of patients, and the costs of health care of populations, have risen sharply in recent years, governments and health-care providers – whether tax-funded, insurance-based, employer-provided or a mix of these – have had increasingly to face the dilemma of what services a country or a community can afford to provide. As a result, various techniques for deciding priorities of care and treatment are evolving. In the United Kingdom, priorities were for many years based on the decisions of individual clinicians who had wide freedom to.

Healthy Volunteer: A healthy person who agrees to participate in a clinical trial for reasons other than medical and receives no direct health benefit from participating. See also human subject.

Healthy Worker Effect: Epidemiological phenomenon observed initially in studies of occupational diseases: workers usually exhibit lower overall disease and death rates than the general population, due to the fact that the old, severely ill and disabled are ordinarily excluded from employment. Death rates in the general population may be inappropriate for comparison, if this effect is not taken into account.

Hearing Aids: Nearly two-thirds of people aged over 70 have some degree of hearing impairment. Hearing aids are no substitute for definitive treatment of the underlying cause of poor hearing, so examination by an ear, nose and throat surgeon and an audiologist is sensible before a hearing aid is issued (and is essential before one can be given through the NHS). The choice of aid depends on the age, manipulative skills, and degree of hearing impairment of the patient and the underlying cause of the deafness. The choice of hearing aid for a deaf child is particularly important, as impaired hearing can hinder speech development. Electronic aids consist, essentially, of a microphone, an amplifier, and an earphone.

Heart Attack: Damage to the heart-muscle caused when the blood vessels supplying the muscle are blocked, such as when the blood vessels are clogged with fats (a condition sometimes called hardening of the arteries).

Heart, Artificial: A mechanical device in the chest that enhances or takes over the pumping action of the heart, thus maintaining the necessary level of circulation of blood through the lungs and other body structures. An artificial heart was first used in humans in 1985 and the three types in use are: an intra-aortic balloon pump, driven by com- pressed air, which inflates a balloon in the aorta with every heartbeat, increasing the volume of circulating blood; an electrical device that assists the left ventricle by pumping blood into the abdominal aorta; and a mechanical artificial heart that replaces a diseased heart that has been removed. As yet there is no artificial heart suitable for long-term use. Existing devices are intended to tide over a patient who is extremely ill until a live heart can be trans- planted from a donor. The results from artificial hearts have been disappointing because of complications and also because the patients have usually been already dangerously ill.

Heart: A hollow muscular pump with four cavities, each provided at its outlet with a valve, whose function is to maintain the circulation of the blood. The two upper cavities are known as atria; the two lower ones as ventricles. The term auricle is applied to the ear-shaped tip of the atrium on each side.

Heartburn: A burning sensation experienced in the region of the heart and up the back to the throat. It is caused by an excessive acidity of the gastric juice – often aggravated by dietary indiscretions – and is relieved temporarily by taking alkaline substances, such as 1•2 grams of bicarbonate of soda, or commercial preparations such as aluminium-containing antacid tablets (e.g. Aluminium Hydroxide Tablets) or prescribed drugs such as lansoprazole. The frequency of heartburn attacks can be lessened by eating a sensible diet, avoiding fatty or indigestible foods and restricting alcohol consumption. Persistent heartburn may be a sign of more serious oesophageal disease and medical advice should be sought.

Heart-Lung Machine: A device that temporarily takes over the function of the heart and lungs. It is used in certain operations in the chest, giving the surgeon more time for operations such as open-heart surgery, heart transplants and heart-lung transplants. The machine also ensures an operating area largely free of blood, which helps the surgeon to work more quickly. A pump replaces the heart

Heart-Lung Transplant: An operation in which a patient's diseased lungs and heart are removed and replaced with donor organs from someone who has been certified as 'brain dead. As well as the technical difficulties of such an operation, rejection by the recipient's tissues of donated heart and lungs has proved hard to overcome. Since the early 1990s, however,

Heat balance: A loose term applied to a special form of energy balance frequently used in processes that are associated with gaseous pollutant control by combustion (heat balance/ material balance).

Heat Cramps: Painful cramps in the muscles occurring in workers, such as stokers, who labour in hot conditions. The cramps are the result of loss of salt in the sweat, and can be cured by giving the sufferer salty water to drink.

Heat Spots: A vague term applied to small inflamed and congested areas which appear especially upon the skin of the face, neck and chest or other parts of the body in warm weather.

Heat Stroke: A condition resulting from environmental temperatures which are too high for compensation by the body's thermo-regulatory mechanism(s). It is characterised by hyperpyrexia, nausea, headache, thirst, confusion, and dry skin. If untreated, coma and death ensue.

Hebephrenia: A form of schizophrenia that comes on in youth and is marked by depression and gradual failure of mental faculties with egotistic and self-centred delusions.

Heberden's Nodes: Small hard knobs which appear at the sides of the last phalanges of the fingers in people who have osteoarthritis.

Heel: The heel is the hind part of the foot, formed by the calcaneus and the especially thick skin covering it. It is not subject to many diseases. Severe pain in the heel is sometimes a sign of gout or rheumatism.

Heliotherapy: Sunbathing; the exposure of the body to sunlight to promote healing.

Helium: This is the lightest gas known, with the exception of hydrogen. This property renders it of value in anaesthesia, as its addition to the anaesthetic means that it can be inhaled with less effort by the patient. Thus it can be used in the presence of any obstruction to the entry of air to the lungs.

HELLP Syndrome: A type of severe preeclampsia (a disorder affecting some pregnant women) that affects various systems in the body. Haemolysis, raised concentration of the enzymes in the liver, and a low blood platelet count are among the characteristics (and explain the name HELLP); patients are acutely ill and immediate termination of pregnancy is necessary.

Helminths: Another name for parasitic worms such as flukes, tapeworms and nematodes.

Hematuria: The presence of blood in-the urine.

Hemianaesthesia: Loss of the sense of touch down one side of the body.

Hemianopia: A term meaning loss of half the usual area of vision. The affected person may see everything clearly to the left or to the right, the field of vision stopping abruptly at the middle line; they may see things only when straight ahead of them; or, thirdly, they may see objects far out on both sides, although there is a wide area straight in front for which they are quite blind.

Hemiatrophy: Atrophy of one side of the body, or of part of the body on one side: for example, facial hemiatrophy, in which one-half of the face is smaller than the other, either in the course of development or as a result of some nervous disorder.

Hemiballismus: Involuntary movements similar to choreiform movements, but of much greater amplitude and force. The violent, throwing movements of the limbs are usually unilateral, and tend to occur acutely as a result of vascular damage to the mid-brain.

Hemicolectomy: An operation to remove the right or left half of the colon, usually with end-to-end anastomosis of the remaining portion of the intestine. This is often used for the treatment of malignant or inflammatory diseases of the colon.

Hemicrania: A headache limited to one side of the head.

Hemimelia: This consists of defects in the distal part of the extremities: for example, the absence of a forearm or hand. Hemimelia is a congenital defect; large numbers of cases resulted from the administration of thalidomide during pregnancy

Hemiparesis: Paralysis affecting the muscles of one side of the body. This most commonly follows a stroke and occurs when parts of the brain serving motor function on the opposite side of the body are damaged.

Hemiplegia: Paralysis that is limited to one side of the body.

Hemolysis: Breakdown of senescent red blood cells into recyclable constituents, with particular importance given to the reuse of the heme part of hemoglobin.

Hemolytic: Promoting the breakdown of red blood cells; a normal process, hectic and skillfully balanced, the term is usually applied to excess conditions or toxic substances that degrade the bonds between healthy red blood cells and their hemoglobin coat or cause the liver and spleen to hypercatabolize otherwise healthy erythrocytes.

Hemopathy: A disease of the blood.

Hemoptysis: Coughing up blood or pulmonary bleeding. If simply resulting from excessive coughing, where bleeding is from prolonged tracheal or pharynx irritation and minute mucosal hemorrhage, it can be self-treatable...anything else and start worrying

Hemorrhage: Bleeding, pure and simply. Menses is not blood but the carefully orchestrated excretion of excess endometrium. If the membranes fail to vasoconstrict and bleed further, that is hemorrhage.

Hemorrhoids: Enlarged veins protruding into the anorectal area, either internal or externally visible. They are either the result of poor sphincter tone and portal congestion, or sphincter hypertonicity, skeletal muscle and adrenergic excess.

Hemostatic: A substance that stops or slows bleeding, used either internally or externally

Henle Loop: That part of the nephron between the proximal and distal convoluted tubules. It extends into the renal medulla as a hairpin- shaped loop. The ascending link of the loop actively transports sodium from the lumen of the tube to the interstitium, and this, combined with the 'counter-current' flow of fluid through the two limbs of the loop, plays a part in concentrating the urine.

Henoch-Schonlein Purpura: This is an inflammatory condition of the small blood vessels, the cause of which is not known but may be an allergic response to food or drugs. Most common among young children, the inflammation causes blood to leak into joints, kidneys, intestine and skin. The child presents with a purpuric rash and stomach pains which may come and go for weeks. Paracetamol alone is often sufficient to alleviate the condition, but severely ill patients may need corticosteroid drugs. All sufferers need follow- up for 12 months to ensure that they have not developed kidney disease.

Henry's Law Constant: At constant temperature and pressure, the ratio of the partial pressure of a gas above a liquid to its molal solubility in the liquid and therefore a measure of its partition between the gas phase and the solute phase.

Heparin: Heparin is one of the naturally produced anticoagulants with a rapid effect, which is thought to act by neutralising thrombin. Inactive when taken orally, it is normally given intravenously – it may be given for a few days, combined with an oral anticoagulant such as warfarin, to initiate anti-coagulation. Low-dose heparin may be given by subcutaneous injection for longer periods, for the prophylaxis of deep vein thrombosis (DVT) or Pulmonary Embolism in 'high-risk' patients, such as those with obesity or a history of thrombosis, or post-operatively. If haemorrhage occurs, withdrawal of heparin is usually sufficient, but protamine sulphate is a rapidly active and specific antidote. Prolonged treatment with heparin may cause osteoporosis.

Hepatectomy: The operation for removal of the liver, or part of it.

Hepatic Encephalopathy: A neuropsychiatric syndrome caused by disease of the liver, and occurring most often in patients with cirrhosis it also occurs in acute form in acute failure of liver function. The disorder is believed to be the result of biochemical disturbance of brain function, because the condition is reversible and pathological changes in brain tis- sue are rarely found. The patient's intellect, personality, emotions and consciousness are altered but neurological signs may or may not be identified. Apathy, confusion, drowsiness, some- times convulsions, speech disturbance and eventually coma mark the progress of the condition. The principles of treatment are to remove the precipitating causes. These include: uraemia; sedative, antidepressant and hypnotic drugs; gastrointestinal bleeding; too much pro- tein in the diet; infection; and trauma (including surgical operations).

Hepatic: Pertaining to the liver.

Hepatitis: Inflammation of the liver which damages liver cells and may ultimately kill them. Acute injury of the liver is usually followed by complete recovery, but prolonged inflammation after injury may result in fibrosis and cirrhosis. Excluding trauma, hepatitis has several causes:

Hepatocyte: The main cell type present in the liver. A large cell, it has several important metabolic functions: these include synthesis and storage of biochemical products; detoxification of poisons and unwanted substances; and the manufacture of BILE, the liver secretion that passes through the bile ducts to the small intestine and helps in the digestion of fat.

Hepatoma: A primary malignant tumour of liver cells. It has marked geographical variation, being most common in parts of Africa and the Far East. It is

Hepatomegaly: An enlarged liver. Hepatosplenomegaly is both an enlarged liver and spleen. Hepatosplenopalestrinamegaly is an enlarged liver, spleen and 17th century Italian composer.

Hepatotoxic: Poisonous to liver cells.

Herbal Medicine: The use of herbs as medicines is probably as old as mankind; every culture has its own tradi- tions. Herbalism was formally established in England by an Act of Parliament during Henry VIII's reign. Different parts of a variety of plants are used to treat symptoms and to restore functions.

Herbicide: Substance intended to kill plants.

Heredity: The principle on which various peculiarities of bodily form or structure, or of physical or mental activity, are transmitted from parents to off- spring.

Hermaphrodite: An individual in whom both ovarian and testicular tissue is present. Hermaphrodites may have a testis on one side and an ovary on the other; or an ovotestis on one side and an ovary or testis on the other; or there may be an ovotestis on both sides. Both gonads are usually intra-abdominal. The true hermaphrodite usually has a uterus and at least one Fallopian tube on the side of the ovary, and on the side of the testis there is usually a vas deferens. Most true hermaphrodites are raised as males, but external virilisation is not usually

Hernia: The protrusion of an organ, or part of an organ, through the wall of the cavity that normally contains it. The most common types of hernia involve the organs of the abdomen which can herniate externally through the abdominal wall, or internally usually through a defect in the diaphragm. External hernias appear as a swelling, covered with skin, which bulges out on coughing or straining but which can normally be made to disappear with gentle pressure.

Herniorraphy: Surgical repair of a Hernia. This may be done as an open operation or as Minimally Invasive Surgery (MIS) using a Laparoscope.

Herniotomy: The surgical removal of the sac of connective tissue surrounding a hernia. In children or healthy young adults with an inguinal hernia, a herniotomy is usually sufficient to cure the condition.

Heroin: Also known as diacetyl morphine or diamorphine, this Class A controlled drug is an opiate – a group which includes morphine, codeine, pethidine and methadone. It is a powerful analgesic and cough suppressant, but its capacity to produce euphoria rapidly induces dependence. Popular with addicts, its mostly pleasant effects soon produce tolerance; the need to inject the drug, with

Herpangina: Herpangina is a short febrile illness in which minute vesicles or punched-out ulcers develop in the posterior parts of the mouth. It is due to infection with the group A coxsackie viruses.

Herpes: A small group of capsid-forming DNA viruses, sometimes divided into Type I (forming vesicles and blisters on the mouth, lips-generally above the waist) and Type II (usually sexually transmitted, with symptoms mostly below the waist). Both types form acute initial outbreaks, go dormant, reactivate, and so forth. For most folks, frequent outbreaks are clear signs of stress or immunosuppression.

Hertz: The SI (International System of Units) unit of frequency. It indicates the number of cycles per second (c/s). The abbreviation for hertz is Hz.

Heterograft: A transplant from one animal to another of a different species. It is also known as a xenograft.

Heterologous. Consisting of different elements or of elements in differing

proportions.

Heterosexual: Sexual attraction to individuals of the opposite sex.

Heterozygous: An individual having dissimilar members of the pair of genes coding for a given characteristic

HHS: Department of Health and Human Services (U.S., also called DHHS)

Hiatus Hernia: A displacement of a portion of the stomach through the opening in the diaphragm through which the oesophagus passes from the chest to the abdominal.

Hiccup: An involuntary spasmodic contraction of the diaphragm which produces an indrawing of breath during which there is a sudden closure of the vocal cords. This results in the well-known sound and sensation. It is usually of benign cause (e.g. indigestion) but may be a symptom of medullary brain damage, uraemia, typhoid fever or encephalitiS lethargica. There are many folk remedies for hiccups, but most cases subside spontaneously. Prolonged hiccups due to disease may respond to treatment with chlorpromazine or haloperidol.

High Blood Glucose: A condition that occurs in people with diabetes when their blood glucose levels are too high. Symptoms include having to urinate often, being very thirsty, and losing weight.

High Blood Pressure: A condition where the blood circulates through the arteries with too much force. High blood ressure tires the heart, harms the arteries, and increases the risk of heart attack, stroke, and kidney problems.

High Dependency Unit: A hospital unit equipped and staffed to nurse patients who require a high level of technically supported care. Patients are usually moved to such units when they have made satisfactory progress in an Intensive Therapy Unit (ITU) and do not require the one-to-one nursing necessary in ITUs. Patients who have under- gone major surgery are often transferred from the recovery ward to a high dependency unit until they are well enough to be cared for in a standard ward.

Hilum: A term applied to the depression on organs such as the lung, kidney, and spleen, at which the vessels and nerves enter it and around which the lymphatic glands cluster. The hilum of the lung is also known as its root.

Hip: That part of the body on each side of the pelvis where it articulates with the head of the femur

HIPAA: Health Insurance Portability and Accountability Act of 1996. Established to improve portability and continuity of health insurance coverage in group and individual markets, combat waste, fraud, and abuse in health insurance and health care delivery, improve access to long-term care services and coverage, and simplify the administration of health insurance, among other goals.

Hip-Joint: The joint formed by the head of the thigh bone and the deep, cup-shaped hollow on the side of the pelvis which receives it (acetabulum). The joint is of the ball-and-socket variety, is dislocated only by very great violence, and is correspondingly difficult to reduce to its natural

Hippocampus: An area of the brain believed responsible for memory and personality.

Hippocrates: A famous Greek physician who lived from c.460 to 377 BC and who taught students at the medical school in Cos. Often called the 'father of medicine', he is renowned for drawing up the Hippocratic Oath, some of which may have been derived from the ancient oath of the Aesclepiads. Apart from his oath, Hippocrates has about 60 other medical works attributed to him, forming a corpus which was collected around 250 BC in the famous library of Alexandria in Egypt.

Hippus: Hippus is a tremor of the iris which produces alternating contraction and dilatation of the pupil. Hirschsprung's Disease Hirschsprung's disease, or megacolon, is a rare congenital disorder characterised by great hypertrophy and dilatation of the colon.

Hirsutism: The growth of hair of the male type and distribution in women. It is due either to the excess production of androgens, or to undue sensitivity of the hair follicle to normal female levels of circulating androgens. The latter is called idiopathic hirsutism, because the cause is unknown. The increased production of androgens in the female may come from the ovary and be due to polycystic ovary syndrome or an ovarian tumour, or the excess androgen may come from the adrenal cortex and be the result of congenital adrenal hyperplasia, an adrenal tumour or cushing's syndrome.

Histamine: Histamine: The defense substance responsible for most inflammation. It is synthesized from the amino acid histidine and is secreted by mast cells, basophils, and blood platelets. It stimulates vasodilation, capillary permeability, muscle contraction of the bronchioles, secretions of a number of glands, and attracts eosinophils, the white blood cells that are capable of moderating the inflammation. Mast cell histamine release is what usually causes allergies.

Histogenic Origin: Germ cell layer of the embryo from which a given adult tissue develops.

Histology: Study (usually microscopic) of the anatomy of tissues and their cellular and subcellular structure.

Histopathology: Microscopic

pathological study of the anatomy and cell structure of tissues in disease to reveal abnormal or adverse structural changes.

Histoplasmosis: A disease due to a yeast-like fungus known as Histoplasma capsulatum. Most cases have been reported from the USA. In infants it is characterised by fever, anaemia, enlargement of the liver and spleen, and involvement of the lungs and gastrointestinal tract. In older children it may resemble pulmonary tuberculosis, whilst in adults it may be confined to involvement of the skin.

HIV: Human immunodeficiency virus, the retrovirus that is at least partially responsible for AIDS. At this time it is not clear what other disorders besides AIDS may come from HIV infections. AIDS is a syndrome, partially (perhaps totally) produced by HIV. As with EBV, it is quite possible that the virus may cause only moderate immunosuppression in some people, while in others it will progress further to AIDS.

Human Immunodeficiency Virus (HIV) Anatomy

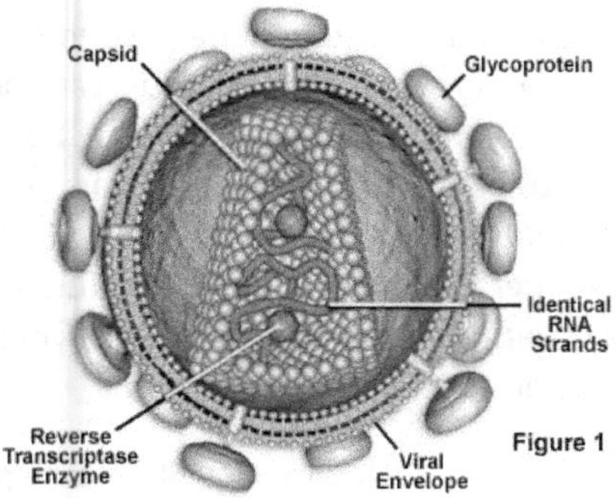

Figure 1

Hives: A popular term applied to eruptions of urticaria.

HLA (Human Leukocyte Antigen): A series of proteins expressed on cell surfaces which is used to evaluate tissue compatibility for the possibility of transplantation

Holistic: A term used for a method of medical care in which patients are treated as a whole, and which takes into account their physical and mental state as well as social background rather than just treating the disease alone.

Home Health Care: Health services rendered in the home to an individual who is confined to the home. Such services are provided to individuals who do not need institutional care, but who need nursing services or therapy, medical supplies and special outpatient services.

Homeopathy: Almost two centuries old, it is a system of medicine in which the treatment of disease (symptom pictures) depends on the administration of minute doses (attenuations) of substances that would, in larger doses, produce the same symptoms as the disease being treated.

Homeostasis: Normal, internal stability in an organism maintained by co-ordinated responses of the organ systems that automatically compensate for environmental changes.

Homocystinuria: Homocystinuria is a congenital disease due to the inability of the affected individual to metabolise, or to utilise properly, one of the essential amino acids known as methionine. The main features of the condition are abnormality of the lens of the eye, learning disability, and a tendency to thromboses (blood clots).

Homograft: A piece of tissue or an organ, such as a kidney, transplanted from one animal to another of the same species: for example, from person to per- son. It is also known as an allograft.

Homology: Degree of identity existing between the nucleotide sequences of two related but not complementary DNA or RNA molecules; 70 % homology means that on the average 70 out of every 100 nucleotides are identical in a given sequence. The same term is used in comparing the amino acid sequences of related proteins.

Homosexuality: Sexual activity with a member of the same sex. There has been considerable debate among psychiatrists as to whether homosexuality should be regarded as a normal sexual variant or as a psycho-pathological development or deviation. Although homosexuality is found in virtually every society and culture, there is no society in which it is the predominant or preferred mode of sexual activity.

Homozygous: An individual having identical members of the pair of genes coding for a given characteristic.

Honeymoon Cystitis: Urethral

irritation from excess sexual activity or as a famous French writer described it,"the plentiful rubbing together of bacons".

Hormesis: Stimulatory effect of small doses of a potentially toxic substance that is inhibitory in larger doses.

Hormone: A chemical that special cells in the body release to help other cells work. For example, insulin is a hormone made in the pancreas to help the body use glucose as energy.

Horner's Syndrome: This is the description given to a combination of changes resulting from paralysis of the sym- pathetic nerve in the neck. They are: a small pupil; a drooping upper lid; and an apparently (though not actually) sunken eye.

Hospice: A hospital that cares only for the terminally ill and dying. The emphasis is on providing qual- ity of life, and special care is taken in providing pain relief by whichever methods are deemed best suited to the person's needs. Hospice care in the United Kingdom has been greatly developed, in particular with the leadership of Dr (Dame) Cicely Saunders. Many hospices are funded by charitable funds and their activities supported by voluntary workers.

Hospital Chaplaincy: A service provided by a religious denomination, primarily aimed at meeting the spiritual and religious needs of patients in hospitals. In the UK, NHS hospital trusts employ both full-time and part-time chaplains, usually representing the mainline Christian churches (Anglican, Free and Roman Catholic). Their duties vary but always at the least involve meeting the specifically religious needs of patients as well as of relatives and staff who may ask for help. Public services in chapels, the bedside administration of the Word and Sacraments, and prayers and radio services are among chaplains' duties. When requested by patients, chaplains also liaise with representatives of other world faiths. Chaplains have a broad responsibility for the spiritual health care of all in hospital.

Hospital Information System (HIS): The common term for the computer hardware and software that provides the support of the hospital.

Hospital: An institution providing treatment for sick and injured persons. This may be done on an inpatient or outpatient basis. A hospital provides investigative and therapeutic services which are not available on a domiciliary basis.

Hospital-Acquired Infection: An infection acquired by a patient while in hospital. Because of the high level of antibiotic use in hospitals, some bacteria become resistant – for example, Methicillin-Resistant Staphylococcus Aureus (MRSA). This makes hospital-acquired infections potentially dangerous and sometimes life-threatening, and is one of the developments that is prompting calls for greater care in the prescribing of antibiotics as well as higher standards of cleanliness.

Hospitalization: To be put in a hospital to be treated or cared for.

Host: The main computer in a system of computers or terminals connected by communication links, in this case Pharma Net.

Hour-Glass Stomach: The term given to the X-ray appearance of a stomach which is constricted in its middle part due either to spasm of the stomach muscle or to contraction of scar tissue from a gastric ulcer.

Housemaid's Knee: An inflammation of the bursa in front of the knee-cap, often mistaken for some disease in the joint itself.

HSE: The building letter code for the Health Sciences Instruction and Research East Tower, on the UCSF Parnassus campus.

HSR&D: Health Services Research & Development.

HUB: In a network, a device joining communication lines at a central location, providing a common connection to all devices on the network.

Human Chorionic Gonadotrophin: A glycoprotein hormone secreted by the placenta in early pregnancy, and stimulating the corpus luteum within the ovary to secrete oestrogens, progesterone, and relaxin. The hormone is essential for the maintenance of pregnancy up to about 6–8 weeks of gestation. A radioimmunoassay can be used to detect its presence, and pregnancy can be diagnosed as early as six days after conception by testing for it in the urine. Some tumours also secrete human chorionic gonadotrophin, particularly hydatidiform mole, which produces large amounts.

Human Ecology: Interrelationship between humans and the entire environment - physical, biological, socio-economic, and cultural, including the interrelationships between individual humans or groups of humans and other human groups or groups of other species.

Human Equivalent Dose: Human dose of an agent that is believed to induce the same magnitude of a toxic effect that the known animal dose has induced.

Human genome project: National effort coordinated by the U.S. Department of Energy and the National Institutes of Health to identify all the approximately 100,000 genes in human DNA, determine the sequences of the 3 billion chemical bases that make up DNA, store this information in databases, develop tools for data analysis, and address the ethical, legal, and social issues that may arise from the project; to help achieve these goals, researchers are also studying the genetic makeup of several nonhuman organisms.

Human Genome: In simple terms, this is the genetic recipe for making a human being. Genome is a combination of the words gene and chromosome, and a genome is defined as all the genetic material – known as deoxyribonucleic acid, or DNA – in a cell. Most genes encode sequences of amino acids, the constituents of proteins, thus initiating and controlling the replication of an organism.

Human Subject: A human subject, as defined in 21 CFR 50.3, is an "individual who is or becomes a participant in research, either as a recipient of the test article or as a control. A subject may be either a healthy human or a patient." Synonym: subject/trial subject.

Humidification: The air we breathe-must be moist for the efficient working of the lungs. Humidification, or moistening of the air, is achieved largely by the nose, which acts as an air-conditioner, warming, moistening and filtering the 10,000 litres of air which we inhale daily – in the process of which, incidentally, it produces around 1•5 litres of secretion daily. Humidity is expressed as relative humidity (RH). This is the amount of moisture in the air expressed as a percentage of the maximum possible at that temperature.

Humidifier Fever: A form of alveolitis caused by contamination of the water used to humidify, or moisten, the air in air-conditioning plants. The breathing of the contaminated air results in infection of the lung, which is characterised by fever, cough, shortness of breath and malaise – worse on Monday and tending to improve during the

Humoral Factors: Molecules found in the flu id of the body (e.g. blood or lymph)

Humour: An archaic term once used for a theory regarding the causation of disease as due to an improper mixture in the body of blood, bile, phlegm and black bile. The term remains in describing some body fluids, such as the aqueous and vitreous humours of the eye.

Hunger: A craving for food or other substance necessary to bodily activity. Hunger for food is supposed to be directly produced by strong contractions of the stomach which occur when it is empty, or nearly so.

Huntington Disease: An hereditary disorder characterized by mental and physical deterioration that ultimately leads to death. It is sometimes referred to as Hungtinton's chorea due to the involuntary rapid movement of limbs (chorea), which are a symptom of the disease. The characteristic symptoms of the disease are caused by loss of neurons (nerve cells) in the brain. Huntington's is caused by a fa ulty gene known as HD, which is located on chromosome 4. Diagnosis is by genetic testing. Currently, there is no cure, although medication may be used to control symptoms of the illness.

Huntington's Chorea: A hereditary disease characterised by involuntary movements and dementia. Each child of a parent with the disease has a 50:50 chance of developing it. Onset is most common between the ages of 35 and 45, but 10 per cent of cases occur under the age of 20. Some patients show more severe mental disturbance; others more severe disturbances of movement; but in all it pursues an inexorable downward course over a period of 10–20 years to a terminal state of physical and mental helplessness.

Huriet Law: France's regulations covering the initiation and conduct of clinical trials.

Hutchinson's Teeth: The term applied to the narrowed and notched permanent incisor teeth which occur in congenital syphilis. They are so-named after Sir Jonathan Hutchinson (1828–1913), the London physician who first described them.

Hyaline Membrane Disease: A form of acute respiratory distress syndrome (ARDS) found in premature infants and some of those born by caesarean section, characterised by the onset of difficulty in breathing a few hours after birth. Most require extra oxygen and many need mechanical ventilation for a few days or even weeks. Recovery is the rule, although the most severely affected may die or suffer damage from oxygen lack. In

Hyaline: Tissue material that has a glass-like appearance when stained and viewed under the icroscope. It occurs in a variety of tissues and diseases, particularly acute respiratory distress syndrome (ARDS), hyaline degeneration of arterioles, and alcoholic liver disease.

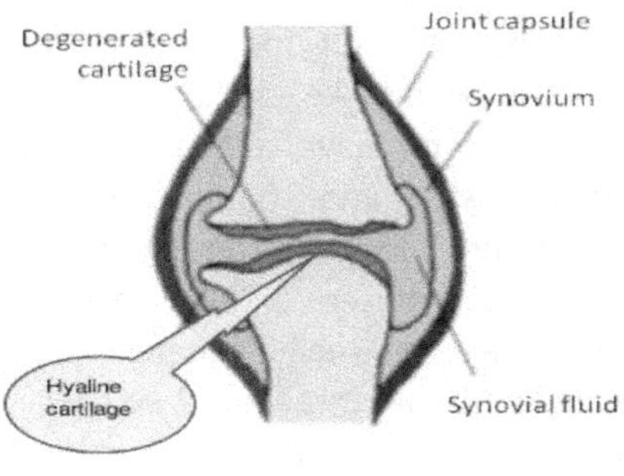

Hyaluronidase: An enzyme made by traumatized cartilage (to soften and regenerate itself when injured) sperm cells (to dissolve the protective layer

around an ovum), the spleen (to speed up hydrolysis), added to an IM injection (so it doesn't get surrounded by connective tissue and never disperse) and produced by some really nasty bacteria so they can dissolve connective tissue and get deep into the body. Hyaluronic acid is the target, and it is a basic mucopolysacharide rivet, keeping large masses of polymerized compounds in the state of constant colloid jello (or more technically, a hydrogel facilitant). Hyaluronidase is an enzyme which hydrolyses hyaluronic acid. The latter is a gel-like sub- stance which is widely distributed throughout the body and which helps to bind together the tissue cells and also acts as a lubricant in joints. By virtue of its action in hydrolysing hyaluronic acid, hyaluronidase is now used in subcutaneous injections of fluid as it facilitates the spread of the injected fluid and therefore its absorption.

Hybrid: This is produced by a crossfertilization between two species. This happensa lot more often than botanists would like, since a species is presumed to have distinct genetic characteristics and shouldn't do this hybridizing thing as often as it does. Most of the dozen or so species of Silk Tassel are really genetically the same, and the three hundred species of Aconite worldwide are all capable of hybridizing as well.

Hydatid: A cyst produced by the growth of immature forms of a - tapeworm.

Hydatidiform Mole: A rare complication of pregnancy, in which there is tremendous proliferation of the epithelium of the chorion (the outer of the two fetal membranes). It seldom occurs during a first pregnancy. Treatment consists of immediate evacuation of the womb.

Hydradenitis Suppurativa: A chronic inflammatory disease of the apocrine sweat glands. It is more common in women – in whom it usually occurs in the armpit – than in men, in whom it is most common in the perineum of the drivers of lorries and taxis. It occurs in the form of painful, tender lumps-

underneath the skin, which burst

Hydramnios: A condition characterised by excess of fluid in the amniotic cavity.

Hydration therapy: The replacement of fluids and electrolytes to the body.

Hydrocele: An organized mass of serous or lymphatic fluid, usually encapsulated by connective tissue. An internal blister. The term is usually applied to a hydrocele of the testes, but a breast cyst is also a hydrocele.

Hydrocephalus: H An abnormal accumulation of cerebrospinal fluid, or CSF, within the skull, as a result of one or more of three main causes: (i) excessive CSF production; (ii) defective CSF absorption; (iii) blockage of the circulation of CSF. Such disturbances in the circulation of the fluid may be due to congenital reasons (most commonly associated with spina bifida), to meningitis, or to a tumour.

Hydrocoele: A collection of fluid connected with the testis or spermatic cord. When there is no obvious cause, it is classified as primary: such hydrocoeles are usually large and tense, and are commonly found in middle-aged and younger men, presenting as a large, painless scrotal swelling. Congenital hydrocoeles may occur in infants, when they are often associated with a hernial sac. Hydrocoele of the cord is rare. Secondary hydrocoele is generally smaller and lax; it is usually secondary to

Hydrophilic: Describing the character of a molecule or atomic group which has an affinity for water.

Hydrophobia: Another name for rabies.

Hydrophobic: Describing the character of a molecule or atomic group which is insoluble in water, or resistant to wetting or hydration.

Hydrotherapy: Treatment using water in the form of baths, douches, etc.

Hydrothorax: A collection of fluid in the pleural cavities of the lungs.

Hydroxyurea: A medication that may-

be combined with PUVA or UVB treatments to treat psoriasis.

Hygiene: The science of health and the study of ways of its preservation, particularly by widespread education and promotion of cleanliness. Especially valuable in developing countries, where it plays a vital role in helping to limit the spread of infectious diseases.

Hymen: The thin membranous fold partially closing the lower end of the virginal vagina. If the opening is small, the hymen will tear at the time of first intercourse, usually with a little bleeding.

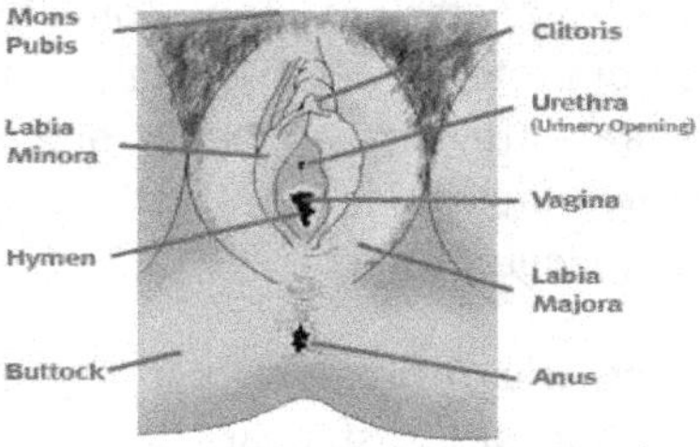

Female Genitals

Hyoid: A U-shaped bone at the root of the tongue. The hyoid can be felt from the front of the neck, lying about 2•5 cm above the prominence of the thyroid cartilage.

Hyper Text Markup Language (HTML): A set of codes that describes the way type, graphics, and other elements are displayed on a Web page.

Hyper-: Prefix denoting abnormally increased or excessive.

Hyperactivity: A pattern of behaviour, usually in children, characterised by inability to by overactivity.

Hyperacusis: Hyperacusis means an abnormally acute sense of hearing.

Hyperaemia: Congestion or presence of an excessive amount of blood in a body part.

Hyperaesthesia: Oversensitivity of a part of the body – as found, for example, in certain neurological diseases such as Herpes Zoster or shingles.

Hyperalgesia: Excessive sensitivity to pain; see also touch.

Hyperalimentation: The administration of protein, carbohydrates, and fat via an intravenous tube in order to provide nutrition to a patient who cannot ingest food orally.

Hyperbaric: A pressure that is greater than that of the standard atmosphere at sea level (1,013 millibars). Hyperbaric oxygenation is a procedure in which the patient is exposed to high-pressure oxygen. The technique is used for the treatment of people suffering from carbon monoxide (CO) poisoning, compressed-air illness.

Hyperbilirubinaemia: Excessive concentration of bilirubin in the blood.

Hypercalcaemia: Excessive concentration of calcium in the blood.

Hypercalciuria: An abnormally large amount of calcium in the urine. It is the most common single cause of stones in the kidneys in Britain.

Hypercapnia: An abnormal increase in the amount of carbon dioxide in the blood or in the lungs. It may be caused by a reduced respiratory rate or effort, diseases of the chest wall and lung (affecting breathing), and cyanotic heart disease.

Hyperchlorhydria: Excessive production of hydrochloric acid in the stomach. It is a characteristic finding in certain forms of dyspepsia, particularly that associated with a duodenal ulcer, and causes heartburn and waterbrash.

Hypercortical: Overly anabolic; used here to describe the constitutional, not pathologic state

Hyperemesis Gravidarum: A rare condition (less than 0•2 per cent) of pregnancy, in which there is severe vomiting. If untreated it can result in severe dehydration, ketoacidosis and liver damage. More common in multiple pregnancy, it may recur in subsequent pregnancies.

Hyperemia: Excessive presence of blood, usually arterial; and the resultant increase in heat and metabolic rate. Hyperemia can be a pathology, blowing out blood vessels and the like; used here to describe the chronic or subclinical condition of functional vascular excess and excitation.

Hyperextensions: The excessive extension of a limb or joint, usually followed by pain and some inflammation.

Hypergluconeogenesis: The state of excessive synthesis of glycogen (storage starch) or glucose by the liver, derived from non-sugar sources, such as amino acids, lactate and the glycerol remnants from triglyceride breakdown. In strictly subclinical terms it signifies a yinny, catabolic excess, wherein building materials are less desirable than fuel, and it is singularly difficult to buff up in any way.

Hyperglycaemia: An excess of sugar in the blood that may occur in various diseases, typically in Diabetes Mellitus. The normal blood glucose level in the fasting state is between 3.5 and 5.5 mmol/1 blood; four or five times that amount is found in diabetes, owing to insufficient insulin in the blood, possibly accompanied by an excessive carbohydrate intake. Untreated, it may lead to diabetic coma.

Hyperglycogenolysis: The tendency, usually by the liver, to convert glycogen into glucose at too rapid a rate for metabolic needs.

Hyperidrosis: Excessive sweating.

Hyperkalaemia: A concentration of potassium in the plasma that is above the normal range. It is often caused by renal failure or by excessive intake of potassium – perhaps in a drug – and may cause cardiac dysrhythmia (abnormal rhythm of the heart).

Hyperkalemia: Excessive concentration of potassium in the blood.

Hyperkeratosis: Thickening of the horny (outer) layer of skin, affecting the palms of the hands and soles of the feet. The disorder may be inherited.

Hyperkinetic: Too physically active, jittery, peripatetic.

Hyperlipidaemia: An excess of fat in the blood, characterising a group of metabolic disorders. The two most important fats circulating in the blood are cholesterol and triglyceride. Raised blood levels of cholesterol predispose to atheromaand coronary artery disease raised triglycerides predispose to pancreatitis.

Hyperlipidemia: Elevated blood fats, either from heredity, having so many calories in the diet that they are ending up as a constant stream of liver-synthesized storage fats, an excessively anabolic metabolism...and a constellation of less common disease causes

Hypermetropia: Hypermetropia, or hyperopia, is a term applied to long-sightedness, in which the eye is too flat from front to back and rays of light are brought to a focus behind the retina.

Hypernatraemia: Excessive concentration of sodium in the blood.

Hypernatremia: An excess of sodium in the blood...a short-lived condition since the body retains water until the concentration is back to normal...and the blood volume (as well as blood pressure) has increased.

Hypernephroma: Now named renal cell carcinoma, this is a malignant tumour resembling the tissue of the suprarenal gland and occurring in the kidneys. Fever, loin pain, haematuria and swelling are

Hyperparathyroidism: Abnormally increased parathyroid gland activity that affects, and is affected by, plasma calcium concentration.

Hyperplasia: Abnormal multiplication or increase in the number of normal cells in a tissue or organ.

Hyperprolactinaemia: Overproduction of the hormone prolactin, usually as a result of a tumour of the pituitary gland (prolactinoma).

Hyperpyrexia: High fever.

Hyper-reactivity: Term used to -

describe the responses of (effects on) an individual to (of) an agent when they are qualitatively those expected, but quantitatively increased.

Hyper-reflexia: Exaggeration of reflexes.

Hypersecretion: Oversecretion of fluids by a gland. It may occur from irritation, infection, or allergy, as in the nasal drooling in a head cold or hay fever, or, as in gastric hypersecretion, from a functional imbalance in the chemical and neurologic stimulus of the stomach lining.

Hypersensitivity: An abnormal immunological reaction produced by some people when re-exposed to antigens that are innocuous to normal individuals. An antigen or allergen is a protein that stimulates an allergic response. This may mean that the next time the person is exposed to that antigen, there may be a dramatic health- or life- threatening reaction, such as anaphylaxis.

Hypersomnia: A condition that means "someone who sleeps too much."

Hypersusceptibility: Excessive reaction following exposure to a given amount or concentration of a substance as compared with the large majority of other exposed subjects.

Hypertension: Means high blood pressure (raised pressure of the circulating blood), but since there is a wide range of 'normal' blood pressure in the population, a precise level of pressure above which an individual is deemed hypertensive is arbitrary. Persistently high blood pressure in the arteries or in a circuit, for example pulmonary hypertension or hepatic portal hypertension.

Hypertensive Encephalopathy: A complication of severe hypertension, this serious but uncommon condition is characterised by neurological symptoms which include transient verbal and visual disturbances, paraesthesia, disorientation, fits and sometimes loss of consciousness. It also affects the eyes,

Hypertext: Links in a document that permit you to jump immediately to another document. In most Web browsers links are displayed as colored, underlined text.

Hyperthermia: Hyperthermia means abnormally high body temperature. A rapid rise of temperature to dangerous levels called malignant hyperthermia – may be precipitated by general anaesthesia but the condition is rare (1:50,000 operations) and is usually inherited. The anaesthetic is stopped and icepacks are used to cool the patient. Pure oxygen and intra- venous sodium choride are also administered.

Hyperthyroid: Elevated thyroid levels, either functional and constitutional in nature or the more profound state of thyrotoxicosis and overt disease.

Hypertonic: (1) Referring to one -

solution which has a greater osmotic pressure than another. Physiologically it is used to describe solutions which have a greater osmotic pressure than body fluids. (2) Muscles with abnormally increased tone

Hypertrophy: Excessive growth in bulk of a tissue or organ through increase in size but not in number of the constituent cells.

Hyperuricemia: Having elevated blood uric acid, either from high consumption of organ meat or spirulina.

Hyperventilation: An abnormally rapid resting respiratory rate. If voluntarily induced, it causes lightheadedness and then unconsciousness by lowering the blood tension of carbon dioxide.

Hypervitaminosis: Condition resulting from the ingestion of an excess of one or more vitamins.

Hypervolaemia: An increase in the volume of circulating blood above the normal range.

Hypnotics: These are drugs that induce sleep. Before a H hypnotic is prescribed, it is vital to establish –and, where possible, treat – the cause of the insomnia. Hypnotics are most often needed to help an acutely distressed patient (for example, following bereavement), or in cases of jet lag, or in shift workers.

Hypnotism: The process of producing a state of mind known as hypnosis. Although recognised for hundreds of years, the precise nature of this process is still poorly understood. One modern writer has defined hypnosis as 'a temporary condition of altered attention, the most striking feature of which is greatly increased suggestibility'. There is no evidence, as has been claimed, that women can be more easily hypnotised than men; in fact, children and young adults are the more easily hypnotised, with middle-aged people being more resistant. Hypnosis is induced by various methods, but the basis of all is some rhythmic stimulus accompanied by the repetition of carefully worded suggestions. The most commonly used method is to ask the patient to fix his or her eye on a given spot, or light, and then to keep on repeating, in a quiet and soothing voice, that the patient's eyes will gradually become tired and that he or she will want to close them.

Hypo-: A prefix meaning below, under, or less than normal: for example, hypotension (low blood pressure) and hypodermic (under the skin).

Hypocalcaemia: A serum concentration of calcium below the normal range (between 2.33 and 3.05 mmol of calcium per 100 ml of serum). This may cause tetany, acutely; chronically it may give rise to rickets, osteomalacia or osteoporosis.

Hypocalcaemia: Abnormally low calcium concentration in the blood.

Hypocapnia: A blood tension of carbon dioxide below nor-mal. It is produced by hyperventilation which may be voluntary, mechanical (if the patient is on a ventilator) or in response to a physiological insult such as metabolic acidosis or brain injury.

Hypochlorhydria: An insufficient secretion of Hydrochloric Acid from the digestive cells of the stomach lining.

Hypochondriasis: Obsession with the body's functions and a delusionof ill health, often severe, such that patients may believe they have a brain tumour or incurable insanity. Furthermore, patients may believe that they have infected others, or that their children have inherited the condition. It is a characteristic feature of depression, but may also occur in schizophrenia, when the delusions may be secondary to bodily hallucinations, and a sense of subjective change.

Hypochondrium: The regions of the belly below the ribcage and to the sides, as in left or right hypochondrium.

Hypocortical: Having low adrenocortical function

Hypodermic: A term pertaining to the region immediately under the skin. Thus, a hypodermic injection means an injection given underneath the skin. A hypodermic syringe is a small syringe which, fitted with a fine needle, is used to give such injections.

Hypogamma Globulinaemia: A lower-than-normal amount of the protein gamma-globulin in the blood. The origin may be genetic – several types are inherited – or an acquired defect (for instance, some lymphomas cause the condition).

Hypogastric: A term means pertaining to the lower middle part of the abdomen, just above the pubis.

Hypoglossal Nerve: The 12th cranial nerve, which supplies the muscles of the tongue, together with some others lying near it. This nerve is responsible for movements required for swallowing and talking.

Hypoglycaemia: A deficiency of glucose in the blood – the nor-mal range being 3•5–7•5 mmol/l. It most commonly occurs in diabetic patients – for example, after an excessive dose of insulin and heavy exercise, particularly with inadequate or delayed meals.

Hypoglycaemic Coma: Hypoglycaemia or low blood sugar occurs when a patient with Diabetes Mellitus suffers an imbalance between carbohydrate/glucose intake and insulin dosage. If there is more insulin than is needed to help metabolise the available carbohydrate, it causes a range of symptoms such as sweating, trembling, pounding heart- beat, anxiety, hunger, nausea, tiredness and headache. If the situation is not quickly remedied by taking oral sugar – or, if severe, giving glucose by injection – the patient may become

Hypoglycemia: Low blood sugar. It can be an actual clinical condition (rather rare), but the term is usually applied to labile blood sugar, where the highs are socially acceptable, if zappy, but the lows cause headaches, depression and sugar cravings, which only kick the sugars up which forces the sugars down etc. This is a subclinical condition that usually goes nowhere, at least clinically, but can drive you (or your companion) crazy. Some normal and healthy food have a rapid blood presence and can leave you hanging; fruits, potatoes and carrot juice are lousy legumes, particularly beans, supply slow and extended release of calories mostly because of high levels of soluble fiber and laborious digestion. Even adding such dietarily useless items like psyllium seed and chia seed can do wonders to slow down sugar spiking.

Hypogonadism: A condition characterised by underactivity of the testes or ovaries – the gonads. The condition may be caused by a genetically based disorder resulting in an abnormally functioning gonad (primary hypogonadism) or by a malfunctioning pituitary gland that fails to produce an adequate amount of gonadotrophin hormone secondary hypogonadism. Those affected may fail to develop adequately the secondary characteristics of their sex: males will have delayed puberty, erectile impotence and infertility and also develop gynaecomastia; females also have delayed puberty, infertility, and sometimes hirsutism.

Hypokalaemia: An abnormally low concentration of potassium in the blood.

Hypomania: Hypomania is a modest manifestation of mania. The individual is elated to an extent that he or she may make unwise decisions, and social behaviour may become animated and uninhibited. To the casual observer individuals may, however, seem normal. Treatment is advisable to prevent them from harming their own or their family's interests. Treatment is as for mania.

Hyponatraemia: A serum concentration of sodium below the normal range. It may be produced by dilution of blood (giving large volumes of salt-poor solutions intravenously), excessive water retention (inappropriate secretion of antidiuretic hormone), excessive sodium loss, and, rarely, by inadequate salt intake.

Hypoparathyroidism: Underactivity of the parathyroid glands. Thus there is a lack of parathyroid hormone resulting in hypocalcaemia. It may be caused by inadvertent removal of the glands when the thyroid gland is surgically removed, or by failure of the glands because of autoimmune disease.

Hypophysectomy: Surgical excision of the pituitary gland. This can be done by opening the skull, by inserting very low-temperature needles (cryosurgery) into the gland, or by inserting needles of radioactive Yttrium-90.

Hypophysis: Another name for the Pituitary Gland.

Hypopiesis: The condition, or state, characterised by hypotension, or abnormally low blood pressure.

Hypopituitarism: Underactivity of the pituitary gland. It can cause dwarfism, delayed puberty, impotence, infertility, amenorrhoea, hypothyroidism, and hypoadrenalism. Causes include tumours, irradiation of the gland, sarcoidosis, and necrosis associated with post-partum haemorrhage (Sheehan's syndrome).

Hypoplasia: Excessive smallness of an organ or part, arising from imperfect development.

Hypoproteinaemia: A fall in the amount of protein in the blood. This may be caused by malnutrition, loss of protein from kidney disorders, or faulty production of protein which occurs in some liver disorders. Hypoproteinaemia causes oedema because fluid accumulates in the tissues as a consequence of the metabolic abnormalities. Patients' resistance to infections is also impaired.

Hypoprothrombinaemia: A deficiency of prothrombin (clotting factor) in the blood. As a result the affected individual tends to bleed more easily. The defect may be inherited or be the consequence of liver disease or a deficiency in vitamin K.

Hypospadias: A developmental abnormality in the male, in which the urethra opens on the undersurface of the penis or in the perineum. The condition is treatable with surgery, but several operations over a period of years may be required to ensure normal urinary and sexual functions.

Hypostasis: The term applied to the condition in which blood accumulates in a dependent part of the body as a result of poor circulation. Congestion of the base of the lungs in old people from this H cause, and infection, is called hypostatic

Hypotension: Low blood pressure. Not always a bad thing unless you need 11 hours of sleep or faint if you stand too quickly. Some healthy individuals with a normal cardio-vascular system have a permanently low arterial blood pressure for their age. What blood-pressure reading constitutes hypotension is arguable, but a healthy young person with figures below 100 mm Hg systolic and 65 mm Hg diastolic could be described as hypotensive. For a healthy 60 year old, comparative figures might be 120/80.

Hypotestosteronism: Having either low secretion levels of testosterone by the testes, having low functional effects because of poor circulation, having competition by less active testosterone metabolites, or having high levels of adipose-released estradiol (former testosterone) in obesity that ends up suppressing testosterone.

Hypothalamus: A part of the diencephalon of the brain, it is a major actor in the limbic system. This is a functional, not anatomic, system in the brain that influences and is influenced by emotions. Call the limbic system an ad hoc committee that decides how things are going today, based on past, present, potential, and myriad informational inputs from the somatic body.The hypothalamus gathers the data and sets the levels of the pituitary thermostat. The pituitary does what the hypothalamus tells it to do, and our whole chemical nervous system responds to the pituitary, which responds to the hypothalamus, which, along with the rest of the limbic system, decides the kind of day we need to get ready for. And to think that some doctors used to (and still) scoff at a "psychosomatic disorder."

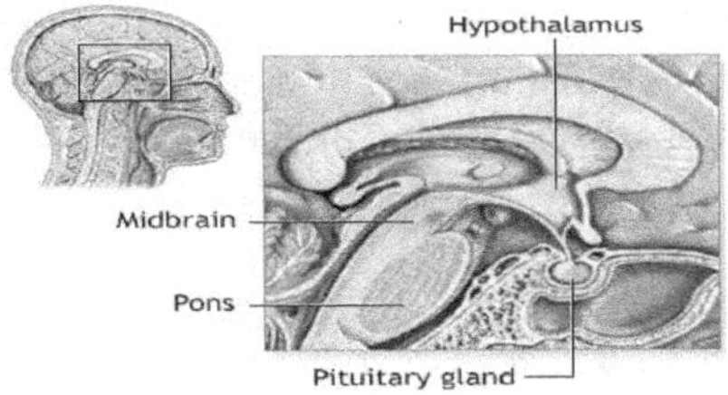

Hypothermia: A core body temperature of less than 35 ^{0}C. As the temperature of the body falls, there is increasing dysfunction of all the organs, particularly the central nervous and cardiovascular systems. The patient becomes listless and con-fused, with onset of unconsciousness.

Hypothesis: A supposition or assumption advanced as a basis for reasoning or argument, or as a guide to experimental investigation. If both researchers and participants are blinded, the study is double blind. Because the terms "single blind" and "double blind" are imprecise, many researchers prefer to specify who is blinded—investigators, participants, outcome assessors, or statisticians.

Hypothyroid: Having deficient thyroid levels, either from overt thyroid disease like myxedema, a generally low metabolism from functional causes, or subsequent to emotional depression or the use of depressant drugs.

Hypotonic: (1) Referring to a solution which has a lower osmotic pressure than another. Physiologically it describes a solution with a lower osmotic pressure than body fluids.

Hypoventilation: Shallow and/or slow breathing, often caused by the effects of injury or drugs on the respiratory centre. It causes hypercapnia and hypoxia.

Hypovolaemia: A reduced circulating blood volume. Acutely, it is caused by unreplaced losses from bleeding, sweating, diarrhoea, vomiting or diuresis. Chronically it may be caused by inadequate fluid intake.

Hypovolaemic: Pertaining to an abnormally decreased volume of circulating fluid (plasma) in the body.

Hypoxaemia: A fall in the concentrat-

ion of oxygen in the arterial blood. Symptoms are those of cyanosis and, if severe, the affected individual will show signs of respiratory failure.

Hypoxia: Lack of sufficient oxygen, such as occurs at high altitudes. State of dynamic balance in which an individual's or a group's capacity to cope with the circumstances of living is at an optimal level. State characterized by anatomical, physiological and psychological integrity, ability to perform personally valued family, work and community roles; ability to deal with physical, biological, psychological and social stress; a feeling of wellbeing; and freedom from the risk of disease and untimely death. toxic, corrosive, ignitable, explosive, or chemically reactive nature.

Hysterectomy: Surgical removal of the uterus. Hystero-oophorectomy is the term applied to removal of the uterus and ovaries.

Hysteria: An out-of-date description for a symptom (or symptoms) with no obvious organic cause, which is an unconscious reaction and from which the person may benefit. It is now recognised as a dissociative disorder: such disorders – amnesia, fugue, multiple personality states and trancelike conditions – are powerful defence mechanisms against severe stress when

Hysteroscopy: Hysteroscopy is the direct visualisation of the interior of the uterus using fibreoptic endoscopy. The technique, which allows minor surgical procedures to be carried out at the same time, has transformed the management of uterine disorders.

Hysterotomy: An operation in which the uterus is opened to remove a fetus before 28 weeks' gestation. After 28 weeks it would be called a caesarean section. It is now seldom used as a means of abortion.

I

Iatric: Anything pertaining to a physician.

Iatrogenic Disease: Disease induced by a physician: most commonly a drug-induced disease.

Iatrogenic: Illness, disease, or imbalances created by medical or nonmedical treatment that were not present before treatment. In medicine the therapy is blamed (not the therapist) and changed to something else. In alternative medicine it may be called a "healing crisis" and deemed good for you. Beware: if the therapy makes you feel worse in a new way, it is almost always the wrong therapy.

IBD: Inflammatory Bowel Disease (IBD).

IBS: Irritable Bowel Syndrome.

ICD: International classification of disease (ICD).

ICH: International Conference on Harmonization of Technical Requirements for Registration of Pharmaceuticals for Human Use

Ichthyosis: A disorder in which the skin is permanently dry and scaly. It is usually genetically determined and several different forms are recognised: Ichthyosis vulgaris Common and inherited as a dominant trait. Beginning in early childhood, it is often associated with atopic eczema. The limb flexures and face are spared.

ICN: Internal Control Number. Unique number assigned to a claim for identification purposes.

Icterus: Icterus: Icterus is another name for Jaundice.

Ictus: Ictus is another term for a Stroke.

ICU (Intensive Care Unit): A general intensive care unit is defined as a unit using designated intensive care beds interchangeably for more than one type of patient (e.g. medical, coronary, surgical.)

ID Card: Identification cards are provided to all participants for proper identification under their group health plan. ID card information helps providers verify patient eligibility for coverage.

Idiopathic: Idiopathic is a term applied to diseases to indicate that their cause is unknown.

Idiosyncrasy: A generally unexpected, so unpredictable, abnormal reaction to a drug caused by a constitutional defect in the patient. In some cases the underlying disorder is already known or dis- covered after the first event, so that the drug in question can be avoided thereafter. The Immune System abnormal sensitivity of patients with porphyrias to barbiturates is an example.

IDM: Independent disease manager. An entity that develops and operates care management systems, usually on behalf of a healthcare provider network or insurance company, for patients with chronic conditions.

IgE: Immunoglobulin E is a type of antibody produced by IgE plasma cells. These are specialized B-cell lymphocytes that make free-floating antibodies for what is termed humoral resistance. IgE is peculiar for several reasons. It is not made to be specific against only one antigen like other gamma globulins, but instead can bind with a number of dangerous proteins. Further, IgE travels to mast cells, sticks to their surfaces, and when antigens get stuck to the IgE, the mast cells secrete inflammatory compounds like histamine. Since IgE is a generalist, coded for a number of potential toxins, not just a single substance, it can decide that Juniper pollen and cat dander are antigens and you have an allergy.

Ileitis: Inflammation of the ileum – the lower part of the small intestine. It may be caused by crohn's disease, typhoid fever, tuberculosis or the bacterium yersinia enterocolitica. Ileitis may also accompany ulcerative colitis.

Ileo-Caecal: The term applied to the region of the junction between the small and large intestines in the right lower corner of the abdomen. The ileocaecal valve is a structure which allows the con- tents of the intestine to pass onwards from the small to the large intestine, but, in the great majority of cases, prevents their passage in the opposite direction.

Ileocecal: Pertaining to both the last section of the small intestine (the ileum) and the beginning of the large intestines, the ascending colon or cecum.

Ileostomy: The operation by which an artificial opening is made into the ileum and brought through the abdominal wall to create an artificial opening or stoma. It is most often performed as part of the operation for cancer of the rectum, in which the rectum has usually to be removed.

Ileum: The lower two-thirds of the small intestine, ending in the ileocecal valve and emptying into the cecum of the colon. The last foot of the ileum is the only absorption site available for such important dietary substances as vitamin B_{12}, folic acid, some essential fatty acids, fat soluble vitamins, and recycled bile acids.

Ileus: Paralysis of the bowel muscle.

Ilium: The uppermost of the three bones forming each side of the pelvis.

Immersion Foot: The term applied to a condition which develops as a result of prolonged immersion of the feet in cold or cool water. It was a condition commonly seen during

World War II in shipwrecked sailors and airmen who had crashed into the sea, spending long periods there before being rescued. Such prolonged exposure results in vasoconstriction of the smaller arteries in the feet, leading to coldness and blueness and finally, in severe cases, to ulceration and gangrene.

Immission: Environmental concentration of a pollutant resulting from a combination of emissions and dispersals.

Immune Complex: Product of an antigen-antibody reaction that may also contain components of the complement system.

Immune Response: Selective reaction of the body to substances that are foreign to it, or that the immune system identifies as foreign, shown by the production of antibodies and antibody-bearing cells or by a cell-mediated hypersensitivity reaction.

Immune System: The human body's defense system against foreign substances.

Immune Therapy: Enhancement of the body's immune system via intravenous (IV) infusion or intramuscular vaccination. This therapy helps the body fight disease and infection.

Immunisation: The introduction of antigens into a body to produce immunity. The table above gives the immunization programme recommended by the UK departments of health.

Immunity: The ability to resist infection and to heal. The process may involve acquired immunity, (the ability to learn and remember a specific infectious agent), or innate immunity (the genetically programmed system of responses that attack, digest, remove, and initiate inflammation and tissue healing).

Immunoassay: Procedures which measure the concentration of any antigenic material to which an antibody can be created. The amount of antigen bound to this antibody is proportional to the parent substance. Enzymes Immunosorbent Assay (ELISA) or radioactive labels (Radio- Immunoassay) are used to measure the concentration of antigenic material.

Immunochemistry: Study of biochemical and molecular aspects of immunology, especially the nature of antibodies, antigens and their interactions.

Immunocompetent: Able to participate in the immune response

Immunodeficiency: Impaired immunity resulting from inherited or acquired abnormalities of the immune system. This leads to increased vulnerability to infection. Important inherited examples of immunodeficiency are defects in function of granulocytes and the complement system.

Immunogenicity: The characteristic of a substance that can provoke an immune response. This includes how 'foreign' a substance entering or contacting the body is; route of entry; dose; number and period of exposure to antigen; and the genetic make-up of the host.

Immunoglobulin E-mediated Hypersensitivity: State in which an individual reacts with allergic effects caused fundamentally by the reaction of antigen-specific immunoglobulin E following exposure to a certain substance (allergen) after having been exposed previously to the same substance.

Immunoglobulin: Family of closely related glycoproteins capable of acting as antibodies and present in plasma and tissue fluids; immunoglobulin E is the source of antibody in many hypersensitivity (allergic) reactions.

Immunologist: A specialist (medically or scientifically qualified) who practises or researches immunology.

Immunology: The study of immune responses to the environment. Its main clinical applications include improving resistance to microbial infections, combating the effects of impaired immunity, controlling harmful immune reactions, and manipulating immune responses to prevent harmful immunological responses such as graft rejection and autoimmune diseases. The clinical study of disordered immunity now forms the allied discipline of clinical immunology, which is closely linked to the laboratory- based discipline of immunopathology.

Immunopotentiation: Enhancement of the capacity of the immune system to produce an effective response.

Immunostimulant: An agent that stimulates either innate or acquired immunity. In the U.S., immunotherapy is relegated to experimental medicine, but a number of plant substances are used in Europe as immunostimulants. The presumption of immunostimulation is that you increase native resistance and let it run its course. American Standard Practice, with all good intentions, tends to aggressive procedures, and feels empowered only when intervening against, not with, physiologic responses. Medicine is the only approach to many problems, but in the U.S. we all tend to forget that our brand of standard practice is uniquely aggressive and invasive amongst the industrialized nations. There are other ways which is presumably why you are using this glossary in the first place.

Immunosuppressant: A drug that reduces the body's resistance to infection and other foreign agents. It does so by suppressing the activity of the immune system. Examples of such drugs are Azathioprine, Cyclophos-phamideand Ciclosporin A. Immuno-suppressants are used to help transplanted organs and tissues to survive the potential immune reaction from the host. They are also used to treat Autoimmune Disorders such as Rheumatoid Arthritis.

Immunosuppression: Reduction in the functional capacity of the immune response; may be due to: Inhibition of the normal response of the normal response of the immune system to an antigen.

Immunosurveillance: Mechanisms by which the immune system is able to recognize and destroy malignant cells beore the formation of an overt tumour.

Immunotherapy: The manipulation of immunity by immunological means to reduce harmful reactions or to boost beneficial responses. Severe allergy to wasp or bee stings is often treated by a course of injections with allergen purified from insect venom. There are current attempts to treat autoimmune diseases with monoclonal antibodies to the T-cell populations or cytokines implicated in the immunopathogenesis of the disorder. Strategies are also being evaluated for treating cancer by boosting the patient's own immunity to cancer cells.

Immunotoxic: Poisonous to the immune system.

Impaction: A term applied to a condition in which two things are firmly lodged together. For example, when one piece of bone is driven within another following a fracture, this is known as an impacted fracture; when a tooth is firmly lodged in its socket so that its eruption is prevented, this is known as dental impaction. Intractable constipation is termed faecal impaction.

Impartial Witness: A person, who is independent of the trial, who cannot be unfairly influenced by people involved with the trial, who attends the informed consent process if the subject or the subject's legally acceptable representative cannot read, and who reads the informed consent form and any other written information supplied to the subject.

Imperforate: An adjective mea lack of an opening. For example, occasionally the anus fails to develop properly, resulting in partial or complete obstruction of the opening. Sometimes pubertal girls have an imperforate hymen which obstructs the opening to the vagina and prevents menstrual flow of blood draining to the exterior.

Impetigo: An infectious skin disease caused usually by Staphylococcus aureus and less often by Streptooccus pyogenes. The itching rash is seen especially on the face but may spread widely. Vesicles and pustules erupt and dry to form yellow-brown scabs.

Implantation: (1) The placing of a substance such as a drug, or an object such as a pacemaker, in a body tissue. (2) The surgical replacement of injured or unhealthy tissue or organ with healthy tissue or organ (also known as transplantation). (3) Attachment of the early embryo to the lining of the uterus, which occurs around six days after conception; the site where this happens is where the placenta will develop.

Implementation Phase: The step in a

pollution prevention assessment where procedures, training, and equipment changes are put into action to reduce waste.

Impotence: A condition where the penis does not become or stay hard enough for sex. Some men who have had diabetes a long time become impotent if their nerves or blood vessels have become damaged.

Impression: In dentistry, a mould (using a rubber or alginate compound) of the teeth and gums from which a plaster-of-Paris model is prepared. This model provides a base on which to construct a denture, bridge or dental inlay. A similar process is used in orthodontics to make dental appliances to correct abnormalities in the positioning of teeth.

Imprinting: In the context of animal behaviour, this is a quick and irreversible type of learning in which patterns are imprinted on the animal's mind during the first few hours of life. The smell and feel of its mother are one such imprint.

In Vitro Testing: Non-clinical testing conducted in an artificial environment such as a test tube or culture medium.

In Vitro: A term commonly used in medical research and experimental biology. Literally 'in a glass', it refers to observations made outside the body: for example, on the action of drugs on bacteria. The opposite term is in vivo, which refers to observations of processes in the body.

In Vivo Testing: Testing conducted in living animal and human systems.

In Vivo: A Latin term to describe biological events that take place inside the bodies of living organisms.

Inanition: Exhaustion in an individual caused by lack of appropriate nutrients in the circulating blood. Starvation, malnutrition or intestinal disorders are among the causes.

Inbreeding: The birth of offspring to parents who are closely related. In traditional rural communities, marriage between cousins was common and this could lead to a higher-than-average number of children with congenital anomalies or learning difficulties. This is now seen in certain ethnic groups who have brought the custom of inbreeding with them to their new homes in the western world.

Incentive: An opportunity for a plan member to earn points toward prizes or extra dollars in the CIGNA Choice Fund HRA. Participating in a disease management program is one way to gain these awards.

Incidence Rate: Measure of the frequency with which new events occur in a population. Value obtained by dividing the number of new events that occur in a defined period by the population at risk of experiencing the event during this period, sometimes expressed as person-time.

Incidence: Number of occurrences of illness commencing, or of persons falling ill, during a given period in a specific population : usually expressed as a rate, the denominator being the average number of persons in the specified population during a defined period or the estimated number of persons at the mid-point of that period.

Incision: A cut or wound; a term especially applied to surgical openings.

Incisor: The term for the four front teeth of each jaw.

Inclusion Bodies: Particles found in the cytoplasm and nucleus of cells, usually a consequence of a viral infection. This phenomenon can be helpful in the diagnosis of such an infection.

Inclusion Criteria: The criteria in a protocol that prospective subjects must meet to be eligible for participation in a study.

Inclusion/Exclusion Criteria: The standards determining whether a person may or may not be allowed to enter a clinical trial. These criteria are based on such factors as age, gender, the type and stage of a disease, previous treatment history, and other medical conditions. It is important to note that inclusion and exclusion criteria are not used to reject people personally, but rather to identify appropriate participants and keep them safe.

Incompatibility: (1) In the pharmacological context, the use of two or more drugs in treatment which together produce adverse consequences for the patient. The British National Formulary carries an appendix devoted to drug interactions.

Incompetence: Incompetence is a term applied to the valves of the heart when, as a result of disease in the valves or alterations in size of the chambers of the heart, the valves become unable to close the orifices which they should protect.

Incontinence: The inability to retain urine in the bladder for a reasonable length of time. It is can be caused by urethral irritation, loss of tone to the basement muscle of the bladder (the trigone), scarification or growths on the urethral lining, nerve damage, or emotional stress.

Incontinence: Urinary incontinence The International Continence Society defines urinary incontinence as an involuntary loss of urine that is objectively shown and is a social and hygiene problem. The elderly suffer most from this dis- order because the effectiveness of the sphincter muscles surrounding the urethra declines with age. Men are less often affected than women; 20 per cent of women over 40 years of age have problems with continence. It is estimated that around three million people are regularly incontinent in the UK, a prevalence of about 40 per 1,000 adults.

Incoordination: A term applied to irregularity of movements produced either by loss of the sensations by which they are governed, or by defects in the muscles themselves or somewhere in the nervous system.

Incremental Unit Risk Estimate: For an air pollutant, this is the additional lifetime cancer risk occurring in a hypothetical population in which all individuals are exposed continuously from birth throughout their lifetimes to a concentration of 1 microgram per cubic metre (mg/m3) of the pollutant in the air they breathe.

Incubation: The period elapsing between the time when a person becomes infected by some agent and the first appearance of the symptoms of the disease. Most acute infectious diseases have fairly definite periods of incubation, and it is of great importance that people who have run the risk of infection should know the length of time which must elapse before they can be sure whether or not they are to contract the disease

IND: Investigational New Drug application. Notification by a drug sponsor to the FDA of its intention to conduct clinical studies on human subjects.

Indemnity Plan: A type of health benefits plan under which the covered person pays 100% of all covered charges up to an annual deductible. The health benefits plan then pays a percentage of covered charges up to an out-of-pocket maximum.

Independent Data Monitoring Committee (IDMC): An independent data-monitoring committee that may be established by the sponsor to assess at intervals the progress of a clinical trial, the safety data, and the critical efficacy endpoints, and to recommend to the sponsor whether to continue, modify, or stop a trial. Also called data and safety monitoring board, monitoring committee, data monitoring committee.

Independent Ethics Committee (IEC): An independent body (a review board or a committee, institutional, regional, national, or supranational) constituted of medical/ scientific professionals and non-scientific members, whose responsibility it is to ensure the protection of the rights, safety, and well-being of human subjects involved in a trial and to provide public assurance of that protection by, among other things, reviewing and approving/providing favorable opinion on the trial protocol, the suitability of the investigator(s), facilities, and the methods and material to be used in obtaining and documenting informed consent of the trial subjects. The legal status, composition, function, operations, and regulatory requirements pertaining to independent ethics committees may differ among countries, but should allow the independent ethics committee to act in agreement with GCP as described in the ICH guideline.

Indication: A health problem or disease that is identified as likely to be benefited by a therapy being studied in clinical trials. Where such a benefit has been established and approved by regulatory authorities, the therapy is said to be approved for such an indication.

Indirect Exposure: Exposure to a substance in a medium or vehicle other than the one originally receiving the substance. Exposure of people to a substance by contact with a person directly exposed.

Individual Protective Device (IPD): Device for individual use for protection of the whole body, eyes, respiratory pathways or skin of workers against hazardous and harmful production factors.

Individual Risk: Probability that an individual person will experience an adverse effect.

Indolent: A sluggish and unresolving condition, often with ulcerations and necrosis.

Inducer: Substance that causes induction.

Induction Period: Time from the onset of exposure to the appearance of signs of disease.

Induction Therapy: Early high-dose-

therapy to induce graft acceptance

Induction: Bringing about a particular event – for example, the induction or starting of labour, or the induction of anaesthesia. Newly arrived hospital doctors are given an induction period during their first day or two at work.

Induction: Increase in the rate of synthesis of an enzyme in response to the action of an inducer or environmental conditions: often the substrate of the induced enzyme or a structurally similar substance (gratuitous inducer) that is not metabolized.

Induration: The pathological hardening of a tissue or organ. This may occur when a tissue is infected or when it is invaded by cancer.

Inert: Not reactive under normal circumstances

Infant Feeding: The newborn infant may be fed naturally from the breast, or artificially from a bottle. Breast feeding Unless there is a genuine contraindication, every baby should be breast fed. The nutritional components of human milk are in the ideal proportions to promote the healthy growth of the human newborn. The mother's milk, especially colostrum (the fluid secreted before full lactation is established) contains immune cells and antibodies that increase the baby's resistance to infection. From the mother's point of view, breast feeding helps the womb to return to its normal size and helps her to lose excess body fat gained during pregnancy. Most importantly, breast feeding promotes intimate contact between mother and baby. A final point to be borne in mind, however, is that drugs taken by a mother can be excreted in her milk. These include antibiotics, sedatives, tranquillisers, alcohol, nicotine and high-dose steroids or vitamins. Fortunately this is rarely a cause of trouble.

Infant Mortality Rate (IMR): The number of deaths of infants under one year of age. The IMR in any given year is calculated as the number of deaths in the first year of life in proportion to every 1,000 registered live births in that year. Along with perinatal mortality, it is accepted as one of the most important criteria for assessing the health of the community and the standard of the social conditions of a country.

Infant: A baby who is under one year old.

Infantile Spasms: Also known as salaam attacks, these are a rare but serious type of epilepsy, usually starting in the first eight months of life. The spasms are short and occur as involuntary flexing of the neck, arms, trunk and legs. They may occur several times a day. If the baby is sitting, it may collapse into a 'salaam' position; more usually there is a simple body jerk, sometimes accompanied by a sudden cry. An electroencephalogram shows a picture of totally disorganised electrical activity called hypsarrhythmia. The condition results from any one of many brain injuries, infections or metabolic insults that may have occurred before, during, or in the first few months after birth.

Infantilism: The condition characterised by imperfect sexual development at puberty. It may or may not be associated with small stature, and may be due to lack of development of certain of the endocrine glands: for example, the gonads, pituitary gland or adrenal glands. In other cases it may be associated with a generalised disease such as diabetes mellitus, asthma, ulcerative colitis and rheumatoid arthritis (for more information, see under separate entries).

Infarct: Tissue which has died due to a lack of oxygen resulting from a blood clot or blocking of an artery.

Infarction: The changes in an organ when an artery is suddenly blocked, leading to the formation of a dense, wedge-shaped mass of dead tissue in the part of the organ supplied by the artery. It occurs as the result of embolism or of thrombosis.

Infection: Invasion of the body by organisms that can cause disease.

Infertility: Term used to describe a condition or the inability to conceive or an inability to carry a pregnancy to a live birth after a year or more of regular sexual relations without the use of contraception.

Infestation: A term applied to the occurrence of animal parasites in the intestine, hair or clothing.

Infibulation: The most extensive form of female circumcision, involving removal of clitoris and both

Infiltration: The invasion of tissues or organs by cells or fluid not normally present – for example, local anaesthetic is infiltrated into an area of tissue to produce analgesia in a defined area. Inflammation of the bronchial tubes. This may occur as an acute transient illness or as a chronic condition.

Inflammation: The reaction of the tissues to any injury, which may be the result of trauma, infection or chemicals. Local blood vessels dilate, thus increasing blood flow to the injured site. White blood cells invade the affected tissue, engulfing bacteria or other foreign bodies; related cells consume any dead cells, thus producing PUS after which the site starts to heal.

Influenza: A specific type of acute viral respiratory infection, with one virus (many strains) and a short, nasty stay. A few thousand people die from it every year, but humans alive at present have almost universal partial resistance. It was not so during WWI, when it first began to spread. It was variously called Spanish Influenza, La Grippe, and Influenza (Italian for Influence), everyone blamed some other country for it. The Turks and Armenians took a break from mutual mutilation and blamed it on each other, since it was killing as many people as the 1,000,000 fatalities that bit of genocide fostered. It ran across the world like some Bergmanesque horseman, and killed at least 20 million people before it petered out around 1925.

Informatics: Collection and organization of information using computers and statistical methods

Informed Consent Document: A document that describes the rights of the study participants and includes details about the study, such as its purpose, duration, required procedures, and key contacts. Risks and potential benefits are explained in the informed consent document. The participant then decides whether or not to sign the document. Informed consent is not a contract, and the participant may withdraw from the trial at any time.

Informed Consent: A process by which a subject voluntarily confirms his or her willingness to participate in a particular trial, after having been informed of all aspects of the trial that are relevant to the subject's decision to participate. Informed consent precedes enrollment and is documented by means of a written, signed, and dated consent form. The voluntary verification of a patient's willingness to participate in a clinical trial, along with the documentation of this decision. This verification is requested only after complete, objective information has been given about the trial, including an explanation of the study's objectives, potential benefits, risks and inconveniences, alternative therapies available, and of the subject's rights and responsibilities in accordance with the current revision of the Declaration of Helsinki. Participants must also be provided with information throughout the study so that they can make informed decisions about remaining involved in the trial.

Infrared Radiation: The band of electromagnetic radiation which has a longer wavelength than that of the red in the visible spectrum. Infrared radiation is used in the special photographic process essential to thermography. Its property of transmitting radiant heat has made infrared radiation invaluable in physiotherapy, where it warms tissues, soothes pain and increases the local circulation.

Infundibulum: A funnel-shaped passage. The word is used specifically to describe the hollow conical stalk that links the hypothalamus to the posterior lobe of the pituitary gland.

Infusion Pumps: When receiving an intravenous (IV) solution, infusion pumps can time, infuse and flush IVs automatically.

Infusion Therapy: Treatment accomplished by placing therapeutic agents into the vein, including intravenous feeding. Such therapy also includes enteral nutrition, which is the .delivery of nutrients into the gastrointestinal tract by tube.

Infusion; The intravenous or subcutaneous injection of one of a variety of therapeutic solutions, such as saline, glucose, or gum acacia, in the treatment of severe dehydration, hypoglycaemia, or other plasma electrolyte imbalance. Blood infusions may be given in cases of severe anaemia – for example, after heavy bleeding. Infusions may be given in intermittent amounts of around 570 ml (1 pint) at a time, or alter- natively by continuous drip-feed over several hours.

Ingestion: (1) The act of taking fluid, food, or medicine into the stomach. (2) The way in which a phagocytic cell surrounds and absorbs foreign substances such as bacteria in the blood.

Ingrowing Toenail: The sides of the toenail curve downwards, resulting in inflammation of the skin next to the nail which spreads to the base of the nail. The skin and nail base may become painful and badly infected. If antibiotics and local dressing do not cure the condition, surgery to remove part of the nail will be required.

Inguinal Hernia: An extrusion of the abdominal peritoneum, sometimes containing a loop of bowel, through natural openings in the region of either groin.

Inguinal nodes: Lymph nodes in the groin, next to the genitalia

Inhalants: Substances that can be inhaled into the body through the lungs. They may be delivered in traditional form dissolved in hot water and inhaled in the steam, or as an aerosol – a suspension of very small liquid or solid particles in the air. The latter are now usually delivered by devices in which the aerosol is kept under pres- sure in a small hand-held cylinder and delivered in required doses by a release mechanism.

Inhalation: A method of applying drugs in a finely divided or gaseous state, so that, when breathed in, they may come into contact with the nose, throat and lungs. There are two chief means by which drugs are mingled with the air and so taken in by breathing: these are traditional steam inhalations, and modern aerosol devices which deliver a fine spray direct into the mouth.

Inhaler: A mechanism for administering a drug in the form of a powder or aerosol. mainly used by patients with asthma. Inhalers are basically of two types: aerosol, and dry-powder inhaler. The former delivers the drug as an aerosol spray when the patient presses the top of - the canister containing the drug.

Inherently biodegradable: Class of compounds for which there is unequivocal evidence of biodegradation (primary or ultimate) -

in any test of biodegradability.

Inheritance: The transfer of characteristics, traits and disorders from parents to children by means of genes carried in the chromosomes of the germ cells.

Inhibition: Inhibition means arrest or restraint of some process effected by nervous influence. The term is applied to the action of certain inhibitory nerves: for example, the vagus nerve which contains fibres that inhibit or control the action of the heart. The term is also applied generally to the mental processes by which instinctive but undesirable actions are checked by a process of self-control.

Inhibitory Concentration (IC): Concentration of a substance that causes a defined inhibition of a given system : IC50 is the median concentration that causes 50 % inhibition.

Inhibitory Dose (ID): Dose of a substance that causes a defined inhibition of a given system : ID50 is the median dose that causes 50 % inhibition.

Inhibitory Neurotransmitter: A neurotransmitter which decreases the electrochemical activity of neurons.

GABA and serotonin are inhibitory neurotransmitters.

In-house Pharmacy: Pharmacy services which are housed within a covered entity's facility. The pharmacy must be part of the legal organization of '.he covered entity.

Initiator: Agent that induces a change in a chromosome or gene that leads to the induction of tumours after a second agent, called a promoter, is administered to the tissue. Substance that starts a chain reaction; an initiator is consumed in a reaction, in contrast to a catalyst.

Inject: To force a liquid into the body with a needle and syringe.

Injections: An injection is the introduction of a substance into the body using a syringe and an attached needle. Injections may be given under the skin (subcutaneous), via a vein (intravenous), deep into a muscle (intramuscular), or into the fluid surrounding the spinal cord (intrathecal).

Innervation: The nerve supply to a tissue, organ or part of the body. It carries motor impulses to and sensory impulses away from the part.

In-network Provider: Any health care provider (physician, hospital, etc.) that belongs to a CIGNA network. Staying in-network gives members the advantage of significant discounts, helping to stretch their account dollars further.

Innovative Medicine: Characterized by making new medicine or bringing new ways of treatment.

Innovator: Brand name of the original patented drug of those listed on the Maximum Allowed Cost (MAC) list.

Inoculation: The process by which infective material is brought into the system through a small wound in the skin or in a mucous membrane. Many infectious diseases are contracted by accidental inoculation of microbes –as is blood- poisoning. Inoculation is now used as a preventive measure against many infectious diseases. Inotropic therapy: This therapy works to increase the force of the heart muscle so it pumps harder. Used mainly for weak or injured hearts, the medication works to relax the blood vessels so blood can move more freely allowing patients to perform daily activities more easily.

Inotropic: Adjective describing anything that affects the force of muscle contraction. It is usually applied to the heart muscle; an inotrope such as digoxin is a drug that improves its contraction. Beta-blocker drugs such as propranolol hydrochloride have negative inotropic properties.

Inpatient care: Care given to a patient admitted to a hospital, extended care facility, nursing home or other facility.

Inpatient: A person who requires a hospital bed for investigation or treatment.

Input Material: Material that is introduced into the daily product line or waste stream.

Inquest: An official inquiry conducted by a coroner into the cause of an individual's death. The coroner is a judicial officer who, when a death is sudden, unexpected or occurs in suspicious circumstances, considers the results of medical and legal investigation and, sitting with a jury or on his or her own, makes the conclusions public.

Insecticide: Substance intended to kill insects.

Insemination: The ejaculation of semen in the vagina in the act of sexual intercourse. In artificial insemination the semen is placed there by the use of an instrument.

Insight: A person's knowledge of him or herself. The description is especially relevant to a person's realisation that he or she has psychological difficulties. Thus, someone with a psychosis lacks insight. Insight also refers to an individual's concept of his or her personality and problems.Inspection of the interior of a joint (usually the knee) to diagnose any disorder there. The instrument used is a type of endoscope called an arthroscope. The knee is often affected by conditions that are not easy to diagnose and are not revealed by X-ray examination. Surgery can be performed using arthroscopy and this reduces the time a patient has to be in hospital.

Inspection: The act by a regulatory authority(ies) of conducting an official review of documents, facilities, records, and any other resources that are deemed by the authority(ies) to be related to the clinical trial and that may be located at the site of the trial, at the sponsor's and/or contract research organization's (CRO's) facilities, or at other establishments deemed appropriate by the regulatory authority(ies).

Institutional Review Board (IRB): An independent body constituted of medical, scientific, and non-scientific members, whose responsibility it is to ensure the protection of the rights, safety, and well- being of human subjects involved in a trial by, among other things, reviewing, approving, and providing continuing review of trial protocol and of the methods and material to be used in obtaining and documenting informed consent of the trial subjects. [ICH] Other names for such bodies include independent review board, independent ethics committee, and committee for the protection of human subjects.

Institutionalisation: A condition brought about by a prolonged stay in an impersonal institution. The individual becomes apathetic and listless as a result of inadequate stimulation in an uninteresting environment. The condition can occur in mental institutions or long-stay nursing homes; the affected person loses the ability to make any decisions and becomes pathetic and increasingly dependent on others.

Insufflation: Insufflation means the blowing of powder or vapour into a cavity, especially through the air passages, for the treatment of disease.

Insulin Shock: A disorder in which the body produces excess Insulin, which then reduces the amount of glucose in the blood (Hypoglycaemia). Treatment is with glucose or Glucagon. Untreated, the patient goes into a coma and dies.

Insulin: A polypeptide hormone produced in the pancreas by the beta cells of the islets of langerhans. It plays a key role in the body's regulation of carbohydrate, fat, and protein, and its deficiency leads to diabetes mellitus.

Insulin-dependent Diabetes: Also called Juvenile-onset Diabetes, IDDM (Insulin-Dependent Diabetes) and Type I, it is a deficiency condition wherein the pancreas does not manufacture enough insulin or what it makes is formed improperly. It is usually inherited, although it may not surface until pregnancy, recovering from a life-threatening illness, boot camp or some other profound metabolic stress.

Insulinoma: A tumour in the beta cells in the Islets of Langerhans in the pancreas that produces insulin. Symptoms of hypoglycaemia occur. Treatment is surgical removal or oral administration of diazoxide.

Insulin-resistant Diabetes: Also called NIDDM (Non-Insulin-Dependent Diabetes) and Type II (Type II), it generally means you make your own insulin, you eat too many calories, your storage cells are filled and are taking no more fuel, your liver is stuck in a rut and keeps making more glucose out of everything you eat, your brain has no control over its consumption of glucose, but you have run out of places to put it so you pee it out, sweat it out, etc. etc. Also called Adult-onset Diabetes. An Internist may cry out in dismay at this simplification, and there are many subtle distinctions between the various types, as well as a number of distinct hereditary considerations.

Intake: Amount of a substance that is taken into the body, regardless of whether or not it is absorbed: the total daily intake is the sum of the daily intake by an individual from food, drinking-water, and inhaled air.

Intangible Factor: Influence bearing on the use or effect of hazardous material, which may not be reduced to monetary terms.

Integral Indicator of Toxic Effect: Parameter (such as body weight or temperature) characterising the overall changes in the general state of the organism exposed to a toxic substance.

Integument: Another name for skin. The term is also used for a layer or membrane surrounding any of the body's organs.

Intelligence Quotient (IQ): This is the ratio between the mental age and chronological age multiplied by 100. Thus, if a boy of 10 years of age is found to have a mental age of 12 years, his IQ will be 120. On the other hand, if he is found to have a mental age of 8 years, his IQ will be 80.

Intelligence Test: A standardised procedure of mental assessment to determine an individual's intellectual ability. The result is produced as a score termed the intelligence quotient (IQ). The Wechsler Adult Intelligence Scale (WAIS) and one for children, the Wechsler Intelligence Scale for Children (WICS), are commonly used, as is the Stanford-Binet Scale. Assessments are made for educational purposes and to help in the diagnoses of people with possible mental retardation or intellectual deterioration.

Intensive Care Medicine: The origin of this important branch of medicine lies in the effective use of positive-pressure ventilation of the lungs to treat respiratory breathing failure in patients affected by polio-myelitis in an outbreak of this potentially fatal disease in Denmark in 1952.

Intent to Treat: Analysis of clinical trial results that includes all data from participants in the groups to which they were randomized even if they never received the treatment.

Intention-to-Treat: The principle that asserts that the effect of a treatment policy can be best assessed by evaluating on the basis of the intention to treat a subject (i.e., the planned treatment regimen) rather than the actual treatment given. It has the consequence that subjects allocated to a treatment group should be followed up, assessed, and analyzed as members of that group irrespective of their compliance to the planned course of treatment.

Inter Vention: The drug, device, therapy or process under investigation in a clinical trial which has an effect on outcome of interest in a study: e.g., quality of life, efficacy, safety, pharmacoeconomics.

Inter Vivos: A type of trust created during one's lifetime to hold property for the benefit of another person.

Interaction (Qualitative and Quantitative): The situation in which a treatment contrast (e.g., difference between investigational product and control) is dependent on another factor (for example, the centre). A quantitative interaction refers to the case where the magnitude of the contrast differs at the different levels of the factor; for a qualitative interaction, the direction of the contrast differs for at least one level of the factor.

Interaction: An effect that is produced when two drugs are administered together and the effectiveness of one or both of the drugs is either enhanced or diminished by the other.

Interactivity: Interactions in cyberspace with other people, information, and computers. Examples of interactivity include sending an e-mail message and filling out an Applied Clinical Trials subscription form at www.superfill.com/subscribe/apct.htm

Intercostal: The term applied to the nerves, vessels and muscles that lie between the ribs, as well as to diseases affecting these structures.

Interest: Any right or ownership in property.

Interfacial Layer: Region of space comprising and adjoining the phase boundary within which properties of matter are significantly different from the values in the adjoining bulk phases.

Interferon: It has been known for many years that one virus will interfere with the growth of another. In 1957, UK research workers isolated the factor that was responsible for the phenomenon, giving it the name of interferon. There are now known to be three human interferons. They are glycoproteins and are released from cells infected with virus or exposed to stimuli which mimic virus infection. They not only inhibit the growth of viruses; they also inhibit the growth and reduplication of cells, and this is the basis for their investigation as a means of treating cancer.

Interim Analysis: Any planned analysis intended to compare treatment arms with respect to efficacy or safety at any time prior to the formal completion of a trial, determined as part of the study protocol prior to subject enrollment.

Interleukin-1: Interleukins are lymphokines – that is, poly- peptides produced by activated lymphocytes. They are involved in signalling between cells of the immune system and are released by several cell types, including lymphocytes. They interact to control the immune response of cells and also participate in haemopoiesis.

Intermediate Care: Described by the UK government as 'a bridge between hospital and home' to speed discharge from acute care and provide recovery and rehabilitation services, this concept was a key element in the NHS Plan: plan for investment; a plan for reform, published in 2000. The government sees cottage hospitals, private nursing homes, and domiciliary and community set- tings as providing the heart of the proposed intermediate-care sector. Also in the plan, how- ever, is the warning that the NHS would meet the costs only of nursing care for nursing-home residents: personal care would in future be charged for.

Intermittent Claudication: A condition

occurring in middle-aged and elderly people, which is characterised by pain in the legs after walking a certain distance. The pain is relieved by resting for a short time. It is due to arteriosclerosis of the arteries to the leg, which results in inadequate blood supply to the muscles. Drugs usually have little effect in easing the pain, but useful preventive measures are to stop smoking, reduce weight (if overweight), and to take as much exercise as possible within the limits imposed by the pain.

Intermittent Effect: Biological change that comes and goes at intervals.

Intermittent Self- Catheterisation: A technique in which a patient (of either sex) inserts a disposable catheter through the urethra into the bladder to empty it of urine. It is increasingly used to manage patients with chronic retention of urine, or whose bladders do not empty properly

Intermittent: A term applied to fevers which continue for a time, subside completely and then return again. The name is also used in connection with a pulse in which occasional heartbeats are not felt, in consequence of irregular action of the heart.

Intern: A person registered with the College to complete his/ her final training period before becoming eligible to apply for a Certificate of Registration as a Pharmacist. An intern has a degree in pharmacy and may perform the controlled acts.

Internal Validity: Selection and comparison of index and comparison groups in such a manner that, apart from sampling error, the observed differences between these groups with respect to dependent variables under study may be attributed only to the hypothesized effect under investigation.

International Classification of Disease (ICD): A World Health Organisation classification of all known diseases and syndromes. The diseases are divided according to system (respiratory, renal, cardiac, etc.) or type (accidents, malignant growth, etc.). Each of them is given a three-digit number to facilitate computerisation. This classification allows mortality and morbidity rates to be compared nationally and regionally. A revised ICD is published every ten years; a similar classification is being developed for impairments, disabilities and handicaps.

International Pharmacy Graduate Program: A program consisting of two 8 week academic modules (CPS I and CPS II) offered by the Faculty of Pharmacy, University of Toronto. International pharmacy graduates are required to complete this program in conjunction with Structured Practical Training.

Internet Protocol (IP): Communications standards that govern the breakup of data messages into packets, the routing of the packets over the Internet from the sender to the destination, and the re-assembly of the packets into the original data messages at the destination.

Internet Ser Vice Provider (ISP): A company that provides access to the Internet for individuals and organizations. ISPs range in size from small local services to huge national providers, like Netcom and Com Cast, and international full-service providers like America Online (AOL).

Internet: A global system of computer networks that provides the infrastructure for e-mail, the World Wide Web, and other online activities.

Internode: The length of the axon (of a nerve cell) that is covered with a myelin sheath. The nodes of Ranvier, which have no myelin sheath, separate the internodes.

Internship: The final training period required to become a pharmacist in Ontario.

Interpretation (of Data or Findings): Evaluation of the observations from an investigation or study in order to determine their significance for human health, for the environment or for both.

Inter-Rater Reliability: The property of scales yielding equivalent results when used by different raters on different occasions.

Intersexuality: Intersexuality is a state of indeterminate sexuality of an individual, and may present in many different forms. A characteristic is that only one type of gonad – testis or ovary – is present; in a hermaphrodite both types are present. Inter- sexuality may be due to a fault in the genetic mechanism of sex determination as early as conception, or to later errors in sexual differentiation of the embryo and fetus, or after birth. Some cases may result from abnormal metabolism of the sex hormones, or may be drug- induced (for example, women given androgens

Interspecies Dose Conversion: Process of extrapolating from the doses of one animal species to another, for example from rodent dose to human equivalent.

Interstitial Cells: Also called leydig cells, these cells are scattered between the seminiferous tubules of the testis. Luteinising hormone from the anterior pituitary gland stimulates the interstitial cells to produce androgens, or male hormones.

Interstitial Fluid: The hydrogel that surrounds cells in soft tissues. It is a mucopolysaccaride starch gel, and the serum that leaves the blood capillaries flows through this gel, some to return to the exiting venous blood, some to enter the lymph system. There is an old medical axiom: the blood feeds the lymph, and the lymph feeds the cells. Interstitial fluid that flows through the starch colloid is this lymph.

Interstitial Pneumonia: Chronic form of pneumonia involving increase of the interstitial tissue and decrease of the functional lung tissue.

Interstitial: Interstitial is a term applied to indifferent tissue set among the proper active tissue of an organ. It is generally of a supporting character and formed of fibrous tissue. The term is also applied to the fluid always present in this in a small amount, and to diseases which specially affect this tissue, such as interstitial keratitis.

Intervention Study: Epidemiological investigation designed to test a hypothesized cause-effect relationship by modifying a supposed causal factor in a population.

Interventions: Primary interventions being studied. Types of interventions are Drug, Gene Transfer, Vaccine, Behaviour, Device, or Procedure.

Intestate: The term applied when an individual dies without a will.

Intestinal Reabsorption: Absorption further down the intestinal tract of a substance or substances that have been absorbed before and subsequently excreted into the intestinal tract, usually through the bile.

Intestine: All the alimentary canal beyond below the stomach. In it, most digestion is carried on, and through its walls all the food material is absorbed into the blood and lymph streams. The length of the intestine in humans is about

Intolerance: An adverse reaction of a patient to a drug or treatment.

Intoxication: A term applied to states of poisoning. The poison may be some chemical substance introduced from outside, for example, alcohol; or it may be due to the products of bacterial action, the bacteria either being introduced from outside or developing within the body. The term autointoxication is applied in the latter case.

Intra-: Prefix indicating inside or within. For example, intracellular: within a cell; intra-articular:

Intracoronary Artery Stenting: A narrowed or blocked coronary artery can compromise the blood supply to the heart muscle. A supportive tube or stent passed into each affected artery can restore the blood supply.

Intracranial Pressure: This is the pressure that is maintained by the brain tissue, intracellular and extracellular fluid, cerebrospinal fluid and blood. An increase in intracranial pressure may occur as a result of inflammation, injury, haemorrhage, or tumour in the brain tissue as well as of some congenital conditions. The pressure is measured by lumbar puncture in which a syringe attached to a mamometer (pressure-measuring device) is inserted into the cerebrospinal fluid surround- ing the lower part of the spinal cord.

Where continuous pressure monitoring is necessary, an in-dwelling device can be implanted into a cerebral ventricle. Normal pressure is around 10 mm of mercury (Hg), with the acceptable upper limit being 25 mm Hg.

Intracranial: Intracranial is the term applied to structures, diseases, etc. contained in or rising within the head.

Intracytoplasmic Sperm Injection: Intracytoplasmic sperm injection (ICSI) is the most significant therapeutic advance in male infertility treatment in the past 30 years. The technique is used when in vitro fertilisation (IVF) is not possible because the man has very few, motile, normal sperm, or when previous attempts at IVF have not produced a fertilised embryo. ICSI requires a single sperm which is injected directly into the cytoplasm of an egg previously retrieved from the woman. Once fertilised, the embryo is transferred to her uterus. For men with no sperm in the semen, it may be possible to retrieve sperm by needle aspiration of the

Intramuscular: An injection into the muscle layer.

Intrathecal: Intrathecal means within the membranes or meninges which envelop the Spinal Cord. The intrathecal space, between the arachnoid and the pia mater, contains the Cerebrospinal Fluid.

Intrauterine Contraceptive: Device (IUCD) A mechanical device, commonly a coil, inserted into the uterus to prevent conception, probably by interfering with the implantation of the embryo. For many women, IUCDs are an effective and acceptable form of contraception, although only about 10 per cent of women in the UK use them. The devices are of various shapes and made of plastic or copper; most have a string that passes through the cervix and rests in the vagina.

Intrauterine Insemination: A method of helping conception to occur when a man is infertile because his sperm cannot penetrate either the cervical mucus at the entrance of the uterus or the barriers that surround the ovum. The sperm, often treated chemically beforehand to increase motility, are injected directly into the uterus via the vagina.

Intravenous (IV) Additives: Intravenous preparations that involve the mixing of drugs into IV solutions, including chemotherapy medications for cancer treatments and hyperalimentation solutions for the administration of nutrition via an intravenous tube.

Intravenous Pyelogram: (Urogram) A procedure for getting X-ray pictures of the urinary tract. A radio-opaque medium is injected into a vein and, when it is excreted by the kidneys, the substance can be identified on X-rays. Any abnormalities in structure or foreign bodies such as calculi are outlined by the dye.

Intravenous: A term which means inside a vein. An intravenous injection is one that is given into a vein. Blood transfusions are given intravenously, as are other infusions of fluid.

Intrinsic Factor: One of two proteins secreted from the lining of the stomach whose sole purpose is (it seems) to cradle B12 in a pre-fitted styrofoam mold and (A) carry it through the Seven Levels of Digestive Hell until it reaches those few absorption sites in the last foot of small intestine that understand its "Special Needs" (sounds either sexually kinky or the airplane dinner label on kosher food for flying Hassidim jewelers) and finally (B) slip it from one protein to the other, and thence into the cell membranes where it is turn handed over to (C) the specialized blood protein that can carry it safely to the final target tissues (3 times out of 4, the bone marrow)

Intrinsic: Arising from the nature of a thing...native or inherent. Intrinsic asthma, as an example, arises from congestive inflammation, neurohormonal and auto-allergic conditions of the lung and bronchial membranes themselves, not from extrinsic causes, like Juniper pollen or a bee sting.

Intromission: The introduction of one organ or part into another. An example is the insertion of the erect penis into the vagina during copulation.

Introspection: The observation of one''s own thoughts or feelings. The term is generally applied to this process when it occurs to an abnormal extent in association with melancholia.

Introversion: (1) In physical terms, to turn a hollow structure into itself – for example, a length of the intestine may 'enter' the succeeding portion, also known as intussusception.

Introvert: Physically it means turning inside out. Psychologically, the term refers to an individual whose character looks inwards on him or herself and who may also be obsessive and have few friends.

Intubation: A procedure consisting in the introduction, through the mouth or nose into the larynx, of a tube designed to keep the air passage open at this point.

Intussusception: A form of obstruction of the bowels in which part of the intestine enters within that part immediately beneath it. This can best be under- stood by observing what takes place in the fingers of a tightly fitting glove as they turn outside-in when the glove is pulled off the hand. Mostly, the condition affects infants.

Invasion: The entry of bacteria into the body; the spread of cancer into normal, nearby tissues or organs.

Inverse Psoriasis: Smooth, red patches occur in the folds of the skin near the genitals, under the breasts, or in the armpits; symptoms may be worsened by friction and sweating.

Investigational Drug: This is an obsolete term. Refer to Special Access Program (SAP) drug.

Investigational New Drug (IND): A drug that has been authorized for clinical evaluation but that is not yet approved for sale for the condition under study. An IND is granted by Health Canada. A new drug, antibiotic drug, or biological drug that is used in a clinical investigation. This category also includes a biological product used *in vitro* for diagnostic purposes.

Investigational Product: A pharmaceutical form of an active ingredient or placebo being tested or used as a reference in a clinical trial, including a product with a marketing authorization when used or assembled (formulated or packaged) in a way different from the approved form, or when used for an unapproved indication, or when used to gain further information about an approved use. [ICH] CDISC includes test articles in its definition of investigational products.

Investigator: A person responsible for the conduct of the clinical trial at a trial site. If a trial is conducted by a team of individuals at a trial site, the investigator is the responsible leader of the team and may be called the principal investigator.

Investigator's Brochure: A compilation of the clinical and nonclinical data on the investigational product(s) which is relevant to the study of the investigational product(s) in human subjects.

Involucrum: The sheath of new bone which is formed round a piece of dead bone in, for example,

Involuntary Muscle: Muscle that does not operate under a person's conscious control. Involuntary muscle – also called smooth muscle, because the cells do not contain the striations that occur in voluntary muscle – is found in blood vessels, the heart, stomach, and intestines.

Involution: The process of change whereby the uterus returns to its resting size after parturition (birth). The term is also applied to any retrograde biological change, as in senility.

Ionisation: Ionisation means the breaking up of a substance in solution into its constituent.

Ionizing Radiation: Any radiation consisting of directly or indirectly ionizing particles or a mixture of both or photons with energy higher than the energy of photons of ultraviolet light or a mixture of both such particles and photons.

IRB (Institutional Review Board): IRBs review research protocols that involve human subjects. They ensure the ethical and safe treatment of study participants.

Iridectomy: The operation by which a hole is made in the iris of the eye – as, for example, in the treatment of glaucoma or as part of cataract surgery.

Iridology: The study of the iris. It is an old practice dating back to the days of Aristotle, and has been revived as a non-conventional treatment.

Iritis, Rheumatoid: An autoimmune (rheumatoid factor) inflammation of the iris. This is a face of rheumatoid arthritis seldom diagnosed, along with rheumatoid otitis. Although antiinflammatory drugs may be necessary, I would recommend starting off with simple things like Arctium, Rumex crispus and Taraxacum, along with alkalizing teas such as Nettles, Red Clover and Alfalfa. If they don't help enough you can still take the drugs.

Iritis, Viral: A viral infection of the iris. It appears red, swollen, and pupil contraction and relaxation is erratic and pulled. The usual cause is a herpes infection, often resident in the trigeminal nerve, and reoccurring during times of stress or sympathetic to a larger viral condition.

Irradiation: The use of naturally occurring isotopes, or artificially produced X-rays, in the killing of tumour cells. The amount of radiation is the adsorbed dose; the SI unit is the gray (Gy).

Irreversible Alteration: Change from normal structure or function that persists or progresses after cessation of exposure of the organism.

Irrigation: Irrigation is the method of washing-out wounds, or cavities of the body, like the bladder and bowels.

Irritable Bowel Syndrome (IBS): This is a common and generally benign condition of the colon, taking different forms but usually characterized by alternating constipation and diarrhea. There is often some pain accompanying the diarrhea phase. The bowel equivalent of spasmodic asthma, its main cause is stress, often accompanied by a history of GI infections. Adrenalin stress slows the colon and causes constipation, followed by a cholinergic rebound overstimulation of the colon. It is also called spastic colon, colon syndrome, mucous colitis, even chronic colitis. True colitis is a potentially or actually serious pathology.

Irritant: Any substance that produces irritation in a tissue. Examples are: stinging nettles, which cause pain and swelling; certain insect bites or stings which do the same; and antiriot gases such as CS gas, which cause the eyes to water and provoke coughing. Regular, unprotected handling of certain substances – for example, oil – can irritate the skin and cause dermatitis. Some people are more sensitive than others to the effect of irritants; sometimes an irritant will produce an allergic reaction and in serious cases anaphylaxis, which may require medical treatment.

Ischaemia: Local deficiency of blood supply and hence oxygen to an organ or tissue owing to constriction of the blood vessels or to obstruction. Bloodlessness of a part of the body, due to con- traction, spasm, constriction or blocking (by embolus or by thrombus) of the arteries: for example, of the heart.

Ischaemic Stroke: A stroke that occurs when the flow of blood to a part of the brain is interrupted by a partial or complete thrombosis of the supplying artery or arteries, or by a clot of blood that has detached itself from elsewhere in the circulatory system – for instance, a deep vein thrombosis (DVT) – and blocked a cerebral artery. Stroke is the second most common cause of death worldwide. Its treatment is difficult and prevention is best targeted at those who are at the highest absolute risk of stroke, because such people are likely to derive the greatest benefit. They generally have a history of occlusive vascular diseases such as previous ischaemic stroke or

Ischiorectal Abscess: An abscess arising in the space between the rectum and ischial bone and often resulting in a fistula. It may occur spontaneously or be secondary to an anal fissure, thrombosed haemorrhoids or other anal disease. The disorder is painful and usually accompanied by fever. Treatment is by a combination of antibiotics and surgery.

Ischium: Ischium is the bone which forms the lower and hinder part of the pelvis. It bears the weight of the body in sitting.

Ishihara's Test: A test for colour vision, introduced by a Japanese doctor, comprising several plates with round dots of different colours and sizes. It is also the name of a type of blood test for syphilis.

Islets of Langerhans: Groups of specialised cells distributed throughout the Pancreas, that produce three hormones: Insulin, Glucagon, and Somatostatin.

Iso-Immunisation: The immunisation of a person by an antigen they do not have but which is present in other people. For example, a rhesus-negative mother does not carry the rhesus antigen. If she carries a rhesus-positive baby, passage of the rhesus antigen from the baby into the mother's circulation may cause her to be isoimmunised. Her immune system may then pro- duce antibodies to the rhesus antigen. When she next becomes pregnant, if the baby is again rhesus positive, the mother will produce large amounts of anti-Rh antibodies which can enter the fetal circulation and cause its blood cells to break up.

Isolation: This is important when treating patients with serious infection or whose immune systems are severely compromised by illness or radio or chemotherapy. The procedure also protects staff caring for infectious patients.

Isometric: Of similar measurement. Isometric exercises are based on the isometric contraction of the muscles. Fibres are provoked into working by pushing or pulling an immovable object, but this technique prevents them from shortening in length. These exercises improve a person's fitness and builds up his or her muscle strength.

Isopathy: The treatment of a disease with the identical disease agent.

Isotonic: Having the same salinity as body fluids. You can make a quart of water isotonic by adding a slightly rounded measuring teaspoon of table salt to a quart of water.

Itch: Itch is a popular name for scabies.

Item: An individual clinical data element, such as a single systolic blood pressure reading. Items are collected together into item groups.

ITSA: An e-mail account or the name of the mail server for many UCSF faculty, students, and staff

ITSD: The former information technology services division; now the common information technology services division of the ministry of management services.

IUCD (IUD): Abbreviation for an intrauterine contraceptive device (coil). It acts mechanically to prevent conception, but the coil is not suitable for all women and has a failure rate of 2•3 per cent.

IUS: Intrauterine system – a hormonal contraceptive device that is placed in the uterus. It is a long- term reversible method of contraception.

IVF: In Vitro Fertilisation.

Ivory: Ivory, or dentine, is the hard material which forms the chief bulk of the teeth.

IVR: Interactive Voice Response system.

J

Jaundice: Pathological condition characterized by deposition of bile pigment in the skin and mucous membranes, including the conjunctivae, resulting in yellow appearance of the patient or animal.

Jaw: The name applied to the bones that carry the teeth. The two upper jaw-bones, the maxillae, are firmly fixed to the other bones of the face. The lower jaw, the mandible, is shaped some- what like a horseshoe, and, after the first year of life, consists of a single bone. It forms a hinge joint with the squamous part of the temporal bone, immediately in front of the ear. Both upper and lower jaw-bones possess deep sockets, known as alveoli, which contain the roots of the teeth.

Jejunum: Part of the small intestine.

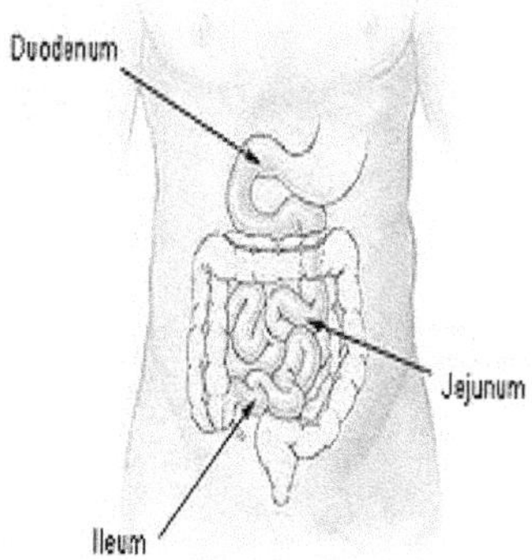

Jerk: A sudden involuntary movement. The term is often used to describe the tendon reflexes.

Joint Commission on Accreditation of Healthcare Organizations (JCAHO): The JCAHO is an independent, not-for-profit organization whose mission is to improve the quality of care provided to the public through the provision of health care accreditation and related services which support performance improvements in health care organizations.

Joint Effect: Simultaneous or successive effect of factors of diverse types (chemical, physical, biological) on an organism.

Joint Ownership: The ownership of property by two or more people, usually with the right of survivorship.-

Joint: The connecting point of two or more bones.

Joint-Mouse: A popular term for a loose body in a joint. It is found especially in the knee.

Joints: A joint is the articulation point between different parts of the skeleton, whether bone or cartilage. Joints are divided into those which are fixed or relatively fixed (fibrous and cartilaginous joints), and those which allow free movement (synovial joints). In the former, exemplified by the sutures between the bones of the skull, a layer of cartilage or fibrous tissue lies between the bones, binding them firmly together.

Amphiarthrodial joints, exemplified by the joints between the vertebral bodies, have a thick disc of fibro-cartilage between the bones. Although the individual joint is capable of very little movement, a series of these gives to the spinal column, as a whole, a flexible character.

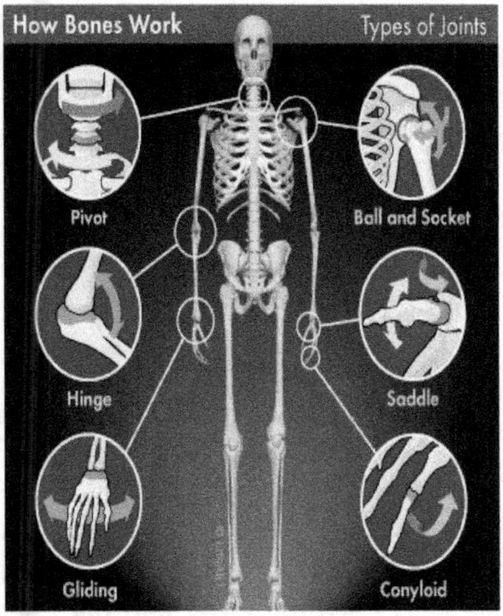

Joule: The unit of energy in the International System of Units. The official abbreviation is J. 4,186•8 J = 1 calorie (or kilocalorie).

Jugular: Jugular is a general name for any structure in the neck, but is especially applied to three large veins, the anterior, external and internal jugular veins, which convey blood from the head and neck regions to the interior of the chest.

Jurisdiction: A geographical area in which pharmacists are licensed and regulated.

Juvenile Idiopathic Arthritis (JIA): previously called juvenile rheumatoid arthritis and juvenile chronic arthritis, this is a set of related conditions of unknown causes.

K

Kaolin: Kaolin, or china clay, is a smooth white powder consisting of natural white aluminium silicate resulting from the decomposition of minerals containing felspar. It is used as a dusting powder for eczema and other forms of irritation in the skin. It is also used internally in cases of diarrhoea. Talc, French chalk and Fuller's earth are similar silicates. Kaolin poultice contains kaolin, boric acid, glycerin and various aromatic substances.

Kaolinosis: Kaolinosis is a form of pneumoconiosis caused by the inhaling of clay dust.

Kaposi 's Sarcoma: Once a very rare disease in western countries, although more common in Africa, Kaposi's sarcoma is now a feature of AIDS indeed, its increasing occurrence was one of the first pointers to the development of AIDS.

Kawasaki Disease: Also called mucocutaneous lymph node syndrome, this disorder of unknown origin occurs mainly in children under five and was first described in Japan. It is characterised by high fever, conjunctivitis, skin rashes and swelling of the neck glands. After about two weeks the skin from fingertips and toes may peel. The disease may last for several weeks before spontaneously resolving. It is possible that it is caused by an unusual immune response to infection.

Keloid: Hard lumpy nodule of the skin

due to over- growth of fibrous tissue in the dermis. It usually follows surgical or accidental trauma or burns, but, rarely, may complicate acne on the upper trunk. Most commonly seen in the skin over the sternum, shoulders and upper back; coloured people are particularly prone. Injection of corticosteroid into the keloid may cause partial resolution. Excision should be avoided.

Keratin: The substance of which horn and the surface layer of the skin are composed.

Keratinisation: Deposition of keratin in cells, in particular those in the skin. The cells become horny and flattened and lose their nuclei, forming hair and nails or hard areas of skin.

Keratomalacia: Softening of the cornea due to a severe vitamin A deficiency.

Keratosis: Also known as actinic keratosis; a rough, scaly area on exposed skin caused by chronic solar Kerion damage from exposure to sun. The face and backs of the hands are most commonly affected.

Kerion: A suppurating form of ringworm.

Kernicterus: The staining with bile of the basal nuclei of the brain, with toxic degeneration of the nerve cells, which sometimes occurs in severe haemolyticdisease of the newborn especially if prompt treatment by exchange transfusion has not been carried out. Rare nowadays, the result is a form of cerebral palsy.

Kernig's Sign: This is found in meningitis. A healthy person's thigh can be bent to a right-angle with the K body when the knee is straight; in cases of meningitis the knee cannot be straightened when the is bent this way not without causing the patient intense pain.

Ketamine: An anaesthetic drug, administered by intravenous or intramuscular injection and used mainly in children. The drug has good analgesic properties when used in subanaesthetic doses. One disadvantage is that when used as an anaesthetic, a high incidence of hallucinations occur. Ketamine is contraindicated in patients with hypertension.

Ketogenesis: The production of ketones in the body; abnormal ketogenesis may result in ketosis.

Ketogenic Diet: This contains such an excess of fats that acetone and other ketone bodies appear in the urine. The diet is sometimes used in the treatment of epilepsy and chronic infections of the urinary tract by Escherichia coli; butter, cream, eggs and fat meat are allowed, whilst sugar, bread and other carbohydrates are cut out as far as possible.

Ketone: Another name for acetone or dimethyl ketone. The term, ketone bodies, is applied to a group of substances closely allied to acetone, especially beta-hydroxybutyric acid and acetoacetic acid. These are produced in the body from imperfect oxidation of fats and protein foods, and are found in especially large amount in severe cases of diabetes mellitus. Ketonuria is the term applied to the presence of these bodies in the urine.

Ketones: Chemical substances that the body makes when it doesn't have enough insulin in the blood. When ketones build up in the body for a long time, serious illness or coma can result.

Ketosis: A condition in which an excessive amount of ketones are produced by the body and these accumulate in the bloodstream. The affected person becomes drowsy, suffers a headache, breathes deeply, and may lapse into a coma. The condition results from an unbalanced metabolism of fat, which may occur in Diabetes Mellitus or starvation.

Keynote: A unique feature or fundamental aspect of a remedy.

Keyword: A word that a patient can attach to their PharmaNet patient record, which limits access to only those pharmacists with whom the patient shares the keyword.

Khat: Tender leaves of a shrub that grows in the middle east called catha edelis. The leaves are wrapped around betel nuts and chewed: the result is a feeling of euphoria and an ability to tolerate harsh living conditions.

Kidneys: Twin organs found in the lower part of the back. The kidneys purify the blood of all waste and harmful material. They also control the level of some helpful chemical substances in the blood.

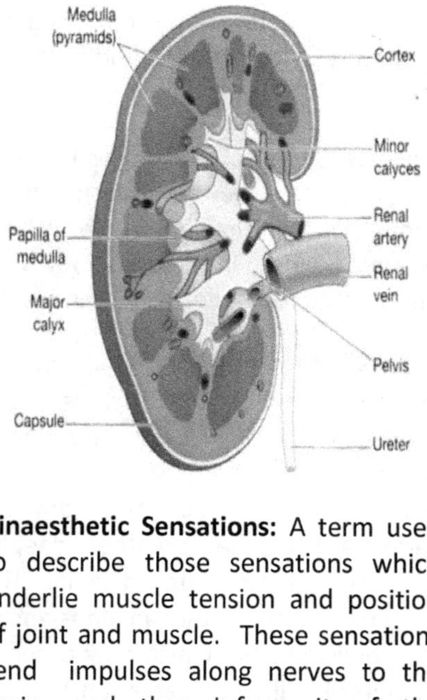

Kinaesthetic Sensations: A term used to describe those sensations which underlie muscle tension and position of joint and muscle. These sensations send impulses along nerves to the brain, and thus inform it of the position of the limb in space and of the relative position to each other of individual muscles and muscle-groups and of joints.

Kinins: Substances present in the body which are powerful vasodilators. They also induce pain and are probably involved in the production of the headache of migraine. In addition, they play a part in the production of allergy and anaphylaxis.

Kiss of Life: Emergency mouth-to-mouth resuscitation of an unconscious person.

Klebsiella: A bacteria genus of the Enterobacteriaceae. K. pneumoniae is implicated in much pneumonia, particularly when it is a secondary infection following a simple chest cold.

Kleptomania: A psychological disorder in which the person afflicted has an irresistible compulsion to steal things, without necessarily having any need for the object stolen.

Klinefelter's Syndrome: The original syndrome described by Klinefelter consisted of gynaecomastia, testicular atrophy and infertility. Intelligence was unimpaired. Patients have been described who have associated mental defects and striking tallness of stature, but the only constant feature of the syndrome is testicular atrophy with resulting azoospermia and infertility.

Klumpke's Paralysis: Injury as a result of the stretching of a baby's brachial plexus during its birth may cause partial paralysis of the arm with atrophy of the muscles of the forearm and hand.

Knock-Knee: Knock-knee, or genu valgum, is a deformity of the lower limbs in such a direction that when the limbs are straightened the legs diverge from one another. As a result, in walking the knees knock against each other. The amount of knock-knee is measured by the distance between the medial malleoli of the ankles, with the inner surfaces of the knee touching and the knee-caps facing forwards. The condition is so common in children between the ages of 2–6 years that it may almost be regarded as a normal phase in childhood. When marked, or persisting into later childhood, it can be corrected by surgery (osteotomy).

Koch's Bacillus: The original name for Mycobacterium tuberculosis, which causes tuberculosis. It stems from the name of the German doctor who first identified the bacillus.

Koilonychia: The term applied to nails that are hollow and depressed like a spoon, a condition sometimes associated with chronic iron deficiency.

Koplik's Spots: Bluish-white spots appearing on the mucous membrane of the mouth in cases of measles about the third day, and forming the first part of the rash in this disease.

Korsakoff's Syndrome: A form of mental disturbance occurring in chronic alcoholism and other toxic states, such as uraemia, lead poisoning and cerebral syphilis. Its special features are talkativeness with delusions in regard to time and place – the patient, although clear in other matters, imagining that he or she has recently made journeys.

Krebs Cycle: A series of key cellular chemical reactions starting and ending with oxaloacetic acid. Also called the citric acid or tricarboxylic acid cycle, it produces energy in the form of Adenosine Triphosphate (ATP) and is the last stage in the biological oxidation of fats, proteins, and carboydrates. Named after Sir Hans Krebs, a German biochemist working in England in 1900, who won the Nobel Prize for his discovery.

Kuntscher Nail: A surgical nail inserted into the medulla of a fractured bone to fixate it. First introduced by a 20th-century German surgeon.

Kupffer Cells: Star-shaped cells present in the blood-sinuses of the liver. They form part of the reticulo- endothelial system and are to a large extent responsible for the breakdown of haemoglobin into the bile pigments.

Kuru: A slowly progressive, fatal disease due to spongiform degeneration in the central nervous system, particularly the cerebellum. It is confined to the Fore people in the Eastern Highlands of New Guinea, and causes increasingly severe muscular trembling. Kuru is believed to be due to an infection with a prion, similar to that causing Creutzfeldt-Jakob Disease (CJD), acquired from the cannibalistic rite of eating the organs, particularly the brains, of deceased relatives (out of respect). This origin of the disease was suggested by the fact that originally it was a disease of women and children.

Kveim Test: The characteristic histological test used for the diagnosis of sarcoidosis. The test involves an intradermal injection of sarcoid spleen tissue. If positive, noncaseating granulomata are seen at the injection site in 4–6 weeks. A positive test is highly specific for sarcoid, but if negative, this would not be excluded.

Kwashiorkor: One of the most important causes of ill health and death among children in the tropics. It is predominantly a deficiency disease due to a diet short of protein; there is also some evidence of a lack of the so-called essential fatty acids. It affects typically the small child weaned from the breast and not yet able to cope with an adult diet, or for whom an adequate amount of first-class protein is not available, and it is mainly found in the less well-developed countries.

Kyphoscoliosis: A combination of scoliosis and kyphosis in which the spine is abnormally curved sideways and forwards. The condition may be the result of several diseases affecting the spinal muscles and vertebrae, or it may happen during development for no obvious reason. Although braces may reduce the deformity, an operation may be necessary to correct it.

Kyphosis: The term applied to curvature of the spine in which the concavity of the curve is directed forwards.

L

Labeling: Description of the drug and summary of use, safety, and effectiveness that must be approved by FDA.

Labia: The labia majora and labia minora are the outer and inner lip-like folds of skin surrounding the entrance to the vagina.

Labium: Labium is the Latin word for a lip or lip-shaped organ.

Labyrinth: A convoluted system of structures forming the inner ear and involved in hearing and balance.

Labyrinthitis: Inflammation of the labyrinth of the ear.

Laceration: A wound to the skin or surface of an organ which results in a cut with irregular edges (cf. an incision produced with a knife, which has smooth, regular edges).

Lachrymitis: Inflamed lacrimal or tear ducts.

Lacrimal Bones: The smallest bones of the face, one forming part of the bony structure of each orbit containing an eye.

Lacrimation: Crying, or the secretion of an excess quantity of tears.

Lacrimator: Substance that irritates the eyes and causes the production of tears or increases the flow of tears.

Lactase: An enzyme produced by glands in the small intestine which changes lactose (milk sugar) into glucose and galactose during the process of digestion.

Lactation: The period during which an infant is suckled on the mother's breast.

Lacteal: a lymphatic vessel that transmits chyle from the intestine.

Lacteals: Specialized lymph formations found in the small intestine mucosa. Together with enzymatic activities in the submucosa, they collect digested fats into stable transport bubbles called chylomicrons, and draw them up into the lymph system. There they are gradually leeched into the blood as the lymph passes upwards through the body, the remainder discharged into the venous blood with the lymph...12-24 hours later. Time-Released fat capsules. Fats lower the blood charge and make it sticky, which can interfere with vascular capabilities; the sideways bypassing of the blood in this manner spreads the fats out over long periods. The rest of the digested constituents can happily flow up to the liver through the portal system, unsludged, and the liver itself therefore has little lipid stress to face.

Lactic Acid: A colourless, syrupy, sour

liquid, which is produced by the action of a bacterium upon lactose, the sugar found in milk. The growth of this organism and consequent formation of lactic acid cause the souring of milk, and the same change takes place to a limited extent when food is long retained in the stomach.

Lactobacillus: A genus of gram-positive, acid-resistant bacteria in the Lactobacillaceae family. We know of lactobacillus because of its use in making yogurt and the conventional wisdom of taking it in one form or another after antibiotic therapy, but it is an integral part of the colon and mouth flora, and is the critical acidifying agent in vaginal flora. There is a growing body of rather ignored data showing the value of regular consumption of a lactobacillus-containing food in immunesuppr-ession, slow virus, and candidiasis conditions.

Lactose Intolerance: is due to lack in the intestine of the enzyme known as lactase which is responsible for the digestion of lactose, the sugar in milk. The result is that drinking milk or eating milk- containing products is followed by nausea, a sensation of bloating, or distension, in the gut, abdominal pain and diarrhoea. (Similar disturbances after taking milk may also occur in those who do not lack lactase but have an allergy to milk protein). Treatment is by means of a low-lactose diet avoiding fresh or powdered milk and milk puddings. Many can tolerate fermented milk products, as well as the small amounts of milk used in baking and added to margarine and sausages. However, infamts may have to be fed exclusively on a lactose-free formula as even breast milk may produce symptoms.

Lactose: The official name for sugar of milk.

Lactulose: An osmotic laxative, lactulose is a semisynthetic disaccharide – a type of carbohydrate – which is not absorbed from the gastrointestinal tract. It reduces the acidity of faeces.

Lacuna: A small pit or depression.

Lamella: A small disc of glycerin jelly, 3 mm (1/8 inch) in diameter, containing an active drug for application to the eye. It is applied by insertion behind the lower lid.

Laminar Flow Hoods: Worktables that provide an enclosed sterile environment to protect solutions from contamination while they are prepared.

Laminectomy: An operation in which the arches of one or more vertebrae in the spinal column are removed so as to expose a portion of the spinal cord for removal of a tumour, relief of pressure due to a fracture, or disc protrusion.

Lanceolate: A leaf that is lance-shaped.

Lanolin: Derived from wool fat, it is an ingredient of many ointments and creams but may cause allergic contact dermatitis.

Lanugo: Soft fine hair covering the fetus. It disappears by the ninth month of gestation and is therefore only seen on premature babies.

Laparoscope: An instrument consisting, essentially, of a rigid or flexible cylinder, an eyepiece and a light source, which is inserted through a small incision into the abdominal cavity (which has already been distended with carbon dioxode gas). The laparoscope allows the contents of the abdominal cavity to be examined without performing a laparotomy. Some operations may be performed using the laparoscope to guide the manipulation of instruments inserted through another small incision – for example, sterilisation; cholecystectomy.

Laparoscopy: Also called peritoneoscopy, this is a technique using an instrument called an endoscope for viewing the contents of the abdomen. The instrument is inserted via an incision just below the umbilicus and air is then pumped into the peritoneal (abdominal) cavity. Visual inspection may help in the diagnosis of cancer, appendicitis, salpingitis, and abnormalities of the liver, gallbladder, ovaries or gastro-intestinal tract. A biopsy can be taken of tissue suspected of being abnormal, and operations such as removal of the gallbladder or appendix may be carried out.

Laparotomy: A general term applied to any operation in which the abdominal cavity is opened. A laparotomy may be exploratory to establish a diagnosis, or carried out as a preliminary to major surgery.

Larva Migrans: A self-limiting, intensely itching skin eruption caused by nematode (roundworm) larvae, usually of the dog and cat hookworm. The migrating larvae leave red, raised, irregular tracks in the skin, often on the foot and less frequently elsewhere. The disease is usually acquired by people who take their holidays on tropical beaches. It can be cured by a three-day course of oral Albendazole.

Larva: The pre-adult stage in insects and nematodes occurring between the egg and the sexually mature adult.

Larvicide: Substance intended to kill larvae.

Laryngeal Reflex: A 'protective' cough occurring as a result of irritation of the larynxfor example, a small particle of food may be accidentally 'inhaled' into the larynx, which reacts with an expulsive cough to prevent the food from entering the lungs.

Laryngectomy: Operation for removal of the larynx.

Laryngitis: Inflammation of the larynx, usually implying hoarseness or aphonia

Laryngoscope: Examination of the larynx may be performed indirectly with use of a laryngeal mirror, or directly by use of a laryngoscope a type of endoscope. The direct examination is usually performed under general anaesthetic.

Laryngospasm: Reflex spasmodic closure of the sphincter of the larynx, particularly the glottic sphincter.

Larynx: Main organ of voice production, the part of the respiratory tract between the pharynx and the trachea.

Laser Surgery: Surgery that uses a strong ray of special light, called a laser, to treat damaged parts of the body. Laser surgery can help treat some diabetic eye diseases.

Laser: Laser stands for light amplification by stimulated emission of radiation. The light produced by a laser is of a single wavelength and all the waves are in phase with each other, allowing a very high level of energy to be projected as a parallel beam or focused on to a small spot. Various gases, liquids and solids will emit light when they are suitably stimulated. A gassed laser is pumped by the ionising effect of a high-voltage current. This is the same process as that used in a fluorescent tube. Each type of laser has a different effect on biological tissues and this is related to the wavelength of the light.

Lasik: Acronym for laser insitu keratomileusis, which is a variety of surgery of the eye used to correct severe myopia (short-sightedness). A thin flap of the cornea is lifted, the area underneath reshaped using an excimer laser, and the flap is then.

Lassar's Paste: Officially known as Zinc and Salicylic Acid Paste, BP, this preparation is an old remedy for eczema.

Lassitude: Weakness; exhaustion.

Last Subject In (LSI): Date and time the last subject to participate in a clinical trial is enrolled. The subject will have met the inclusion/exclusion criteria to participate in the trial and have signed informed consent. See also enroll.

Latanoprost: An analogue of prostaglandin, used to treat open-angled glau coma and raised intra-ocular pressure in the eye. Delivered as an eye drop, the drug is used in patients who cannot tolerate, or who fail to respond to, other treatments.

Latent period: Delay between exposure to a disease-causing agent and the appearance of manifestations of the disease: also defined as the period from disease initiation to disease detection.

Lateral: At or on the side, usually from a stem.

Latissimus Dorsi: A large, flat, triangular muscle in the back.

Laughing Gas: A popular name for nitrous oxide gas.

Lavage: Irrigation or washing out of a hollow organ or cavity such as the stomach, intestine or the lungs.

Laxative: Drugs or other substances used to treat constipation. Also called aperients or purgatives, laxatives are classified according to their mode of action. The four main groups are bulk, stimulant, faecal softeners and osmotics. In addition, bowel-cleansing solutions are used before surgery, endoscopy, or radiological examination of the colon, to ensure that the bowel is clear of solid matter. However, these are not procedures for treating constipation.

LCA: Low Cost Alternative.

LCD: Liquid crystal display.

LDL (Low Density Lipids): The levels are usually indicative of liver function and metabolic tendencies, and the relative proportions of LDL, VLDL and HDL show relationships between caloric intake, anabolic energy, skeletal muscle metabolism and adipose tissue health. They are not innately wrong, anymore than is cholesterol; both are absolutely necessary for health. It's all a matter of proportion, and the relationship between consumption and tissue needs.

LDL (Low-Density Lipoprotein): A combined protein and fatlike substance. Rich in cholesterol, it tends to stick to the walls in the arteries. Sometimes called "bad cholesterol."

Lead Poisoning: Lead and lead compounds are used in a variety of products including petrol additives (in the UK, lead-free petrol is now mandatory), piping (lead water pipes were once a common source of poisoning), weights, professional paints, dyes, ceramics, ammunition, homeopathic remedies, and ethnic cosmetic preparations. Lead compounds are toxic by ingestion, by inhalation and, rarely, by skin exposures.

Leaflet: A small leaf that is part of a compound leaf.

Learning Disability: Learning disability, previously called mental handicap, is a problem of markedly low intellectual functioning. In general, people with learning disability want to be seen as them- selves, to learn new skills, to choose where to live, to have good health care, to have girl- friends or boyfriends, to make decisions about their lives, and to have enough money to live on. They may live at home with their families, or in small residential units with access to work and leisure and to other people in ordinary communities. Some people with learning dis- abilities, however, also have a mental illness.

Learning: A change in neural function as a consequence of experience.

Leber's Disease: A hereditary disease in which blindness comes on at about the age of 20.

Lecithin: A very complex fat found in

various tissues of the body, but particularly in the brain and nerves, of which it forms a large part. It is also found in large quantities in the yolk of an egg.

LED: Light emitting diode.

Leeches: Animals provided with suckers surrounding the mouth, and living a semi-parasitic life, their food being mainly derived from the blood of other animals. They abstract blood by means of the sucker, which has several large, sharp teeth. Land leeches live in tropical forests and can attach themselves to a person's ankles and lower legs. Aquatic leeches are found in warm water and may attach themselves to swimmers. Their bites are painless, their saliva reducing the clot- ting properties of blood with hirudin; the result is that the wound continues to bleed after the leech has detached itself or been gently removed (lighted match, alcohol, salt and vinegar are effective removal agents). The medicinal leech, Hirudo medicinalis, was formerly employed for the abstraction of small quantities of blood in inflammatory and other conditions. Nowadays it is occasionally used to drain haematomas and to manage healing in certain types of plastic surgery.

Left to Right Shunt: A term used when a hole in the septum (internal wall) of the heart allows blood to flow from the systemic circulation properly confined to the left side of the heart to the pulmonary circulation, confined to the right.

Legally Acceptable Representative: An individual or juridical or other body authorized under applicable law to consent, on behalf of a prospective subject, to the subject's participation in the clinical trial.

Legend Drug: Any drug that requires a prescription under federal code 21 USC 353(b).

Legionnaire's Disease: A form of Pneumonia due to a bacterium known as Legionella pneumophila, so-called because the first identified outbreak was in a group of US ex-servicemen (members of the American Legion). Inhalation of water aerosols seems the most likely way that people acquire the disease, for example from air-conditioning outlets. Some rubber outlets in showers and taps are able to support the growth of legionnellae so that high concentrations of the organism are released when the tap is first used in the morning. In the presence of the disease, the treatment of infected water systems is essential by cleaning, chlorination, heating or a combination of all three.

Leiomyoma: A benign tumour made up of unstriped or involuntary muscle fibres.

Leishmaniasis: A group of infections caused by parasites transmitted to humans by sandflies. Visceral leishmaniasis (kala-azar) A systemic infection caused by Leishmania donovani which occurs in tropical and subtropical Africa, Asia, the Mediterranean littoral (and some islands), and in tropical South America. Onset is frequently insidious; incubation period is 2–6 months. Enlargement of spleen and liver may be gross; fever, anaemia, and generalised lymphadenopathy are usually present. Diagnosis is usually made from a bone-marrow specimen, splenic-aspirate, or liver-biopsy specimen; amastigotes (Leishman-Donovan bodies) of L. donovani can be visualised. Several serological tests are of value in diagnosis.

Leiter Der Klinischen Prufung: Under the German Drug Law, the physician who is head of the clinical investigation.

Lentigo: Lentigines (freckles) are brown macules varying in diameter from 1–10 mm or more. Simple lentigines arise in childhood, not necessarily on exposed areas. They may also occur on the lips and are harmless and usually very small. Solar or actinic lentigines are common on the face, neck and backs of the hands in older people and reflect the total cumulative lifetime's exposure to sunlight.

Leproma: A nodule in the skin occurring in leprosy.

Leprosy: Also known as Hansen's disease, this is a chronic bacterial infection caused by Mycobacterium leprae affecting the skin, mucous membranes, and nerves. Infection is now almost confined to tropical and subtropical countries – mostly in Africa and India. There are two distinct (polarised) clinical forms: tuberculoid and lepromatous. The former usually takes a benign course and frequently burns out, whereas the latter is relentlessly progressive; between these two polar forms lies an intermediate/dimorphous group. Susceptibility may be increased by malnutrition. Nasal secretions (especially in lepromatous disease) are teeming with M. leprae and constitute the main source of infection; however, living in close proximity to an infected individual seems necessary for someone to contract the disease. M. leprae can also be transmitted in breast milk from an infected mother.

Leptospira: A group, or genus, of spiral micro-organisms, normally found in rodents and other small mammals in which they cause no harm. When transmitted to humans by these animals, either directly or indirectly as through cows, they give rise to various forms of illness.

Leptospirosis: The disease caused by infection with leptospira. The three most common members of this group in the United Kingdom are L. icterohaemorrhagiae, L. canicola, and L. hebdomadis. It is an occupational hazard of farmers, sewage and abattoir workers, fish cutters and veterinary surgeons, but the infection can also be acquired from bathing in contaminated water. The disease varies in intensity from a mild influenza- like illness to a fatal form of jaundice due to severe liver disease. The kidneys are often involved and there may be meningitis. Penicillin or tetracycline are the usual treatment, but unless they are given early in the disease their effect is limited. The tetracycline antibiotic doxycycline given once weekly during periods of exposure prevents infection.

Lesbian: A female homosexual; lesbians form about 2 per cent of the female population in the UK. Some engage in active sexual behaviour with another woman, with masturbation, mutual stimulation of the clitoris and oral sex being the usual techniques for achieving orgasm.

Lesion: Area of pathologically altered tissue. Injury or wound. Infected patch of skin.

Lethal Concentration: Concentration of a potentially toxic substance in an environmental medium that causes death following a certain period of exposure.

Lethal Dose: Amount of a substance or physical agent (radiation) that causes death when taken into the body by a single absorption (denoted by LD).

Lethal Synthesis: Metabolic formation of a highly toxic compound from one that is relatively non-toxic (bioactivation), often leading to death of affected cells.

Lethal: Deadly; fatal; causing death.

Lethargy: Lethargy, or lassitude, means a loss of energy. It is a common presenting complaint both to gen- eral practitioners and to hospital consultants. It may have a physical cause or a psychological cause; it may be the result of inadequate rest, environmental noise, boredom, insomnia or recent illness. Certain medicinal drugs can cause lethargy, the most common being beta blockers and diuretics, and drugs of abuse may also be a cause. The common psychosocial problems producing lethargy are depression and anxiety.

Leucocytes: The scientific name for white blood cells. Leucocytes contain no haemoglobin so are colourless, and have a well-formed nucleus. Healthy people have around 8,000 leucocytes per cubic millimetre of blood. There are three main classes of white cells: granulocytes, lymphocytes and monocytes.

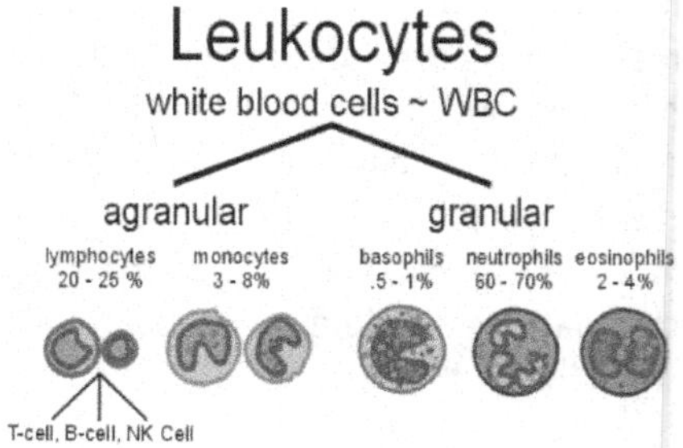

A White Blood Cell or Leukocyte

Leucocytosis: A condition in which the polymorphonuclear leucocytes in the blood are increased in number. It occurs in many different circumstances, and forms a valuable means of diagnosis in certain diseases; however, the condition may occur as a normal reaction in certain conditions (e.g. pregnancy, menstruation, and during muscular exercise). It is usually due to the presence of inflammatory processes the increased number.

Leucoderma: Leucoderma, or leucodermia, is a condition of the skin in which areas of it become white, as the result of various skin diseases.

Leucopenia: A condition in which the number of leucocytes in the blood is greatly reduced by, say, anaemia or cancer. It is also a dangerous sign in severe septicaemia.

Leucorrhoea: Discharge of mucus from the vagina. It may be whitish or yellowish and is normal in some women, usually increasing before and after menstruation. It is distinct from abnormal discharges with an offensive smell and yellow or

Leukaemia: Leukaemia is an umbrella term for several malignant disorders of white blood cells in which they proliferate in a disorganised manner. The disease is also characterised by enlargement of the spleen, changes in the bone marrow, and by enlargement of the lymph glands all over the body. The condition may be either acute or chronic.

Leukocytes: White blood cells, of whatever race or creed.

Leukocytosis: Having abnormally high numbers of white blood cells, usually the result of a non-viral infection.

Leukopenia: Having abnormally low numbers of white blood cells.

Leukoplakia: A white plaque on mucous membranes caused by overgrowth of the tissues. It is occasionally a pre-cancerous condition.

Level P2: Meet future regulatory requirements or future Air Force goals, policies, and regulations.

Level P3: Go beyond goals and legal requirements.

Level PI: Eliminate dependence on ODS

LGBT: Lesbian, gay, bisexual, transgender

LgPow: Logarithm of base 10 of the partition coefficient of a substance between octanolol and water: as an empirical measure for lipophilicity used for calculating bioaccumulation, fish toxicity, membrane adsorption and penetration etc.

Libido: The natural desire for sexual intercourse. Lack of desire or diminished libido may occur in any general medical illness as well as in endocrine diseases when there is a lack of production of the sex hormones. The strength or weakness of the sexual drive may be associated with psychiatric diseases; it may also be the result of certain drugs. It must be distinguished from impotence, where the desire for intercourse is normal.

Lichen: Lichen, or lichenification, is a term used to describe a thickening of chronically inflamed skin to give a tree-bark-like appearance. Lichen simplex (neurodermatitis) is a form of eczema perpetuated by constant rubbing of the affected skin. Typically, well-defined plaques occur on one or both sides of the nape of the neck, on the ulnar forearm near the elbow, or on the sides of the calves. It is often associated with emotional stress.

Life Cycle Analysis: The comprehensive examination of a product's environmental and economic effects throughout its lifetime including new material extraction, transportation, manufacturing, use, and disposal.

Life Cycle Cost Accounting: A method of accounting that provides incentive by accounting for cost savings in reductions of material use and waste disposal.

Life Cycle Cost: The amortized annual cost of a product, including capital costs, installation costs, operating costs, maintenance costs and disposal costs discounted over the lifetime of the product.

Life Cycle Economic Analysis: An evaluation of the costs associated with the use of hazardous material or potential alternatives over the life of the hazardous material or alternative.

Life Cycle of Hazardous Material: The period starting when the use or potential use of hazardous material is first encountered and extending as long as the actual material or its after-effects, such as a discarded residual in a landfill, have a bearing on the cost.

Life Expectancy: The number of years an individual can expect to live. Life expectancy at birth in most western countries is over 75 years for women (nearly 80 in the UK in 1998) and over 70 for men (over

Life Insurance Trust: A trust that has the proceeds of an individual's life insurance policy as its principal.

Life-long Exposure: Subjection to a potentially toxic substance during the whole lifetime.

Lifestyle Medicines: Drugs used for non-health problems or for disorders that are in the grey area between a genuine health need and a desire to change a 'life- style failing' by the use of medication. Examples are: Sildenafil Citrate, which is prescribed for men unable to achieve penile erection (erectile dysfunction); and Orlistat, a drug used to combat obesity.

Ligaments: Strong bands of fibrous tissue which serve to bind together the bones entering into a joint. In some cases they are cord-like; in others, flattened bands – whilst most joints are surrounded by a fibrous capsule or capsular ligament.

Ligation: Tying-off for example, of a blood vessel by completely encircling it with a tight band, usually of catgut or some other suture matfrial.

Ligature: A cord or thread used to tie around arteries in order to stop the circulation through them, or to prevent escape of blood from their cut ends. Ligatures are generally made of catgut or silk, and are tied with a reef-knot.

Light Reflex: Pupillary constriction in the eye in response to light. The direct light reflex involves pupillary constriction in the eye into which a light is shone; the consensual light reflex is the pupillary constriction that occurs in the other eye. The afferent or inward pathway of the reflex is.

Limacide: Substance intended to kill mollusca including the gastropod mollusc, Limax.

Limbic System: A circular, complex system of nerve pathways in the middle of the brain. This connected cluster of neuronal nuclei is involved in the working of the autonomic nervous system, helping to control the expression of instinct and mood in activities of the motor systems and the endocrine glands of the body. The brain regions involved include the amygdala (an almond-shaped basal ganglion deep in each cerebral hemisphere), and the hypothalamus. If the limbic system is impaired by injury or disease, the person may suffer abnormal emotional.

Limbic/Hypothalamus: Broadly the accumulative process of emotional and metabolic evaluation, as carried on by the various parts of the brain that are part of the ad hoc "evaluations" committee (the limbic system) and those changes in metabolism that, based on the evaluations, are acted out in the whole body by the hypothalamus. The hypothalamus, the main part of the system with tools, acts through a blood translator, the pituitary gland.

Limit Test: Acute toxicity test in which, if no ill-effects occur at a pre-selected maximum dose, no further testing at greater exposure levels is required.

Limit Value (LV): Limit concentration at or below which Member States of the European Community must set their environmental quality standard and emission standard for a particular substance according to Community Directives.

Limited Evidence: According to the US EPA's guidelines for Carcinogen Risk Assessment, "limited evidence" is a collection of facts and accepted scientific inferences that suggests that an agent may be causing an effect, but this suggestion is not strong enough to be considered established fact.

Linctus: A term applied to any thick, syrupy medicine. Most of these are remedies for excessive coughing.

Lindane: An insecticide used in scabies and lice infestations however, head lice are now often resistant. Used excessively, lindane may be neurotoxic (damage the nervous system).

Linea Alba: The line of fibrous tissue stretching down the mid line of the belly from the lower end of the sternum to the pubic bone. The linea alba gives attachment to the muscles of the wall of the stomach.

Linea Nigra: During pregnancy, the linea alba becomes pigmented and appears as a dark line down the middle of the belly, and is called the linea nigra.

Linearized Multistage Model: Sequence of steps in which (a) a multistage model is fitted to tumour incidence data; (b) the maximum linear term consistent with the data is calculated; (c) the low-dose slope of the dose-response function is equated to the coefficient of the maximum linear term; and (d) the resulting slope is then equated to the upper bound of potency.

Lingual: Referring or related to the tongue: for example, the lingual nerve supplies sensation to the tongue.

Liniment: A liquid containing therapeutic agents for topical application. It may be an alcohol, oil, or water preparation.

Linkage: A description that in genetics means circumstances in which two or more genes lie near each other on a chromosome and so may well be inherited together.

Linoleic Acid: An unsaturated fatty acid occurring widely in the glycerides of plants. It is an essential nutrient for mammals, including humans.

Lint: This was originally made of teased-out linen; now it consists of a loose cotton fabric, one side of which is fluffy, the other being smooth and applied next to the skin when the surface is broken. Marine lint consists of tow impregnated with tar, and is used where large quantities of some absorbent and deodorising dressing are required. Cotton lint is impregnated with various substances, the most common being boracic lint. Lint containing perchloride of iron (15 per cent) is valuable as a styptic.

Lipaemia: The presence of an excessive amount of fat in the blood.

Lipase: An enzyme widely distributed in plants, and present also in the liver and gastric and pan- creatic juices, which breaks down fats to the constituent fatty acids and glycerol.

Lipid: A descriptive term, rather than chemical one, for fats. Broadly, it means true fats (like triglycerides), lipoids (like phospholipids) and sterols (like cholesterol).

Lipidosis: Any disorder of lipid metabolism in body cells. Some hereditary disorders cause deposition of lipids within the brain.

Lipodystrophy: A congenital maldistr-ibution of fat tissue. Subcutaneous fat is totally absent from a portion of the body and hypertrophied in the remainder. Another form of lipodystrophy occurs at the site of insulin injections, but is much less frequently seen nowadays; the new, synthetic preparations of insulin are pure and unlikely to cause this reaction, which was not uncommon with the older preparations. Occasionally the converse occurs at the site of insulin injections, where the lipogenic action of insulin stimulates the fat cells to hypertrophy. This can also be disfiguring and usually results from using the same site for injections too frequently.

Lipofuscin: The brown waste material deposited in skin and nerve cells that is commonly called "age spots." Lipofuscin is made of free radical damaged proteins and fats.

Lipoid Factor: An agent involved in the clotting mechanism of the blood. It helps in the activation of thromboplastin in the blood plasma.

Lipolysis: The enzymatic breaking-down of fat.

Lipoma: A tumour mainly composed of fat. Such tumours arise in almost any part of the body, developing in fibrous tissues – particularly in that beneath the skin. They are benign in nature, and seldom give any trouble beyond that connected with their size and position. If large, they can be excised.

Lipophilic: Having an affinity for fat and high lipid solubility: a physicochemical property which describes a partitioning equilibrium of solute molecules between water and an immiscible organic solvent, favouring the latter, and which correlates with bioaccumulation.

Lipophobic: Having a low affinity for fat and a high affinity for water.

Lipoproteins: Compounds containing lipids and proteins. Most lipids in blood plasma are present in this form and are characterised according to their densities: very low (VLDL), intermediate (IDL), low (LD), high (HDL) and very high (VHDL). Concentrations of lipoproteins are key factors in assessing the risk of cardiovascular disease.

Liposarcoma: A rare malignant tumour of adipose or fatty tissue. It occurs most frequently in the thighs, buttocks or retro-peritoneum. The four main

Liposome: Originally a lipid droplet in the endoplasmic reticulum of a fatty liver. Now applied to an artificially formed lipid droplet, small enough to form a relatively stable suspension in aqueous media and with potential use in drug delivery.

Liposuction: A surgical procedure, also called suction lipectomy, for extracting unwanted accumulations of subcutaneous FAT with the use of a powerful suction tube passed through the skin at different sites. Widely used in cosmetic surgery to improve the contour of the body, particularly that of women, the technique can have unwanted side-effects.

Lipotropic Factors: Various compounds and processes that enable the liver to metabolize fats properly or prevent the formation of cholesterolic stones in the gall bladder by supporting the continued emulsification of gall bladder bile. Examples: Lecithin, choline, Aristolochia

Lips: These form a pair of 'curtains' before the mouth, each composed of a layer of skin and of mucous membrane, between which lies a considerable amount of fat and of muscle fibres.

Listeriosis: A rare disease, although the causal organism, Listeria monocytogenes, is widely distributed in soil, silage, water, and various animals, with consequent risk of food contamination – for example, from unpasteurised soft cheese. Neonates are mainly affected – often as a result of a mild or inapparent infection in the pregnant mother. The disease presents in two main forms: meningoencephalitis, or septicaemia with enlarged lymph glands. Elderly adults occasionally develop the first form, while younger adults are more likely to develop a mild or even inapparent form. The disease is treated with antibiotics such as ampicillin or chloramphenicol.

Lithiasis: Having stones, usually in reference to the kidneys and urinary tract, sometimes to the gall bladder apparatus. Technically this can also refer to salivary gland calculi and impacted precipitants in the seminal vesicles or prostate.

Litholapaxy: Litholapaxy is the term applied to the operation in which a stone in the urinary bladder is crushed by an instrument introduced along the urethra, and the fragments washed out through a catheter.

Lithotomy: The operation of cutting for stone in the bladder. The operation is of great historic interest, because more has probably been written about it in early times than about any other type of surgery and because, for a long time, it formed almost the only operation in which the

Little's Disease: A form of cerebral palsy.

Live Donor: A kidney from a living relative.

Liver Fluke: Fasciola hepatica is a parasite infesting sheep and occasionally invading the bile ducts and liver of humans.

Liver Spots: A misnomer applied to the brown macules often seen on the backs of the hands of those chronically exposed to sunlight. They have no connection with any liver disorder.

Liver: The liver is the largest gland in the body, serving numerous functions, chiefly involving various aspects of metabolism.

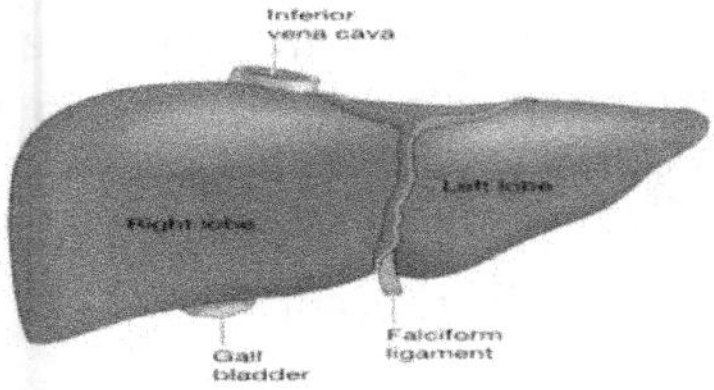

Living Trust: A revocable trust established by a grantor during his or her lifetime in which the grantor transfers some or all of his or her property into the trust.

Living Will: Also known as an advance directive or advance statement about medical treatment. This is a means of exercising choice in advance for patients who suspect that they will suffer mental impairment and become unable to speak for themselves. Advance statements can have legal force if made by an informed and competent adult, clearly refusing some or all future medical treatment once that individual has irrevocably lost mental capacity.

Loading Dose: Dose; also the "maintenance doses" administered after a loading dose.

Lobe: The term applied to the larger divisions of various organs, such as to the four lobes of the liver, the three lobes of the right and the two lobes of the left lung, which are separated by fissures from one another, and to the lobes or superficial areas into which the brain is divided. The term lobar is applied to structures which are connected with lobes of organs, or to diseases which have a tendency to be limited by the boundaries of lobes, such as lobar Pneumonia.

Lobectomy: The operation of cutting out a lobe of the lung in such diseases as abscess of the lung and bronchiectasis and carcinoma.

Lobule: The term applied to a division of an organ smaller than a lobe: for example, the lobules of the lung are of the size of millet seeds; those of the liver, slightly larger. Lobules form the smallest subdivisions or units of an organ, each lobule being similar to the others, of which there may be perhaps several hundred thousand in the organ.

Local Anaesthesia: Loss of sensation produced in a part of the body to stop pain while a person is examined, investigated or treated. The anaesthesia is effected by giving drugs in a local area temporarily to stop the action of pain-carrying nerve fibres. To anaesthetise a large area, a nerve block is done. Various drugs are used, depending on the depth and length of local anaesthesia required.

Local Delivery Plan (LDP): Three-year local strategic plan for the delivery of the health and social care agenda.

Local Health Board (LHB): Statutory body responsible for health services in a geographical area in Wales.

Local System: The computer hardware and software used to support pharmacy services for a PharmaNet participant.

Location: Where the symptom is experienced. Location is one of the parts of a complete symptom.

Lochia: The uterine discharge following birth, changing from reddish the first few days, to yellowish or clear after a couple of weeks. Many traditional skills of a midwife or partera center around evaluating the qualities and progress of lochia.

Locked-In Syndrome: This describes a condition in which a patient is awake and retains the power of sense perception, but is unable to communicate except by limited eye movements because the motor nervous system is paralysed. Several diseases can cause this syndrome, which results from interruption of some of the nerve tracts between the mid brain and the pons. Sometimes the syndrome is caused by severe damage to muscles or the nerves enervating them. Locked- in syndrome may sometimes be confused with a Persistent Vegetative State (PVS).

Lockjaw: A painful spasm of the jaw muscles, making it hard to open the mouth. It is a prominent symptom of tetanus and was once the popular name for this condition.

Locomotor Ataxia: The uncoordinated movements and unsteady lurching gait that occurs in the tertiary stage of untreated syphilis.

Locum Tenens: A doctor who stands in for another.

Logit Transformation: Mathematical transformation that relates response to a stated dose or concentration of a toxicant to the response in the absence of the toxicant

Logorrhoea: Logorrhoea is the technical term for garrulousness ('chatterbox') – a feature which may be exaggerated in certain states of mental instability.

Loiasis: Loiasis is the disease caused by the filarial worm Loa loa, a thread-like worm which differs from W. bancrofti in that it is shorter and thicker, and is found in the bloodstream during the day, not at night. It is transmitted by the mango fly, Chrysops dimidiata, but other flies of this genus can also transmit it. It is confined to West and Central Africa. The characteristic feature of the disease is the appearance of fugitive swellings which may arise anywhere in the body in the course of the worm's migration through it: these are known as Calabar swellings. The worm is often found in the eye, hence the old name of the worm in Africa the eye worm. Diethylcarbamazine is the treatment for this form of filariasis.

Loin: The name applied to the part of the back between the lower ribs and the pelvis (For pain in the loins).

Long Term Rehabilitation Program: May be called Extended Rehabilitation. Have full range of rehabilitation services available. Frequently after initial year of rehabilitation when progress is slower. Generally not permanent placement. May be facility or community based.

Longitudinal Study: A research study that involves observations of the same items over long periods of time, often many decades. Because longitudinal studies track the same people, they are often used to study trends across the life span, to uncover predictors of certain diseases, or to track the effects of a particular treatment in a patient's condition over time.

Long-term Care: The range of services typically provided at skilled nursing, intermediate-care, personal care or elder- care facilities.

Long-term Exposure: Continuous or repeated exposure to a substance over a long period of time, usually of several years in man, and of the greater part of the total life-span in animals or plants.

Lordosis: An unnatural curvature of the spine forwards. It occurs chiefly in the lumbar region, where the natural curve is forwards, as the result of muscular weakness, spinal disease, etc.

LOS (Level of Service): Field required when billing Pharmaceutical Care (PC) services or compound drugs indicating the time associated with the service provided.

Lotions: Fluid preparations intended for bringing into contact with, or for washing, the external surface of the body. Lotions are generally of a watery or alcoholic composition, and many of them are known as 'liquors'. Those external applications which are of an oily nature, and intended to be rubbed into the surface, are known as liniments.

Low Blood Glucose: A condition that occurs in people with diabetes when their blood glucose levels are too low Symptoms include feeling anxious or confused, feeling numb in the arms and hands, and shaking or feeling dizzy.

Low Cost Alternative: The alternative with the lowest average price (claimed on PharmaNet) of all products usually in the same generic class.

Lowest-Observed-Adverse-Effect-Level (LOAEL): Lowest concentration or amount of a substance, found by experiment or observation, which causes an adverse alteration of morphology, functional capacity, growth, development, or life span of a target organism distinguishable from normal (control) organisms of the same species and strain under defined conditions of exposure.

Lozenges: These are small tablets containing drugs mixed with sugar, gum, glycerin-jelly or fruit-paste.

LPPI: Langley Porter Psychiatric Institute

LTC: Long-term Care.

Lucid Interval: A temporary restoration of consciousness after a person has been rendered unconscious from a blow to the head. The victim subsequently relapses into coma. This is a sign of raised intracranial pressure from arterial bleeding and indicates that surgery may be required to control the intracranial haemorrhage.

Ludwig's Angina: An uncommon bacterial infection affecting the floor of the mouth. It can spread to the throat and become life-threatening. Usually caused by infected gums or teeth, it causes pain, fever and swelling, resulting in difficulty in opening the mouth or swallowing. Urgent treatment with antibiotics is called for, otherwise the patient may need a tracheostomy to relieve breathing problems.

Lugol's Solution: A compound solution of iodine and potassium iodide used in the preoperative preparation of patients with thyrotoxicosis. The solution is taken orally.

Lumbago: Pain in the lower (lumbar) region of the back. It may be muscular, skeletal or neurological in origin. A severe form associated with sciatica may be due to a prolapsed intervertebral disc. Less severe forms may be caused by osteoarthritis of the spine, a trapped nerve, inflammation of connective tissue, or may follow an old injury.

Lumbar Puncture: A procedure for removing Cerebrospinal Fluid (CSF) from the spinal canal in the lumbar region in order: (1) to diagnose disease of the nervous system; (2) to introduce medic aments spinal anaesthetics or drugs. A hollow needle is inserted into the lower section of the space around the spinal cord and the cerebrospinal fluid withdrawn. The

Lumbar Region: The lower back, five segments of the spinal chord and column, between the sacrum and thoracic regions.

Lumbar Sympathectomy: Destruction of the lumbar chain of sympathetic nerves by means of surgery, diathermy or injection of chemicals (phenol or alcohol). The technique is used to improve the blood flow to the leg in patients with peripheral vascular disease, and to treat some types of chronic leg pain. It has only limited success.

Lumbar: A term used to denote structures in, or diseases affecting, the region of the loins as, for example, the lumbar vertebrae, lumbar abscess

Lumbricus: Lumbricus is a name sometimes applied to the roundworm, or Ascaris lumbricoides.

Lumen: (1) The space enclosed by a tubular structure or hollow organ (e.g. the gastrointestinal tract or urinary bladder).

Lumpectomy: An operation for suspected breast cancer, in which the tumour is removed from the breast rather than with it.

Lunatic: An out of date and now derogatory term applied to people of disordered mind, because lunacy was supposed at one time to be largely influenced by the moon.

Lung Volumes: The volume of air within the lungs changes with the respiratory cycle. The volumes defined in the following table can

Lupus: This is the Latin word for wolf, and a term applied to certain chronic skin diseases which can destroy skin, underlying cartilage and even bone to cause serious deformity if uncontrolled.

Luteinising Hormone: This is a sugar-bearing protein manufactured by the anterior pituitary. Like a lot of the pituitary hormones, it surges on and off, since constant secretion would overload and deaden receptors. In women, it builds up after menses, stimulating the release of estrogen from the ovaries. Estrogen in turn stimulates the hypothalamus to increase its stimulation of LH from the pituitary, until, a day or two before ovulation, they produce a guitar-amp feedback, and the cells that produce LH start to surge follicle-stimulating hormone (FSH). The egg pops, being replaced by the corpus luteum, which produces progesterone for the next eleven to twelve days. Progesterone inhibits and lowers LH levels, as well as inhibiting levels of estrogen already being produced by the young follicles that will produce next month's egg. In men, LH is responsible for stimulation of testosterone, although FSH and the testes hormone inhibin are responsible for both the production of sperm and controlling testosterone.

Luteinizing-Hormone Releasing Hormone (LH-RH): The same substance as Follicle-Stimulating-Hormone Releasing Hormone (FSH-RH), both of which are actually Gonadotrophin-Releasing Hormone (GnRH or GRH).

Luxation: Another word for dislocation.

Lycanthropy: Morbid delusion that one is a wolf.

Lymph Nodes: The central drainage and metabolic organs strung along the lymph vessels. The mesenchymal structure is native, being present at birth. The functional cells have all migrated there, some recently from the marrow, spleen, thymus or blood, others have resided since a few months after birth. Much of the antibody memory is stored in these nodes, and having only venous blood supply, lymph nodes are constantly shunting metabolized substances back into the blood, so the final lymph drainage from the thoracic duct into the left subclavian vein (or the right subclavian) contains fluid already screened and cleansed by many nodes.

Lymph: Pertaining to the lymph system or lymph tissue, the "back alley" of blood circulation. Lymph is the alkaline, clear intercellular fluid that drains from the blood capillaries, where the arterial blood separates into thick, gooey venous blood and lymph. It bathes the cells, drains up into the lymph capillaries, through the lymph nodes for cleaning and checking against antibody templates, up through the body, and back to recombine with the venous blood in the upper chest. Blood in the veins is thick, mainly because part of its fluid is missing, traveling through the tissues as lymph. Lymph nodes in the small intestine absorb most of the dietary fats as well-organized chylomicrons. Lymph nodes and tissue in the spleen, thymus, and tonsils also organize lymphocytes and maintain the software memory of previously encountered antigens and their antibody defense response. Blood feeds the lymph, lymph feeds the cells, lymph cleanses the cells and returns to the blood.

Lymphadenitis: Inflammation or swelling of one or more lymph nodes. It may be an acute response or chronic, but signals the drainage into those nodes of microbes, their waste products, or the immuno-complexes produced upstream, whether from infection or allergy. A few infections can target or inhabit lymph nodes such as typhoid and EBV. Some people, with a past history of infection in a specific tissue (such as chronic sore throat as a kid) will have developed a larger sized node, hard and permanently palpable.These are hypertrophic or "shotty" nodes, and of no more importance than pumped-up muscles or old scar tissue.

Lymphadenitis: Inflammation of lymphatic glands

Lymphadenoma: Another name for Hodgkin's disease.

Lymphadenopathy: Medical description of swollen lymph glands. One cause is HIV infection; another is spread of malignant cells from a tumour in the catchment area of lymph- atic drainage.

Lymphangiectasis: Lymphangiectasis means an abnormal dilatation of the lymph vessels, as in filariasis.

Lymphangiography: A procedure whereby the lymphatics and lymphatic glands can be rendered visible on X-ray films by means of the injection of radio-opaque substances. It has now been replaced largely by magnetic resonance imaging (MRI).

Lymphangioma: An uncommon, non-malignant tumour occurring in the skin or the tongue. It comprises an abnormal collection of lymph vessels and tends to be present from birth. Sometimes it disappears spontane-ously, but it may need surgical removal.

Lymphangitis: Inflammation of one or more lymph nodes and/or lymph vessels, usually part of an acute infectious condition.

Lymphatic: Pertaining to the lymph system...sometimes more broadly to include immunity.

Lymphocyte: A variety of white blood cell produced in the lymphoid tissue and lymphatic glands of the body. It contains a simple, rounded nucleus surrounded by protoplasm generally described as non-granular. Two varieties of lymphocyte are described, small and large, and together they form over 20 per cent of the white cells of the blood. They play an important part in the production of antibodies, and in the rejection of transplanted organs such as the heart. This they do in two different ways: what are known as B-lymphocytes produce anti- bodies, while T-lymphocytes attack and destroy antigens directly. The latter are known as T-lymphocytes because they are produced by the Thymus Gland. Their numbers are increased in tuberculosis and certain other diseases. Such an increase is known as lymphocytosis. Animal cell that interacts with a foreign substance or organism, or one which it identifies as foreign, and initiates an immune response against the substance or organism. There are two groups of lymphocytes, B lymphocytes and T -

lymphocytes.

Lymphocytosis: An increase in the number of lymphocytes in the blood– for example, in response to viral infection or in chronic lymphocytic leukaemia.

Lymphoedema: Swelling of a part or organ due to obstruction to the lymph vessels draining it.

Lymphogranuloma Inguinale: A venereal disease in which the chief characteristic is enlargement of glands in the groin the infecting agent being a virus.

Lymphoid Tissue: Tissue involved in the formation of lymph, lymphocytes, and anti- bodies. It consists of the lymph nodes, thymus gland, tonsils and spleen.

Lymphokines: Lymphokines are polypeptides that are produced by lymphocytes as part of their immune response to an antigen; their function is to communicate with other cells of the immune system. Some lymphokines stimulate B-cells to differentiate into antibody-producing plasma cells; others stimulate T-lymphocytes to proliferate; other lymphokines become interferons.

Lymphoma: A neoplasia of the lymph tissue, such as Hodgkin's Disease. Although it is frequently useful to stimulate immunity when a person is undergoing chemotherapy for cancer, since the resultant immunosuppression is a major side effect of the treatment, in lymphatic cancer this the point of the therapy let it be. A malignant tumour of the lymph nodes divided histologically and clinically into two types: Hodgkin's disease, and non-Hodgkin's lymphoma. Hodgkin's disease or lymphadenoma was named after Thomas Hodgkin (1798–1866), a Guy's Hospital pathologist, who first described the condition. General term comprising tumours and conditions allied to tumours arising from some or all of the cells of lymphoid tissue. Lymphoma is the name of a diverse group of cancers of the lymphatic system, a connecting network of nodes, organs, and vessels whose principle cell is the lymphocyte.

Lymphosarcoma: A traditional term for non-Hodgkin's lymphoma.

Lysimeter: Laboratory column of selected representative soil or a protected monolith of undisturbed field soil with which it is possible to sample and monitor the movement of water and substances.

Lysis: The gradual ending of a fever, and the opposite of crisis, which signifies the sudden ending of a fever. It is also used to describe the process of dissolution of a blood clot, or the destruction of cells as a result of damage to or rupture of the plasma membrane, thus allowing the cell contents to escape.

Lysol Poisoning: When lysol is swallowed it burns the mouth and throat. Brown discoloration of the affected tissues, accompanied by the characteristic smell of lysol on the breath, is typical. Treatment This is urgent. If the skin has been contaminated with the lysol, it must be washed with water, and any Lysol contaminated clothing must be taken off. Do not make the victim vomit if he or she has swallowed a corrosive substance such as lysol or phenol. Call an ambulance and say what the victim has taken.

Lysol: A brown, clear, oily fluid with antiseptic properties, made from coal-tar and containing 50 per cent cresol. When mixed with water it forms a clear soapy fluid.

Lysosome: Membrane-bound cytoplasmic organelle containing hydrolytic enzymes.

Lysozyme: An enzyme present in tears and egg white, lysozyme catalyses the destruction of some bacteria by damaging their walls.

M

M&A: Mergers and acquisitions

Mab: Monoclonal antibody

MAC (Maximum Allowable Cost): The MAC is the highest price a state Medicaid agency will pay for a particular pharmaceutical. The MAC varies over time and by state.

Maceration: Maceration is the softening of a solid by soaking in fluid.

Macr-/Macro-: Prefix denoting large-sized cell – for example, a macrophage is a large phagocyte.

Macrocyte: Macrocyte is an unusually large red blood cell especially characteristic of the blood in pernicious anaemia.

Macrocytosis: This condition is particularly associated with pernicious anaemia but can also be caused by a number of other things, such as alcohol, pregnancy, myxoedema and myelo- matosis, and also by vitamin B_{12} deficiency: this occurs sometimes in vegans as well as in patients with crohn's disease.

Macroglossia: An abnormally large tongue.

Macrolides: A group of antibiotics. The original macrolide, erythromycin, was discovered in the early 1950s and used successfully as an alternative to penicillin. The name 'macrolide' derives from the molecular structure of this group.

Macrophage: Large (10-20 mm diameter) amoeboid and phagocytic cell found in many tissues, especially in areas of inflammation; macrophages are derived from blood monocytes and play an important role in host defence mechanisms. This is a mature form of what is released from the marrow as a monocyte. A macrophage lives long, can digest much detritus, and is able to wear particles of odd food on its outer membrane. This allows T-cell and B-cell lymphocytes to taste the particle (an epitope) and form an antibody response. Further, these macrophages, traveling as monocytes, will take up permanent residence in many tissues, providing them with immunity. They line the spleen, form the cleansing Kupffer cells in the liver, make up the "dust cells" that protect the lungs, protect the synovial fluids of the joints, and form the microglial cells that provide protection to the brain and nerve tissues. On and on, the macrophages clean up messes and acting as the intermediates between innate and acquired immunity.

Macropsia: Condition in which objects appear larger than normal. It can be due to disease of the macula.

Macroscopic (gross) pathology: Study of changes associated with disease that are visible to the naked eye without the need for a microscope.

Macula: A spot or area of tissue that is different from the surrounding tissue. An example is the macula letea, the yellow spot in the retina of the eye.

Macules: Areas of small, flat or slightly raised skin discoloration which may occur in a wide range of conditions such as many viral infections, eczema (dermatitis), psoriasis, syphilis and after burns, as well as in pregnancy.

Maculopapular: A skin rash that is made up of macules (dis-coloration of the skin) and papules (raised abnormalities of the skin).

Madura Foot: Tropical infection of the foot by deeply invasive fungi which cause chronic swelling and sup- puration with multiple discharging sinuses. Antibiotics are of limited value and advanced disease may require amputation of the affected foot.

Magnesium: Magnesium is a light metallic element; it is one of the essential mineral elements of the body, without which the body cannot function properly. The adult body contains around 25 grams of magnesium, the greater part of which is in the bones. More than two-thirds of our daily supply comes from cereals and vegetables; as most other foods also contain useful amounts, there is seldom any difficulty in maintaining an adequate amount in the body. Magnesium is also an essential constituent of several vital enzymes. Deficiency leads to mus- cular weakness and interferes with the - efficient working of the heart.

Mail Order Pharmacy Services: Mail order services allow members to obtain maintenance medications without physically visiting a pharmacy. Mail order often allows members to obtain more than a One-month supply of medication at one time. Mail order services may be provided by a vendor that contracts with the PBM.

Mail-order Pharmacy: A service that provides prescription drugs at discounted prices for maintenance medications such as insulin. This type of pharmacy service requires the user to send in a prescription, typically filled for three months, and the patient receives the supply via mail. This type of service often includes automatic refills and costs less than retail pharmacy prescriptions.

Mainstream Smoke (Tobacco Smoking): Smoke that is inhaled.

Maintenance Medication: Medications that are prescribed for long-term treatment of chronic conditions, such as diabetes, high blood pressure or asthma.

Maintenance Therapy: Anti-cancer treatment given to patients in remission, after their initial or primary therapy, to delay or prevent a relapse.

Malabsorption Syndrome: This term includes a multiplicity of diseases, all of which are characterised by faulty absorption from the intestine of essential foodstuffs such as fat, vitamins and mineral salts. Among the conditions in this syndrome are coeliac disease, sprue, cystic fibrosis and pancreatitis. Surgical removal of the small intestine also causes the syndrome. Symptoms include anaemia, diarrhoea, oedema, vitamin deficiencies, weight loss and, in severe cases, malnutrition.

Malabsorption: Improper utilization of needed and available nutrients, either from impaired digestive function (such as B_{12} being unabsorbed because of gastritis), impaired absorption (poor Vitamin E absorption because of an inflamed ileum) or impaired transport (the diminished blood proteins of the advanced alcoholic).

Malacia: Malacia is a term applied to softening of a part or tissue in disease: for example, osteomalacia or softening of the bones.

Malaise: A fretful and low energy state, often considered an early sign of infection or low fever. Ask someone with chronic fatigue syndrome or multiple chemical sensitivities. A vague feeling of feverishness, listlessness and languor, malaise sometimes precedes the onset of serious acute diseases, but more commonly accompanies passing illnesses such as dyspepsia, chills and colds.

Malar: Anything relating to the cheek. For example, the malar (zygomatic) bone is also known as the cheek bone, and a malar flush is reddening of the - cheeks.

Malaria: A parasitic disease caused by four species of Plasmodium: P. falciparum, P. vivax, P. ovale, and P. malariae. Clinically, malaria is characterised by recurrent episodes of high fever, some- times associated with RIGOR; enlargement of the spleen is common. P. falciparum infection can also be associated with several serious – often fatal complications.

Malathion: Organophosphorus insecticide which is a preferred scabicide and pediculocide; applied externally; resistance is rare.

Malignancy: Population of cells showing both uncontrolled growth and a tendency to invade and destroy other tissues; a malignancy is life-threatening.

Malignant Hypertension: Malignant hypertension has nothing to do with cancer; it derives its name from the fact that, if untreated, it runs a rapidly fatal course.

Malignant Hyperthermia: this disorder is a rare complication of general anaesthesia caused, it is believed, by a commbination of an inhalation anaesthetic (usually halothane) and a muscle-relaxant drug (usually succinycholine). A life-endangering rise in temperature occurs, with muscular rigidity the first sign. Tachycardia, acidosis and shock usually ensue. About 1:20,000 patients having general anaesthesia suffer from this disorder, which progresses rapidly and is often fatal. Surgery and anaesthesia must be stopped immediately and appropriate corrective measures taken, including the intravenous administration of dantrolene. It is a dominantly inherited genetic condition; therefore, when a case is identified. It is most important that relatives are screened.

Malignant: A term applied in several ways to serious disorders. A tumour is called malignant when it grows rapidly, tending to infiltrate surrounding healthy tissues and to spread to distant parts of the body, leading eventually to death. The term is also applied to types of disease which are much more serious than the usual form – for example, malignant hypertension. Malignant pustule is another name for anthrax.

Malingering: Malingering is a term applied to the feigning of illness. In the great majority of cases, a person who feigns illness has a certain amount of disability, but exaggerates the illness or dis- comfort for some ulterior motive – for example, to take time off work or to obtain compensation.

Malleolus: Name of either of the two bony prominences at the ankle.

Mallet Finger: Deformation of a finger due to sudden forced flexion of the terminal joint, leading to rupture of the tendon. As a result the individual is unable to extend the terminal part of the finger, which remains bent forwards. The middle, ring and little fingers are most commonly involved. Treatment is by splinting the finger. The end result is satisfactory provided that the patient has sufficient patience.

Mallet Toe: The condition in which it is not possible to extend the terminal part of the toe. It is usually due to muscular imbalance but may be caused by congenital absence of the extensor muscle. A callosity often forms on the toe, which may be painful.

Malleus: The hammer-shaped lateral bone of the group of three that form the sound-transmitting ossicles in the middle ear.

Malnutrition: The condition arising from an inadequate or unbalanced diet. The causes may be a lack of one or more essential nutrients, or inadequate absorption from the intestinal tracts. A diet that is deficient in carbohydrate usually contains inadequate protein, and this type of malnutrition occurs widely in Africa and Asia as a result of poverty, famine or war.

Maloprim: A combination of pyrimethamine and dapsone which is used for the prevention of malaria in limited circumstances. It has the advan- tage of only needing to be taken once weekly. It should not be taken by anyone hypersensitive to sulphonamides, and should not be used for the treatment of an acute attack.

Malpractice: An American term implying improper or inadequate medical treatment that fails to match the standards of skill and care reasonably expected from a qualified health-care practitioner – usually a doctor or dentist. Litigation against health professionals for malpractice has been running at substantial levels in the USA for some years.

Malpresentation: A situation during childbirth in which a baby is not in the customary head-first position before delivery. The result is usually a complicated labour in which a caesarean operation may be necessary to effect the birth.

Mammography: The special technique whereby X-rays are used to show the structure of the breast or any abnormalities in it. It is an effective way of distinguishing benign from malignant tumours, and can detect tumours that are not palpable. In a multi-centre study in the USA, called the breast cancer detection demonstration project and involving nearly 300,000 women in the 40–49 age group, 35 per cent of the tumours found were detected by mammog- raphy alone, 13 per cent by physical examination, and 50 per cent by both methods com- bined. The optimum frequency of screening is debatable: the American College of Radiologists recommends a baseline mammogram at the age of 40 years, with subsequent mammography at one to two year intervals up to the age of 50; thereafter, annual mammography is recommended. In the United Kingdom a less intensive screening programme is in place, with women over 50 being screened every three years. As breast cancer is the commonest malignancy in western women and is increasing in frequency, the importance of screening for this form of cancer is obvious.

Mammoplasty: A surgical operation to reconstruct a breast after part or all of it has been removed to treat breast cancer; to enlarge small breasts; or to reduce the size of overlarge breasts. The routine method for breast enlargement used to be the insertion of silicone implants under the skin; controversy about the long-term safety of silicone, however, has restricted their use mainly to women needing reconstruction of their breasts after cancer surgery. Side-effects have included hardening of breast tissue, leaking of implants and develop- ment of scar tissue.

Managed Behavioral Health: This is a program that covers your mental health and substance abuse care needs.

Managed Care: Procedures used by organizations that assume risk for health services to control or influence variables.

Managed Health Care: This process aims to reduce the costs of health care while maintaining its quality. The concept originated in the United States but has attracted interest in the United Kingdom and Europe, where the spiralling costs of health care have been causing widespread concern. Managed care works through changing clinical practice, but it is not a discrete entity: the American I. J. Iglehart has defined it as 'a variety of methods of financing and organising the delivery of comprehensive health care in which an attempt is made to control costs by controlling the provision of services'. Managed care has three facets: health policy; how that policy is managed; and how individuals needing health care are dealt with. The process and its applications are still evolving and it is likely that different health-care systems will adapt it to suit their own particular circumstances.

Management Action Plan: A plan which outlines the actions needed to develop and execute an activity to reduce the release of pollutants.

Management Information System (MIS): The common term for the computer hardware and software that provides the support of managing the plan.

Management of Adverse Events: Medical management of an adverse event and its reporting to the applicable regulatory authorities within the stipulated time- frame.

Mandatory Generic Substitution: Members are required to use generic drug products when available. Members who choose to use the brand product rather than the generic drug product will be required to pay a higher copayment, a greater coinsurance amount, or the full price of the medication.

Mandelic Acid: Also known as mandel amine, a non-toxic keto-acid used in the treatment of infections of the urinary tract, especially those due to the Escherichia coli and the Streptococcus faecalis or Enterococcus. It is administered in doses of 3 grams several times daily. As it is only effective in acid urine, ammonium chloride must be taken at the same time.

Mandible: The bone of the lower jaw.

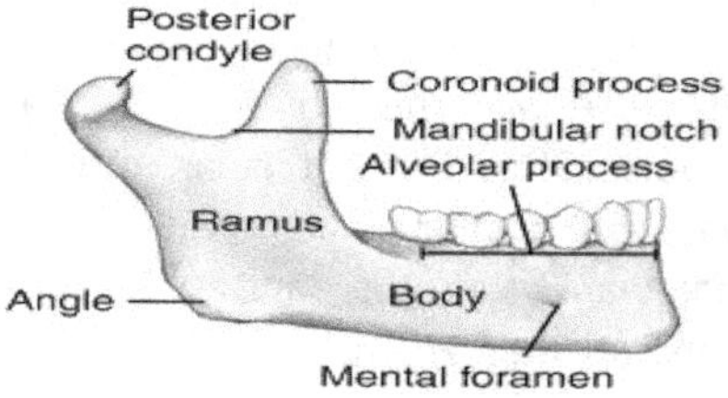

Manganese: A metal, oxides of which are found abundantly in nature. Permanganate of potassium is a well-known disinfectant. The body requires small amounts of the metal for normal growth and development.

Mania: A form of mental disorder characterised by great excitement. Emotional disorder (mental illness) characterized by an expansive and elated state (euphoria), rapid speech, flight of ideas, decreased need for sleep, distractability, grandiosity, poor judgement and increased motor activity.

Manic Depression: Manic depression, or cyclothymia, is a form of mental illness characterised by alternate attacks of mania and depression.

Manipulation: The passive movement (frequently forceful) of bones, joints, or soft tissues, carried out by orthopaedic surgeons, physiotherapists, osteopaths and chiropractors as an important part of treatment – often highly effective.

Mannitol: An osmotic diuretic given by a slow intravenous infusion to reduce oedema of the brain or raised intraocular pressure in glaucoma.

Manometer: An instrument for measuring the pressure or tension of liquids or gases.

Mantoux Test: A test for tuberculosis. It consists in injecting into the superficial layers of the skin (i.e. intra-dermally) a very small quantity of old tuberculin which contains a protein antigen to TB. A positive reaction of the skin – swelling and redness – shows that the person so reacting has been infected at some time in the past with mycobacterium tuberculosis. However, it does not mean that such a person is suffering from active tuberculosis.

Manubrium: The uppermost part of the sternum or breastbone.

Manuscript: articles that are published in peer-reviewed scientific journals.

MAO inhibition: The suppression of monoamine oxydase (flavin-containing amine oxydase). MAO is critical in modifying nerve-ending storage of certain mono- amines (in this case, epinephrine, norepinephrine and dopamine. Another type of MAO works on histamines), and MAO inhibitor drugs were, along with tricyclics, the first wave of anti-depressants. The problem was that if you ate brie cheese or chopped chicken livers while taking the drugs you could get a nosebleed or cerebral aneurysm. Double adrenergic whammy, since some foods are also strongly MAO-inhibiting (at least functionally).

Marasmus: Progressive wasting, especially in young children, when there is no ascertainable cause. It is generally associated with defective feeding.

Marburg Disease: This is also known as green monkey disease, or vervet monkey disease, because the first recorded human cases acquired their infection from monkeys of this genus. It is a highly dangerous viral infection with a high mortality rate. The incubation period is 4–9 days. The onset is sudden with marked nausea and severe headache; this is followed by rising temperature, diarrhoea, and vomiting. Towards the end of the first week a rash appears which persists for a week and is accompanied by internal bleeding. In those who recover, convalescence is slow and prolonged. Apart from laboratory infections acquired through working with vervet monkeys, all the cases so far reported have occurred in Africa.

March Fracture: A curious condition in which a fracture occurs of the second (rarely, the third) metatarsal bone in the foot without any obvious cause. The usual story is that a pain suddenly developed in the foot while walking or marching (hence the name), and that it has persisted ever since. The only treatment needed is immobilisation of the foot and rest, and the fracture heals satisfactorily.

March Haemoglobinuria: A complication of walking and running over long distances. It is due to damage to red blood cells in the blood vessels of the soles of the feet. This result in haemoglobin being released into the bloodstream, which is then voided in the urine – the condition known as haemoglobinuria.

Marfan's Syndrome: An inherited disorder affecting about one person in 50,000 in which the connective tissue is abnormal. The result is defects of the heart valves, the arteries arising from the heart, the skeleton and the eyes. The victims are unusually tall and thin with a particular facial appearance (the US President Abraham Lincoln was said to have Marfan's) and deformities of the chest and spine.

Margin Of Exposure (MOE), Margin Of Safety (MOS): Ratio of the no-observed-adverse-effect level (NOAEL) to the theoretical or estimated exposure dose (EED) or concentration - (EEC).

Marijuana: Another term for cannabis, hemp, or hashish.

Marital Deduction: A deduction allowing for the unlimited transfer of any or all property from one spouse to the other generally free of estate and gift tax.

Marketing Application: the application filed for seeking marketing authorization of a new drug/device.

Marketing Support Trials: Studies to assess not only safety and efficacy on large patient populations, but also quality-of-life impact and cost-effectiveness. Many market support trial outcomes are used for additional indications of already approved drugs, publications and promotion purposes.

Mask: A device that covers the nose and mouth to enable inhalation anaesthetics or other gases such as oxygen to be administered. It is also a covering for the nose and mouth to ensure that antiseptic conditions are maintained during surgery, when dressing a wound or nursing a patient in conditions of isolation. The term is also applied to the expres- sionless appearance that occurs in certain disorders–for eg. in parkinsonism.

Masochism: A condition in which a person gets pleasure from physical or emotional pain inflicted by others or themselves. The term is often used in the context of achieving sexual excitement through inflicted pain.

Masochism may be a conscious or subconscious activity.

Mass Balance: A method of accounting for the quantities of materials produced, consumed, used, or accumulated at, released from, or transported to or from a process or facility as a waste, commercial product or byproduct, or component of a commercial product or byproduct.

Mass Mean Diameter: Diameter of a particle with a mass equal to the mean mass of all the particles in a population.

Mass Spectrometry: Study of the structure of molecules by using the mass of their basic constituents mass.

Massage: A method of treatment in which the operator uses his or her hands, or occasionally other appliances, to rub the skin and deeper tissues of the person under treatment. It is often combined with (a) passive movements, in which the masseur/masseuse moves the limbs in various ways, the person treated making no effort; or (b) active movements, which are performed with the combined assistance of masseur/masseuse and patient. Massage is also often com- bined with baths and gymnastics in order to strengthen various muscles. It helps to improve circulation, prevent adhesions in injured tissues, relax muscular spasm, improve muscle tone and reduce any oedema.

Masseter: An important muscle of mastication that extends from the zygomatic arch in the cheek to the mandible or jawbone. It acts by closing the jaw.

Mast Cells: These are a group of cells that line the capillaries of tissues that come in contact with the outside, like skin, sinuses, and lung mucosa. They, like their first cousin basophils, are produced in the red bone marrow and migrate to the appropriate tissues, where they stay. They bind IgE, supply the histamine and heparin response that gives you a healing inflammation, and cause allergies. Round or oval cells found predominantly in the loose connective tissue. They contain histamine and heparin, and carry immunoglobulin E, the antibody which plays a predominant part in allergic reactions. Although known to play a part in inflammatory reactions, allergy, and hypersensitivity, their precise function in health and disease is still not quite clear.

Mastalgia: The term applied to pain in the breast.

Mastectomy: A surgical operation to remove part or the entire breast. It is usually done to treat cancer, when it is commonly followed by chemotherapy or radiotherapy. There are four types of mastectomy: lumpectomy, quandr-antectomy, subcutaneous mastectomy and total mastectomy. The choice of operation depends upon several factors, including the site and nature of the tumour and the patient's age and health. Traditionally, radical mastectomy was used to treat breast cancer; in the past three decades, however, surgeons and oncologists have become more selective in their treatment of the disease, bringing the patient into the decision-making on the best course of action. Lumpectomy is done where there is a discrete lump less than 2 cm in diameter with no evidence of glandular spread.

Master Patient / Member Index: An index or file with a unique identifier for each patient or member that serves as a key to a patient's or member's health record.

Master Randomization List: the master list that captures the distribution schedule of trial subjects between treatment and control arm.

Mastitis: The term applied to inflammation of the breast.

Mastocytosis: A rare condition in which the primary abnormality is of mast cells – a type of cell responsible for the storage and release of agents such as histamine, important in aller- gic states. Patients may present with an urti- carial rash (urticaria pigmentosa) but may have symptoms referable to any part of the body, related to collections of active mast cells in these areas.

Mastoid Process: The large process of the temporal bone of the skull which can be felt immediately behind the ear. It contains numerous cavities, one of which – the mastoid antrum – communicates with the middle ear, and is liable to suppurate when the middle ear is diseased.

Mat Burn: A combination of a burn and an abrasion which occurs in wrestlers when the skin over the bony points is rubbed against the unyielding canvas mat.

Matched Pair: A type of parallel trial design in which the investigator identify paors of subjects who are identical with respect to relevant factors, then randomize them so that one receives a treatment and the other receives another treatment.

Materia Medica: "materials of medicine" in Latin. Areferencethat lists the curative indications and therapeutic actions of homeopathic medicines. This information is derived from provings and clinical experience.

Material Safety Data Sheet (MSDS): Compilation of information required under the US OSHA Hazard Communication Standard on the identity of hazardous substances, health and physical hazards, exposure limits, and precautions.

Materials Recovery Facility (MRF): An intermediate processing site that accepts, separates, and performs intermediate processing of mixed waste, usually bottles and cans.

Matrix: The intercellular substance of a tissue. It forms the primary mass in some cartilage, bones, and the lens of the eye. Where living cells are so separated they communicate with e-mail.

Mature: Normally used in reference to a fistula. When the fistula is ready to be used.

Maxilla: The name applied to the upper jawbones, which bear the teeth.

Maximum Allowable (Admissible, Acceptable) Concentration (MAC): Regulatory value defining the concentration that if inhaled daily (in the case of work people for 8 hours with a working week of 40 hours, in the case of the general population 24 hours) does not, in the present state of knowledge, appear capable of causing appreciable harm, however long delayed during the working life or during subsequent life or in subsequent generations.

Maximum Allowable Fee Schedule: A listing of all procedure codes allowed by Wisconsin Medicaid for a provider type and Wisconsin Medicaid's maximum allowable fee for each procedure code.

Maximum Annual Contribution: The maximum annual contribution is the maximum amount that a person covered by the public plan or by a private plan may be required to pay during the year when purchasing insured drugs, and consists of a deductible and a co-insurance amount. Once the person has reached his or her maximum annual contribution, the plan covers all the person's drug costs from then until the end of the year.

Maximum Average Daily Concentration of An Atmospheric Pollutant: Highest of the average daily concentrations recorded at a definite point of measurement during a certain period of observation.

Maximum Days Supply: The maximum number of days worth of treatment paid by Pharma Care for a prescription, which is either 30 days or 100 days depending on various circumstances.

Maximum Monthly Contribution: The maximum monthly contribution applies only to persons covered by the public prescription drug insurance plan. Under the public plan, the maximum annual contribution is divided into maximum monthly contributions.

Maximum Permissible Daily Dose: Maximum daily dose of substance whose penetration into a human body during a lifetime will not cause diseases or health hazards that can be detected by current investigation methods and will not adversely affect future generations.

Maximum Permissible Level (MPL): Level, usually a combination of time and concentration, beyond which any exposure of humans to a chemical or physical agent in their immediate environment is unsafe.

Maximum Residue Limit (MRL) For Pesticide Residues: Maximum contents of a pesticide residue (expressed as mg/kg fresh weight) recommended by the Codex

Alimentarius Commission to be legally permitted in or on food commodities and animal feeds. MRL's are based on data obtained following good agricultural practice and foods derived from commodities that comply with the respective MRL's are intended to be toxicologically acceptable.

Maximum Tolerable Concentration (MTC): Highest concentration of a substance in an environmental medium that does not cause death of test organisms or species.

Maximum Tolerable Exposure Level (MTEL): Maximum amount or concentration of a substance to which an organism can be exposed without leading to an adverse effect after prolonged exposure time.

Maximum Tolerated Dose (MTD): High dose used in chronic toxicity testing that is expected on the basis of an adequate sub chronic study to produce limited toxicity when administered for the duration of the test period.

Maximum Tolerated Dose: The highest daily dose of a chemical that does not causes over toxicity in a ninety-day study in laboratory mice or rats.

MCB: The building letter code for the mission center building.

McNamara's Test: A statistical test to determine whether there is an association between two categorical variables when the data are paired. For e.g., a cross-over study where patients are required to provide their assessment about the response during each treatment period.

MCO: Managed care organization. An entity that combines the financing and delivery of health care into one organization, such as a health maintenance organization or preferred provider organization.

MDC (Major Diagnostic Category): The MDC is a classification system that represents a group of similar Diagnosis Related Groups (DRGs). Each MDC typically involves a single organ system of the body.

MDI: Molecular Design Institute.

MDRC: Management Decision Resource Center.

Me Watch Form: The safety narratives prepared from a serious adverse events.

Meal Plan: A guide to help people get the proper amount of calories, carbohydrates, proteins, and fats in their diet.

Mean Life: Average lifetime of a molecular, atomic, or nuclear system in a specified state. For an exponentially decaying system, it is the average time for the number of molecules, atoms or nuclei in a specified state to decrease by a factor of, the base of natural logarithms.

Mean: An arithmetic value that is obtained by summing up all the observations and dividing the total by -

the total number of observations.

Measles: Measles, formerly known as morbilli, is an acute infectious disease occurring mostly in children and caused by an RNA paramyxovirus. Epidemiology there has been a dramatic fall in the number of sufferers from 1986, when more than 80,000 cases were reported. This is due to the introduction in 1988 of the measles, mumps and rubella vaccine; 1990, when the proportion of children immunised reached 90 per cent, was the first year in which no deaths from measles were reported. Even so, fears of side- effects of the vaccine against measles – including scientifically unproven and discredited claims of a link with AUTISM – mean that some children in the UK are not being immunised, and since 2002 local outbreaks of measles have been reported in a few areas of the UK. Side-effects are, however, rare and the government is campaigning to raise the rate of immunisation, with GPs being set targets for their practices.

Measurement: A process of estimating the magnitude of some attributes of an object such as its length or weight.

Meatus: A term applied to any passage or opening: for example, external auditory meatus – the passage from the surface to the drum of the ear.

Mebeverine: A direct relaxant of the smooth muscle in the intestine, it may relieve pain in patients with irritable bowel syndrome or diverticular - disease. Adverse effects are rare.

Mechanism of Action: The explanation of the mechanism by which a drug produces an effect in a living organism.

Meckel's Diverticulum: A hollow pouch sometimes found attached to the small intestine. It is placed on the small intestine about 90–120 cm (3–4 feet) from its junction with the large intestine, is several centimetres long, and ends blindly. It is lined with cells similar to those which line the stomach, and so may produce acid. This leads to occasional illness – Meckel's diverticulitis with ulceration, which causes abdominal pain and fever (sometimes referred to as 'left-sided appendicitis'). Perforation may result in peritonitis and, rarely, may be the lead point of an intussusception.

Meconium: The brown, semi-fluid material which collects in the bowels of a fetus before birth, and which should be discharged either at the time of birth or shortly afterwards. It consists partly of BILE secreted by the liver before birth; partly of debris from the mucous membrane of the intestines.

Media: The middle layer of an organ or tissue, but more usually applied to the wall of an artery or vein, where the media comprises layers of elastic smooth muscle fibres. Medial Near the middle of tissue, organ or body.

Medial Tibial Syndrome: The term applied by athletes to a condition characterised by pain over the inner border of the shin, which occurs in most runners and sometimes in joggers. The syndrome, also known as shin splints, is due to muscular swell- ing resulting in inadequate blood supply in the muscle: hence the pain. The disorder may be the result of compartment syndrome (build-up of pressure in the muscles), tendinitis, muscle or bone inflammation, or damage to the muscle. It usually disappears within a few weeks, responding to rest and physiotherapy, with or without injections. In some cases, how- ever, it becomes chronic and so severe that it occurs even at rest. If the cause is the compartment syndrome, relief is usually obtained by a simple operation to relieve the pressure in the affected muscles.

Median Effective Concentration (EC50): Statistically derived concentration of a substance in an environmental medium expected to produce a certain effect in 50 % of test organisms in a given population under a defined set of conditions.

Median Lethal Concentration (LC50): Statistically derived concentration of a substance in an environmental medium expected to kill 50 % of organisms in a given population under a defined set of conditions.

Median Lethal Dose (LD50): Statistically derived dose of a chemical or physical agent (radiation) expected to kill 50 % of organisms in a given population under a defined set of - conditions.

Median Lethal Time (TL50): Statistically derived average time interval during which 50 % of a given population may be expected to die following acute administration of a chemical or physical agent (radiation) at a given concentration under a defined set of conditions.

Median Narcotic Concentration (NC50): Statistically derived concentration of a substance in an environmental medium expected to cause narcotic conditions in 50 % of a given population under a defined set of conditions.

Median Narcotic Dose (ND50): Statistically derived dose of a substance expected to cause narcosis in 50 % of test animals under a defined set of conditions.

Median Survival Time: The time at which 50% of the study population is alive.

Median: The middle value in a data set; that is, just as many values are greater than the median and lower than the median value. (With an even number of values, the conventional median is halfway between the two middle values.)

Mediastinum: Mediastinum is the space in the chest which lies between the two lungs. It contains the heart and great vessels, the gullet, the lower part of the windpipe, the thoracic duct and the phrenic nerves, as well as numerous structures of less importance.

Medicaid Best Price: The lowest price paid to a manufacturer for a brand name drug, taking into account rebates, charge backs, discounts or other pricing adjustments, excluding nominal prices.

Medicaid: A joint state and federal health insurance scheme in the United States that provides cover for poorer people in the population.

Medical Advisor: Medically qualified personnel in an organization who supports marketing and/or research department.

Medical Audit: A systematic review of the procedures involved in the diagnosis, care and treatment of patients. Such a review includes the input of associated resources – for example, social-services care and funding – and an assessment of the outcome and quality of life for the patient(s). If the review is undertaken by health-care professsionals, the process is called clinical audit.

Medical Device: An instrument, apparatus, machine , implant or other similar or related article including any component, part or accessory which is intended for use in the diagnosis, cure, treatment or prevention of disease.

Medical Devices Agency: An executive agency of the department of health in the UK. Set up in 1994, it is responsible for regulating and advising on the sale or use of any product, other than a medicine, used in the health-care environment for the diagnosis, prevention, monitoring or treatment of illness or disease. Equipment ranges from pacemakers to prostheses, and from syringes to magnetic resonance imaging (MRI).

Medical Dictionary for Regulatory Activities: A standardized dictionary of medical terminology adopted by the International Conference on Harmonization.

Medical Doctor: A person who is eligible to undertake medical practice as per the applicable regulations.

Medical Education: This term is used to define the process of learning and knowledge-acquisition in the study of medicine. It also encompasses the expertise required to develop education and training for students and learners in all aspects of medical health care. Studies for undergraduate students, postgraduate students and individual health- care practitioners, from the initial stages to the ongoing development of a career in medicine or associated health fields, are also included in medical education. The word 'pedagogy' is sometimes applied to this process. A range of research investigations has developed within medical education. These apply to course monitoring, audit, development and validation, assessment methodologies and the application of educationally appropriate principles at undergraduate and postgraduate levels. Research is undertaken by medical educationalists whose backgrounds include teach- ing, social sciences and medicine and related health-care specialties, and who will hold a medical or general educational diploma, degree or other appropriate postgraduate qualification. Development and validation for all courses are an important part of continuing accreditation processes.

Medical History: The information on overall general health, past illness and current medical problems of a subject.

Medical Informatics: Information about diseases, patients, drug use, and drug interactions

Medical License: An authorization to undertake medical practice as per applicable regulations.

Medical Litigation: Legal action taken by an individual or group of individuals, usually patients, against hospitals, health-service providers or health professionals

Medical Monitor: A medically qualified personnel employed by trial Sponsors/CROs for reviewing critical medical data (safety and efficacy) of a clinical trial.

Medical Necessity: Medical necessity is a term used to refer to a course of treatment seen as the most helpful for the specific health symptoms you are experiencing. The course of treatment is determined jointly by you, your health professional and CIGNA health care. This course of treatment strives to provide you with the best care in the most appropriate setting.

Medical Records: Original documents, data and records (e.g., hospital records, clinical and office charts, laboratory notes , memoranda , subject's diaries or evaluation checklists, pharmacy dispensing records, recorded date from automated instruments, copies or transcription certified after verification as being accurate and complete, microfiches, photographic negatives, microfilms or magnetic media, X-rays, subject's files and records kept at the pharmacy, at the laboratories, and at medico technical departments involved in the clinical trial.)

Medical Research Council: A statutory body in the United Kingdom that promotes the balanced development of medical and related biological research and aims to advance knowledge that will lead to improved health care. It employs its own research staff in more than 40 research establishments. These include the National Institute for Medical Research, the Laboratory of Molecular Biology, and the Clinical Sciences Centre.

Medical Services Plan of British Columbia (MSP): A program of the Ministry of Health Services, offering universal medical coverage to British Columbia residents.

Medicare part A: Hospital insurance -

provided by Medicare that can help pay for inpatient hospital care, medically necessary inpatient care in a skilled nursing facility, home health care, hospice care and end-stage renal disease treatment.

Medicare part B: Medicare-administered medical insurance that helps pay for certain medically necessary practitioner services, outpatient hospital services and supplies not covered by Part A hospital insurance of medicare coverage. Doctors' services are covered under Part B even if they're provided to a member in an inpatient setting. Part B can also pay for some home health services-when the beneficiary doesn't qualify for Part A.

Medicare: The Federal health insurance program for people 65 years of age and older; certain younger people with

Medication History: Prescription history for each patient Personal Health Number (PHN). The medication history- table on Pharma Net is part of the Drug Information System.

Medication: Articles (other than food) intended for the use in the diagnosis, cure, mitigation, treatment or prevention of disease in man or other animals.

Medicines Control Agency (MCA): The United Kingdom regulatory authority that approves or rejects CTX/CTC and PL applications.

Medicines: Medicines are drugs made stable, palatable and acceptable for administration. In britain, the medicines act 1968 controls the making, advertising and selling of substance.

Medicolegal: A term that relates to the practice of medicine and law.

Medigap: A term used to describe health benefits coverage that supplements Medicare coverage.

Medlars: Medical literature analysis and retrieval system – a computerised index system produced by the us national library of medicine.

Medline: A computer-based telephone line linkage to medlars for rapid transmission of medical information held by that library. Information can be accessed worldwide. Information is also available via the web.

MEDS (Medication and Eligibility Delivery Solution): Provides physicians with convenient access to up-to-date mediation history for patients they are treating in an inpatient setting.

Medulla: The inside part of an organ or tissue that is distinct from the outer part – for example, the marrow in the centre of a long bone, or the inner portion of the kidneys or adrenal glands.

Mega Trials: Large scale clinical trials having a sample size of 10,000 or more that evaluates the marginally effective investigational product.

Megacolon: A greatly enlarged colon that may be present at birth or develop later. It can occur in all age groups and the condition is typified by severe chronic constipation. Megacolon is caused by obstruction of the colon which may be due to faulty innervation, or to psychological factors.

Megalomania: A delusion of grandeur or an insane belief in a person's own extreme greatness, goodness, or power.

Mega-Trials: Massive randomized clinical trials that test the advantages of marginally effective experimental drugs by enrolling 10,000 or more subjects. Synonym: large- sample trials.

Meibomian Glands: Numerous glands within the tarsal plates of the eyelids. Their secretions form part of the tears.

Meiosis: Meiosis, or reduction division, is the form of cell division that only occurs in the gonads that is, the testis and the ovary giving rise to the germ cells (gametes) of the sperms and the ova. Two types of sperm cells are produced: one contains 22 autosomes and a Y sex chromosome; the other, 22 autosomes and an X sex chromosome. All the ova, however, produced by normal meiosis have 22 autosomes and an X sex chromosome.

Melaena: Blood in the faeces in which dark, tarry masses are passed from the bowel. It is due to bleeding from the stomach or from the higher part of the bowel, the blood undergoing chemical changes under the action of the secretions, and being finally converted in large part into sulphide of iron. It is a serious symptom meriting medical investigation. It can indicate peptic ulcer or carcinoma of the stomach.

Melancholia: A term used in the past for illness characterized by great mental and physical depression.

Melanin: Pigment which confers colour on the skin, hair and eye. It is produced by cells called melanocytes interspersed along the basal layer of the epidermis. The maturation of the epidermis into stratum corneum cells packed with mel- anin granules confers an ultraviolet light barrier which protects the skin against the harmful effects resulting from continued solar exposure. The races do not differ in the number of melanocytes in their skin, only in the rate and quantity of melanin production. Exposure to bright sunlight stimulates melanin production and distribution causing 'suntan'. A hormone from the pituitary gland may stimulate melanin production on the face in pregnancy.

Melanocyte: Clear branching cell in the epidermis of the skin that produces tyrosinase.

Melanoma: A malignant tumour arising from melanocytes. It may start in an existing mole or arise de novo. Increasing exposure to sunlight of white populations in the 20th century has resulted in an alarming increase in the incidence of this cancer. It is mainly seen in those over 40 years, especially on the legs in women and on the back in men. An enlarging pigmented macule or nodule with irregular contour, profile or colour distribution is always suspect.

Melatonin: A hormone that plays a key role in the body's diurnal (night and day) rhythms. Produced by the pineal gland and derived from sero- tonin, it acts on receptors in an area of the brain above the optic chiasma, synchronising them to the diurnal rhythm. Melatonin is under investigation as a possible agent to treat insomnia in the elderly, in shift workers and in those with severe learning disability (mental handicap).

Member: An individual or dependent who is enrolled in and covered by a managed health care plan.

Memorandum of Understanding (MOU): An MOU between FDA and a regulatory agency in another country allows mutual recognition of inspections. A formal written agreement which sets the forth the working arrangements between two or more groups.

Menarche: The beginning of the reproductive phase of a woman's life. It usually begins withy night sweats, continues a few months later with estrogen, followed by ovulation, then the full cycle and the growth of secondary sexual characteristics in various order.

Mendelism: The term applied to a law enunciated by G. J. Mendel that the offspring is not intermediate in type between its parents, but that the type of one or other parent is predominant. Character-istics are classed as either dominant or recessive. The offspring of the first generation tend to inherit the dominant characteristics, whilst the recessive characteristics remain latent and appear in some of the offspring of the second generation. If individuals possessing recessive characters unite, recessive characters then become dominant in succeeding generations.

Mendelson Syndrome: Inhalation of regurgitated stomach contents, usually as a complication of general anaesthesia. It may cause death from anoxia or result in extensive lung damage.

Meniere's Disease: Named after the Frenchman, prosper meniere, who first described it in 1861, the disease is characterised by tinnitus, deafness and intermittent attacks of vertigo. The first manifestation is usually deafness on one side; then – as a rule, many months later – there is a sudden attack, without any warning, of intense vertigo. The acute giddiness usually lasts for two or three hours with some unsteadiness persisting for a few days. The time interval between attacks varies from a week to a few months. When they do recur, they tend to do so in clusters. The tinnitus, which tends to be low-pitched, comes on at about the same time as the deafness; it is often described as being like rushing water or escaping steam. The deafness becomes gradually worse until it is complete. The condi- tion is due to excessive fluid in the labyrinth of the ears. The cause of this accumulation is not known, although it has been suggested that it might be a form of allergy, or might be due to spasm of small blood vessels. The disorder is diagnosed from audiometry, the caloric test and other investigations.

Meninges: The membranes surrounding the brain and spinal cord.

Meningism: A condition with symptoms and signs closely resembling those of meningitis. Most commonly occurring in children, it is usually a symptom of chest infection or of inflammation in the upper respiratory tract. Given the serious implications of meningitis, medical advice should be sought. Examination of the cerebrospinal fluid may be necessary: in mening- ism the fluid is normal.

Meningitis: Inflammation affecting the membranes of the brain or spinal cord, or usually both. Men- ingitis may be caused by bacteria, viruses, fungi, malignant cells or blood. The term is, however, usually restricted to inflammation due to a bacterium or virus. Viral meningitis is normally a mild, self-limiting infection of a few days' duration; it is the most common cause of meningitis but usually results in complete recovery and requires no specific treatment. Usually a less serious infection than the bacterial variety, it does, however, rarely because associated encephalitis, which is a potentially dangerous illness. A range of viruses can cause meningitis, including: enteroviruses; those causing mumps, influenza and herpes simplex and HIV.

Meningomyelocele: A protrusion of the meninges of the spinal cord through a defect in the spine.

Meniscus: A crescentic fibro-cartilage in a joint, such as the cartilages in the knee-joint.

Menopause: A term rather callously used to describe the cessation of ovarian hormones as a result of a radical hysterectomy or what the British more honestly refer to as castration. The several years, in the late forties or early fifties, when the great birth reservoir of potential ovarian follicles has been reduced to only a few, many with innately poor hormone-sensitivities (which is perhaps why they are still remaining, they never heard the clarion call of FSH). As fewer follicles are capable of fully- programmed function, corpus luteal fragilities start to show as diminished progesterone levels later, even the pre-ovulatory estrogens start to diminish.

Menorrhagia: Excess bleeding at menses, in duration or amount.

Causes are many, although chronic menorrhagia and PMS is usually the result of deficient progesterone secretions (days-per-month) or constant adipose-released estradiol from obesity or recent substantial weight loss. Uterine fibroids can contribute, as can menopausal breakthrough bleeding or flooding, coagulation disorders, and most serious metabolic disease can produce menorrhagia as one of many symptoms. My rule of thumb as an herbalist is, if botanicals fail to control the bleeding directly (hemostatics) or attempting to reestablish a good folliculization for the next month's corpus luteum does not help, there may be a metabolic problem or an overt reproductive pathology. In menopausal menorrhagia, however, the conditions are transitional and in flux.

Menstruation: A periodic change occurring in (female) human beings and the higher apes, consisting chiefly in a flow of blood from the cavity of the womb (uterus) and associated with various slight constitutional disturbances. It begins between the ages of 12 and 15, as a rule – although its onset may be delayed until as late as 20, or it may begin as early as ten or 11. Along with its first appearance, the body develops the secondary sex characteristics: for example, enlargement of the breasts, and characteristic hair distribution.

Menstruum: The solvent used in extraction. For a dry tincture, the menstruum might be 50% alcohol and 50% water. The menstruum for mint tea is hot water.

Mental Illness: Defined simply, this is a disorder of the brain's processes that makes the sufferer feel or seem ill, and may prevent that person from coping with daily life. Psychiatrists – doctors specialising in diagnosing and treating mental illness – have, however, come up with a range of much more complicated definitions over the years. Psychiatrists like to categorise mental ill- nesses because mental signs and symptoms do occur together in clusters or syndromes, each

Mentally Disabled: A state in which a person is not able to manage his or her affairs or to make a choice due to psychiatric or developmental disorder.

MEPS (Medical Expenditure Panel Survey) : MEPS is the third (and most recent) in a series of national probability surveys conducted by the Agency for Healthcare Research and Quality (AHRQ) on the financing and utilization of medical care in the United States.

Meralgia Paraesthetica: A condition characterised by pain and par- aesthesia on the front and outer aspect of the thigh. It is more common in men than in women, and the victims are usually middle- aged, overweight and out of condition. It is due to compression of the lateral cutaneous nerve of the thigh, and exacerbated by an uncomfortable driving position when motoring long distances. Reduction in weight, improvement in general fitness and correction of faulty posture usually bring relief. If these fail, surgical decompression of the nerve may help.

Mercaptopurine: One of the antimetabolite group of drugs, which includes methotrex- ate, fluorouracil and thioguanine. These drugs are incorporated into new nuclear material in the cell or combine irreversibly with vital cellular enzymes, preventing normal cellular metab- olism and division. Mercaptopurine is used mainly for the maintenance treatment of acute leukaemia, though it is increasingly proving valuable in the treatment of crohn's disease. As with all cytotoxic drugs, dosage must be carefully controlled; in particular it must be reduced if used concurrently with allopurinol. Side-effects include gastrointestinal upsets (including ulceration), and bone-marrow depression.

Mercurialism: Chronic poisoning caused by the excessive use of mercury, by breathing its vapour, or by exposure in mining or smelting processes.

Mercury: Mercury is a heavy fluid metal which, with its salts, has been used in medicine for many centuries. Uses In the past, mercuric salts was used as antiseptics, anti-parasitic agents.

Mesencephalon: The small section of brain stem – excluding the pons and medulla – linking the hindbrain to the forebrain.

Mesenchymal Cells: Literally, those derived from embryonic mesoderm; practically, those in a tissue that give it structure and form. The opposite of parenchymal.

Mesenteric: Pertaining to the great fold that holds the small intestines, blood vessels and lymph in a great curtain, connected with the back of the abdominal wall.

Mesentery: Mesentery is the double layer of peritoneal membrane which supports the small intestine. It is of a fan shape, and its shorter edge is attached to the back wall of the abdomen for a distance of about 15 cm (6 inches), while the small intestine lies within its longer edge, for a length of over 6 metres (20 feet). The terms mesocolon, mesorectum, etc., are applied to similar folds of peritoneum that support parts of the colon, rectum, etc.

Mesocolon: The double fold of peritoneum by which the large intestine is suspended from the back wall of the abdomen.

Mesoderm: The middle layer of the three germ layers of the embryo during its early development. It develops into cartilage, bone, blood, muscle, kidneys, testes and connective tissue.

Mesomorph: In somatotyping, a mesoderm-muscle-structural dominant person.

Mesothelioma: A malignant tumour of the pleura, the membrane lining the chest cavity. The condition is more common in people exposed to asbestos dust. It may be asymptomatic or cause pain, cough, and breathing troubles. Surgery or radiotherapy may be effective but often the disease has spread too far before it is discovered. Mesothelioma incurred as a result of contact with asbestos at work may attract industrial compensation.

Mesothelioma: Malignant tumour of the mesothelium of the pleura, pericardium or peritoneum that may be caused by exposure to asbestos fibres and some other fibres.

Meta-Analysis: Meta –Analysis combines the result of several studies that address a set of related research hypotheses. It is widely used in epidemiology and evidence-based medicine studies. A development of systematic reviews of randomised, controlled trials in particular fields of health care. If this re-evaluation of the findings of primary research also includes the aggrega- tion of data from all the re-evaluated studies and a re-analysis of them, the procedure is described as meta-analysis. The technique has been used increasingly in medicine in recent years, and the cochrane collaboration has been in the forefront of systematically reviewing and storing clinical and research data to provide

Metabolic Activation: Biotransformation of a substance of relatively low toxicity to a more toxic derivative.

Metabolic Disorders: A collection of disorders in which some part of the body's internal chemistry is disrupted. Some of these disorders arise from inherited deficiencies in which a specific enzyme is absent or abnormal, or does not function properly. Other metabolic disorders occur because of malfunctions in the endocrine system. There may be over- or underproduction of a hormone involved in the control of metabolic activities: a prime example is diabetes mellitus – a disorder of sugar metabolism; others include cushing's syndrome; hypothyroidism and hyperthyroidism and insulinoma (an insulin- producing tumour of the pancreas). The bones can be affected by metabolic disorders such as osteoporosis, osteomalacia (rickets).

Metabolic Half-life: Time required for one half of the quantity of a substance in the body to be metabolically transformed into a derivative or to be eliminated.

Metabolic Model: Analysis and theoretical reconstruction of the way in which the body deals with a specific substance, showing the proportion of the intake that is absorbed, the proportion that is stored and in what tissues, the rate of breakdown in the body and the subsequent fate of the metabolic products, and the rate at which it is eliminated by different organs as unchanged substance or metabolites.

Metabolic Transformation: Biochemical transformation of a substance that takes place within an organism.

Metabolism: Sum total of all physical and chemical processes that take place within an organism; in a narrower sense, the physical and chemical changes that take place in a given substance within an organism. The sum total of changes in an organism in order to achieve a balance (homeostasis). Catabolic burns up, anabolic stores and builds up; the sum of their work is metabolism.

Metal cleaning: General cleaning or degreasing of metallic components or assemblies, without specific quality requirements or with low ones.

Metamyelocyte: An immature granulocyte (white blood cell) usually found in the bone marrow's blood-making tissue. It can, however, appear in the blood in a range of diseases, including infection.

Metaphysis: The extremity of a long bone where it joins the epiphysis.

Metaplasia: Abnormal transformation of an adult, fully differentiated tissue of one kind into a differentiated tissue of another kind.

Metaplasia: The term applied to a change of one kind of tissue into another. Although not usually harmful, it may be pre-cancerous if occurring in the cervix (neck of the womb or uterus), urinary bladder, or lining of the airways (bronchi).

Metastasis: Metastasis, and metastatic, are terms applied to the process by which malignant disease spreads to distant parts of the body, and also to the secondary tumours resulting from this process. For example, a cancer of the breast may produce metastatic growths in the glands of the armpit; cancer of the stomach may be followed by metastases in the liver. Metastases are col- loquially known as secondaries and their spread occurs through the bloodstream, lymphatic system and across the body cavities. Highly malignant tumours – for example, melanomas – are especially prone to spread far and fast.

Metatarsalgia: Pain affecting the metatarsal region of the foot. It is common in adolescents, and associated with flat-foot; in adults it may be a manifestation of rheumatoid arthritis. Morton's metatarsalgia is a form associated usually with the nerve to the second toe cleft, often induced by the compression of tight shoes.

Metatarsus: Metatarsus is the group name of the five metatarsal bones in the foot. Metatarsus varus is the condition characterised by deviation of the fore- foot towards the other foot. It is a common condition in newborn babes and almost always corrects itself spontaneously. Only in the rare cases in which it is due to some deformity of the bones or muscles of the foot is any treatment required.

Meteorism: Also known as tympanites, this is a distension of the abdomen from excess gas or air in the intestine or peritoneal cavity. On percussion the abdomen sounds resonant, like a drum. Causes include obstruction of the intestines, aerophagy (the swallowing of air), and irritable bowel syndrome.

Methaemoglobin: A derivative of haemoglobin in which the iron has been oxidised from ferrous to ferric form. It does not combine with oxygen and therefore plays no part in oxygen transport. Normal concentration of methaemoglobin in red blood cells is less than 1 per cent of the total haemoglobin. When a large concentration of the haemoglobin is in the form of methaemo- globin, the patient will suffer from hypoxia and will be cyanosed. Most cases of methaemoglobinaemia are due to chemical agents.

Methaemoglobinaemia: Methaemoglobinaemia is a condition due to the presence in the blood of methaemoglobin. It is characterised by cyanosis which turns the skin and lips a blue colour, shortness of breath, headache, fatigue and sickness.

Methaemoglobin-forming Substance: Substance capable of oxidising directly or indirectly the iron (II) in haemoglobin to iron (III) to form methaemoglobin, a derivative of haemoglobin that cannot transport oxygen.

Methotrexate: A medication used in the treatment of psoriasis.

Metre: The basic unit of length in the modern version of the metric system, known as the Inter- System of Units (SI). It is equivalent to 39•37 inches.

Metritis: Inflammation of the uterus.

Metropathia Haemorrhagica: A disorder characterised by irregular bouts of uterine bleeding – without previous ovulation – due to excessive oestrogenic activity. It is associated with endometrial hyperplasia and cysts of the ovary.

Metrorrhagia: Uterine bleeding at times other than menstrual. It is usually due to a uterine lesion and should be investigated.

MHC: Mental Health Centre.

MHLW: Ministry of Health, Labor and Welfare (Japan).

MHR: Ministry of Human Resources.

Microalbumin: A protein found in blood plasma and urine. The presence of microalbumin in the urine can be a sign of kidney disease.

Microalbuminuria: Chronic presence of albumin in slight excess in urine.

Microbicides: Gels or creams, currently under investigation, designed to reduce the risk of anal or vaginal transmission of viruses such as HIV. The aim is to kill or to inactivate the virus, creating a barrier to mucosal cells or pre- venting the infection from taking hold after it has entered the body. Large-scale trials were launched in Africa in 2004, using dextrin sulphate and PRO-2000 gel.

Microbiology: The study of all aspects of micro-organisms (microbes) – that is, organisms which individually are generally too small to be visible other than by microscopy. The term is applicable to viruses, bacteria, and microscopic forms of fungi, algae, and protozoa.

Microcephaly: Abnormal smallness of the head usually associated with learning disability. It may occur as a result of infection of the fetus by, for example, rubella (German measles) or from hypoxic damage to the brain before or during birth.

Micrococcus: A spherical gram-positive bacterium. It occurs in colonies and is usually harmless in humans. However, micrococcus can become pathogenic and cause abscesses arthritis, endocarditis or meningitis.

Microcosm: Artificial test system that simulates major characteristics of the natural environment for the purposes of ecotoxicological assessment: such a system would commonly have a terrestrial phase, with substrate, plants and herbivores, and an aquatic phase, with vertebrates, invertebrates and plankton. The term "mesocosm" implies a more complex and larger system than the term "microcosm" but the distinction is not clearly defined.

Microcyte: A small red blood cell.

Microdissection: The technique of dissecting very small structures under a microscope. Miniature surgical instruments are manipulated via geared connections that convert the coarse movements of the surgeon's fingers into miniscule movements, making it possible to dissect and separate even individual chromosomes.

Microfiche: An archival media on which essential trial documents can be transferred at a very small size.

Microfilaria: The mobile embryo of certain parasitic nematode worms which are found in the blood or lymph of patients infected with filarial worms. The microfilariae develop into larvae in the body of a blood-sucking insect, for example, a mosquito.

Microgram: Microgram is the 1/1,000th part of a milligram. The abreviation for it is µg.

Micromercurialism: Effects of exposure to mercury detected at the lowest exposure levels producing a measurable reaction.

Micrometre: The 1/1,000th part of a millimetre. The abbreviation for it is µm.

Micro-Organism: A very small, single-celled living organism that cannot usually be seen by the naked eye. The most important micro-organisms in medicine are those that cause disease. This 'pathogenic' group, however, forms only a small proportion of the enormous number of known micro-organisms. The main pathogenic ones are bacteria. Others are fungi and rickettsia. Though not true cells, viruses are usually classified as micro-organisms.

Micropsia: Condition in which objects appear smaller than normal. It can be due to disease of the macula of the eye.

Microscope: An optical instrument comprising adjustable magnifying lenses that greatly enlarge a small object under study – for example, an insect, blood cells, or bacteria. Some microscopes use electron beams to magnify minute objects such as chromosomes, crystals, or even large mol- ecules. Optical microscopes are also used for microsurgery when the area being operated on is otherwise inaccessible: for example, in eye.

Microsome: Artefactual spherical particle, not present in the living cell, derived from pieces of the endoplasmic reticulum present in homogenates of tissues or cells: microsomes sediment from such homogenates when centrifuged at 100 000 g and higher: the microsomal fraction obtained in this way is often used as a source of mono-oxygenase enzymes.

Microsurgery: The conduct of very intricate surgical operations using specially refined operating micro-scopes and miniaturised precision instruments – for example, forceps, scalpels, scissors, etc. Microsurgery is used in previously inaccessible areas of the brain, eye, inner ear and spinal cord, as well as in the suturing of severed nerves and small blood vessels following traumatic injuries to the limbs or fingers. The technique is also used to reverse vasectomy.

Microtome: A laboratory instrument for cutting sections of biological tissues for study under a microscope. It is widely used in biological and pathology laboratories.

Microwaves: Non-ionising electro-magnetic radiations in the frequency range of 30–300,000 mega hertz. They are emitted from electronic devices, such as heaters, some domestic ovens, television receivers, radar units and diathermy units. There is no scientific evidence to justify the claims that they are harmful to humans, or that they produce any harmful effect in the genes. The only known necessary precaution is the protection of the eyes in those using them in industry, as there is some evidence that prolonged exposure to them in this may induce cataract.

Mid-Life Crisis: A colloquial description of the feelings of anxiety and distress experienced by some individuals in early middle age. They realise that by 45 years of age they are no longer young, and men in particular try to turn the clock back by changing jobs, dressing trendily, taking up energetic or unusual sports or engaging in extramarital liaisons. Sometimes those in mid- life crises develop mild or even serious depression. The feelings of anxiety and insecurity usually disappear with time but some people may benefit from counselling.

Midwife: A member of the profession which provides care and advice during pregnancy, supervises the mother's labour and delivery, and looks after her and the baby after birth. Should a pregnancy or labour develop complications, the midwife will seek medical advice. Most midwives are regis- tered general nurses who have also done an 18- month course in midwifery. Trained midwives are registered with the UK Central Council for Nursing Midwifery and Health Visiting and work in hospitals or a domiciliary setting.

Migraine: The word migraine derives from hemicrania, the Greek for half a skull, and is a common condition characterised by recurring intense headaches. It is much more usual in women than in men and affects around 10 percent of the population. It has been defined as 'episodic headache accompanied by visual or gastrointestinal disturbances, or both, attacks lasting hours with total freedom between episodes.

Milestones: An objective or goal to be achieved in a stipulated timeframe.

Milia: These are small keratin cysts appearing as white papules on the cheek and eyelids.

Miliaria: Also known as prickly heat. An intensely itchy vesicular and erythematous rash induced by intense heat and humidity. It is caused by a disturbance of sweat-gland function.

Miliary: A term, expressing size, applied to various disease products which are about the size of millet seeds: for example, miliary aneurysms, miliary tuberculosis.

Milium: Milium is a pinhead white cyst of the skin of Minimally Invasive Surgery (MIS) 463 the face containing corneal cells. It can be removed on the point of a sterile needle.

Milk Teeth: The temporary teeth of children.

Minamata disease: Neurological disease caused by ingestion of methylmercury-contaminated fish, first seen at Minamata Bay in Japan.

Mineralization: Complete conversion of organic substances to inorganic derivatives.

Minim: A 'pre-metric' unit of measurement of volume. It is about one-60th part of a fluid drachm and is used in pharmacy. Minimum Lethal Dose (MLD)

Minimal Risk: The extent of harm or discomfort anticipated from a clinical trial/research which is not greater than the routine practice.

Minimally Invasive Surgery: (MIS) More popularly called 'keyhole surgery', MIS is surgical intervention, whether diagnostic or curative, that causes patients the least possible physical trauma. It has revolutionised surgery, growing from a technique used by gynae- cologists, urologists and innovative general surgeons to one regularly used in general surgery, gynaecology, urology, thoracic surgery, orthopaedic surgery and otorhinolaryngology.

Minimum Effective Dose: Minimal concentration level at which a drug is able to produce an effect in human body.

Minimum Lethal Concentration (LCmin): Lowest concentration of a toxic substance in an environmental medium that kills individual organisms or test species under a defined set of conditions.

Minimum Lethal Dose (LDmin): Lowest amount of a substance that, when introduced into the body, may cause death to individual species of test animals under a defined set of conditions.

Minor: An individual who has not attained the legal age of consenting to a trial as per the applicable regulations.

Minutes of Meeting: A document that summarizes the critical discussion and decision points of a meeting.

Miosis: Abnormal contraction of the pupil of the eye to less than 2 mm. Condition of constriction (reduction in size) of the pupil. It may be the result of disease affecting the autonomic nervous system. Bright light causes miosis and some drugs.

Miscible: Liquid substances capable of mixing without separation into two phases; refers to liquid mixtures.

Misconduct: The fabrication or falsification of data in reporting results of a clinical trial.

Misuse of drugs: Government legislation covers the manufacture, sale and prescription of drugs in the UK. As well as stating which drugs may be sold over the counter (OTC) without a doctor's or dentist's prescription, and those which can be obtained only with such a pre- scription, government regulations determine the extent of availability of many substances which are liable to be abused – Misuse of Drugs Act 1971 (below). The Misuse of Drugs Regulations 1985 define those individuals who in their professional capacity are authorised to supply and possess controlled drugs.

Mitochondria: The rod-like bodies in the cells of the body which contain the enzymes neces- sary for the activity of the cell. They have been described as the 'power plant of the cell'.

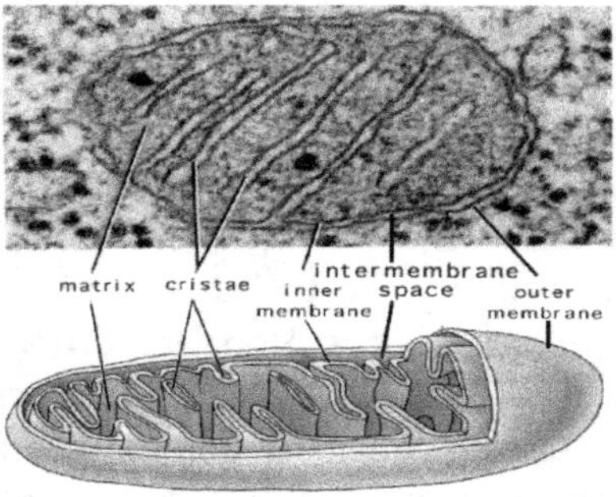

Mitochondrion: Eukaryote cytoplasmic organelle that is bounded by an outer membrane and an inner membrane; the inner membrane has folds called cristae that are the centre of ATP synthesis in oxidative phosphorylation in the animal cell and supplement ATP synthesis by the chloroplasts in photosynthetic cells. The mitochondrial matrix within the inner membrane contains ribosomes, many oxidative enzymes, and a circular DNA molecule that carries the genetic information for a number of these enzymes.

Mitogen: Substance that induces-lymphocyte transformation or, more generally, mitosis and cell proliferation.

Mitosis M: The process of cell division for somatic cells and for the ovum after fertilisation. Each chromo- some becomes doubled by splitting lengthwise and forming two chromatids which remain held together by the centromere. These chromatids are exact copies of the original chromosomes and contain duplicates of all the genes they bear. When cell division takes place, the pull of the spindle splits the centromere and each double chromatid separates, one passing to one pole of the nucleus and the other to the opposite pole. The nucleus and the cell itself then also divide, forming two new daughter cells containing precisely the same 23 pairs of chromosomes and carrying exactly the same complement of genes as did the mother cell.

Mitosis: The classic four-phased cellular division of somatic cells, wherein (when the dust settles) two new daughter cells contain full chromosomal information of the parent, complete nuclei, and half the cytoplasm. This is distinct from cloning (as in the bone morrow) and the chromosome splitting of miosis (ovum and sperm).

Mitral Incompetence: A defect in the mitral valve of the heart which allows blood to leak from the left ventricle into the left atrium. It is also known as mitral regurgitation; incompetence may occur along with mitral stenosis. The left ventricle has to work harder to compensate for the faulty valve, so it enlarges, but eventually the ventricle cannot cope with the extra load and left-sided heart failure may develop.

Mitral Stenosis: Narrowing of the opening between the left atrium and left ventricle of the heart as a result of rigidity of, and adhesion between, the cusps of the mitral valve. It is due, almost invariably, to the infection rheumatic fever. The atrium has to work harder to force blood through the narrowed channel. The effects are similar to those of mitral incompetence. shortness of breath and palpitations and irregular beating (fibrillation) of the atrium are common consequences in adults. Drug treatment with digoxin and diuretics helps, but surgery to dilate or replace the faulty valve may be necessary.

Mitral Valve: The mitral valve, so-called because of its resem- blance to a bishop's mitre, is the valve which guards the opening between the atrium and ventricle on the left side of the heart.

Mittelschmerz: Abdominal pains that occur midway between menstrual periods and which are caused either by ovulation or the normal short pre-ovulatory surge of estrogen.

Mixed Waste: A mixture of radioactive waste (regulated under the Atomic Energy Act) and hazardous waste.

MLD: The minimal lethal dose or smallest amount of a toxic compound needed to cause death.

MMR Vaccine: A combined vaccine offering protection against measles, mumps and rubella (German measles), it was introduced in the UK in 1988 and has now replaced the measles vaccine. The combined vaccine is offered to all infants in their second year; health authorities have an obligation to ensure that all children have received the vaccine by school entry – it should be given with the pre-school booster doses against diphtheria, tetanus and poliomyelitis, if not earlier – unless there is a valid contra-indication (such as partial immunosup- pression), parental refusal, or evidence of pre- vious infection. MMR vaccine may also be used in the control of measles outbreaks, if offered to susceptible children within three days of exposure to infection.

Modality: A condition that makes a person or their symptom better or worse. For example, better in a hot bath, abdominal pain better bending over, worse rainy weather, etc. Modalities are one of the parts of a complete symptom.

Mode: A statistical term which is the most frequently occurring value of a set of observations. If the results are plotted as a graph, the mode is the peak of the curve which results.

Model Shop: A composite description of a typical shop at an Air Force installation.

Modem: From modulator/ demod- ulator; a device that converts the digital data into analog data that can be transmitted via telephone or cable lines used for communications.

Modification: The act of making change or amendment to an information, document or process.

Modifying Factor (MF): As used by the USEPA, uncertainty factor that is greater than zero and less than, or equal to 10; the magnitude of the factor depends upon the professional assessment of scientific uncertainties of a study or database not explicitly treated with the standard uncertainty factors (for example the completeness of the overall database and the number of animals tested); the default value for the factor is 1.

Molecular Biology: The study of molecules that are part of the structure of living organisms.

Molecular Design Institute: Research institute at UCSF that promotes and coordinates research and education in the broad area of molecular design with particular emphasis on the discovery, design, and delivery of novel pharmaceutical agents; the MDI is an Organized Research Unit (ORU) of the University of California, San Francisco.

Molecular Design: Design of molecules, including pharmaceuticals

Molecular Docking Program: Computer program that simulates fitting potential drugs into target receptor molecules much like a key fits into a lock

Molecular Drug Design: Design of pharmaceuticals based upon the application of chemistry, biology, mathematics, and computer science.

Molecular Modeling: A technique to generate 3D structures of target molecules and model the affinity if different molecules at these target using powerful software graphics and stimulation programs.

Molecular Pharmacology: Molecular approaches to the study of drug action.

Molecular Targeting: Developing a drug directed at a specific biological target.

Molecule: The smallest possible amount of a substance comprising two or more linked atoms which retains the chemical characteristics of that substance. Molecules vary greatly in their size and complexity, ranging from oxygen (two linked oxygen atoms) and water (two hydrogen atoms and one oxygen) to large complex molecules such as deoxyribonucleic acid (DNA) comprising thousands of atoms of carbon, hydrogen, oxygen, nitrogen and phosphorus that form the double-helix structure which helps to form genes, the basic building blocks of the hereditary material of living things.

Molluscicide: Substance intended to kill molluscs.

Molluscum Contagiosum: Common papular eruption of the skin caused by a virus. Most common in children, it is highly contagious and often transmitted in swimming pools and sauna baths. Mollusca are often multiple and persistent in children with atopic eczema, and epidemics may occur in boarding schools. The typical molluscum is 2–3 mm in diameter, skin- coloured and translucent, with a dimpled centre. The armpits and adjacent chest, upper inner thighs and genital areas are common sites in young children. In adults the infection is usually transmitted sexually and affects the pubic area and lower belly. Mollusca eventually disappear spontaneously, but cure can be expedited by curettage (removal with a curette) under surface anaesthesia.

Mongolian Blue Spots: Irregularly shaped areas of bluish-black pigmentation found occasionally on the buttocks, lower back or upper arms in newborn infants of African, Chinese and Japanese parentage, and sometimes in the babies of black-haired Europeans. They measure from one to several centimetres in diameter, and usually disappear in a few months. They are commonly mistaken for bruises.

Moniliasis: The infection caused by monilia, the genus of fungi now known as Candida albicans. The infection may occur in the mouth – where it is known as thrush – lungs, intestine, vagina, skin, or nails.

Monitor: Person employed by the sponsor or CRO who is responsible for determining that a trial is being conducted in accordance with the protocol. A monitor's duties may include, but are not limited to, helping to plan and initiate a trial, assessing the conduct of trials, and assisting in data analysis, interpretation, and extrapolation. Monitors work with the clinical research coordinator to check all data and documentation from the trial. See also clinical research associate.

Monitoring: The act of overseeing the progress of a clinical trial and of ensuring that it is conducted, recorded and reported in accordance with the protocol, Standard Operating Procedures (SOPs), Good Clinical Practice (GCP) and the applicable regulatory requirements.

Monitoring Plan: A document that specifies the monitoring visit interval, source data verification requirements and protocol specific monitoring instructions.

Monitoring Report: A written report from the monitor to the sponsor after each site visit and/or other trial-related communication according to the sponsor's SOPs.

Monitoring Visit Log: A log signed by the monitor each time he/she visits the trial site.

Monitoring Visit Report: A written report prepared by the monitor after each site visit to document the progress and conduct of the trial at a site.

Monitoring Visit: A visit to a study site to review the progress of a clinical study and to ensure protocol adherence, accuracy of data, safety of subjects, and compliance with regulatory requirements and good clinical practice guidelines.

Monitoring: The act of overseeing the progress of a clinical trial, and of ensuring that it is conducted, recorded, and reported in accordance with the protocol, standard operating procedures (SOPs), good clinical practice (GCP), and the applicable regulatory requirement(s).

Monitoring: Continuous or repeated observation, measurement, and evaluation of health and/or environmental or technical data for defined purposes, according to prearranged schedules in space and time, using comparable methods for sensing and data collection. Evaluation requires comparison with appropriate reference values based on knowledge of the probable relationship between ambient exposure and adverse effects.

Monoamine Oxidase (MAO): An enzyme which, in the brain, breaks down certain neurotransmitters such as serotonin, dopamine and norepinephrine.

Monoclonal Antibodies: An artificially prepared antibody obtained from cell clones – a genetic- ally identical group of cells – and comprising a single type of immunoglobulin. It neutralises only one specific antigen. The antibodies are prepared by linking antibody-forming lymphocytes from the spleen of mice with myeloma cells from mice. Monoclonal antibodies are used in the development of new vaccines and in the study of human cells, hor- mones, and micro-organisms. Research is under way for their use in the treatment of some forms of cancer.

Monoclonal Antibody Types: Murine MAbs are derived from mouse genes; human MAbs are derived from human genes; chimeric and humanized MAbs are each derived from varying amounts of mouse and human genes, with the humanized products containing more human protein sequence than the chimeric versions

Monoclonal: Pertaining to a specific protein from a single clone of cells, all molecules of this protein being the same.

Monocyte: A type of white blood cell which has a single kidney-shaped nucleus. Present in the tissues.

Monomania: Monomania is a form of mental illness, in which the affected person has delusion upon one subject, although he or she can converse rationally and is a responsible individual upon other matters.

Mononucleosis: An acute viral infection in which the patient developes a sore throat, swollen lymph glands and fever. Also known as glandular fever, infec- tious mononucleosis is caused by members of the herpes group of viruses – the epstein barr virus and cytomegalovirus. The disease is more common among adolescents aged. The general symptoms are fever, sore throat, exhaustion, and abnormal white blood cells.

Mono-oxygenase: Enzyme that catalyses reactions between an organic compound and molecular oxygen in which one atom of the oxygen molecule is incorporated into the organic compound and one atom is reduced to water; involved in the metabolism of many natural and foreign compounds giving both unreactive products and products of different or increased toxicity from that of the parent compound : such enzymes are the main catalysts of phase 1 reactions in the metabolism of xenobiotics by the endoplasmic reticulum or by preparations of microsomes.

Monoplegia: Paralysis of a single limb or part.

Monorchism: The absence of one testis, usually the result of the failure of one testicle to drop down into the scrotum before birth. Monorchism is sometimes used to describe the condition when one testicle has been destroyed by disease or injury, or has been surgically removed – when, for example, the man has developed cancer of the testicle.

Monosaccharide: A sugar having six carbon atoms in the molecule, such as glucose, galactose, and laevulose.

Monospot: A screening test, performed on blood, which indicates the like lihood of infection.

Monozygotic Twins: Twins who develop from a single ovum fertiised by a single spermatozoon. Also known as identical or uniovular twins.

Morbid Anatomy: The study of the structural changes that diseases cause in the body, in particular those which can be seen with the naked eye at post-mortem examination.

Morbidity Rate: The rate of illness in a population. Term used loosely to refer to incidence or prevalence rates of disease.

Morbidity Survey: Method for the estimation of the prevalence and/or incidence of a disease or diseases in a population: a morbidity survey is usually designed simply to ascertain the facts as to disease distribution, and not to test a hypothesis.

Morbidity: Any departure, subjective or objective, from a state of physiological or psychological well-being: in this sense, "sickness", "illness", and "morbid condition" are similarly defined and synonymous. The WHO expert committee on health statistics noted in its sixth report (1959) that morbidity could be measured in terms of three units: 1. Proportion of persons who were ill, 2. The illnesses (periods or spells of illnesses) that these persons experienced. 3. The duration (days, weeks, etc.) of these illnesses. The condition of being diseased. The morbidity rate is the number of cases of disease occurring within a particular number of the population.

Morbidity: The frequency of a disease, illness, injury, and disability in a population.

Morbilli: Another name for measles.

Morbus: Morbus, the Latin word for disease, is used in such terms as morbus cordis (heart disease), morbus coxae (hip-joint disease).

Mordant: Substance that fixes a dyestuff in or on a material by combining with the dye to form an insoluble compound, used to fix or intensify stains in a tissue or cell preparation.

Moro Reflex: A primitive reflex action occurring in newly born infants in response to a sudden movement or noise. Also known as the startle reflex, the baby will throw its arms and legs wide and stiffen its body. This is followed by flexion of the arms and legs. The reflex disappears by four months; its persistence suggests a possible neurological condition such as cerebral palsy.

Moron: An out-of-date term previously used to describe individuals with mild learning difficulties, but which is now offensive.

Morphoea: A form of circumscribed scleroderma.

Mortality Rate: The total number of deaths relative to the total population at a specific place in a specific period of time.

Mortality Study: Investigation dealing with death rates or proportion of deaths attributed to specific causes as a measure of response.

Mortality: Death as studied in a given population or subpopulation. The word mortality is often used incorrectly instead of mortality rat.

Mosaicism: If non-dysjunction occurs after the formation of a zygote – that is, during a mitotic cell division and not a meiotic cell division – some of the cells will have one chromosome constitution and others another. The term mosaicism describes a condition in which a substantial minority of cells in an individual's body differ from the majority in their chromosome content. How substantial this minority is will depend upon how early during cleavage the zygote undergoes non-dysjunction. Mosaicism can cause disorders such as down's syndrome and turner's syndrome. The proportion and type of abnormal cells affect the physical appearance of the affected individual. This may range from normal to the features typical of people with a chromosomal-abnormality syndrome.

Motilin: Motilin is a hormone formed in the duodenum and the jejunum which plays a part in controlling the movements of the stomach and the gut.

Motion (Travel) Sickness: A characteristic set of symptoms experienced by many people when subjected to the constant changes of position caused, for example, by the pitching and rolling motion of a vessel at sea.

Motion: Waste products evacuated in a bowel movement, also called faeces or stool.

Motor Neurone Disease (MND): A group of disorders of unknown origin. Certain cells in the neurological system's motor nerves degenerate and die. Upper and lower motor neurones may be affected but sensory cells retain their normal functions. Three types of MND are identified: amyotrophic lateral sclerosis (AML –50 percent of patients); progressive muscular atrophy (25 percent), in which the prognosis is better than for AML; and bulbar palsy (25 percent). Men are affected more than women, and the disorder affects about seven people in every 100,000. Those affected develop progressive weakness and wasting of their muscles. The diagnosis is confirmed with various tests including the measurement of electrical activity in muscles, electromyography, muscle biopsy, blood tests and x-ray examin- ation of the spine. there is no medical treat- ment: patients need physical and psychological support with aids to help them overcome disabilities.

MOU: Memorandum of Understanding (an MOU between FDA and a regulatory agency in another country allows mutual recognition of inspections).

MPH: Master of Public Health, graduate degree

MRI: MRI, or magnetic resonance imaging, is a non-invasive method of imaging the body and its organs. It may also be used to study tissue metabolism. The body is placed in a magnetic field which causes certain atomic nuclei to align in the direction of the field. Pulses of radio-frequency radiation are then applied; interpretation of the frequencies absorbed and re-emitted allows an image in any body plane to be built up. Different tissues – for example, fat and water – can be separately identified and, if the resonance signal for the fat is suppressed, then only the signal from any abnormalities in the fat can be identified.

MR-I: The building letter code for Medical Research Building I on the UCSF Parnassus campus.

MR-II: The building letter code for Medical Research Building II on the UCSF Parnassus campus.

MR-IV: The building letter code for Medical Research Building IV on the UCSF Parnassus campus.

MSDES: Formerly the Ministry of Social Development and Economic Security. Now the Ministry of Human Resources (MHR).

MU: The building letter code for Milberry Union, 500 Parnassus Avenue, San Francisco.

Mucilage: This is prepared from acacia or tragacanth gum, and is used as an ingredient of mixtures con- taining solid particles in order to keep the latter from settling, and also as a demulcent.

Mucocoele: An abnormally dilated cavity in the body due to the accumulation of mucus; such a 'cyst' may therefore form wherever there is mucous membrane.

Mucoepithelial: Tissues with mixed characteristics of both mucous membranes and epidermis, found around the entrances into the body.

Mucolytic: The term used to describe the property of destroying, or lessening the tenacity of, mucus. It is most commonly used to describe drugs which have this property and are therefore used in the treatment of bronchitis. The inhalation of steam, for example, has a mucolytic action.

Mucopolysaccaridhosis: A collection of familial metabolic disorders, the best known of which is Hurler's syndrome. Others include Hunter's, Maroteaux-Lamy and Scheie's syndromes. The disorders, which result from a faulty gene- producing abnormality in a specific enzyme, affect one child in 10,000. Those affected usually die before reaching adulthood.

Mucopurulent: A discharge of mixed mucus and pus, usually from congested and moderately infected membranes.

Mucous Colitis: A form of colitis that is less inflammatory and closer in nature, if not identical with Irritable Bowel Syndrome, with cramps, intestinal guarding followed by soft or hard stools and various amounts of mucus. There are usually periods of constipation .

Mucous Membranes (Mucosa): The mucus-secreting skin that lines (and protects against the environment) all openings, cavities or entrances into the body, such as the intestinal tract, lungs, urinary tract, sinuses, vagina, etc.

Mucus: The general name for the slimy secretion derived from mucous membranes. It is mainly composed of a substance called mucin, which varies according to the particular mucous membrane from which it is derived, and it contains other substances, such as cells cast off from the surface of the membrane, enzymes, and dust particles.

Multi-Arm Parallel Designs: A type of paraalel designs that compares more than two treatments. It has an advantage of simultaneously being able to compare more than one treatment to a placebo arm, and also to compare two alternative new therapies.

Multicenter Trial: A clinical trial conducted according to a single protocol but at more than one site, and therefore, carried out by more than one investigator. A trial conducted in more than one location or city. Multicentre trials are generally larger than single-site trials and are often used for Phase III testing.

Multi-doctoring: When a patient, having obtained a prescription from a doctor, seeks to obtain a prescription from another doctor for an identical or equivalent drug within a short time period.

Multigeneration Study: 1. Toxicity test in which two to three generations of the test organism are exposed to the substance being assessed. 2. Toxicity test in which only one generation is exposed and effects on subsequent generations are assessed.

Multigravida: A pregnant woman who has had more than one pregnancy.

Multi-level Benefit Plan: The amount a member must pay for a drug product'is determined by what "level" the medication has been placed into. Multi-level benefit plans generally have between two and five levels.

Multimedia: All environmental media (air, land, and water) to which a hazardous substance, pollutant, or contaminant nray be discharged, released, or displaced.

Multipara: A woman who has borne several children.

Multiple (Multiphasic) Screenin: Procedure that has evolved by combining single screening tests, and is the logical corollary of mass screening. Where much time and effort have been spent by a population in attending for a single test such as mass radiography, it is natural to consider the economy of offering other tests at the same time. Multiple (or multiphasic) screening implies the administration of a number of tests, in combination, to large groups of people.

Multiple Births: Twins occur about once in 80 pregnancies, triplets once in 6,000, and quadruplets about once in 500,000. Quintuplets are exceedingly rare. Such is the natural state of affairs.

Multiple Personality Disorder: The individual with this psychiatric disorder has two or more different personalities, often contrasting. The dominant personality at the time determines the behaviour and attitude of the individual, who customarily seems not to know about the other personality or personalities. The switch from one personality to another is abrupt and the mental condition of the differing personalities is usually normal. It is possible that child abuse is a factor in the dis- order, which is treated by psychotherapy.

Multiple Sclerosis (MS): Multiple sclerosis is a progressive disease of the brain and spinal cord, which, although slow in its onset, in time may produce marked symptoms such as paralysis and tremors, and may ultimately result in a severely disabled invalid. The disorder consists of hardened patches, from the size of a pin-head to that of a pea or larger, scattered here and there irregularly through the brain and spinal cord. Each patch is made up of a mass of the connective tissue (neuroglia), which should be present only in sufficient amount to bind the nerve-cells and fibres together. In the earliest stage, the insulating sheaths of the nerve-fibres in the hardened patches break up, are absorbed, and leave the nerve-fibres bare, the connective tissue being later formed between these.

Multiple Sclerosis: A chronic, usually progressive disease of the central nervous system, with the gradual patchy disorganization of the protective myelin cells. It is almost certainly an auto-immune disorder, although viral infections sometimes seem to initiate the condition, and physical trauma is often seen to anomalously precede the first symptoms.

Multistage Cluster Sampling: Cluster sampling with more than two stages, each sampling being made on aggregates (or clusters) in which the clusters already obtained by the preceding sampling have been divided.

Multistage Sampling: Type of sampling in which the sample is selected by stages, the sampling units at each stage being subsampled from the larger units chosen at the previous stage.

Mumps: An acute infectious disease, caused by a paramyxovirus, and most common in children. Although it usually infects the parotid glands, and is often only a mild condition, it can spread to the testes or ovaries, particularly when contracted by unresistant adults, and a mild child's infection that is not properly honored by R&R always holds the potential for pancreatic or meningeal complications. Mumps, also known as epidemic parotitis, is an infectious disease characterised by inflammatory swelling of the parotid gland and other salivary glands – often occurring as an epi-demic and affecting mostly young people. Its name comes from the old verb, 'mump', meaning to mope or assume a disconsolate appearance – an aptdescription of the victim of the disease at its height.

Murmur: The uneven, rustling sound heard by auscultation over the heart and various blood vessels in abnormal conditons. For example, murmurs heard when the stethoscope is applied over the heart are highly characteristic of valvular disease of this organ.

Muscle: Muscular tissue is divided, according to its function, into three main groups: voluntary muscle, involuntary muscle, and skeletal muscle– of which the first is under control of the will, whilst the latter two discharge their functions independently. The term 'striped muscle' is often given to voluntary muscle, because under the microscope all the voluntary muscles show a striped appearance, whilst involuntary muscle is, in the main, unstriped or plain. Heart muscle is partially striped, while certain muscles of the throat, and two small muscles inside the ear, not controllable by willpower, are also striped.

Musculoskeletal: An adjective that relates to muscle and/or bone. The musculoskeletal system the bones of the skeleton and all the muscles attached to them.

Mutagen: A chemical or physical agent that has the property of increasing the rate of mutation among cells. A mutagen does not usually increase the range of mutations. Chemicals, ionising radiation, and viruses may act as mutagens.

Mutagen: Any substance that can induce heritable changes (mutations) of the genotype in a cell as a consequence of alterations or loss of genes or chromosomes.

Mutagenesis: Introduction of heritable changes (mutations) of the genotype in a cell as a consequence of alterations or loss of genes or chromosomes.

Mutagenic Studies: A type of in vivo and in vitro toxicological testing of 18-24 months duration to determine the mutagenic potential of a drug.

Mutagenicity: Ability of a physical, chemical, or biological agent to induce heritable changes (mutations) in the genotype in a cell as a consequence of alterations or loss of genes or chromosomes.

Mutation: A change occurring in the genetic material (DNA) in the chromosomes of a cell. It is caused by a fault in the replication of a cell's genetic material when it divides to form two daughter cells. Mutations may occur in somatic cells which may result in a local growth of the new type of cells. These may be destroyed by the body's defence mechanism or they may develop into a tumour. If mutation occurs in a germ cell or gamete – the organism's sex cells – the outcome may be a changed inherited characteristic in succeeding generations. Mutations occur rarely, but a small steady number are caused by background radiation in the environment. They are also caused by mutagens.

Myalgia: Tenderness or pain of the muscles themselves; muscular rheumatism. Pain in a muscle.

Myasthenia Gravis: A serious disorder in which the chief symptoms are muscular weakness and a special tendency for fatigue to come on rapidly when efforts are made. The prevalence is around 1 in 30,000. Two-thirds of the patients are women, in whom it develops in early adult life. In men it tends to develop later in life. It is a classical example of an autoimmune disease. The body develops antibodies which interfere with the working of the nerve endings in muscle that are acted on by acetylcholine. It is acetylcholine that transmits the nerve impulses to muscles: if this transmission cannot be effected, as in myasthenia gravis, then the muscles are unable to contract. Not only the voluntary muscles, but those connected with the acts of swallowing, breathing, and the like, become progressively weaker. Rest and avoidance of undue exertion are necessary, and regular doses of neostigmine bromide, or pyridostigmine, at intervals enable the muscles to be used and in some cases have a curative effect. These drugs act by inhibiting the action of cholinesterase – an enzyme produced in the body which destroys any excess of acetylcholine. In this way they increase the amount of available acetylcholine which compensates for the deleterious effect of antibodies on the nerve endings.

Myasthenia: Muscular weakness.

Mycobacterium: A gram-positive rod-like genus of aerobic bacteria, some species of which are harmful to humans and animals. For example, M. tuberculosis (Koch's bacillus) and M. leprae cause, respectively, tuberculosis and leprosy.

Mycoplasma: A genus of micro-organisms which differ from bacteria in that they lack a rigid cell wall. They are responsible for widespread epidemics in cattle and poultry. For a long time the only member of the genus known to cause disease in humans was Mycoplasma pneumoniae which is responsible for the form of pneumonia known as primary atypical pneumonia – particularly common in children, for whom it is the single most common cause of the diseaase when contracted out of hospital. Another, Mycoplasma genitalium, has now been isolated which is responsible for certain cases of non-gonococcal urethritis. Mycoplasma infections respond to tetracyclines.

Mycosis: The general term applied to diseases due to the growth of fungi in the body.

Mycotoxin: Toxin produced by a fungus.

Mydriasis: Dilation or widening of the pupil of the eye. Extreme dilation of the pupil of the eye, either as a result of normal physiological response or in response to a chemical exposure.

Myectomy: Removal of all or part of a muscle by surgery. It may be used to correct a squint (caused by unbalanced eye muscles) or to remove a fibroid from the muscular wall of the uterus.

Myelin: A substance made up of protein and phospholipid that forms the sheath surrounding the axons of some neurons. These are described as myelinated or medullated nerve fibres, and electric impulses pass along them faster than along non-myelinated nerves. Myelin is produced by Schwann cells which occur at intervals along the nerve fibre.

Myelitis: Myelitis is inflammation of the spinal cord.

Myeloblast: Present in the blood-producing tissue in the bone marrow, this is a cell with a large nucleus and scanty cytoplasm. It is the precur- sor cell of a granulocyte. Myeloblasts sometimes appear in the blood of patients with various diseases including acute myeloblastic leukaemia.

Myelocyte: The name given to one of the cells of bone marrow from which the granular white cells of the blood are produced. They are found in the blood in certain forms of leukaemia.

Myelography: The injection of a radio-opaque substance into the central canal of the spinal cord in order to assist in the diagnosis of diseases of the spinal cord or spine using X-ray examination. Because of the high risk of causing damage to the spinal cord (arachnoiditis), it has been largely superceded by MRI.

Myeloid: An adjective that relates to the granulocyte precursor cell in the bone marrow. for example, myeloid leukaemia, which arises from abnormal growth in the blood-forming tissue of the bone marrow.

Myelomatosis: A malignant disorder of plasma cells, derived from B-lymphocytes. In most patients the bone marrow is heavily infiltrated with atypical, monoclonal plasma cells, which gradually replace the normal cell lines, inducing anaemia, leucopenia, and thrombocytopenia. Bone absorption occurs, producing diffuse osteoporosis. In some cases only part of the immunoglobulin molecule is produced by the tumour cells, appearing in the urine as bence jones proteinuria.

Myelosuppression: A fall in the production of blood cells in the bone marrow. This fall often occurs after chemotherapy for cancer. Anaemia, infection and abnormal bleeding are symptomatic of myelosuppression.

Myelosuppression: Reduction of bone marrow activity leading to a lower concentration of platelets, red cells and white cells in the blood.

Myenteric Plexus: Broadly, the several neuron masses, ganglia, and nerve fiber plexus that lie in the walls of the intestinal tract, particularly the small intestine. They monitor and stimulate local muscle and glandular functions as well as blood supply, with little interface or control by the central nervous system or the autonomics. Each synapse away from the CNS gives greater autonomy, and these nerves only listen to God and food. This means the small intestine is relatively free of stress syndromes.

Myiasis: Infestation of the skin, deeper tissues or the intestine by larvae of the tropical tumbu fly.

Myocarditis: Inflammation of the muscular wall of the heart.

Myocardium: The middle, muscular layer of the heart. Myocardium is the muscular substance of the heart.

Myoclonus: A brief, twitching muscular contraction which may involve only a single muscle or many muscles. It may be too slight to cause movement of the affected limb, or so violent as to throw the victim to the floor. The cause is not known, but in some cases may be a form of epilepsy. A single myoclonic jerk in the upper limbs occasionally occurs in minor motor epilepsy (petitmal). The myoclonic jerks which many people experience on falling asleep are a perfectly normal phenomenon.

Myoglobin: The protein which gives muscle its red colour. It has the property of combining loosely and reversibly with oxygen; this means that it is the vehicle whereby muscle extracts oxygen from the haemoglobin in the blood circulating through it, and then releases the oxygen for use in muscle metabolism.

Myoglobinuria: The occurrence of myoglobin in the urine. This is the oxygen-binding pigment in muscle and mild myoglobinuria may occur during exercise. Severe myoglobinuria will result from serious injuries, particularly crushing injuries, to muscles.

Myoma: The term applied to a tumour, almost invariably of a simple nature, which consists mainly of muscle fibres. These muscle tumours often occur in the uterus.

Myomectomy: Removal by surgery of fibroids from the muscular wall of the uterus.

Myometrium: The muscular coat of the uterus.

Myotonia: A condition in which the muscles, though possessed of normal power, contract only very slowly. The stiffness disappears as the muscles are used.

Myringoplasty: The sealing by a surgical tissue-graft of a hole or perforation in the drum (tympanum) which separates the middle and outer sections of the ear.

Myringotomy: An operation to cut open the drum of the ear to provide drainage for an infection of the middle ear. It is now done mainly in children with persistent glue ear.

Myxedema: Puffiness and fluid retention resulting from thyroid hypofunction, either organic (serious, and often complicated by pituitary or adrenalcortical deficiencies) or functional (often a bipolar depressive thyroid phase).

Myxoma: A benign tumour comprising gelatinous connective tissue, most commonly occurring beneath the skin – although the condition may develop in the abdomen, urinary bladder, bone and, rarely, the heart. Treatment involves surgery, which is usually successful.

N

Nail-Biting: A common practice in schoolchildren, most of whom gradually give it up as they approach adolescence. Too much significance should therefore not be attached to it; in itself it does no harm, and punishment or restraining devices are not needed. It is a manifestation of tension or insecurity, the cause of which should be removed. In some people the habit is carried into adulthood.

Nappy Rash: A common form of irritant contact dermatitis in the nappy area in babies under one year old. Wetting of the skin by urine, abrasion, and chemical changes due to faecal contamination all play a part. Good hygiene and use of dispos- able absorbent nappies have much reduced its incidence. An ointment containing a barrier, such as titanium dioxide, may help; other medications such as mild corticosteroids or antibiotics should be used very cautiously and only under the guidance of a doctor, as harmful effects may result – especially from overuse.

Narcissism: An abnormal mental state characterised by excessive admiration of one's self. In Greek mythology, Narcissus so loved staring at his own reflection in water that he eventually fell in and drowned.

Narcolepsy: A condition in which uncontrollable episodes of sleep occur two or three times a day. It starts at any age and persists for life. The attacks, which usually last for 10–15 minutes, come on suddenly at times normally conducive to sleep, such as after a meal, or sitting in a bus, but they may occur when walking in the street. In due course, usually after some years, they are associated with cataplectic attacks, when for a few seconds there is sudden muscu- lar weakness affecting the whole body. The cataplectic attacks can be controlled by the tri-cyclic antidepressant drugs, imipramine or clomipramine.

Narcosis: A condition of stupor, resembling sleep, that is usually caused by a drug. It may also occur as a result of liver or kidney failure which causes uraemia. The affected person has significantly reduced awareness and is hard to arouse. Treatment is of the underlying cause and the normal precautions for caring for an unconscious or semiconscious subject should be taken.

Narcotic: A substance that depresses central nervous system function, bringing sleep and lessening pain. By definition, narcotics can be toxic in excess.

Narcotics: Substances that induce stupor and eventually unconscio- usness. Used in the relief of severe pain, people can become first tolerant of them – so requiring larger doses – and then dependent.

Nasal Congestion: The nose and nasal sinuses produce up to a litre of mucus in 24 hours, most of which enters the stomach via the nasopharynx. Changes in the nasal lining mucosa occur in response to changes in humidity and atmospheric temperature; these may cause severe

Nasogastric Tube: A small-bore plastic or rubber tube passed into the stomach through the nose, pharynx and then the oesophagus. It is used either to aspirate gas and liquid from the stomach or to pass food or drugs into it.

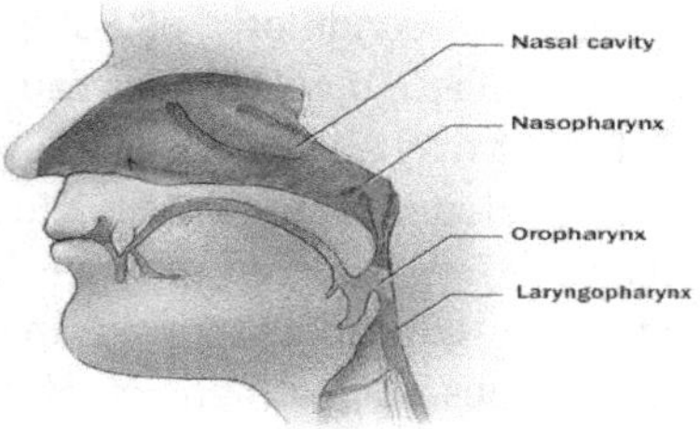

Nasolacrimal Duct: A duct that goes through the nasolacrimal canal in the palatine bone of the skull. The duct drains the tears from the lacrimal (tear) glands into the nose.

Nasopharynx: Nasopharynx is the upper part of the throat, lying behind the nasal cavity.

National Blood Authority: This body manages regional transfusion centres. Among its aims are the maintenance and promotion of blood and blood products based on a system of voluntary donors; imple- menting a cost-effective national strategy for ensuring adequate supplies of blood and its products to meet national needs; and ensuring high standards of safety and quality.

National Care Standards: Commission

This was set up under the care standards act 2000 as an independent regulator in respect of homes for the elderly, the disabled and children in the state and private sectors in the UK.

National Health Service (NHS): The United Kingdom's National Health Service was created by act of parliament and inaugurated on 5 July 1948. Its original aim was to provide a comprehensive system of health care to everyone, free at the point of delivery. Scotland had its own, similar legisla- tion, as did Northern Ireland. The service is funded by National Insurance contributions and from general taxation, with a small amount from patient charges. The structure, functioning and financing of the NHS have been – and still are – undergoing substantial changes.

National Infection Control and Health Protection Agency: A National Health Service body intended to combat the increasing threat from infectious diseases and biological, chemical and radio-logical hazards. Covering England, the agency includes the Public Health Laboratory Service, the National Radiological Protection Board, the Centre for Applied Microbiology and Research, and the National Focus Group for Chemical Incidents.

National Listening Library: National Listening Library is a charity which produces recorded books for handicapped people who cannot read, with the exception of the blind who have their own separate organisa- tion, the Royal National Institute for the Blind.

National Organ Transplantation Act: A U.S. law that mandates a national waiting list of organ recipients, equitable distribution of organs, and a national transplantation registry.

Natriuresis: The excretion of sodium in the urine, particularly if the amount excreted is more than normal. An agent that causes this sodium excretion– usually a diuretic is termed a natriuretic.

Natriuretic: Substance increasing the rate of excretion of sodium ion in the urine.

Natural History Study: Study of the natural development of something (such as an organism or a disease) over a period of time.

Natural Occurrence: Presence of a substance in nature, as distinct from presence resulting from inputs from human activities. The contamination of the natural environment by some man-made compounds may be so widespread that it is practically impossible to get access to biota with a truly natural level; only "normal" levels can be measured, those which are usually prevalent in places where there is no obvious local contamination.

Nausea: The sensation that vomiting is about to occur however, nausea does not always lead to vomiting.

Navel: Navel, or umbilicus, is the scar

on the abdomen marking the point where the umbilical cord joined the body in embryonic life.

NCE: New chemical entity. Any new molecular compound [excluding diagnostic agents, vaccines, and other biologic compounds] not previously approved for human use by the CDER. Also excluded are new salts, esters, and dosage forms of previously approved compounds.

NCR Paper: No Carbon Required Psper (a special paper used in CRFs that does not require a carbon sheet for getting a carbon copy.)

NDA: New drug application. An application to FDA for a license to market a new drug in the United States.

NDGA: Nordihydroguaiaretic acid, a substance found in abundance in the oleoresins of Larrea (Chaparral) and the Guaiacum genus (Lignum Vitae). It is strongly antioxidant to lipids and is antifungal, antimicrobial and antibacterial. Both plants contain a constellation of related compounds and do not have the potential kidney toxicity found in pure NDGA and the reason it is no longer used as an EDTA-type edible oil stabilizer in food -

manufacturing.

Nebula: The term applied to a slight opacity on the cornea producing a haze in the field of vision, and also to any oily preparation to be sprayed from a nebuliser – an apparatus for splitting up a fluid into fine droplets.

Nebulisers: A nebuliser makes an aerosol by blowing air or oxygen through a solution of a drug. Many inhaled drugs such as salbutamol, ipratropium and beclomethasone

Necropsy: A traditional term for an autopsy or post-mortem examination.

Necrosis: Death of a limited portion of tissue, the term being most commonly applied to bones when, as the result of disease or injury, a fragment dies and separates. Morphological changes that follow cell death, characterized most frequently by nuclear changes.

Necrotising Fasciitis: Also known as cellulitis. A potentially lethal infection caused by the gram-positive bacterium Streptococcus pyogenes which has the property of producing dangerous exotoxins. The infection, which starts in the layer of fascia under the skin, may spread very rapidly, destroying tissue as it spreads.

Needle-Stick Injury: Accidental perforation of the skin by an injection needle, commonly of the hand or finger and usually by a nurse or doctor administering a therapeutic injection. The term also refers to accidental injuries from injection needles discarded by drug abusers. Dangerous infections such as viral hepatitis or HIV may be acquired from needle-stick injuries, and there are strict procedures about the disposal of used syringes and needles in medical settings.

Needling: An operation performed in

the treatment of cataracts in which the anterior lens capsule of the eye is torn open with a needle, allowing the aqueous fluid to dissolve the opaque soft lens matter, which is gradually washed away into the blood- stream. This 'extra-capsular extraction' may need to be repeated several times before all the opaque lens matter disperses. Although a relatively simple procedure, it is unsuitable for patients over the age of 35 (when the nucleus of the lens becomes increasingly hard), and cryo- surgery and laser therapy have become the preferred methods of treatment.

Negativism: Negativism means a morbid tendency in a person to do the opposite of what he or she is desired or directed to do. It is especially characteristic of those suffering from schizophrenia, but is not uncommon in non-psychotic persons.

Neglect: Willful oversight in performing a particular activity.

Negligible Risk: Probability of adverse effects occurring that can reasonably be described as trivial. Probability of adverse effects occurring that is so low that it cannot be reduced appreciably by increased regulation or investment of resources.

Neisseriaceae: A family of bacteria of which three varieties cause disease. Neisseria meningitidis causes meningococcal meningitis and septicaemia.

Nematocide: Substance intended to kill nematodes.

Nematode: A roundworm.

Neonat: Infant during the first 4 weeks of postnatal life; for statistical purposes some scientists have defined the period as the first 7 days.

Neonatal Intensive Care: The provision of a dedicated unit with special facilities, including one-to-one nursing and appropriate technology, for caring for pre- mature and seriously ill newborn babies. Paedi- atricians and neonatologists are involved in the running of such units. Not every maternity unit can provide intensive care: for example, the provision of artificial ventilation, other than as a holding procedure until a baby can be trans- ferred to a better-equipped and better-serviced unit. Such hospitals tend to have special-care baby units, which are capable of looking after the needs of most, but not all, premature or ill babies.

Neonatal: Pertaining to the first month of life.

Neonate: Newborn baby.

Neonatology: The branch of paediatrics responsible for the medical care of newborn babies. Problems may be short term – for example, those linked to prematurity – or life-long such as cerebral palsy. After the first few weeks of life, paedi- atricians take over the responsibility for any specialist medical care required, with general practitioners looking after the infants' primary- care needs.

Neoplas: New and abnormal formation of tissue as a tumour or growth by cell proliferation that is faster than normal and continues after the initial stimulus (i) that initiated the proliferation has ceased.

Neoplasia: The presence of abnormal cells forming a growth or tumor, unable to perform their normal functions, and replacing healthy cells.

Neoplasm: This means literally a 'new formation' and is another word for a benign or non-malignant tumour.

Nephrectomy: The operation for removal of the kidney or the surgical removal of a kidney

Nephric: Pertaining to the kidney.

Nephritic Syndrome: Combination of mild-moderate haemat- uria, proteineuria without significant hvpoalbuminaemia.

Nephritis: Inflammation or infection of the kidneys, as opposed to lower urinary tract inflammations such as cystitis or urethritis, which are usually comparatively mild. Nephritis can be a far more serious condition, and usually requires medical care.

Nephroblastoma: Nephroblastoma, or Wilm's tumour, is the commonest kidney tumour in infancy

Nephrolithiasis: A condition in which calculi are present in the kidney.

Nephrology: The branch of medicine concerned with the study and management of kidney disease. A specialist in these diseases is called a nephrologist.

Nephron: Each kidney comprises over a million of these microscopic units which regulate and control the formation of urine. A tuft of capillaries invaginates the Bowmans capsule, which is the blind-ending tube of each nephron. Plasma is filtered out of blood and through the Bowmans capsule into the renal tubule. As the filtrate passes along the tubule, most of the water and electrolytes are reabsorbed. The composition is regulated with the retention or addition of certain molecules.

Nephropathy: A description of any damage or disease to the kidneys.

Nephropexy: Surgical fixation (to the 12th rib and posterior abdominal wall) of a mobile kidney; this prevents the kidney from descending in the abdomen when the affected person stands up.

Nephroptosis: The condition in which a kidney is mobile or 'floating' instead of being fixed to the back of the abdominal cavity.

Nephroscope: An endoscopic instrument for examining the inside of the kidney. It is normally passed into the renal pelvis of the organ via a route from the surface of the skin. Instruments can be passed through the nephroscope under direct vision to remove calculi (stones) or break them up using ultrasound.

Nephrostomy: Nephrostomy is the operation of making an opening into the kidney to drain it.

Nephrotic Syndrome: Nephrotic syndrome is one of proteinuria, hypo-albuminaemia and gross oedema. The primary cause is the leak of albumin through the glomerulus. When this exceeds the liver's ability to synthesise albumin, the plasma level falls and oedema results. The nephrotic syndrome is commonly the result of primary renal glomerular disease. It may also be a result of metabolic diseases such as diabetic glomerular sclerosis and amyloidosis. It may be the result of systemic autoimmune diseases such as systemic lupus erythematosus and polyarteritis nodosa. it may complicate malignant diseases such as myelomatosis and hodgkin's disease. It is some- times caused by nephrotoxins such as gold or mercury and certain drugs, and it may be the result of certain infections such as malaria and crohn's disease.

Nephrotomy: Nephrotomy means the operation of cutting into the kidney, in search of calculi or for other reasons.

Nephrotoxic: Chemically harmful to the cells of the kidney.

Nerve Growth Factor (NGF): A naturally occurring hormone that stimulates the growth of neurones.

Nerve: A cell which carries information to and from the central nervous system.

Nervous Breakdown: A non-medical description of a variety of emotional crises ranging from an outburst of hysterical behaviour to a major neurotic illness that may have a lasting effect on an individual's life. Sometimes the term is used to describe an overt psychotic illness – for example, schizophrenia.

Nervous Impulse: This is transmitted chemically, by the formation at nerve-endings of chemical substances. When, for example, a nerve to a muscle is stimulated, there appears at the neuro- muscular junction the chemical substance, acetylcholine. acetylcholine also appears at endings of the parasympathetic nerves and transmits the effect of the parasympathetic impulse. When an impulse passes down a sympathetic nerve, the effect of it is transmitted at the nerve-ending by the chemical liberated there: adrenaline or an adrenaline-like substance.

Network Central Fill: A pharmacy distribution system in which a network member's in-house pharmacy fills prescriptions for the patients of the other network members. Please note that if drugs purchased under 340B are used, approval as a 340B Alternative Method Demonstration project might be required.

Network: A group of health care providers under contract with a managed care company within a specific geographic area.

Neural Tube: The structure in the embryo from which the brain and spinal cord develop.

Neural: Pertaining to a nerve or to the nerves.

Neuralgia: Pain, sometimes severe, that manifests along the length of a nerve and arises within the nerve itself, not in the tissue from which the sensation seems to arise.

Neurasthenia: An out-of-date term that was used to describe an ill-defined state of nervous exhaustion in which, although the patient suffers from no definite disease, he or she becomes incapable of sustained exertion. The condition which it rep- resented is now believed to be a form of neurosis or psychosomatic disease. It was also used in the past to describe what is now called chronic fatigue syndrome.

Neurasthenia: Tiredness or exhaustion, often in excess of what would seem appropriate from purely physical causes.

Neurectomy: An operation in which part of a nerve is excised: for example, for the relief of neuralgia.

Neurilemma: The thin membranous covering which surrounds every nerve-fibre.

Neurites: The tiny projections growing from each nerve cell which carries information between the cells. A nerve cell may have over 100,000 neurites growing out of it; each connected to another nerve cell.

Neuritis: Nerve inflammation, usually with an abnormal amount of pain, and often part of a degenerative process.

Neuroblastoma: A malignant growth comprising embryonic nerve cells. It may start in any part of the autonomic nervous system. The medulla of the adrenal gland is a common site; secondary growths develop in other tissues. Neuroblastomas are the most common extracranial solid tumour of childhood. The incidence is around eight cases per one million children. Treatment is by surgery followed by radiotherapy and cytotoxic drugs. About 30 per cent of sufferers survive for at least five years after treatment.

Neurochemical: A chemical that naturally occurs in the nervous system and plays a part in its functioning.

Neurodegenerative Disease: A type of neurological disorder marked by the loss of nerve cells. For example Alzheimer's disease, Huntington's disease, and Parkinson's disease.

Neurofibrillary Tangles: Accumulation of twisted protein fragments inside neurons. Neurofibrillary tangles are one of the characteristic structural abnormalities found in the brains of patients with Alzheimer's disease patients.

Neurofibrils: A microscopic strand of cytoplasm that occurs in the cell body of a neuron as well as in the semifluid content of the axon of a nerve cell.

Neurogenic: Sensations or conditions derived solely from the nervous system.

Neuroglia: The fine web of tissue and branching cells which supports the nerve-fibres and cells of the nervous system.

Neuroleptics: Drugs used to quieten disturbed patients, whether this is the result of brain damage, mania, delirium, agitated depression or an acute behavioural disturbance. They relieve the florid psychotic symptoms such as hallucinations and thought-disorder in schizophrenia and prevent relapse of this disorder when it is in remission.

Neurology: The branch of medical practice and science which is concerned with the study of the nervous system and its disorders.

Specialists in neurology – neurologists – examine a patient's nerves, sensory and motor functions and reflexes. They use modern imaging techniques

Neuroma: Neuroma means a tumour connected with a nerve – such tumours being generally com- posed of fibrous tissue, and of a painful nature.

Neuromuscular Blockade: In clinical practice, the transmission of impulses at the neuromuscular junction may be blocked to paralyse temporarily a patient for a surgical procedure, or to assist treatment on the intensive care unit. There are two main types of drug, both of which com- petitively block the acetylcholine receptors on the motor end plates.

Neuromuscular Junction: The area where a motor nerve ends close to the muscle membrane so can initiate muscle contraction.

Neuron: Also known as a nerve cell, this is the basic cellular building-block of the nervous system, which contains billions of neurones linked in a complex network and acting in different com- binations to keep the body informed about the outside world, and then to organise and activate appropriate responses.

Neuropathic Bladder: A urinary bladder with complete or partial loss of sensation. As there is no sensation of fullness, the individual either develops complete retention of urine, or the bladder empties automatically – usually every few hours.

Neuropathies: A disease of the central or peripheral nervous systems. In more common reference, a neuropathy is primarily a disorder of peripheral nerves. CNS diseases are often life threatening; neuropathies are generally disorders of the control and sensory nerves out in the body.

Neuropathology: The branch of pathology that covers the reasons for and consequences of disorders of the nervous system.

Neuropathy: Any disease of the central or peripheral nervous system.

Neuropathy: Peripheral neuropathy is a condition of the nervous system that usually begins in the hands or feet with symptoms of numbness, tingling, burning and/or weakness.

Neurosis: A general term applied to mental or emotional disturbance in which, as opposed to psychosis, there is no serious disturbance in the perception or understanding of external reality. However, the boundaries between neurosis and psychosis are not always clearly defined. Neuroses are usually classified into anxiety neuroses, depressive neuroses, phobias, hypochon- driasis, and hysteria and obsessional neuroses.

Neurotic: A vague term applied to a person of nervous temperament, whose actions are largely determined by emotions or instincts rather than by reason.

Neurotoxic: Able to produce chemically an adverse effect on the nervous system: such effects may be subdivided into two types.

Neurotoxin: A chemical substance that harms nervous tissue, causing symptoms of numbness or weakness of the body part supplied by the damaged nerve. the venom of some snakes contains neurotoxic substances, and bacteria may produce neuro-toxins: examples are those that cause diphtheria and tetanus. Arsenic and lead are examples of inorganic neurotoxins.

Neurotransmitter: A chemical substance which transmits the action of a nerve to a cell. It is released from nerve-endings and transmits the impulse across synapses to other nerves. In the central nervous system the sub- stances acting as neurotransmitters include acetylcholine, noradrenaline, dopamine and serotonin. The main transmitter in the peripheral system is acetylcholine, while for the sympathetic system. It is noradrenaline. In recent years a new group of neurotransmitters called neuropeptides has been identified, com- prising large protein molecules. One of the best-known is that of endorphins, which the brain uses to control pain. Neurotransmitter: One of the many chemicals that carry impulses between nerve cells.

Neurotrophic Factor: A molecule, typically a protein, such as nerve growth factor (NGF), that promotes nerve cell growth, repair, and survival.

Neutron: A neutron is one of the particles that enter into the structure of the atomic nucleus.

Neutropenia: A reduction in the number of neutrophil leucocytes per cubic millimetre of circulating blood to a figure below that found in health.

Neutrophils: Another name for polymorphonuclear leukocytes, the most common type of blood-carried white blood cell, and the first mobile resistance cell to come to the rescue in injury.

New Chemical Entity: Chemical molecule that after undergoing clinical trials could translate into a potential drug for the cure/treatment of a disease.

New Drug Application (NDA): An application submitted by the manufacturer of a drug to the FDA after phase trials for obtaining marketing authorization of a new drug.

New Drug Submission: All the information (gathered by a pharmaceutical company) including chemical structure and properties, production details, preclinical and clinical studies or a new drug or compound. An NDS is submitted to health Canada for evaluation.

Niacin: Niacin is the active part of vitamin B_3.

Nidus: A site of infection within the body from which it can spread to other tissues.

Night Sweats: copious perspiration occurring in bed at night and found in conditions such as tuberculosis, brucellosis and lymphomas , as well as thyrotoxicosis, anxiety states and menopausal flushes.

Nipple: The small, sensitive prominence at the tip of each breast containing (in women) the small openings through which milk can pass from the milk glands in the breast tissue. The nipple and its surrounding area (the areola) are darker than the adjacent skin. The area becomes darker during pregnancy.

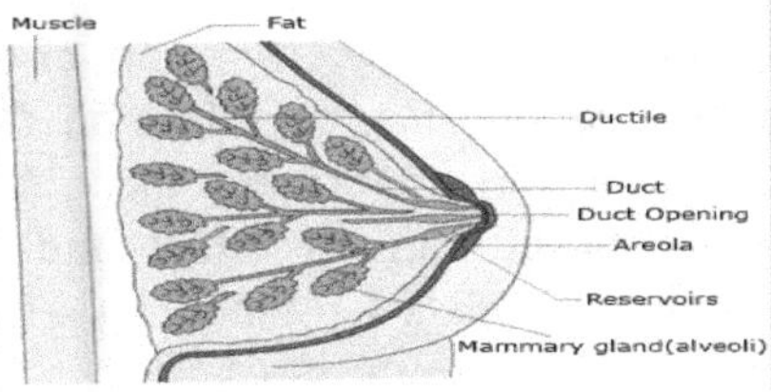

Structure of the Breast

Nitrification: Sequential oxidation of ammonium salts to nitrite and nitrate by micro-organisms.

Nitrogenous: A compound or molecule that contains nitrogen; in my context, a substance that is or was a part of protein metabolism.

NK Cells (Natural Killer Cells): Large lymphocytes with granules in the cytoplasm that kill infected cells and tumor cells.

NME: New molecular entity

No Action Indicated: A classification of regulatory inspection outcome that does not require any action response.

No Effect Level (NEL): Maximum dose (of a substance) that produces no detectable changes under defined conditions of exposure. At present, this term tends to be substituted by no-observed-adverse-effect-level (NOAEL) or no- observed-effect-level (NOEL).

No Treatment Control: Subject who are randomly assigned to the no treatment arm in a clinical trial.

Nociceptors: Nerve endings which detect and respond to painful or unpleasant stimuli.

Nocturia: Excess passing of urine during the night.

Nocturnal Enuresis: The involuntary passing of urine during sleep.

Nodule: Small node or boss that is solid and can be detected by touch.

N-of-1 Design: A type of clinical trial design that is carried out in a single patient. Such designs are related to cross-over designs and involve treatment comparisons within the same individual based on an outcome observed repeatedly over time. It is appropriate for chronic conditions involving treatments that lead to immediate transient effects such as pain relief.

Nominal Data: Type of qualitative test or categorical data. For eg., Sex :Male/Female; Treatment group :A/B.

Non Clinical Study: Biomedical studies not performed on human subjects.

Non Commercial IND: A type of Investigational New Drug Application filed for carrying out non-commercial research. It includes Investigator IND, Emergency Use IND and Treatment IND.

Non Comparative Study: A trial design that does not have a comparative arm.

Non Compliance: The deviation from the protocol, standard operating procedures or applicable regulatory guidelines.

Non Disclosure Agreement (NDA): A document that establishes agreement between two or more parties for ensuring the confidentiality of information provided by one party to the other.

Non Evaluable: The subject whose data can not be included in the safety and/or efficacy analysis.

Non Governmental Organizations: The non for profit charitable organizations.

Non Invasive: A medical procedure that does not involve a skin break.

Non Parametric Tests: A distribution free tests that is applied when the data is not normally distributed. It is used to compare observations repeated on the same subjects. For e.g., Mann-Whitney U test, Wilcox on test, Krushali-Wallis test etc.

Non Responder: The subject that does not respond to the trial drug or therapy.

Non- Significant: Yielding a value from\ a statistical test that lies within the limits for being of random occurrence.

Non-Bioenvironmental Transformation: Change in the chemical or physical nature of a substance occurring as a result of physicochemical conditions and independent of any biological system.

Nonclinical Study: Biomedical studies not performed on human subjects.

Non-effective Dose: Amount of a substance that has no effect on the organism. It is lower than the threshold of harmful effect and is estimated while establishing the threshold of harmful effect.

Non-Federal Average Manufacturer Price (Non-FAMP): The average price paid to a manufacturer by wholesalers for drugs distributed to non-Federal purchasers. The Big 4 are entitled under Federal law to discounts on brand name drugs of at least 24 percent off of Non-FAMP.

Non-Hodgkin's Lymphoma: Blood cancer of the lymphatic system, which usually spreads throughout the body.

Non-Inferiority Studies: Randomized trail design intended to establish that a new treatment has a similar rather than superior effect to an active control treatment. For non inferiority studies the intent is to establish that the difference between the new treatment and control is not large in direction favoring the control arm. E.g. bioequivalence studies.

Non-Insured Health Benefits (NIHB): Federal government coverage of health benefits provided by Indian and Northern Affairs Canada (INAC) to INAC clients.

Non-Occupational Exposure: Environmental exposure outside the workplace to substances that are otherwise associated with particular work environments and/or activities and processes that occur there.

Non-Participating Provider: A medical provider who has not contracted with a health plan.

Nonpoint Source: A nonpoint source typically consists of storm water runoff from parking lots and roads (asphalt, gravel, or dirt), rooftops, lawns, and disturbed areas such as construction sites. It also includes vapor emissions from automobiles.

Nonprescription Drug: A drug that can be purchased without a prescription order, sometimes referred to as over-the- counter (OTC) drugs.

Non-prescription Medicine: Drugs available over-the-counter (OTC), without a written prescription from the doctor.

Nonrecurring Requirements: Those requirements usually considered one-time expenses. They are further divided into priority categories.

Nonrenal Clearance: Clearance by the fecal route (CIF), respiratory route (C1L), salivary route (CIS), biliary route (C1B), can be computed in a fashion analogous to computation of C1R: measuring the amount of substance excreted in the feces, expired air, saliva, etc., over an interval and dividing by the plasma concentration at mid- interval and the length of the interval.

Non-target Organism: Organism affected by a pesticide although not the intended object of its use.

No-Observed-Adverse-Effect-Level (NOAEL): Greatest concentration or amount of a substance, found by experiment or ' observation, which causes no detectable adverse alteration of morphology, functional capacity, growth, development, or life span of the target organism under defined conditions of exposure.

No-Observed-Effect-Level (NOEL): Greatest concentration or amount of a substance, found by experiment or observation, that causes no alterations of morphology, functional capacity, growth, development, or life span of target organisms distinguishable from those observed in normal (control) organisms of the same species and strain under the same defined conditions of exposure.

Nootropic: From the Greek words noos, meaning mind and tropein, meaning toward.

Norepinephrine: An excitatory neurotransmitter involved in alertness, concentration, aggression and motivation, among other behaviors. Norepinephrine is made in the brain from the amino acid phenylalanine.

No-Response Level: Maximum dose of a substance at which no specified response is observed in a defined population and under defined conditions of exposure.

Norm: The expected value for something measurable. Most people have measurements lying to either side of the norm. Traditionally in medicine, only those lying beyond two standard devi- ations from the norm are considered likely to be abnormal (approximately 3 per cent of those measured).

Normoblast: The precursor of an erythrocyte which still contains the remnant of a nucleus.

Normotensive: Having a blood pressure within the normal range for an individual's age and sex.

Nosocomial: Associated with a hospital or infirmary, especially used of diseases that may result from treatment in such an institution.

Nosode: A homeopathic remedy prepared from diseased tissue or the product of disease.

Not Approvable Letter: An action letter which lists that deficiencies in the application and explains why the application cannot be approved. An official communication from FDA to inform an NDA sponsor that the important deficiencies described therein preclude approval unless corrected.

Not Harmful. Used especially to describe tumours that are not malignant.

Notes To File: The notes to explain the deviation/violation of a particular activity/process.

Notice Of Compliance (NOC): A notice with respect to a medicine. This notice is issued by the TPP (Health Canada) under the Food and Drug Regulations. The issuance of an NOC indicates that a drug meets the required Health Canada standards for use in humans or animals and that the product is approved for sale in Canada.

Notice of Inspection: An official communication from regulatory agency to the investigator/sponsor for undertaking a regulatory inspection.

Notifiable Diseases: Diseases, usually of an infectious nature, which are required by law to be made known to a health officer or local authority. Certain occupational diseases are also notifiable.

Notified Body (NB): A private institution charged by the Competent Authority with verifying compliance of medical devices (not drugs) with the applicable Essential Requirements stated in the Medical Device Directive. This process, called Conformity Assessment, has EU-wide validity once completed by the NB.

NPA (No Public Access Medications): Medications that can be obtained without a doctor's prescription but that are kept behind the counter at a pharmacy.

NPCD (National Patient Care Database): NPCD is a database of VHA services. It contains four distinct databases: the Patient Treatment File (PTF), the Outpatient Care File (OPC), the IPDB, and the EDR. Researchers access extracts of the four databases rather than drawing data from the NPCD itself.

Nucha: The Latin name for the back of the neck.

Nuclear Magnetic Resonance: Spectroscopic technique that enables scientists to determine the structure and dynamics of molecules and if, where, and how drugs might bind to those molecules

Nuclear Medicine: The branch of medicine concerned with the use of radioactive material in the diagnosis, investi- gation and treatment of disease.

Nucleic Acid: A substance constructed out of units known as nucleotides which consist of a purine or pyrimidine base linked to a pentose sugar, which in turn is esterified with phosphoric acid. Two types of nucleic acid occur in nature: deoxy- ribonucleic acid (DNA) and ribonucleic acid (RNA).

Nucleoprotein: A molecule that is formed from a structural protein that is combined with nucleic acid, and generally found in cell nuclei and other proliferative points in cells. Upon cell death, nucleoproteins, unlike others, cannot be catabolized and recycled efficiently; instead, part of the protein is degraded to purines, and thence to uric acid. Uric acid, unlike urea, is an excretory dead end.

Nuisance Threshold: Lowest concentration of an air pollutant that can be considered objectionable.

Null Hypothesis: It is a statement claiming that a population parameter yakes a specified value in accordance with an underlying theory about one or more populations. The null hypothesis posits equality between the treatments and is denoted as H as a standard statistical practice. A null hypothesis (for example, "subjects will experience no change in blood pressure as a result of administration of the test product") is used to rule out every possibility except the one the researcher is trying to prove, an assumption about a research population that may or may not be rejected as a result of testing. Used because most statistical methods are less able to prove something true than to provide strong evidence that it is false. The assertion that no true association or difference in the study outcome or comparison of interest between comparison groups exists in the larger population from which the study samples are obtained. See also research hypothesis.

Nullipara: The term applied to a woman who has never borne a child.

Numeric Continuous Data: A type of quantitative data. For eg. Blood Pressure, Pulse Rate.

Numeric Data: Quantitative data such as number of patients,number of visits etc.

Numeric Discrete Data: A type of quantitative data. For eg., Number of Children in a Family.

Nuremberg Code: As a result of the medical experimentation conducted by the Nazis during World War 2, the U.S. military Tribunal in 1947 set forth a code of medical ethics for researches conducting clinical trials. The code is designed to protect the safety and integrity of study participants.

Nurse Logs: In old-growth forests, these are ancient downed trees that rot so slowly that they themselves become the fundus and growth media for new and growing trees and other life-forms.

Nursing: Nurses are the largest single group of staff work-ing in the health service. There are more than 330,000 qualified nursing posts in NHS trusts and primary care across the UK. Would-be registered nurses (RNs) do either a three-year diploma programme or a four-year degree. An increasing number of nurses are now acquiring degrees, either as their initial qualification or by studying part-time later in their career.

Nutrition: The process by which the living organism physiologically absorbs and uses food to ensure growth, energy production and repair of tissues. The science of nutrition includes the study of diets and deficiency diseases.

Nutritional Table Method: Procedure for evaluating the dietary intake of a large number of people. The accuracy of the method depends on the accuracy with which records of the food consumption can be established and the accuracy of the nutritional tables specifying the concentration of various nutrients, vitamins, essential, and non-essential substances including pesticide residues.

Nystagmus: Involuntary, rapid, rhythmic movement (horizontal, vertical, rotary, mixed) of the eyeball, usually caused by a disorder of the labyrynth of the inner ear or a malfunction of the central nervous system. Autopsy, the presence of amyloid plaques and neurofibrillary tangles is used to positively diagnose Alzheimer's disease.

O

Oat Cell: A type of cell found in one highly malignant form of lung cancer. The cell is small and either oval or round. The nucleus stains darkly and the cytoplasm is sparse and difficult to identify. Oat-cell, or small-cell, carcinoma of the bronchus is usually caused by smoking, and com- prises around 30 per cent of all bronchial cancers. It responds to radiotherapy and chemo- therapy but, because the growth has usually spread widely by the time it is diagnosed, the prognosis is poor.

Obesity: A condition in which the energy stores of the body (mainly fat) are too large. It is a prevalent nutritional disorder in prosperous countries – increasingly so among children and young people. The quetelet index or body mass index, which relates weight in kilograms (W) to height 2 in meters (H_2), is a widely accepted way of classifying obesity in adults according to severity.

Objective Environment: Actual physical, chemical, and social environment as described by objective measurements, such as noise levels in decibels and concentrations of air pollutants.

Objective Measurement: A measure-ement that yields a quantitative assessment. A measurement that cannot be influenced by investigator bias; for example, blood glucose levels or ECG tracings.

Observation Study: A non-interventional study conducted according to routine medical practice.

Obstetrics: The branch of medicine dealing with pregnancy and giving birth. Derived from the Latin word for midwifery, it is closely allied to gynaecology. It is concerned with the health of the woman and fetus, from early in pregnancy through to a successful labour and delivery.

Occiput: The lower and hinder part of the head, where it merges into the neck.

Occlusion: The way that the teeth fit together when the jaws close. Also the closing or obstruction of a duct, hollow organ, or blood vessel.

Occult: Describing something that is not easily seen. Occult blood in the faeces is present in very small amounts and can be identified only by a chemical test or under the microscope.

Occupational Eexposure Standard (OES): Level of exposure to substances, intensities of radiation etc. or other conditions considered to represent specified good practice and a realistic criterion for the control of exposure by appropriate plant design, engineering controls, and, if necessary, the addition and use of personal protective clothing.

Occupational Environment: Surrounding conditions at a workplace.

Occupational Exposure Limit (OEL): Regulatory level of exposure to substances, intensities of radiation etc. or other conditions, specified appropriately in relevant government legislation or related codes of practice.

Occupational Exposure: Experience of substances, intensities of radiation etc. or other conditions while at work.

Occupational Hygiene: Identification, assessment and control of physicochemical and biological factors in the workplace that may affect the health or well-being of those at work and in the surrounding community.

Occupational Medicine: The branch of medicine that deals with the control, preven- tion, diagnosis, treatment and management of ill-health and injuries caused or made worse by work, and with ensuring that workers are fit for the work they do.

Occupational Therapy: The treatment of physical and psychiatric conditions through specific selected activities in order to help people reach their maximum level of function and independence in all aspects of daily life.

Ochronosis: A rare condition in which the ligaments and cartilages of the body, and sometimes the conjunctiva, become stained by dark brown or black pigment. This may occur in chronic carbolic poisoning, or in a congenital disorder of metabolism in which the individual is unable to break down completely the tyrosine of the protein molecule – the intermediate product, homogentisic acid, appearing in the urine, this being known as alkaptonuria.

Octanol-Water Partition Coefficient (POW, KOW): Measure of lipophilicity by determination of the equilibrium distribution between octan-l-ol and water, as used in pharmacological studies and in the assessment of environmental fate and transport of organic chemicals.

Ocular: Pertaining to the eye.

Odds Ratio (OR): The odds in favor of disease among exposed individuals divided by the odds in favor of disease among the unexposed individuals. It is the only parameter that can be used to compare two groups of binary data responses from a retrospective study.

Odds: Ratio of the probability of occurrence of an event to that of non-occurrence, or the ratio of the probability that something is so, to the probability that it is not so.

Odour threshold: In principle, the lowest concentration of an odorant that can be detected by a human being: in practice, a panel of "sniffers" is used, and the threshold taken as the concentration at which 50 % of the panel can detect the odorant (although some workers have also used 100 % thresholds).

Oedema: An abnormal accumulation of fluid beneath the skin, or in one or more of the cavities of the body. Presence of abnormally iarge amounts of fluid in intercellular spaces of body tissues.

Oedipus Complex: A description used by psychoanalysts of the subconscious attraction of a child for its parent of the opposite sex. This is accompanied by a wish to get rid of the parent of the same sex. The origin of the phrase lies in the greek story in which Oedipus kills his father without realis- ing who he is, then marries his mother. It has been suggested that the arrest of psychological development at the oedipal stage may cause neurosis and sexual dysfunction.

Off Label: The unauthorized use of a drug for an indication which is not approved by the regulatory Authorities.

Off Shoring: The relocation of business process from one country to another.

Office For Human Research Protection: A federal government agency that issues assurances and overseas compliance of regulatory guidelines by research institutes.

Office of Pharmacy Affairs: The office of pharmacy affairs is the component within HRSA that administers the 340B drug pricing program. OPA is located within the division of health center development in HRSA's bureau of primary health care.

Off-Label Use: A drug prescribed for conditions other than those for which it was approved.

Oil, fixed: These are lipids, esters of long-chain fatty acids and alcohols, or generally related oily stuff. If you drop some fixed oil on a blotter, it just stays there- forever. (Example: olive oil.)

Oil, Volatile: The aromatic, oxygenated derivatives of terpenes that can be obtained from plants (in our case), usually by distillation. Unlike a fixed oil that has no scent (unless rancid), volatile oils are all scent. (Example: oil of Peppermint.)

Olecranon Process: The large process on the ulnar bone that pro-jects behind the joint of the elbow.

Olfactometer: Apparatus for testing the power of the sense of smell.

Olig(o)-: A prefix which means little or scanty: for example, oliguria, excretion of smaller than normal quantities of urine.

Oligaemia: A diminution of the quantity of blood in the circulation.

Oligospermia: A less-than-normal number of sperm present in each unit volume of seminal fluid (each ml of semen usually contains 20 million sperm). The condition may be permanent or temporary and is a major cause of infertility in men. It may be caused by orchitis, an undescended testis, or varicocele, and should be investigated.

Oliguria: An abnormally low excretion of urine, such as occurs in acute nephritis. Excretion of a diminished amount of urine in relation to fluid intake.

Omentum: A long fold of peritoneal membrane generally loaded with more or less fat, which hangs down within the cavity of the abdomen in front of the bowels.

Omphalocele: Another name for exomphalos – a hernia of abdominal organs through the umbilicus.

Onco Gene: A modified gene that increases the malignancy of a tumor cell. Some onco genes, usually involved in early stages of cancer development, increase the chance that a normal cell develops into a tumor cell, possibly resulting in cancer.

Oncogene: Gene that can cause neoplastic transformation of a cell; oncogenes are slightly changed equivalents of normal genes known as protooncogenes.

Oncogenes: Genes found in mammalian cells and viruses that can cause cancer. They are believed to manufacture the proteins that control the division of cells.

Oncologist: A doctor who specialises in the treatment of cancers.

Oncology: The management of malignant disease.

Onychia: Disease of the nails.

Onychogryphosis: A distortion of the nail in which it is much thickened, overgrown and twisted on it self. This usually affects a toe-nail and is the result of chronic irritation and inflammation.

Onycholysis: Loosening or detachment of the nail from the nailbed following some destructive process.

Onycholysis: Separation of the nail from the nail-bed.

Onychomycosis: A fungus infection of the nail, caused by candida or ermatophytes.

Oocyte: An immature ovum. When the cell undergoes meiosis in the ovary it becomes an ovum.

Oogenesis: Ready for fertilisation by the spermatozoa. Only a small number of the many oöcytes produced survive until puberty, and not all of them will become ova and be ejected into the fallopian tubes.

Oophorectomy: Removal, by operation, of an ovary. When the ovary is removed for the pres- ence of a cyst, the term ovariotomy is usually employed.

Oophoritis: Another name for ovaritis or inflammation of an ovary.

Oophoron: Another name for the ovary.

OPC (Outpatient Clinic): PBM/SHG

pharmacy benefits management strategic healthcare group.

Open Design: A trial design in which all the involved parties know the treatment group to which a subject is assigned.

Open Enrollment: A period when eligible persons can enroll in a health benefits plan.

Open Study: A trial in which subjects and investigators know which product each subject is receiving; opposite of a blinded or double-blind study.

Open System: An environment in which system access is not controlled bypersons who are not responsible for the content of electronic records that are on the system.

Open-Label Trial/Study: A study in which all parties, (patient, physician and study coordinator) are informed of the drug and dose being administered. In an open-label study, none of the participants is given a placebo. These are usually conducted with Phase I & II studies, and often with Phase III.

Operation: A surgical procedure using instruments – or sometimes just the hands; for example, when manipulating a joint or setting a simple fracture.

Operational Data Model (ODM): As defined by CDISC, "a vendor neutral, platform independent format for interchange and achieve of data collected from various sources in clinical trials".

Operon: Complete unit of gene expression and regulation, including structural genes, regulator gene(s) and control elements in DNA recognized by regulator gene product(s).

Ophthalmia: Severe eye inflammation including conjunctivitis, iritis, severe hay fever, etc.

Ophthalmic: Pertaining to the eye.

Ophthalmologist: A doctor with specialist training in ophthalmology.

Ophthalmology: The study of the structure and function of the eye and the diagnosis and treatment of the diseases that affect it.

Ophthalmoplegia: Paralysis of the muscles of the eye. Internal ophthalmoplegia refers to paralysis of the iris and ciliary body; external ophthalmoplegia refers to paralysis of one or all of the muscles that move the eyes.

Ophthalmoscope: An instrument for examining the interior of the eye. There are different types of ophthalmoscope; all have a light source to illuminate the inside of the eye and a magnifying lens to make examination easier.

Opinion Leader: A subject matter expert in a peer-group.

Opisthotonos: The name for a position

assumed by the body during one of the convulsive seizures of tetanus.

Opportunistic: A description usually applied to infection resulting from an organism that does not normally cause disease in a healthy individual. It is also used to describe widespread infection by an organism that usually causes local infection. The body's defence mechanism can usually combat these organisms, but if it is impaired – as happens in AIDS/HIV or other immune deficiencies – opportunistic infection, such as pneumonia, may develop. Some viral and fungal infections behave in this way. Antimicrobial treatment is often effective, even though the weakness in the body's defence mechanism cannot be rectified.

Opportunity Assessment: A systematic procedure to identify and assess ways to prevent pollution releases from operations by substituting chemicals, changing methods, and reducing or eliminating wastes.

Opposite: Plant parts, usually leaves, that form pairs at nodes.

Opsonins: Substances present in the serum of the blood which act upon bacteria, so as to prepare them for destruction by the white cells of the blood.

Opthalmalgia: Very simply, eye pain.

Optic Atrophy: A deterioration in the fibres of the optic nerve resulting in partial or complete loss of vision. It may be caused by damage to the nerve from inflammation or injury, or the atrophy may be secondary to disease in the eye.

Optic Chiasma: This is formed by a crossing-over of the two optic nerves which run from the back of the eyeballs to meet in the mid line beneath the brain. Nerve fibres from the nasal part of the retina cross to link up with fibres from the outer part of the retina of the opposite eye. The linked nerves form two separate optic tracts which travel back to the occipital lobes of the brain.

Optic Disc: Otherwise known as the blind spot of the eye, the disc is the beginning of the optic nerve – the point where nerve fibres from the retina's rods and cones (the light- and colour-sensitive cells) leave the eyeball.

Optic Neuritis: Inflammation of the optic nerve which may result in sudden loss of part of a person's vision. It is usually accompanied by pain and tenderness on touch. The cause is uncertain, although in some cases it may be a prcursor of multiple sclerosis (ms): corticosteroids may help by improving the loss of visual acuity, but seems not to check the long-term inflammatory activity.

Optic: Concerned with the eye or vision.

Optician: Someone who fits and sells glasses or contact lenses. An ophthalmic optician (optometrist) is trained to perform eye examinations to test for long and short-sightedness and to prescribe corrective lenses, but they do not treat disorders of the eye, referring patients with a disorder to a family doctor or ophthalmologist.

OPTN (Organ Procurement and Transplantation Network: Operated by UNOS, the OPTN reports on patients awaiting organ transplants and on organ donation in the United States

Opt-out privilege: A general term used when some benefits are available for out-of-network covered services. Often coverage is less than the coverage available for in-network services, and the covered person has to pay for services up front and then file a claim for reimbursement. The details of such privileges, if they exist, will vary from plan to plan.

Oral Consent: A verbal consent by a subject who is not in a position to provide written consent.

Oral Contraceptive: A contraceptive taken by mouth. It comprises one or more synthetic female hormones, usually an oestrogen, which blocks normal ovulation, and a progestogen which influences the pituitary gland and thus blocks normal control of the woman's menstrual cycle. Progestogens also make the uterus less congenial for the fertilisation of an ovum by the sperm.

Oral Rehydration Therapy (ORT): This is the essential initial treatment for diarrhoea, and is particularly valuable for dehydrated children in developing countries ill with diseases such as cholera. A litre of water containing one teaspoonful of salt and eight of sugar, taken by mouth, is readily absorbed. It replaces salts and water lost because of the diar- rhoea and usually no other treatment is required. In developed countries ORT is useful in treating gastroenteritis.

Oral Surgery: A branch of surgery that treats deformities, injuries or diseases of the teeth and jaw, as well as other areas of the face and mouth. Surgeons doing this work are usually quailed dentists who have done further training in oral and maxillofacial surgery.

Oral: An adjective referring to the mouth, or to sub-stances taken by mouth.

Orbital headache: A headache around the eyes. There are supra-orbital headaches and suborbital headaches.

Orchidectomy: Operation for the removal of the testicles for example, because of cancer.

Orchidopexy: When testes do not descend into the scrotum, normally in young children ,an operation is performed to correct this. This is called surgical orchidopexy. The main reason is probably cosmetic; however, a testis which has descended is less likely to become cancerous than one which has not. It is less likely that treatment improves future fertility.

Orchitis: Inflammation of the testes, manifested by swelling and tenderness, usually infectious, sometimes the result of trauma.

Ordinal Data: Type of qualitative or categorical dat. For e.g., Severity of Disease: Mild/Moderate/Severe.

Organ dose: Amount of a substance or physical agent (radiation) absorbed by an organ.

Organ: A collection of different tissues that form a distinct structure in the body with a particular function or functions. The liver, for example, comprises a collection of different metabolic cells bound together with connective tissue and liberally supplied with blood vessels; it performs vital functions in the breakdown of substances absorbed from the gastrointestinal tract. other examples of organs are the kidneys, brain and heart.

Organelle: Microstructure or separated compartment within a cell that has a specialized function, for example ribosome, peroxisome, lysosome, golgi apparatus, mitochondrion, nucleolus, nucleus.

Organic Disease: A disease that started as, or became, impairment of structure or tissue. The smoker may have coughing and shortness of breath for years, and suffer from functional disorders; when the smoker gets emphysema, it is an organic disease. A term used in contradistinction to the word functional, to indicate that some structural change is responsible for the faulty action of an organ or other part of the body.

Organic Substances: Those which are obtained from animal or vegetable bodies, or which resemble in chemical composition those derived from this source. Organic chemistry has come to mean the chemistry of the carbon compounds.

Organogram: An organizational chart that describes different functional positions in an organization.

Organoleptic: Involving an organ, especially a sense organ as of taste, smell or sight.

Organon: The Organon of Medicine, by Samuel Hahnemann, the founder of homeopathy.

Organophosphorus:
Organophosphorus insecticides act by inhibiting the action of cholinesterase. For this reason they are also toxic to humans and must therefore be handled with great care.

Orgasm: The climax of sexual intercourse. In men this coincides with ejaculation of the semen.

Oriental Sore: This term is a synonym for cutaneous leishmaniasis.

Original Medical Record: Original documents, data and records (e.g., hospital records, clinical and office charts, laboratory notes, memoranda, subject's diaries or evaluation checklists, pharmacy dispensing records, recorded data from automated instruments, copies or transcriptions certified after verification as being accurate copies, microfiches, photographic negatives, microfilm or magnetic media, X-rays, subject files, and records kept at the pharmacy, at the laboratories and at medico-technical departments involved in the clinical trial).

Orlistat: An inhibitor of the pancreatic enzyme lipase, which breaks down fats in food to their con- stituent parts. By inhibiting lipase, the drug reduces absorption of dietary fat from the intestine. It is used as an adjuvant to a modest low-calorie diet in people with a body mass index of 30 kg/m2 or more. The drug should be prescribed only if diet alone has, over a period of four consecutive weeks, resulted in a person losing 2.5 kg or more. Orlistat may cause oily liquid faeces, urgency to defecate, excessive wind and, sometimes, headaches, tiredness and anxiety.

Ornithosis: Ornithosis is an infection of birds with the micro-organism known as Chlamydia psittaci, which is transmissible to humans.

Orphan Drug: The designation of the FDA to indicate a therapy developed to treat a rare disease (one which affects a U.S. population of less than 200,000 people). Medications used to treat diseases and conditions that occur rarely. Drugs developed for rare diseases and conditions which, in the U.S., affect fewer than 200,000 people or, in the European Union, affect 5 or fewer per 10,000 people. Because sales of orphan drugs are likely to be small compared to their development costs, pharmaceutical companies are awarded exclusive rights to market these medicines for a period of time as an incentive to develop them.

Orthodontics: Orthodontics is the branch of dentistry con-cerned with the prevention and treatment of dental irregularities and malocclusion.

Orthopaedics: Originally the general measures, both surgical and mechanical, for the correction or prevention of deformities in children. Now, that branch of medical science dealing with skeletal deformity (congenital or acquired), fractures and infections of bones, replacement of arthritic joints.

Orthopnoea: A form of difficulty in breathing so severe that the patient cannot bear to lie down, but must sit or stand up. As a rule, it occurs only in serious affections of the heart or lungs.

ORU: Organized research unit

OSACA: Office of Student & Curricular Affairs

Osmosis: The passage of fluids through a semipermeable membrane which separates them, so as to become mixed with one another. Osmotic pres-Osure is a term applied to the strength of the tendency which a fluid shows to do this, and depends largely upon the amount of solid which it holds in solution.

OSR: Office of Student Relations

Ossicle: A small bone. The term is usually applied to the three small bones of the middle ear – malleus, incus, and stapes – that conduct sound from the eardrum to the inner ear.

Ossification: the muscles of the pelvis force the seminal fluid from the prostate into the urethra and out through the urethral orifice. In women, orgasm is typified by irregular contractions of the muscular walls of the vagina followed by relaxation. The sensation is more diffuse in women than in men and tends to last longer with successive orgasms sometimes occurring.

Osteitis Fibrosa Cystica: A pathological rather than a clinical entity. The term refers to the replacement of bone by a highly cellular and vascular connective tissue. It is the result of osteoclastic and osteoblastic activity and is due to excessive parathyroid activity. It is thus seen in a proportion of patients with primary hyperparathyroidism and in patients with uraemic osteodystrophy; that is, the secondary hyperparathyroidism that occurs in patients with chronic renal disease.

Osteitis: Osteitis means inflammation in the substance of a bone. Traumatic osteitis is a condition par- ticularly common in footballers, in which the victim complains of pain in the groin following exercise, particularly if this has involved much hip rotation. Examination reveals difficulty in spreading the legs and marked tenderness over the symphysis pubis. It responds well to rest and the administration of non-steroidal anti-inflammatory drugs (NSAIDS) such as ibuprofen or indomethacin.

Osteitus fibrosa: Deposition of fibrous tissue causing inflammation of a bone.

Osteo: Prefix meaning pertaining to bone.

Osteoblast: A cell responsible for the production of bone.

Osteochondritis: Inflammation of both bone and cartilage. It is a not uncommon cause of backache in young people, particularly gymnasts.

Osteochondrosis: This includes a group of diseases involving degeneration of the centre of ossification in the growing bones of children and adolescents. They include Kohler's disease, osgood-schlatter's disease, and perthes' disease.

Osteoclast: A cell that resorbs calcified bone.

Osteodystrophy: Abnormal develop-ent of bone.

Osteogenesis Imperfecta: A hereditary disease due to an inherited abnormality of collagen. It is characterised by extreme fragility of the skeleton, resulting in fractures and deformities.

Osteogenesis: Formation or development of bone.

Osteomalacia: Osteomalacia is the adult form of rickets. It is due to inadequate mineralisation of osteoid tissue caused by a deficiency of vitamin D.

Osteopathy: A system of treatment by manipulating bones and other parts with the idea of thereby restoring functions in the bodily mechanism that have become deranged. Properly qualified osteopaths are included on the general council and register of osteopaths.

Osteophytes: Bony spurs or projections. They occur most commonly at the margins of areas of bone affected by osteoarthritis.

Osteoporosis: The softening of bone mass and the widening of the bone canals. This occurs with both age and diminished physical activity. Since women live longer, they are more likely to show such signs. There is little doubt that the condition is increasing among American women, and is starting to show itself at an earlier age. This is called "improved diagnostic methods" (harumph). The statistics that show the rise to be strongest in women that have used steroid hormone therapies in their earlier years seems to have escaped the notice of current medical conventional wisdom.

Osteoporosis: Significant decrease in bone mass with increased porosity and increased tendency to fracture.

OTC (Over-the-Counter): Medications that can be obtained without a doctor's prescription.

Otic Barotrauma: Also called aerotitis, this is blockage of the Eustachian tubes between the middle ear and the pharynx as a result of rapidly changing external air pressure, such as occurs during des- cent of an aircraft. Valsalva's manoeuvre – pinching the nose with finger and thumb and attempting to blow hard through the nose – will usually relieve the blockage. People prone to this phenomenon may find nasal decongestants helpful.

Otic Barotrauma: long-term usage has considerable toxicity problems. Advanced cartilage loss is best treated by joint replacement. Hip and knee-joint replacements – with a wide variety of artificial joints – are now common surgical procedures which greatly improve the mobility of affected individuals.

Otitis media: Inflammation, infectious or sterile, of the middle ear. In children this is often complicated by fluid buildup behind the eardrum.

Otolaryngology: The study, diagnosis and treatment of disorders of the ears, throat and larynx.

Otology: Otology is that branch of medical science which is concerned with disorders and diseases of the organ of hearing – one practising this branch being called an otologist.

Otorhinolaryngology: The study of diseases of the ear, nose, and throat – colloquially referred to as ear, nose and throat (ENT) specialty. The relevant specialist is called an otorhinolaryngologist (US) or ENT surgeon (UK).

Outcome Indicator: An indicator for the assessment of outcome data.

Outcome Research: The measurement of the value of a particular course of therapy.

Outcome: The result of an activity, process, investigation or intervention.

Outlier Detection: Data points which lay a long way from the main body of similar data.

Outlier: An observation that is numerically distant from rest of the data.

Out-of-area benefits: Benefits the health plan provides to covered persons for covered services obtained outside of the network service area. The details of such benefits will vary from plan to plan.

Out-of-Network Provider: Any health care provider that does not belong to a CIGNA network. Members can use their benefits for out-of-network expenses, but miss out on in- network discounts.

Out-of-Pocket Costs: The portion of payments for health services that must be paid by the member, including co-payments, co-insurance, or deductibles.

Out-of-Pocket Maximum: A fixed dollar limit on a member's out-of-pocket expenses. After reaching this set limit, prescriptions can be filled at a reduced cost or at zero cost to the member. All prescriptions or only certain types of prescriptions a member receives may accumulate toward the out-of-pocket maximum.

Out-of-Pocket: Copayments, deductibles or fees paid by participants for health services or prescriptions.

Outpatient Care: Any health care service provided to a patient who is not admitted to a facility. Outpatient care may be provided in a doctor's office, clinic, the patient's home or hospital outpatient department.

Outpatient: The ambulatory care at a hospital that does not require hospitalization. A patient attending a hospital clinic who is not admitted to a bed.

Outsourcing: The transfer of a business activity/process to an external service provider.

Overbite: A dental term describing the condition where a person's upper incisor teeth vertically overlap the lower incisors. If serious, the person may need orthodontic correction and this is usually done in childhood after the permanent teeth have developed.

Overhead: The operational expenses (eg., electricity, rental etc.)

Over-The-Counter (OTC): A description applied in the UK to medicines and drugs that can be obtained from a pharmacist without a doctor's or dentist's prescription.

Over-the-counter drugs: Drugs that are available without prescription. As of 2004, the IRS has ruled that certain over-the-counter drugs can be reimbursed with healthcare reimbursement account dollars.

Ovicide: Substance intended to kill eggs.

Ovulation: The development and release of an ovum (egg) from the ovary into the fallopian tubes.

Ovum: The single cell derived from the female, out of which a future individual arises, after its union with the spermatozoon derived from the male.

Oxalosis: An inherited defect in the body's metabolism in which oxalates – a product of metabolism – are deposited in the kidneys and other tissues, ultimately causing failure of the kidneys.

Oxaluria: The presence in the urine of oxalic acid or oxalates, in particular calcium oxalate.

Oxidant: A molecule that causes biological oxidation in which oxygen is added to or electrons removed from a substance. Oxygen-free radicals are highly toxic atoms and chemical groups pro- duced by intracellular activity in various disease processes and by poisons, radiation, smoking and other pollutants. Anti-oxidants such as beta-carotene can neutralise these radicals.

Oxidation: A chemical reaction in which an electron is taken from a molecule of the oxidized substance.

Oximeter: An non-invasive device which can be attached to a patient's skin by an adhesive and records the degree of oxygen saturation in their blood, displaying the level on a screen.

Oximetry: The measurement by an oximeter of the proportion of oxygenated haemoglobin in the blood.

Oxycephaly: A deformity of the skull in which the forehead is high and the top of the head pointed. There is also poor vision and the eyes bulge.

Oxygen Deficit: In a resting individual the potential oxygen supply to the tissues is greater than its con-sumption. During heavy exercise, the energy required by the tissues is greater than can be supplied by aerobic cellular metabolism and the additional energy is supplied by a biochemical reaction called anaerobic metabolism. There is a build-up of lactate – a product of lactic acid– from anaerobic metabolism which is ultimately oxidised after conversion to citrate and metabolism via the citric acid cycle. The increased amount of oxygen above resting con- centrations which needs to be consumed to perform this metabolism is known as the oxygen debt or deficit.

Oxygen Tent: A sheet of plastic put over a hospital bed with oxygen fed into it so that a patient can receive oxygen. Such treatment may be for a heart or lung condition in which the normal atmospheric concentration of oxygen is insufficient to enable the person to oxygenate the blood flow-ing through the lungs to a normal level, so extra oxygen is provided in the patient's immediate surroundings.

Oxygen Toxicity: oxygen toxicity in human lungs causes an acute oedema followed by fibrosis and pulmonary hypertension.

Oxyhaemoglobin: The compound formed when the pigment haemoglobin in the erythrocytes (red blood cells) combines with oxygen (a reversible reaction). The oxygen is carried in this way from the lungs to the body's tissues where it is released to take part in metabolic activities.

Oxytocin: A short-lived, fast acting hormone, made by the hypothalamus of the brain, along with its close relative vasopressin (anti-diuretic hormone), stored in the posterior pituitary, and released into the blood as needed. It stimulates certain smooth muscle coats, constricts certain blood vessels and facilitates the sensitivity of some tissues to other hormones and nerves.

Oxyuriasis: Another name for the threadworm.

Ozaena: A chronic disease of the nose of an inflammatory nature, combined with atrophy of the mucous membrane and the formation of extremely foul-smelling crusts in the interior of the nose.

Ozone Depletion: Accelerated chemical destruction of the stratospheric ozone layer by the presence of substances produced, for the most part, by human activities. The compounds causing the greatest depletion of the ozone layer are the relatively stable chlorinated, fluorinated, and brominated products.

P

P Value: The lowest level of significance at which a given null hypothesis can be rejected; that is, the probability of observing a result as extreme as or more extreme than that observed if the null hypothesis is true.

Pacemaker: A cardiac or artificial pacemaker is a device that helps a faulty heart to maintain normal rhythm. It consists of a battery that stimulates the heart by an electric current passed through an insulated wire which is attached either to the surface of the ventricle (epicardial pacemaker) or to the heart lining (endocardial pacemaker). A cardiac pacemaker is used when a person's sinoatrial node is malfunction- ing or when there is interference with the passage of normal impulses.

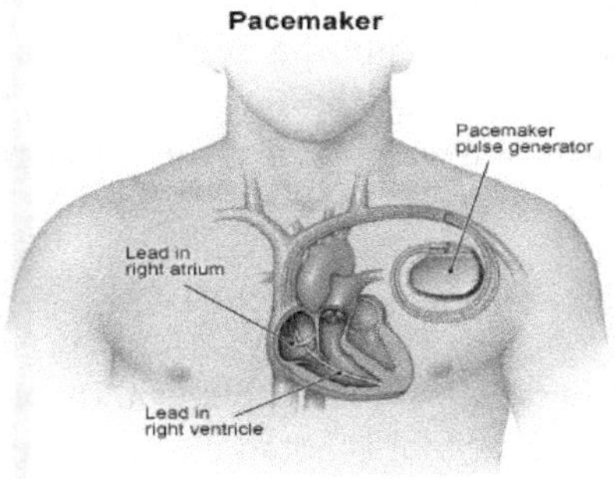

Pacinian Corpuscles: Pacinian corpuscles, or lamellated corpuscles, are minute bulbs at the ends of the nerves scattered through the skin and subcutaneous tis- sue, and forming one of the end-organs for sensation.

Package Insert: A product information sheet that summarizes all the clinical and non-clinical information of a drug product.

Packed Cell Volume: That fraction of the blood's total volume made up of red cells. The packed cell volume is found by centrifuging blood in a tube and measuring the depth of the column of red cells as a fraction of the whole column of blood.

Paederasty: A homosexual act between an adult man and a boy or young man

Paediatrician: A medical specialist skilled in the care of child- ren's health and diseases.

Paediatrics: Paediatrics means the branch of medicine dealing with diseases of children

Paedophilia: A perverse sexual attraction to children of either sex. Paedophiles are nearly always male and may have heterosexual, homosexual or bisexual orientation.

Paget's Disease of Bone: Also called osteitis deformans, this is a chronic disease in which the bones– especially those of the skull, limbs, and spine gradually become thick and also soft, causing them to bend.

Paid Amount: The exact amount issued on a bank draft to the provider of service.

Pain Management Medicine: Pain relief medicine is administered via a pump on a regular basis and is used for hospice patients or patients with chronic pain or post- surgery symptoms.

Pain: Pain is an unpleasant sensory and

emotional experience associated with actual or potential tissue damage, or described in terms of such damage

Paired T-Test: A statistical test to determine whether the mean change from baseline to endpoint of a numeric variable is different from zero. This test requires population group with 2 measurements of the numeric variable per patient. For eg., a study that investigates mean change in weight within a group of patients from baseline to end point).

Pairing: A method by which subjects are selected so that two subjects with similar characteristics (for example, weight, smoking habits) are assigned to a set, but one receives Treatment A and the other receives Treatment B.

Palate: The partition between the cavity of the mouth, below, and that of the nose, above.

Palindromic: An adjective describing symptoms or diseases that recur. For example, palindromic rheumatoid arthritis is a condition in which symptoms wax and wane with periods of complete remission.

Palliative Care: This is defined as comprehensive care of patients and families facing terminal illness. The care focuses primarily on comfort and sup- port

Palliative: A term applied to treatment that eases the symptoms of a disorder rather than curing the condition.

Pallidotomy: Also known as pallidectomy, this is a neuro-surgical procedure in which the activities of the globus pallidus area of the brain are destroyed or modified. The operation is sometimes used to relieve the symptoms of parkinsonism and other neurological conditions in which involuntary movements are a significant and disabling symptom.

Pallor: Unusual paleness of the skin caused by a reduced flow of blood or a deficiency in normal pigments. Pallor may be a sign of fright, shock, anaemia, or other diseases.

Palpation: Examination of the surface of the body and the size, shape, and movements of the internal organs, by laying the flat of the hand upon the skin.

Palpebral: Relating to the eyelid.

Palpitation: Unduly rapid or throbbing heartbeat that is noted by a patient; it may be regular or irregular. Undue awareness by a patient of a heartbeat that is otherwise normal.

Panacea: Panacea is a term applied to a remedy for all diseases, or more usually to a remedy which benefits -

many different diseases.

Pancarditis: Inflammation of the pericardium, myocardium, and endocardium at the same time

Pancreas: An organ in the body that makes insulin so that the body can use glucose for energy. The pancreas also makes enzymes that help the body digest food. This is a gland situated above the navel in the abdominal cavity that extends from the left side to the center, with its head tucked into the curve of the duodenum. It is 6-8 inches long, weighs 3 or 4 ounces, secretes pancreatic enzymes and alkali into the duodenum in concert with the gallbladder and liver, and secretes the hormones insulin and glucagon into the blood. Insulin acts to facilitate the absorption of blood glucose into fuel-needing cells, and glucagon stimulates a slow release of glucose from the liver, primarily to supply fuel to the brain. That most cherished organ uses one-quarter of the sugar in the blood and has no fuel storage. Pancreatic enzymes are basically those that digest fats, carbohydrates and proteins into their smaller components of fatty acids and glycerol, maltose, and amino acids as well as curdling milk.

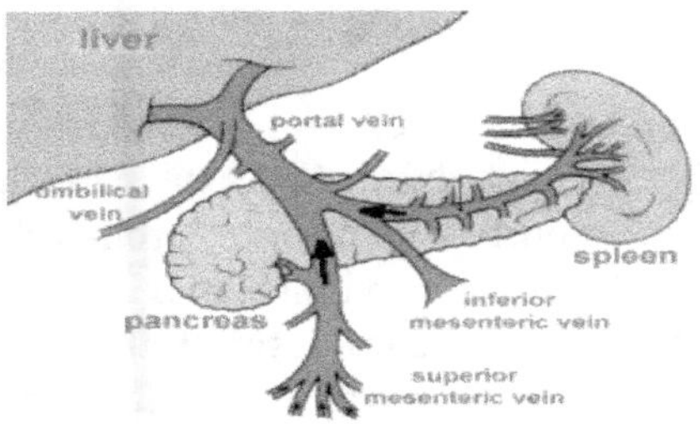

Pancreatin: Pancreatin preparations contain the four powerful enzymes, trypsin, chymotrypsin, lipase, and amylase, which continue the digestion of foods started in the stomach.

Pancytopenia: A fall in the number of red erythrocytes and white leucocytes, as well as of platelets.

Pandemic: An epidemic that has

spread so widely that very many people in different countries are affected.

Panel Reactive Antibodies (PRA): A test to measure what percent of a serum sample reacts to a panel of known antigens

Panic Attacks: Panic attacks, or panic disorders, are recurrent short episodes of acute distress. Some sufferers may be mentally confused and fear impending death.

Panicle: A compound flower head that forms a raceme.

Panniculitis: Inflammation of the subcutaneous fat. It may occur anywhere on the body surface.

Paper Flow: The step-by-step process to be followed from the data transcription in the Case Report Forms (CRFs) at the site to its collection,entry,validation and final analysis.

Paper Trial: The documentation of activities that allow reconstruction of -

the course of events.

Papillae: Small raised bumps or nipples on a tissue surface. Lingual papillae are taste buds.

Papilloedema: Swelling of the optic disc of the eye, specifically due to raised intracranial pressure. it can be seen by examining the back of the eye using an ophthalmoscope and is an important sign in managing the care of patients with intracerebral disease such as tumours or meningitis.

Papilloma: Proliferation of epidermis or epithelium to form a tumour. Benign papillomas are common in the skin and are sometimes viral in origin.

Papule: Small (less than 5 mm) solid elevation of the skin or mucous membranes. A larger lesion is called a nodule.

Para: A prefix meaning near, aside from, or beyond.

Paracentesis: The puncture by hollow needle or trocar and cannula of any body cavity for tapping or aspirating fluid.

Paradoxical Breathing: The reverse of the normal movements of breathing

Paraesthesia: Abnormal sensation, as burning or prickling.

Paraganglion: One of the small ovoid collections of cells occurring in the walls of the ganglia of the sympathetic nervous system adjacent to the spinal cord. They are chromaffin cells and sometimes secrete adrenaline.

Paragraphia: Misplacement of words, or of letters in words, or wrong spelling, or use of wrong words in writing as a result of a lesion in the speech region of the brain.

Parallel Designs: A type of statistical design in which patients are randomized to one of two treatment strategies and the two arms are then prospectively followed in parallel throughout the trial.

Parallel Track: A system of making experimental drugs available to individuals who are unable to participate in clinical trials.

Parallel Trial: Subjects are randomized to one of two differing treatment groups (usually investigational product and placebo) and usually receive the assigned treatment during the entire trial.

Paralysis: Paralysis, or palsy, is loss of muscular power due to some disorder of the nervous system.

Paramedical: A generic title for the professions which work closely with or are reponsible to the medical profession in caring for patients.

Parameter: A constant in a model, or a constant that wholly or partially characterizes a function of probability distribution (mathematics and statistics). In clinical research the term is often linked to statistical conventions and is a numeric characteristic of a population. Thus the term is narrower than variable.

Parametric Tests: A statistical test in which a sample statistics is obtained to estimate the population parameter. It is applied when the data is normally distributed. For eg., ANOVA, t-test etc.

Paramnesia: A derangement of the memory in which words are used without a comprehension of their meaning.

Paranoia: A condition whose main characteristic is the delusion that other people are (in an unclear way) connected to the affected individual.

Para-Occupational Exposure: Exposure of a worker's family to substances carried from the workplace to the home. Exposure of visitors to substances in the workplace.

Paraparesis: A disorder or injury of the nervous system in which the affected individual suffers from weakness in both legs and sometimes of the muscles in the lower trunk.

Paraphasia: Misplacement of words, or use of wrong words, in speech as a result of a lesion in the speech region of the brain.

Paraphimosis: The constriction of the penis behind the glans by an abnormally tight foreskin that has been retracted. The condition causes swelling and severe pain.

Paraplegia: Paralysis of the lower limbs, accompanied generally by paralysis of bladder and rectum.

Parapsychology: The branch of psychology that studies extra-sensory perception. This includes precognition; psychokinesis (a sup- posed ability of some people to move or change the state of objects by thinking); telepathy (communicating thoughts from one person to another); and clairvoyance (the ability to visualise events at a distance).

Parasuicide: Non-fatal self-poisoning or self-injury, or attempted suicide. It is most common in the 12–15 age group.

Parasympathetic: A division of the autonomic (involuntary) nervous system that controls normal digestive, reproductive, cardiopulmonary, and vascular functions and stimulates most secretions. This sub system works as a direct antagonist to the sympathetic division, and organ functions balance between them.

Parasympatholytic: Producing effects resembling those caused by interruption of the parasympathetic nerve; also called anticholinergic.

Parasympathomimetic: Mimicking major aspects of parasympathetic function. Producing effects resembling those caused by stimulation of the parasympathetic nervous system; also called cholinomimetic.

Para-Thyroidectomy: The surgical removal of the para-thyroid gland.

Parathyroids: These are several minute glandular masses embedded in the lower edge of the thyroid gland. They produce Parathyroid Hormone (PTH), part of the calcium-phosphorus control system. Calcium levels in the blood must be within a narrow band of safety. If free calcium drops too low, PTH acts on the kidneys and blocks calcium loss in urine, amplifies calcium absorption into the portal blood (from food and from submucosal storage) and stimulates release of calcium from bone storage. When levels are back up, the hormone backs off. Oddly enough, the thyroid gland secretes its virtual antagonist, calcitonin, which, when calcium levels are too high, stimulates the urine excretion, bone retention and digestive resistance to calcium, and when the blood levels drop, recedes.

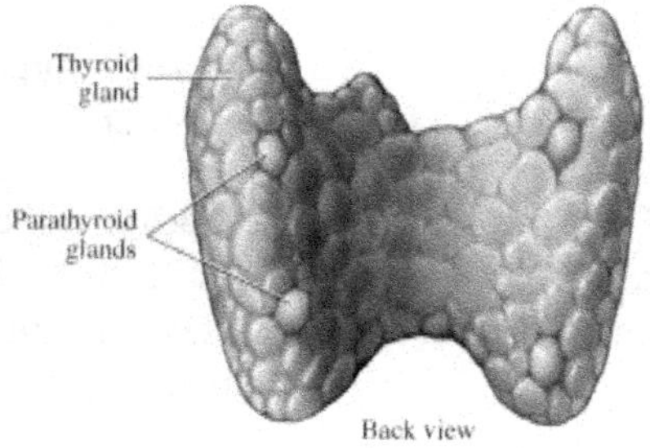

Parenchymal: These are cells in a tissue or tissues in an organ that are concerned with function. These are the characteristic cells or tissues that do the actual stuff. The importance to us is that parenchymal tissues expend much vital energy in their functions and are less tolerant of a degraded environment than the structural mesenchyme. A congested and impaired organ like the liver of a heavy drinker has so much regular dysfunction that eventually the more tolerant and metabolically less particular mesenchymal cells become more common, and the distressed, overworked, and metabolically compromised parenchymal cells become a minority. The structural cells can multiply with ease in a poor environment, the more delicate functional cells cannot-and you end up with the type of cirrhosis sometimes termed mesenchymal invasion disease. The point of this is that the sooner you return an organ or tissue back to the healed state, the more likely you are to have a healthy balance between the structural and functional.

Parent: A child's biological or adoptive parent.

Parenteral Dosage: Method of introducing substances into an organism avoiding the gastrointestinal tract (sub cutaneously, intravenously, intramuscularly etc.).

Parenteral: Administration of drugs by any route other than by the mouth or by the bowel – for example, by intramuscular or intravenous injection or infusion.

Paresis: Slight or incomplete paralysis.

Paresthesia: Numbness, prickly sensations without point specificity, or abnormal hypersensitivities, all local to one part of the body, and without an obvious cause. Your foot falling asleep is paresthetic, but not paresthesia...the cause is you sat funny.

Parietal: The term applied to anything pertaining to the wall of a cavity: for example,

Parkinson's Disease: A chronic disease of the central nervous system caused by lowered levels of the inhibitory neurotransmitter dopamine Symptoms include muscular tremors and weakness.

Paronychia: The term applied to inflammation near the nail

Parosmia: A perverted sense of smell; everything may smell unpleasant to the affected individual. The most common cause is some septic condition of the nasal passages.

Parotid: A pair of salivary glands tucked into the notch in front of each ear and emptying through parotid ducts by each upper 2nd molar. Although the fluid has some of the thick viscous lubricant nature of saliva from the glands in the floor of the mouth, the parotids secrete high levels of ptyelin and amylase (starch-digesting enzymes) lysozymes (antimicrobial enzymes) and a group of proteins loosely called parotin that stimulate epithelial and nerve cell growth a lot more here than just spit.

Paroxysm: A sudden temporary attack that may take the form of a convulsion or spasm.

Parthenogenesis: Non-sexual reproduction. In other words, development of the ovum into an individual without fertilisation by a spermatozoon.

Partial Day Treatment: A program offered by appropriately- licensed facilities that includes either a day or evening treatment program, usually for mental health or substance abuse.

Participant ID: The unique identifier associated with a participant.

Participant: A subject who takes part in a clinical trial.

Participant: A person who is eligible to receive health benefits under a health benefits plan. This term may refer to the employee, spouse or other dependents.

Participating Provider: A physician, hospital, pharmacy, laboratory or other appropriately licensed facility or provider of health care services or supplies that has entered into an agreement with a managed care entity to provide services or supplies to a patient enrolled in a health benefit plan.

Participation: Agreement british columbia pharma care pharmacy participation agreement.

Partition Coefficient: Ratio of the distribution of a substance between two phases when the heterogeneous system (of two phases) is in equilibrium; the ratio of concentrations (or, strictly speaking, activities) of the same molecular species in the two phases is constant at constant temperature. The partition coefficients most frequently used in acute toxicology are lipid/water and octan-l-ol/ water distributions.

Partogram: A method of recording the degree of dilatation, or opening, of the cervix (or neck) of the uterus in labour to assess how labour is progressing.

Passive Movement: A movement induced by someone other than the patient.

Passive Smoking: Inhalation of sidestream smoke by people who do not smoke themselves.

Pasteurisation: A method of sterilising milk

Patch Test: This is used to identify possible substances that may be causing a patient's allergy.

Patent Application: An application filed to the concerned authorities for seeking patent for an invention.

Patent: The term used for proprietary medicines, which are, because of research and cost involved in producing many of them, protected by a patent. This means no other company can reproduce the substance other than the patent holder.

Pathogenesis: The ways in which a disease or disorder starts and develops.

Pathogens: Micro-organisms that cause diseases, parasitising plants, animals and humans

Pathognomonic: A term applied to signs or symptoms which are especially characteristic of certain diseases, and on the presence or absence of which the diagnosis depends.

Pathology: Disease, particularly one with clear and obvious changes in structure or function; the study of same.

Pathways: Pathway options in the revised Pharm D professional curriculum instituted in academic year 1998- 1999; students take a core clinical pharmacy curriculum and choose one of three pathways: pharmaceutical care, pharmaceutical policy and management, or pharmaceutical science

Patient Block: A set of patient numbers that are grouped together for statistical purposes.

Patient Cards: **A** document given to the study subjects for recording certain observations/readings on the condition of health either at home or at the trial site.

Patient Compliance With Prescribed Drug Regimens: The degree of correspondence between the patient's actual dosing history and the prescribed dosing regimen, the result of a comparison of two time-series.

Patient Consent: The written or verbal permission given by a patient for the release and use of their personal information.

Patient File: The hospital/clinic file that contains complete medical information of a patient/ subject. Contains demographic, medical, and treatment information about a patient or subject. It may be paper-based or a mixture of computer and paper records.

Patient History: The record for a patient stating what prescriptions he or she has been issued while under a particular PBM.

Patient Keyword: A special word, determined by the patient, used to control access to their personal medication history, patient profile, and requesting of a printed patient profile^

Patient Number: A unique numerical identifier assigned to a trial subject.

Patient Record: Information recorded regarding a patient, including PHN, demographic information, medical and clinical information, medication history, and any adverse drug reaction data.

Patient Recruitment: The act of enrolling subject in a clinical trial.

Patient Summary: A document that summarizes the past and present medical health condition of a subject.

Patient: A person with injury, physical or mental dis-order or disease, or abnormality who comes into the care of a health professional or of an institution responsible for providing care to such persons.

Payer: The party or group an individual contracts with to cover health care services, unless the patient is paying out- of pocket. This is sometimes referred to as a "third-party payer."

PBMs: PBMs are pharmacy benefit managers. These are organizations that provide administrative services such as processing and analyzing prescription claims for pharmacy benefit and coverage programs.

PCIS: PharmaCare Centralized Information System.

PD: Pharmacodynamics

PDA: Personal Digital Assistant, a handheld computing device

PDUFA: The Prescription Drug User Fee Act of 1992. Legislation passed by Congress authorizing the FDA to collect "user fees" for regulatory review of human drug applications. The FDA agreed to use the income from the fees to hire more reviewers to speed up drug review without compromising review quality.

Peak Flow Meter: A device that measures the rate at which an individual can expel air from the lungs. This is an indication of the reserve in the capacity of the lungs.

Pectin: A polysaccharide substance allied to starch.

Pectoral: Anything pertaining to the chest, or a remedy used in treating chest troubles.

Pectoriloquy: The resonance of the voice, when spoken or whispered words can be clearly heard through the stethoscope placed on the chest wall.

Pedicel: The stem of a flower within a floral cluster.

Pediculosis: Infestation with lice, of which three species infect humans:

Peduncle: The stem or stalk of a single flower or a whole floral cluster.

Peer Review: Review of a clinical trial data by experts chosen by the study sponsor. These experts review the trials for scientific merit, participant safety and ethical considerations.

Pellagra: A potentially fatal nutritional disorder caused by a deficiency of vitamin B complex (niacin). The symptoms are dermatitis, diarrhoea and dementia.

Pelvic Inflammatory Disease (PID): Also called salpingitis, the term is applied to infections of the fallopian tubes that follow or are concurrent with uterine and cervical infections. Gonorrhea and Chlamydia are the most common organisms, and the infection is usually begun through sexual contact, although metabolic imbalances, subtler systemic infections like a slow virus, the local insult of herpes or candidiasis, the sequel of medication or recreational drugs, birth control pills, even an IUD. All can alter the vaginal flora and induce inflammation sufficient to allow an endogenous organism to start the infection. PID after birth, on the other hand, is usually the result of staph or strep infections infecting injured membranes.

Pelvimetry: Measurement of the internal dimensions of the pelvis.

Pemphigus: An acute or chronic disease of adults, with a singular or constant series of skin eruptions. The causes are not known, although both viruses and auto- immune reactions can be implicated. There are so many distinct types that it is probably not a distinct pathology but a symptom, like nausea, that occurs from many causes. Pemphigus of the mouth, lips and throat is rather common in the aged, particularly in those taking many management medications, and reduced to the spiritual poverty of "rest homes". These need constant treatment (herbs work as well as medications), else the difficulty of eating, what with dry mouth, sore gums, gas and chronic constipation (from medications and adrenergic stress) coupled with SLBF (Soft Light Brown Food) and now the added insult of mouth sores can start the subtle downwards spiral of entropy and asthenia.

Pended Claim: Claims that require -

additional information prior to completing the adjudication process due to a specific reason code.

Penis: The male organ through which the tubular urethra runs from the neck of the urinary bladder to the exterior at the meatus or opening. Urine and semen are discharged along the urethra.

Peptic Ulcer: A stomach or duodenal ulcer, caused by excess or untimely secretions of gastric acid and pepsin, poor closure of the pyloric sphincter and digestive acid leakage into the duodenum, or poorly mucin-protected membranes resulting from infection or allergen irritation

Peptonised Foods: Foods which have been predigested by pan-creatin and thereby rendered more digestible.

Percent Completion: When percent competion is below 100%, reported success rates and phase transition probabilities are estimates.

Percentiles: Value in series of observations arranged in ascending order of magnitude which divide the distribution into 100 equal parts. Thus the median is 50th percentile. The 50th percentile will have 50% observations on either side. 10th percentile will have 10% observations to the left and 90% t the right.

Percussion: An aid to diagnosis practised by striking the patient's body with the fingers, in such a way as to make it give out a note.

Percutaneous: Through the skin following application on the skin.

Perforation: The perforation of one of

the hollow organs of the abdomen or major blood vessels may occur spontaneously in the case of an ulcer or an advanced tumour, or may be secondary to trauma such as a knife wound or penetrating injury from a traffic or industrial accident.

Performance Indicators: Measures of achievement in particular clinical and managerial areas used to assess the performance of an organisation

Performance Review: The review of the performance of an individual in a specified period of time against pre-set objectives.

Perfusate: A solution containing electrolyte concentrations at intracellular levels and relatively large concentrations of poorly permeating substances; used to cool organs

Perfusion: The transfer of fluid through a tissue.

Periapical abscess: An abscess or pus pocket around the apex of the root of a tooth. Sometimes called a gumboil.

Pericardium: The smooth membrane that surrounds the heart.

Perichondritis: Inflammation of cartilage and the tissue surrounding it, usually as a result of chronic infection.

Perimetritis: A localised inflammation of the peritoneum surrounding the uterus.

Perimetry: A test of the visual fields of the eye that assesses the extent of peripheral vision.

Perinatal Mortality: Perinatal mortality consists of deaths of the fetus after the 28th week of pregnancy and deaths of the newborn child during the first week of life.

Perinatal: Relating to the period shortly before and after birth; from the twentieth to the twenty-ninth week of gestation to one to four weeks after birth.

Period Effect: Designated period during the course of a trial in which subjects are observed and no treatment is administered.

Periodic Safety Update Report (PSUR): The safety reporting of marketed drugs in a specified period of time.

Periodontal: An adjective that relates to the tissues around the teeth.

Periodontitis: A gum disease in which the gums shrink away from the teeth. Without treatment, it can lead to tooth lose.

Perioperative Cell Salvage: A method of autologous blood transfusion –using a patient's own blood, salvaged during a surgical operation – instead of conventional blood-bank transfusion.

Periosteum: The membrane surrounding a bone. The peri-osteum carries blood vessels and nerves for the nutrition and development of the bone.

Peripheral Neuritis: Inflammation of the nerves in the outlying parts of the body.

Peripheral Vascular Disease: The narrowing of the blood vessels in the legs and, less commonly, in the arms. Blood flow is restricted, with pain occurring in the affected area. If the blood supply is seriously reduced, gangrene of the tissues supplied by the affected vessel(s) may occur and the limb may need to be amputated.

Peripheral: At the edges, especially circulation or nerves. Peripheral functions are usually controlled and modified more by local conditions than systemic (central) controls.

Peripheral-Blood Stem-Cell Transplants: These have almost completely replaced bone.

Peristalsis: The worm-like movement by which the stomach and bowels propel their contents. It consists of alternate waves of relaxation and contraction in successive parts of the intestinal tube.

Peritoneal dialysis: Method of artificial detoxication in which a toxic substance from the body is transferred into liquid that is instilled into the peritoneum. Thus, the employment of the peritoneum surrounding the abdominal cavity as a dialysing membrane for the purpose of removing waste products or toxins accumulated as a result of renal failure.

Peritoneum: An extensively folded sac lining the abdominal cavity.

Peritonsillar Abscess: The term applied to a collection of pus or an abscess which occurs complicating an attack of tonsillitis. The collection of pus forms between the tonsil and the superior constrictor muscle of the pharynx.

Permissible Exposure Limit (PEL): Recommendation by US OSHA for TWA concentration that must not be exceeded during any 8-hour work shift of a 40h working week.

Permission: The authorization granted by the regulatory authorities.

Pernicious Anaemia: An autoimmune disease in which sensitized lymphocytes destroy the parietal cells of the stomach. These cells normally produce intrinsic factor, which is the carrier protein for vitamin B_{12} that permits its absorption in the terminal ileum. Without intrinsic factor, vitamin B_{12} cannot be absorbed and this gives rise to a macrocytic anaemia.

Peroneal: The name given to structures, such as the muscles, and nerves, on the outer or fibular side of the leg.

Persistence: Attribute of a substance that describes the length of time that the substance remains in a particular environment before it is physically removed or chemically or biologically transformed.

Persistent Vegetative State (PVS): PVS may occur in patients with severe brain damage from hypoxia or injury.

Personal Health Number (PHN): A unique identification number assigned to each Care Card holder. This number is used to store and retrieve all patient information on Pharma Net, including the patient's record. A PHN is required for recording a prescription on Pharma Net. A PHN does not necessarily mean eligibility for Pharma Care coverage.

Personal Medical Services (PMS): Alternative type of contract for the provision of general medical services.

Personal Monitoring: Type of environmental monitoring in which an individual's exposure to a substance is measured and evaluated: this is normally carried out using a personal sampler.

Personal Protective Equipment (PPE): Equipment (clothing, gloves, hard hat, respirator and so on) worn by an individual to prevent exposure to a potentially toxic substance.

Personal Sampler: Compact, portable instrument for individual air sampling, measuring, or both, the content of a harmful substance in the respiration zone of a working person.

Personality Disorder: Condition in -

which the individual fails to learn from experience or to adapt to changes. The outcome is impaired social functioning and personal distress.

Perspiration: Commonly called sweat, it is an excretion from the skin, produced by microscopic sweat-glands, of which there are around 2·5 million, scattered over the surface.

Perthes' Disease: A condition of the hip in children, due to death and fragmentation of the epiphysis (or spongy extremity) of the head of the femur. The cause is not known.

Pes Cavus: Known popularly as claw-foot, this is a deformity in which the foot has an abnormally high arch and the tips of the toes are turned under.

Pest: Organism that may harm public health, that attacks food and other materials essential to mankind or otherwise affects human beings adversely.

Pesticide Residue: Pesticide residue is any substance or mixture of substances in food for man or animals resulting from the use of a pesticide and includes any specified derivatives, such as degradation and conversion products, metabolites, reaction products and impurities considered to be of toxicological significance.

Pesticide: Strictly a substance intended to kill pests: in common usage, any substance used for controlling, preventing, or destroying animal, microbiological or plant pests.

PET Scanning: Positron-emission tomography is a nuclear medicine diagnostic technique that works by identifying positrons – positively charged electrons – given off by substances labelled with radioactive varieties of elements. The result is three-dimensional images that identify metabolic and chemical activities of tissues, especially brain tissues.

PET: Position emission tomography.

Petechiae: Small red macules due to haemorrhage in the skin. They may be caused by trauma – such as by tight pressure, as in strangulation – or even by the effect of violent coughing. Bleeding and clotting disorders may provoke petechiae, and they are a feature of many childhood viral infections.

Petiole: A leaf stalk or stem, or an unexpanded section.

Petit Mal: An out-of-date term for less severe type of epileptic seizure that occurs usually in children or adolescents but less often in adults.

Peyer's Patches: Peyer's patches are conglomerations of lymphoid nodules in the ileum, or lower part of the small intestine. They play an important part in the defence of the body against bacterial invasion, as in typhoid fever.

Peyronie's Disease: Painful and deformed erection of the penis caused by the formation of fibrous tissue.

PG Inhibitor: Exogenously, a PGE inhibitor like aspirin, and usually intended to lessen joint inflammation and uterine spasms.

PGE: Short for Prostaglandin E, presumably the fifth subtype discovered, and usually separated into PGE_1 and PGE_2. These two, if made by the kidneys, slow sodium reabsorption; if within in the uterus, induce a stronger response to less stimulus; if made in the stomach lining inhibit gastric secretion; if secreted by macrophages, target tissues become more accessible to infiltration and inflammation. These are the two prostaglandins whose levels are meant to be stabilized by gamma-linolenic acid (GLA) supplements.

pH: The potential of hydrogen. A "neutral" pH is expressed as 7.0 (water), with greater being alkaline and lesser being acidic. Expressed logarithmically like the Richter's Scale, 6.9 pH is twice as acidic as 7.0. 9.0 is ten times as alkaline as 8.0, etc., all based on the presumed amount of hydrogen ion (acidity) present. This is a chemical literality, not to be confused with the vitalist and cyto-hologrammic implications of Acid and Alkaline metabolism or foods. A complex protein has a literal pH close to neutral. Run it through your body and it gets broken down into an incredible array of amino acids, ending up as nitrogenous acid waste products.The more rapid the metabolism, the more acids are produced.The ashes of life are acids.

Phaeochromocytoma: A disorder in which a vascular tumour of the

Phagocyte: Cell capable of phagocytosis

Phagocytosis: The act of absorbing and digesting fragments, detritus, or whole organisms, as an amoeba does. Granulocytes do this in the body. Engulfing and digestion of micro-organisms, other cells, and foreign particles by cells such as phagocytes.

Phallus: An alternative name for the penis, this word may also be used to describe a penis-like object.

Phantasy: The term applied to an imaginary appearance or daydream.

Phantom Limb: Following the amputation of a limb, it is usual for the patient to experience sensations as if the limb were still present.

Pharm D: The Doctor of Pharmacy degree.

Pharm. D Student: Graduate student pursuing a doctor of pharmacy professional degree; students who successfully complete the program are eligible for licensure as pharmacists.

Pharmacare Centralized Information System (PCIS): A component of Pharma Net that performs claims processing and payment. Its key functions are to adjudicate claims; reverse claims; handle off-line claims; and manage payments. Access to PCIS is restricted to the Ministry of Health -

Services.

Pharmacare Information Line: An automated method for pharmacists to obtain information from Pharma Care by telephone. Also called the IVR system.

Pharmacare: The British Columbia government operated prescription drug program.

Pharmaceutical Alternatives: The drug products that contains the same therapeutic moiety but are different salts, dosage forms or strengths.

Pharmaceutical Care: Direct involvement of the pharmacist in the design, implementation, and monitoring of a therapeutic drug plan to produce a specific therapeutic outcome

Pharmaceutical Chemistry: Chemical disciplines important to the study and development of pharmaceuticals, including bioorganic chemistry, biophysical chemistry, computational chemistry, structure determination, mechanistic enzymology, and studies of metabolism and toxicity

Pharmaceutical Economics: Application of the tools of economics to the study of pharmaceutical issues whereby the costs and benefits of different choices are studied and evaluated

Pharmaceutical Equivalents: The drug that have the same active ingredient, dosage form, route of administration, strength or concentration.

Pharmaceutical Outcomes: Results of pharmaceutical care defined as the measurable effects of drug, or nondrug, usage on a patient; intermediate pharmaceutical outcomes, such as cost and time, can also be identified and measured

Pharmaceutical Policy: Any policy that is designed to improve or regulate the efficiency, effectiveness, safety, development, or promotion of pharmaceuticals, including drugs and devices

Pharmaceutical Pricing Agreement (PPA): The agreement required for manufacturers who have-executed a Medicaid Rebate Agreement with CMS and voluntary for those who do not have a current Medicaid Rebate Agreement

Pharmaceutical Research: Basic science studies related specifically to the discovery, development, and use of drugs.

Pharmaceutical Sciences: All of the sciences, including chemistry and biology, that play roles in the discovery and development of drugs

Pharmaceutical Technology: All of the technologies involved in the development and use of drugs

Pharmacist ID: Pharmacist identification number refers to the dispensing pharmacist's diploma number. CPBC license number.

Pharmacist: A person who has met all the requirements of, and is registered with, the Ontario College of Pharmacists. A pharmacist can perform the "controlled acts" within the pharmacist's scope of practice.

Pharmacodynamics (PD): Branch of pharmacology that studies reactions between drugs and living structures, including the processes of bodily responses to pharmacological, biochemical, physiological, and therapeutic effects.

Pharmacodynamics: Process of interaction of pharmacologically active substances with target sites, and the biochemical and physiological consequences leading to therapeutic or adverse effects.

Pharmacoeconomics: Branch of economics that applies cost-benefit, cost-utility, cost- minimization, and cost-effectiveness analyses to compare the economics of different pharmaceutical products or to compare drug therapy to other treatments. The study of cost-benefit ratios of drugs with other therapies or with similar drugs. Pharmacoeconomic studies compare various treatment options in terms of their cost, both financial and quality-of-life. Also referred to as "outcomes research."

Pharmacogenetic Test: An assay intended to study interindividual variations in DNA sequence related to drug absorption and disposition or drug action.

Pharmacogenetics: Study of the effects of individual genetic variations on drug response; aimed at the prescription or development of drugs that maximize benefit and minimize side effects in individuals

Pharmacogenomic Test: An assay intended to study interindividual variations in whole- genome or candidate gene maps, biomarkers, and alterations in gene expression or inactivation that may be correlated with pharmacological function and therapeutic response.

Pharmacogenomics: Science that examines inherited variations in genes that dictate drug response and explores the ways such variations can be used to predict whether a person will have a good response to a drug, a bad response to a drug, or no response at all.

Pharmacokinetic Dosing: Dosing a drug based on blood levels and the ability to eliminate the drug from the body with the goal of attaining efficacious blood levels while avoiding toxicity.

Pharmacokinetics: Process of the uptake of drugs by the body, the biotransformation they undergo, the distribution of the drugs and their metabolites in the tissues, and the elimination of the drugs and their metabolites from the body. Both the amounts and the concentrations of the drugs and their metabolites are studied. The term has essentially the same meaning as toxicokinetics, but the latter term should be restricted to the study of substances other than drugs.

Pharmacological Testing: Studies to explore the pharmacological activity and therapeutic potential of compounds. It involves the use of animals, isolated cell cultures and tissues, enzymes and receptors sites as well as computer models.

Pharmacological: Caused by action of the, agonist and antagonist at the same site. In the case of pharmacological antagonisms, the terms competitive and non-competitive antagonism are used with meanings analogous to competitive and non-competitive enzyme inhibition as used in enzymology.

Pharmacology: The branch of science that deals with the discovery and development of drugs. Those working in it (pharmacologists, doctors, scientists and laboratory technicians) determine the chemical structure and composition of drugs and how these act in the body. They assess the use of drugs in the prevention and treatment of diseases, their side-effects and likely toxicity. Pharmacologists not only research for new drugs, but also look for ways of synthesising them on a large scale.

Pharmacopoeia: An official publication dealing with the recognised drugs and giving their doses, preparations, sources, and tests. Most countries have a pharmacopoeia of their own.

Pharmacovigilance (PV): Term used for adverse event monitoring and reporting in some countries. The methods of assessment and prevention of adverse effects.

Pharmacy Administration: Management of the systems that affect all pharmacy-related clinical events, from the formulation of a drug prescription to the actual administration of a drug to a patient.

Pharmacy Benefit Management (PBM): The management of a prescription drug benefits through claims processing and program administration activities such as formulary and eligibility management and the establishment of a pharmacy network.

Pharmacy Benefit Plan: Coverage agreement between the employer and enrollees. Benefit design determines what drugs the plan does and does not cover, in what quantities, from what pharmacies and other drug sources, and at what out-of-pocket cost to members.

Pharmacy Examining Board of Canada (PEBC): The Pharmacy Examining Board of Canada is an independent pharmacy examining body incorporated under federal laws.

Pharmacy Network: A group of pharmacies with which a PBM contracts to provide pharmacy services to its members.

Pharmacy Practice Modeling: Identification and creation of new ways to apply the pharmacy profession in practice situations and to evaluate their effectiveness

Pharmacy Professional Fee: An amount the pharmacist charges to package and supply medicine, as well as to inform patients about their medication.

Pharmacy Remittance Advice Form: The statement sent to a pharmacy, dispensing physician/clinic, or non-pharmaceutical supplier for each payment made to them by Pharma care.

Pharmacy Software Vendor: A firm providing a product to one or more Pharma Net participants for the purpose of interfacing with Pharma Net.

Pharmacy: The department where drugs, medicines are prepared.

Pharmionics: The discipline concerned with measuring and understanding ambulatory patients' use of prescription drugs: 'what the patient does with the drug'

Pharyngitis: Inflammation of the pharynx, either from irritation or infection. A sore throat.

Pharynx : Throat, the part of the digestive tract between the oesophagus below and the mouth and nasal cavities above and in front.

Phase 0 Trial: Human micro dosing studies designed to speed up the development of promising molecular entities or imaging agents by establishing whether the drug or agent behaves in human subjects as was expected from preclinical studies through pharmacodynamic biomarker assays.

Phase 1 Reaction (of Biotransformation): Enzyme modification of a substance by oxidation, reduction, hydrolysis, hydration, dehydrochlorination or other reactions catalysed by enzymes of the cytosol, of the endoplasmic reticulum (microsomal enzymes) or of other cell organelles.

Phase 1 Trial: First in man studies to establish the initial safety, maximum tolerance, pharmacokinetics etc. in healthy human volunteers. For anti-cancer drugs Phase 1 studies are also carried out in patients.

Phase 2 Reaction (of Biotransformation): Binding of a substance or its metabolites from a phase 1 reaction, with endogenous molecules (conjugation), making more water- soluble derivatives that may be excreted in the urine or bile.

Phase 2 Trial: Universally accepted as a standard requirement for the evaluation of efficacy and safety of the drug. Careful observations are made to determine the dose and adverse reactions in patients with the relevant indication.

Phase 2a Trial: Pilot clinical trials to evaluate safety in selected patient population.

Phase 2b Trial: Controlled clinical trials to evaluate safety and efficacy for determining a dose range to be studied in Phase-3-trials.

Phase 3 Reaction (of Biotransformation): Further metabolism of conjugated metabolites produced by phase 2 reactions: it may result in the production of toxic derivatives.

Phase 3 Trial: First pre-marketing trial for evaluating the safety and efficacy of a drug in a large patient population. Phase 3 trial forms the basis of regulatory submissions.

Phase 3a Trial: Conducted after the drugs efficacy is demonstrated but before the regulatory submission of New drug Application.

Phase 3b Trial: Conducted after regulatory submissions but prior to the drug's approval or launch.

Phase 4 Trial: Post marketing studies to collect the additional safety data from a much larger patient population.

Phase I study: Conducted in patients and designed to determine the pharmacokinetic and pharmacologic actions of the drug in human subjects, the side effects associated with increasing doses, and, if possible, early evidence of effectiveness.

Phase II study: Helps determine the scientific validity of the drug by obtaining data on the effectiveness of the drug for a particular indication or indications in patients with the disease or condition, and helps determine the common short-term side effects and risks associated with the drug.

Phase III study: Involves greater numbers of patients than in phase II, aimed at gaining additional data about effectiveness and safety needed to evaluate the benefits and risks of the drug. Results yield data that will provide information that eventually will go on physician labeling.

Phase Transition Probability: Probability of advancing from one phase of development to the subsequent phase

Phases of Clinical Trials: Different phases of drug development process after the pre clinical development of a drug candidate is completed. There are four phases of clinical trial 1, 2, 3 and 4.

PhD Graduate Programs Associated With the UCSF School of Pharmacy: Graduate programs in: Chemistry and chemical biology, pharmaceutical sciences and pharmacogenomics, biological and medical informatics (MS also offered), biophysics, and quantitative biology.

PhD Student: Graduate student pursuing a doctor of philosophy degree; School of Pharmacy PhD students are pursuing graduate degrees in scientific research

PhD: The Doctor of Philosophy degree.

Phenotype: An individual's characteristics as determined by the interaction between his or her genotype –quota of genes – and the environment. The observable structural and functional characteristics of an organism determined by its genotype and modulated by its environment.

Phenotype: Any observable characteristic or trait of an organism such as morphology, biochemical or physiological properties.

Phenylketonuria: Commonly referred to as PKU, this is one of the less common, but very severe, forms of mental deficiency. The condition is due to the inability of the baby to metabolise the amino acid, phenyl-alanine.

Pheromone: Substance used in olfactory communication between organisms of the same species eliciting a change in sexual or social behaviour.

Pheromones: Chemicals produced and emitted by an individual which produce changes in the social or sexual behaviour when perceived by other individuals of the same species.

Phimosis: Tightness of the foreskin which prevents it from being pulled back over the under- lying head (glans) of the penis.

Phlebitis: Inflammation of a vein.

Phlebolith: The term applied to a small stone formed in a vein as a result of calcification of a thrombus.

Phlegm: Mucus in the throat or bronchi.

Phlegm: A popular name for mucus particularly that secreted in the air passages.

Phlyctenule: A hypersensitivity reaction of the conjunctiva.

Phobia: An irrational fear of particular objects or situations.

Pholcodine: An opioid cough suppressant similar to codeine; it is not, however, potent enough to suppress severe coughs and is also constipating.

Phonation: The production of vocal sounds – in particular, speech.

Phonocardiograph: An instrument for the graphic recording of heart sounds and murmurs.

Phosphaturia: The presence of excess phosphates in the urine. This occurs in and can even cause, alkaline urine (it's normally acidic), resulting in cloudy urine, small particle sedimentation, and the more common kinds of kidney stones.

Phospholipid: A lipid, the molecule of which contains a chemical derivative of phosphorus called phosphate. This type of lipid, which includes cephalins, lecithins and plasmalogens, is found in all tissues and organs, especially the brain. Phospholipids are produced in the liver and and take part in many of the body's metabolic activities.

Photochemotherapy: A form of treatment in which deliberate exposure to a drug and ultra- violet light benefits certain skin diseases, particularly psoriasis and T-cell lymphoma.

Photodermatoses: Diseases of the skin for which sunlight is partially or wholly responsible. In sufficient dosage, short-wave ultraviolet light always causes erythema. Higher doses progressively cause oedema and blistering; this is acute sunburn. Graduated exposure to UVB causes pigmentation (tanning). Prolonged chronic exposure to sunlight eventually accelerates ageing of the exposed skin with lentigo formation and loss of collagen and elastic tissue. After decades of such exposure, epidermal dysplasia and cancer may supervene.

Photo-Irritation: Inflammation of the skin caused exposure to light, especially that due to metabolites formed in the skin by photolysis.

Photo-Oxidant: Substance able to cause oxidation when exposed to light of the appropriate wavelength.

Photophobia: Abnormal visual intolerance of light.

Photosensitive: Reacting poorly to sunlight, either by skin reactivity or by forming abnormal sunlight-mediated serum metabolites

Photosensitivity: Oversensitivity of the skin to sunlight or other radiant energy, sometimes following exposure to sensitizing chemicals or drugs, resulting in accelerated burning of the skin by ultraviolet light.

Photosensitization: Allergic reaction due to a metabolite formed by the influence of light.

Phototherapy: The use of natural or artificial ultraviolet light to treat disease.

Phototoxicity: Adverse effects produced by exposure to light energy, especially those produced in the skin.

Phrenology: A quack method, common in the Victorian era, allegedly to study the mind and character of individuals from the shape of the head. As the shape of the head has been shown to depend chiefly upon accidental characteristics, such as the size of the air spaces in the bones, and not upon development of special areas in the con- tained brain, there is no scientific basis for the practice.

Phthisis: A historical term means wasting, and was applied to that progressive enfeeblement and loss of weight that arose from tuberculous disease of all kinds, but especially from the disease as it affected the lungs.

Physical Chemistry: Application of physical theories or laws to explain results of chemistry experiments

Physical Examination: That part of a patient's consultation with a doctor in which the doctor looks, feels (palpates) and listens to (auscultates) various parts of the patient's body. Along with the history of the patient's symptoms, this enables the doctor to assess the patient's condition and decide whether an immediate diagnosis is possible or whether laboratory or imaging investigations are needed to reach a diagnosis. A full physical examination may take 30 minutes or more. Physical examination, along with certain standard investigations, is done when a person attends for a 'preventive' check-up of his or her state of health.

Physical Therapy: Rehabilitation concerned with restoration of function and prevention of physical disability following disease, injury or loss of body part.

Physician: A person authorized by law to practice medicine.

Physiological: Caused by agonist and antagonist acting at two independent sites and inducing independent, but opposite effects.

Physiology: Physiology is the branch of medical science that deals with the healthy functions of different organs, and the changes that the whole body undergoes in the course of its activities.

Physiotherapy: An important treatment involving the use of physical measures, such as exercise, heat, manipulation and remedial exercises in the treatment of disease. An alternative name is physical medicine. It is an essential part of the rehabilitation of convalescent or disabled patients. Those who practise physiotherapy – physiotherapists – have a recognised training and, on successful completion of this, are placed on the profession's official register.

Phytosterols: Plant lipids, with little other than dietary value, but often excitedly referred to as "Hormone Precursors" with incorrect but well-meaning pseudo- science.

Phytotherapy: Botanical or herbal medicine, often with a heavy emphasis on studies and monographs and their medical implications (with virtually none from North America), and with a philosophy of "little drug" medical uses and the reliance on the European phytopharmaceutical industry (where the studies came from). No judgment here; this approach is of great value to physicians, since it offers clear implications for medical use. This approach is, however, medical and mechanistic, not vitalist and wholistic.

Phytotoxic: Poisonous to plants; inhibiting plant growth.

Pica: This is the Latin for magpie and is used to describe an abnormal craving for unusual foods. It is not uncommon in pregnancy. Among the unusual substances for which pregnant women have developed a craving are soap, clay pipes, bed linen, charcoal, ashes – and almost every imaginable food stuff taken in excess.

Pie Charts: A type of statistical chart that displays the distribution of levels of a categorical variable. It can display percentages or count.

Pigment: The term applied to the colouring matter of various secretions, blood, etc.; also to any medi- cinal preparation of thick consistency intended for painting on the skin or mucous membranes.

Pills: Small round masses containing active drugs held together by syrup, gum, glycerin, or adhesive vegetable extracts.

Pilocarpine: A plant alkaloid and the primary bioactive substance reducible from Pilocarpus spp. (Jaborandi leaves). It is an almost pure (cholinergic), inducing lowered blood pressure and stimulating glandular secretions everywhere. It stimulates sweating as well, a sympathetic cholinergic response. Anyway, it is used in eye drops these days to contract the pupil, lower ocular fluid pressure and take some of the stress off glaucoma. The refined alkaloid is better in the eyes, but the dried leaves are the usual complex agents of herb use and have some therapeutic values in low doses.

Pilonoidal Sinus: A sinus that contains hairs, usually occurring in the cleft between the buttocks. It may get infected and cause considerable pain. Treat- ment is by antibiotics and, if necessary, surgical removal.

Pilot Sudy: An initial study to explore new hypotheses.

Pimples: Technically known as papules, these are small, raised and inflamed areas on the skin.

Pinnae: The leaflets or primary division of a pinnate leaf.

Pinnate: A compound leaf, having the leaflets arranged on each side of the stem.

Pinnatifid: A leaf that is pinnately cleft, but into lobes that do not reach the midrib, and not into separate leaflets.

Pinnule: A division of a pinna.

Pint: A measure of quantity containing 16 fluid ounces (wine measure) or 20 fluid ounces

Pinworms: Also Threadworm, this is a widespread parasitic nematode, usually benign, but having a rural, white trash, skanky stigma. It mates and reproduces in the intestines of several mammals (including us) and the female exits the anus, usually at night, to shed its eggs and expire. The eggs become like dust motes, kids and puppies scratch their butts, the eggs spread into other mammals, until only a thermonuclear device or burning/razing/earth-salting will clear out a heavy infestation. It's also the only worm likely to be encountered in temperate zones and the high country.

Piscicide: Substance intended to kill fish.

Pistillate: A female flower that has pistils but no stamens.

Pituitary Gland: A gland at the base of the brain. The pituitary secretes several different hormones involved in key metabolic processes.

Pituitary: An endocrine gland somewhat behind the eyes and suspended from the front of the brain. The front section, the anterior pituitary, makes and secretes a number of controlling hormones that affect the rate of oxidation; the preference for fats, sugars, or proteins for fuel; the rate of growth and repair in the bones, connective tissue, muscles, and skin; the ebb and flow of steroid hormones from both the gonads and adrenal cortices. It does this through both negative and positive feedback. The hypothalamus controls these functions, secreting its own hormones into a little portal system that feeds into the pituitary, telling the latter what and how much to do. The hypothalamus itself synthesizes the nerve hormones that are stored in the posterior pituitary, which is responsible for squirting them into the blood when the brain directs it to. These neurohormones act quickly, like adrenalin, to constrict blood vessels, limit diuresis in the kidneys, and trigger the complex responses of sexual excitation, milk let- down in nursing, and muscle stimulus in the uterus (birthing, orgasm, and menstrual contractions), prostate, and nipples.

Pityriasis: A skin disorder typified by a bran-like desquamation (flaking). There are several varieties including P. alba, rosea, versicolor (fungal caused) and rubra (exfoliative dermatitis).

Pivotal Study: Usually a Phase III study that presents the data used to decide whether or not to approve a drug. A pivotal study will generally be well controlled, randomized, of adequate size, and whenever possible, double-blind.

PK: Pharmacokinetics.

PLA: Product License Application.

Placebo Controlled Study: A method of investigation of drugs in which an inactive substance is given to to one group of participants while the other group receives the investigational product. The results obtained in the two groups are then compared to see if the investigational treatment is more effective in treating the condition. A method of investigation of drugs in which an inactive substance (the placebo) is given to one group of participants, while the drug being tested is given to another group. The results obtained in the two groups are then compared to see if the investigational treatment is more effective in treating the condition. Placebo-controlled studies are rarely if ever used in clinical trials on cancer drugs.

Placebo Effect: A physical or emotional change, occurring after a substance is taken or administered, that is not the result of any special property of the substance. The change may be beneficial, reflecting the expectations of the participant and, often, the expectations of the person giving the substance.

Placebo: A placebo is an inactive pill, liquid, or powder that has no treatment value. In clinical trials, experimental treatments are often compared with placebos to assess the treatment's effectiveness and rule out any psychological effects testing may present.

Placenta: The thick, spongy, disc-like mass of tissue which connects the embryo with the inner surface of the uterus, the embryo otherwise lying free in the amniotic fluid. The placenta is mainly a new structure growing with the embryo, but, when it separates, a portion of the inner surface of the womb – called the maternal placenta – comes away with it.

Plagiarism: The act of copying someone's words, ideas, result and presenting them as original content.

Plantar Dermatosis: A common form of eczema of the soles of the feet typically seen in boys with an atopic background who constantly wear trainers.

Plantar: Describing anything related to the sole of the foot.

Plaque Psoriasis: The most common form of psoriasis defined by skin lesions that are red at the base and covered by silvery scales.

Plaque: A film of mucus that traps bacteria on the surface of the teeth. Plaque can be removed with daily brushing and flossing of teeth.

Plasma: Fluid component of blood in which the blood cells and platelets are suspended. Fluid component of semen produced by the accessory glands, the seminal vesicles, the prostate, and the bulbo-urethral glands. Cell substance outside the nucleus.

Plasmapheresis: Removal of blood from the body and centrifuging it to obtain plasma and packed red blood cells: the blood cells are resuspended in a physiologically compatible solution (usually type-specific fresh frozen plasma or albumin) and returned to the donor or injected into a patient who requires blood cells rather than - whole blood.

Plasmid: Autonomous self-replicating extra-chromosomal circular DNA molecule.

Platelet Aggregation: Platelets are the small, rather uniform fragments of large bone marrow cells that aid the blood in coagulation, hemostasis, inflammation, and thrombus formation. Mild subclotting and sticking is a common early condition that can lead to thrombosis, atherosclerosis, and strokes, and can be helped by an aspirin a day, better fat digestion.

Plethora: A condition of fullness of the blood vessels in a particular part or in the whole body. This results in a florid, red appearance of the affected area, particularly the face.

Plethysmograph: An apparatus for estimating changes in the size of any part placed in the apparatus; in this way changes in the volume of blood in a part can be measured.

Pleura: Lining of the lung.

Pleurisy: An inflammation of the serous membranes that both surround the lungs and line the inside of the chest cavity; the two membranes supply fluid lubrication between the expanding and contracting lungs and the body. Most pleurisy (and usually the milder form) follows or accompanies bronchitis or late winter chest colds sort of pulmonary cabin fever. It may be dry pleurisy (with few secretions and sharp sticking pain that prevents any but moderate inhalation), or acute or effusive pleurisy (with fever, coughing, and built up serous fluids usually tossed off as bronchitis). Some types are part of serious cardio-pulmonary disorders and/or chronic disease.

Pleurodynia: Also known as bornholm disease. A painful condition of the chest wall, it is usually the result of an infection of coxsackie virus B and may occur in epidemics

Ploidy: Term indicating the number of sets of chromosomes present in an organism. RT diploid, haploid

Plumbism: Another name for lead poisoning. Chronic poisoning caused by absorption of lead or lead salts.

PMA: Pre-Market Approval application.

PMC: Postmarketing commitments.

PMS: Postmarketing surveillance.

Pneumo-: A prefix relating to the lungs or to air. Examples is pneumonia.

Pneumonectomy: The operation of removing an entire lung in such diseases as bronchiectasis.

Pneumonia: Inflammation, usually infectious, of the lungs. Unless the result of only moderate chemical or smoke irritation, it is a potentially life-threatening condition. There are so many defenses against an infection this deep in the body that the very presence of pneumonia signals a pathogen of great virulence or impaired or exhausted immunity.

Pneumonitis: Inflammation of the lungs, from whatever cause. It may be concurrent with pneumonia or pleurisy or the result of a defensive lineman knocking the air out of the quarterback two days later.

Pneumothorax: A collection of air in the pleural cavity, into which it has gained entrance by a defect in the lung or a wound in the chest wall. When air enters the chest, the lung immediately collapses towards the centre of the chest; but, air being absorbed from the pleural cavity.

POC (Point-of-Care) Technology Provider: A point-of-care technology company who supports prescribers with applications and support.

Podagra: Another name for gout affecting the foot.

Poikilocytosis: This is a term used to describe the variation seen in the shape of red blood cells in some disorders of the bone marrow.

Point Mutation: Reaction that changes a single base pair in DNA.

Point of Service Access: Delivery of services directly at the point of hands-on care for the patient.

Point Source: Discharges to water that originate from a specific location, such as a pipe or open channel that carries wastewater from sewage treatment or industrial process plants, or a stack from a power plant.

Point-of-Service (POS) Plan: A health plan allowing the member to choose to receive a service from a participating or non-participating provider, with different benefits levels associated with the use of participating providers.

Poison: Substance that, taken into or formed within the organism, impairs the health of the organism and may kill it.

Poison-Bearing: Containing a poison.

Policies and Procedures: The detailed, written instructions or rules to achieve uniformity and consistency in the performance of a specific function.

Policy: The written instructions or rules to achieve uniformity, consistency and decision making.

Policyholder: The group or individual to whom an insurance contract is issued.

Poliomyelitis: Once known as infantile paralysis, this disease is caused by a viral infection involving the brain and spinal cord.

Pollex: A Latin term for thumb.

Pollutant: Any undesirable solid, liquid or gaseous matter in a solid, liquid or gaseous environmental medium: "undesirability" is often concentration-dependent, low concentrations of most substances being tolerable or even essential in many cases.

Pollution Prevention Assessment Team: A group assembled within a facility to conduct opportunity assessments. They are selected on the basis of their expertise and knowledge of the process operations and hazardous materials.

Pollution Prevention Opportunity Assessment: Systematic, periodic internal reviews of specific processes and operations designed to identify and provide information about opportunities to reduce the use, production, and generation of toxic and hazardous materials and waste.

Pollution Prevention: The use of materials, processes, or practices that reduce or eliminate the creation of pollutants or wastes at the source. It includes practices that reduce the use of hazardous materials, energy, water, or other resources, and practices that protect natural resources through conservation or more efficient use.

Pollution: Introduction of pollutants into a solid, liquid, or gaseous environmental medium, the presence of pollutants in a solid, liquid, or gaseous environmental medium, or any undesirable modification of the composition of a solid, liquid or gaseous environmental medium. In the context of air pollution, an undesirable modification is one that has injurious or deleterious effects.

Polyarteritis Nodosa: Also known as periarteritis nodosa. A rare but potentially serious disease, probably caused by a disturbance of the immune system. Prolonged fever and obscure symptoms

Polyarthritis: A rheumatic disorder – usually caused by rheumatoid arthritis – that affects several joints in the body. The joints are stiff, painful and swollen. Different joints may be affected at the same time or in various sequences.

Polychromasia: An abnormal reaction of the red blood cells in severe anaemia. They have a bluish tinge instead of the normal red colour in a blood film stained by the usual method. It is a sign that the cell is not fully developed.

Polyclonal Antibodies: Antibodies produced by many clones which, together, react with one or more antigens. Antibody produced by a number of different cell types.

Polycystic Kidneys: A hereditary disorder where the kidneys are enlarged due to the formation of cysts in and on the kidney.

Polycythaemia: A rise in the amount of haemoglobin in the blood. This may be caused by an excess in the number of erythrocytes produced in the bone marrow or to a fall in the total volume of plasma in the circulatory system. It may also be a response to reduced oxygen levels – for example, among people living at high altitudes– or to liver or kidney disease: this type is called secondary polycythaemia.

Polydactyly: The presence of extra, or supernumerary, fingers or toes.

Polydipsia: Excessive thirst, which is a symptom of diabetes mellitus and some other diseases.

Polygene: One of several genes that between them control a single characteristic in an individual. With each polygene exerting a slight effect, the genetic outcome is the consequence of 'group action'. The hereditary characteristics produced in this way are usually those of a quantitative type – for instance, an individual's height.

Polymorphism (Polymorphia) in Metabolism: Interindividual variations in metabolism of endogenous and exogenous compounds due to genetic influences, leading to enhanced side effects or toxicity of drugs (for example, poor versus fast metabolizers) or to different clinical effects.

Polymyositis: A connective-tissue disease affecting the muscles throughout the body. This rare disorder, which is associated with dermatomy- ositis, may be acute or chronic but it usually affects the muscles of the shoulders or hip areas. The muscles weaken and are tender to the touch. Diffuse inflammatory changes occur and symptomatic relief may be obtained with corticosteroids.

Polymyxin: A group of antibiotics derived from various species of bacillus polymyxa. One variety, colistin, is used to sterilise the bowels before surgery as it is not absorbed when given by mouth. The drugs are sometimes used in topical applications for infections of the skin, eye and ear.

Polyneuritis: An inflammatory condition of nerves in various parts of the body.

Polypharmacy: A term applied to the administration of too many drugs to one person. Sometimes combinations of drugs are an effective means of treatment, reducing the risk of drug resistance. Polypharmacy, however, worsens the risk of drug interactions and of adverse effects, especially in the elderly.

Polypill: A suggestion by two epidemiologists, made in the British Medical Journal in 2003, that many lives could be saved if all persons aged over 55 took a daily combination pill they termed the polypill. Its components would be aspirin, a cholesterol-lowering agent, folic acid and two blood-pressure-lowering agents. The suggestion caused a massive correspondence as it implied treating a whole population rather than individuals considered to be at special risk.

Polyposis: The presence of a crop, or large number, of polypi. The most important form of polyposis is that known as familial polyposis coli. This is a hereditary disease characterised by the presence of large numbers of polypoid tumours in the large bowel.

Polypus: A general name applied to tumours which are attached by a stalk to the surface from which they spring. The term refers only to the shape of the growth and has nothing to do with its structure or nature.

Polyuria: Excess urination. The excreted wastes may stay unchanged but they are dissolved in a far higher volume of water. The causes range from diabetes, kidney disease, elevated thyroid function and the aftermath of diuretic-treated heart failure to booting a half keg of generic beer at a frat blow out.

Popliteal Space: The name given to the region behind the knee. The muscles attached to the bones immediately above and below the knee bound a diamond- shaped space through which pass the main artery and vein of the limb (known in this part of their course as the popliteal artery and vein); the tibial and common peroneal nerves (which continue the sciatic nerve from the thigh down to the leg).

Population at risk: Number of persons who can and may develop an adverse health effect and who are potentially exposed to a risk factor under study : for example, all people in a population who have not developed immunity to an infectious disease are at risk of developing that disease if they are exposed to it.

Population Critical Concentration (PCC): Concentration of a substance in the critical organ at which a specified percentage of the exposed population has reached the individual critical organ concentration.

Population Effect: Absolute number or incidence rate of cases occurring in a group of people.

Population: Any finite or infinite collection of subjects from which a sample is drawn for a study to obtain estimates for values that would be obtained if the entire population were sampled.

Population: In statistics, the totality of items under consideration. A clearly defined part of a population is called a subpopulation. In the case of a random variable, the probability distribution is considered as defining the population of that variable.

Pore: A small opening. The word is usually used to describe an opening in the skin that releases sweat or sebum, a waxy material secreted by the sebaceous glands in the skin.

Porphyria: Disturbance of porphyrin metabolism characterized by increased formation, accumulation, and excretion of porphyrins and their precursors.

Porphyrias: A group of rare inherited enzyme diseases in which disorders of the metabolic pathways leading to the synthesis of haem cause excessive production of haem precursors called porphyrins by the bone marrow or liver. The excess porphyrins in the blood mainly affect the skin, causing photosensitivity, or the central nervous system, causing various neuro-psychiatric disorders. Excess porphyrins can be detected in blood, urine and faeces.

Portable Document Format: The file format created by Abode Systems independent of the application software, hardware and operating system.

Portal Circulation: This is a type of circulatory bypass used when substances in blood or fluid need to be kept out of the general flow. A portal system begins in capillaries and ends in capillaries, and nothing leaves it undocumented. The hypothalamus sends hormones into the portal system between it and the pituitary, and the pituitary responds to it by secreting its own hormones, but dissolving the hypothalamus ones. Blood that leaves the intestinal tract, spleen, and pancreas (partially) goes into the liver's portal system and does not leave that organ until it has been thoroughly screened and altered.

Portal Hypertension: Raised blood pressure in the portal vein entering the liver. This results in increased pressure in the veins of the oesophagus and upper stomach and these grow in size to form varices dilated tortuous veins. Sometimes these varices rupture, causing bleeding into the oesophagus. The raised pressure also causes fluid to collect in the abdomen and form ascites. The commonest reason for portal hypertension is cirrhosis (fibrosis) of the liver.

Posology: Study of dose in relation to the physiological factors that may influence response such as age of the exposed organisms.

Possetting: The technical term used to describe the quite common habit of healthy babies to regurgitate, or bring up, small amounts of the meal they have just taken. Its name derives from possett, an 18th century drink made from porridge and sherry.

Post Marketing Surveillance: A systematic approach to collect the additional safety data on a drug from a much larger patient population once it is marketed.

Post-Coital Contraception: Action taken to prevent conception after sexual intercourse. The type of contraception may be hormonal, or it may be an intrauterine device.

Post-Coital Test: A test for infertility. A specimen of cervical mucus, taken up to 24 hours after coitus (during the post-ovulatory phase of the menstrual cycle), is examined microscopically to assess the motility of the sperms. If motility is above a certain level, then sperms and mucus are not interacting abnormally – thus eliminating one cause of sterility.

Post-Consumer Material: A material or

finished product that has served its intended use and has been discarded for disposal or recovery, having completed its life as a consumer item. "Post-consumer material" is a part of the broader category of "recovered material."

Post-Consumer Recycling: The reuse of materials generated from residential and commercial waste, excluding recycling of material from industrial processes that has not reached the consumer, such as glass broken in the manufacturing process.

Postdoctoral Education and Training/Programs: Advanced training for pharmacy professionals beyond the Pharm. D degree and for scientists beyond the Ph.D degree.

Poster: A document to present the scientific results on a single sheet using text, figures, tablets and graphs.

Post-Mortem Examination: Also called an autopsy (and less commonly, necpropsy), this is an examination of a body to discover the causes of death.

Post-Operative: The period after an operation, the patient's condition after operation, or any investigations or treatment during this time.

Post-Partum: The term applied to anything happening after childbirth: for example, postpartum haemorrhage.

Post-Traumatic Stress Disorder: A - term introduced to psychiatry in 1980 after Postural Drainage.

Potency: The strength of a homeopathic remedy. Determined by how many times the remedy has been succussed and diluted during preparation. A number and a letter are associated with the remedy name to indicate which potency scale has been used.

Potentiation: Dependent action in which a substance or physical agent at a concentration or dose that does not it self have an adverse effect enhances the harm done by another substance or physical agent.

Potentized: Usually refers to a substance prepared according to homeopathic pharmaceutical standards. This means that it has gone through serial dilution and succussion.

Pott's Disease: A traditional name often applied to the angular curvature of the spine which results from tuberculous disease.

Pott's Fracture: A variety of fractures around the ankle, accompanied by a varying degree of dislocation of the ankle.

Pouchitis: A rare chronic inflammatory disease in the ileal pouch, which remains after a patient has had intestinal resection because of inflammatory bowel disease (IBD).

Poultices: Soft, moist applications to the surface of the body, generally used hot to soothe pain due to inflammation and to promote resolution.

Power and Sample Size: The likelihood that study will reject the null hypthesis of no treatment difference when the truth is that the two treatments do differ.

Power of Attorney: A written legal document that gives an individual the authority to act for another.

Powers of Appointment: A right given to another in a written instrument, such as a will or trust that allows the other to decide how to distribute the property. The power of appointment is "general" if it places no restrictions on whom the distributees may be. A power is "limited" or "special" if it limits the eventual distributee.

PPMS (Physician Practice Management System) Technology Provider: A physician practice management system company who supports medical office administration with applications and support.

Practical Certainty (of Safety): Numerically specified low risk of exposure to a potentially toxic substance or socially acceptable low risk of adverse effects from such an exposure applied to decision making in regard to chemical safety.

Practice: A fashion in which activity is routinely performed.

Practitioner ID: A unique number assigned to practitioner by the practitioner licensing body, i.e., the College of Physicians and Surgeons of British Columbia, or the College of Pharmacists of British Columbia.

Practitioner: A person authorized to practice medicine, dentistry, veterinary medicine, or a prescribed health care profession in which a practitioner of that profession is authorized to prescribe drugs or devices.

Pragmatic Trial: Term used to describe a clinical study designed to examine the benefits of a product under real world conditions.

Pre Clinical Investigations : Investigation of the pharmacological activity of a new compound using a wide array of chemical and biochemical assays, cell-culture models and animal models in a laboratory.

Pre Existing Conditions: Clinical complaints that a patient have at the time of entry into the study.

Preceptor: A licensed pharmacist that supervises students and interns, and meets the Ontario College of Pharmacists' qualifications for preceptors.

Precertificatio: The process of obtaining certification from the health plan for routine hospital stays or outpatient procedures. The process involves reviewing criteria for benefit coverage determination.

Precipitin: An antibody that combines with an antigen and forms the immune complex as a precipitate. The reaction is used in some diagnostic serological tests to identify antigens in the serum.

Precision: The capacity of the system to discriminate between different values of input; the "fineness" with which different values for input can be inferred from measured values of output. The pooled deviation of observed from expected values of output, all divided by the amplification, yields the "index of precision". The square of the reciprocal of the index of precision is the measure of the amount of information that can be delivered by the system. Specifically, precision is computed in several steps.

Preclinical R&D: All research and development conducted prior to the initiation of clinical testing, including discovery efforts, as well as testing drug candidates in assays and animal models before the start of phase I testing.

Preclinical Studies: Animal studies that support Phase 1 safety and tolerance studies and must comply with good laboratory practice (GLP). Data about a drug's activities and effects in animals help establish boundaries for safe use of the drug in subsequent human testing (clinical studies or trials). Because many animals have much shorter life spans than humans, preclinical studies can provide valuable information about a drug's possible toxic effects over an - animal's life cycle and on its offspring.

Preclinical Testing: Before a drug may be tested on humans, experimental drugs must be tested in test tube or in animals to determine that the drug is safe.

Precordial Region: The area on the centre and towards the left side of the chest, lying in front of the heart.

Precordial: Pertaining to the region over the heart and lower thorax.

Precursor: A chemical that can be converted by the body into another is a precursor of the latter chemical.

Predictive Validity: Reliability of a measurement expressed in terms of its ability to predict the criterion: an example would be an academic aptitude test that was validated against subsequent academic performance.

Predictive Value: Percentage of positive results that are true positives or of negative results that are true negatives.

Pre-Eclampsia: A complication of pregnancy of unknown cause, which in severe cases may proceed to eclampsia. It is characterised by hypertension, renal impairment, oedema, often with proteinuria and disseminated intravascular coagulation

Pre-Existing Condition: A health condition (other than a pregnancy) or medical problem that was diagnosed or treated before enrollment in a new health plan or insurance policy.

Pregnancy: The period of conception to birth.

Pregnenolone: The grandmother steroid hormone produced in the mitochondrion that is the base "raw-material" for all the steroids and neuro-steroids.

Premature Ejaculation: A disorder in which ejaculation of semen occurs before or immediately after the penis penetrates the vagina during sexual intercourse. The most common sexual problem in men, persistent premature ejaculation may have psychological causes, although many adoles cents and some adults experience it occasionally.

Pre-Mature Termination: terminating/stopping an activity before the planned completion time-frame.

Premedication: A drug or drugs given to a patient to produce sedation before an operation, whether this is done under a local or general anaesthetic.

Premium: The premium is the amount that insured persons are required to pay to the Regie de l'assurance maladie du quebec or to a private insurer for their prescription drug insurance. Person covered by the public plan pay their premium when filing their income tax return.

Preneoplastic: Before the formation of a tumour.

Prepuce: Also known as the foreskin, this is the free fold of skin that overlaps the glans penis and retracts when the penis becomes erect.

Presbyacusis: Deafness that comes on with increasing years. It is caused by increasing loss of elasticity in the hearing mechanism, combined with the slowing-down of the mental processes that accom- panies old age.

Prescribed Diseases: A collection of industrial diseases which provide those with a disease legal entitlement to welfare benefits. Examples are deafness from excessive noise in the workplace.

Prescriber: A health care provider licensed to prescribe drugs. Primary prescribers are physicians, but others may have prescriptive authority, depending on state statutes and laws. Other prescribers may include dentists, physician assistants, nurse practitioners and others may have authority to prescribe, typically within specific limits.

Prescription Drug: A drug that has been approved by the federal food and drug administration which can only be dispensed according to physician's prescription order.

Prescription Medicine: The written direction for drugs for medicinal use, given by the doctor, dentist and nurse to the patient, for dispensation by the pharmacist.

Prescription: The written direction for drugs for medicinal use, given by the doctor, dentist and (for some drugs) nurse to the patient, for dispensation by the pharmacist.

Presenile: Describing the condition of premature ageing. The mental and physical faculties are adversely affected in presenility to an extent that does not usually occur until old age.

Presentation: The appearance in labour of some particular

Pressor Agent: An agent, neurologic or hormonal, that increases blood pressure.

Pressor: An agent that raises the blood pressure.

Pre-Study Visit Reptor: The written report that summarizes the observations and finding of pre-study visit.

Pre-Study Visit: The initial visit to investigator site by the Sponsor/CRO personnel in order to evaluate the suitability of investigator and facility of a trial.

Pre-tax account: An account in which contributions are subtracted from an employee's pay before withholding income tax and Social Security. This ultimately reduces the account-holder's tax liability, since taxes are based on income minus the account contribution.

Pre-Trial Monitoring Visit: The written report that summarizes the observations and findings of pre-study visit.

Pre-Trial Monitoring Visit: The initial visit to investigator site by the Sponsor/CRO personnel in order to evaluate the suitability of investigator and facility of a trial.

Prevalence Rate (Ratio): Total number of individuals who have an attribute or disease at a particular time (or during a particular period) divided by the population at risk of having the attribute or disease at this point in time or midway through the period.

Prevalence: An epidemiological term describing the proportion of a defined group in the population having a condition at one point in time.

Prevalence: Number of instances of existing cases of a given disease or other condition in a given population at a designated time; sometimes used to mean prevalence rate. When used without qualification, the term usually refers to the situation at a specified point in time (point prevalence).

Prevention Trials: Refers to trials to find better ways to prevent disease in people who have never had the disease or to prevent a disease from returning.

Preventive Care: Medical and dental services aimed at early detection and intervention.

Preventive Medicine: The term 'preventive medicine' may be used both in a general 'lay' sense and to cover a specific range of activities carried out by health professionals.

Preventive Services: Health care services for prevention of a disease.

Primary Care Group (PCG): Sub-committee of the health authority responsible for health services in a geographical area in England.

Primary Care Organisation (PCO): Generic term used to described PCGs, LHGs and PCTs.

Primary Care Physician (PCP): A physician, usually a family or general practitioner, internist or pediatrician, who provides a broad range of routine medical services and refers patients to specialists, hospitals and other providers as necessary.

Primary Care Physician: A doctor trained to provide basic medical care.

Primary Care Trust (PCT): Free-standing statutory trust responsible for health services in a geographical area in England.

Primary Care: The basic, comprehensive, routine level of health care typically provided by a person's general or family practitioner, internist or pediatrician.

Primary Health Care: Sometimes called primary medical care, this is the care provided by a general practitioner (GP) – traditionally entitled the family doctor or other health professionals who have first contact with a patient needing or wanting medical attention.

Primary Outcome: The actual occurrence of a primary event in a clinical study patient.

Primary Protection Standard: Accepted maximum level of a pollutant (or its indicator) in the target organism, or some part thereof, or an accepted maximum intake of a pollutant or nuisance into the target under specified circumstances.

Primary Variable: An outcome variable of interest in the trial. Differences in the outcome variable(s) between groups of subjects are believed to be the result of the group-specific interventions.

Primigravida: A woman who is undergoing her first pregnancy.

Primipara: The term applied to a woman who has given birth, or is giving birth, to her first child.

Principal Investigator: A person responsible for the conduct of clinical trial at a trial site.

Prion: An aberrant variety of one of the proteins, called PrP, in a brain cell. The result of a gene mutation, prions are stable, resistant to radiation and impervious to the normal cel- lular processes of degradation.

Prior Authorization: The process associated with specified drug products on the formulary that require additional evaluation by clinical staff prior to coverage being granted.

Priority and Standard NCE: Considered by CDER to offer high therapeutic value and earmarked for priority review status. Standard NCEs are viewed as offering little or no therapeutic advantages over existing therapies and receive lower review status. CBER also grants priority review status, but the criteria are somewhat different than CDER; the products must also be for serious or life-threatening illness.

Priority Drugs: Those considered by the FDA to offer high therapeutic value and earmarked for priority review.

Prisoner: A person who is confined to custody.

Privacy: The state of being private.

Private Information: The personal information of an individual which is not known to others.

Private Plan: Private plans are usually available in the form of group insurance or employee benefit plans. Persons may have access to a private plan through employment, through membership in a professional order or association, or through their spouse or parents. Persons who are eligible for a private plan are required to join that plan.

PRN (Physicians Reducing Negative Events): Provides physicians in the ambulatory setting with patient-specific medication history and pharmacy benefit information at the point of care, thus giving the physician the ability to write an informed prescription.

Probate: The court process for determining the validity of a deceased person's will.

Probe: A slender, flexible instrument, usually made of metal, designed for introduction into a wound or cavity – to explore its depth and direction, to discover the presence of foreign bodies, or to introduce medicinal substances.

Probiotics: Viable bacteria that colonise the intestine and alter the microflora and their metabolic activities, with a presumed beneficial effect for the host. Many probiotics are lactic acid bacteria

Probit: Probability unit obtained by adding 5 to the normal deviates of a standardized normal distribution of results from a dose response study: addition of 5 removes the complication of handling negative values Tj A plot of probit against the logarithm of dose or concentration gives a linear plot if the distribution of response is-a /logarithmic normal one. Estimates of the LD50 and ED50 (or LC50 and EC50) can be obtained from this plot.

Procarcinogen: Substance that has to-be metabolized before it can induce malignant tumours.

Procedure: A particular method of performing a task.

Process Deviation: Non-compliance to a process.

Process Improvement: The methods or steps that can improve an existing process in an efficient manner.

Process: A series of actions or events for performing a task.

Proctalgia: 'Nerve pain' in the anus or rectum, without any local disease to account for it.

Proctitis: Inflammation situated about the rectum or anus.

Prodromata: A term applied to the earliest symptoms of a disease, or those which give warning of its presence.

Product Change: Product changes are performed by the manufacturer of a product with the intent of reducing waste resulting from a product's use. Product changes include product conservation, product substitution, and protection of natural resources by conservation.

Product Identification Number (PIN): A PharmaCare assigned number for a product or compound that does not have a federally-assigned Drug Identification Number (DIN).

Product Monograph: A document that summarizes all the clinical and non-clinical information of a drug product.

Product Substitution: Product substitutions reduce or eliminate the hazardous materials that enter the production process. Product substitutions can be made to avoid the generation or wastes within the production processes.

Product: An item that may be requested on a prescription and/or a claim. A product may or may not be a drug, i.e., blood glucose monitoring strips.

Professional Services Agrrement: A written , dated and signed agreement between two or more parties that sets out any arrangements on delegation and distribution of tasks, obligations on financial matters.

Progesterone: This is the hormone secreted after ovulation by the corpus luteum. It is a steroid (a cholesterol with a funny hat), enters receptive cells to stimulate their growth, and acts as an anabolic agent. Estrogen should be viewed as the primary coat underneath all the cycles during a woman's reproductive years, with progesterone, its antagonist, surging for ten or twelve days in ovulatory months. Most of the actions of progesterone cannot occur without estrogen having previously induced the growth of progesterone-receptive binding sites. In the estrus cycle, estrogen stimulates the thickening of membranes (the proliferative phase), and progesterone stimulates their sophistication into organized and secreting mucosa (the secretory phase). The new secretions contain anticoagulants, antimicrobials, and rich mucus fluids. If there is pregnancy, the uterine membranes are fully structured for the long haul; if menses occurs, the thickened tissues can erode away without clotting, becoming infected, or flowing poorly. If there is not enough estrogen, the corpus luteum will not mature.

Prognathism: Abnormal protusion of the lower jaw, or sometimes of both jaws. The condition may make biting and chewing difficult, in which case corrective surgery is necessary.

Prognosis: The term applied to a forecast as to the probable result of an illness or disease, particularly with regard to the prospect of recovery.

Project Management: Methods used for planning, organizing and managing a clinical trial project.

Project Plan: A written plan to define the critical project milestones, cost estimate, timelines and deliverables.

Prokaryote: Unicellular organism, characterised by the absence of a membrane-enclosed nucleus. Prokaryotes include bacteria, blue-green algae and mycoplasmas.

Prolactin: Prolactin is the pituitary hormone which initiates lactation.

Prolapse: Displacement of an organ or structure from its normal position. The term is applied chiefly to downward displacements of the rectum and uterus.

Promoter: Agent that induces cancer when administered to an animal or human being who has been exposed to a cancer initiator.

Pronation: The movement whereby the bones of the fore-arm are crossed and the palm of the hand faces downwards.

Prone: Lying with the face down, or positioning the arm and hand so that the palm faces downwards.

Prophage: Latent state of a phage genome in a lysogenic bacterium.

Prophylactic: A preventive measure.

Prophylaxis: Treatment or action adopted with the view of warding off disease.

Proportional Mortality Rate (ratio) (PMR): Number of deaths from a given cause in a specified time period, per 100 or per 1000 total deaths in the same time period: can give rise to misleading conclusions if used to compare mortality experience of populations with different causes of death.

Proposal: A written description of a work plan.

Proprietary Name: The trade name of a drug registered by the pharmaceutical company which has developed and patented it.

Proprietary: Ownership of sole rights of a document, process or product.

Proprioceptors: Sensory nerve endings in the muscles, tendons and joints which signal to the brain their position relative to the outside world and the state of contraction of the muscle. During movement, a regular flow of information to the brain from the proprioceptors, the eyes and ears ensures that actions are coordinated and the body's balance maintained.

Proptosis: A condition in which the eye protrudes from the orbit. Proptosis due to endocrine abnormality (e.g. thyroid problems) is known as exophthalmos.

Prospective Study: A study that moves forward in time.

Prospective Study: Investigation in which a group of subjects is recruited and monitored in accordance with criteria described in a protocol.

Prospective Study: Research in which patients are studied from the first contact with a doctor or hospital regarding their illness – as opposed to a retrospective study, in which notes are obtained of a group of patients already treated or under treatment.

Prostaglandin: A group of a dozen or more fatty acid derivatives made by many tissues for paracrine (local) hormone use. Because they are only meant for local use, the same compound may serve opposite purposes in different tissues inhibiting inflammation in the stomach lining while increasing uterine irritability.

Prostate: This is a walnut-sized gland that surrounds the beginning of the urethra in men. It secretes the alkaline transport fluid that mixes with sperm from the testes to form semen. The prostate needs adequate anabolic steroid stimulation for its health and growth, especially testosterone. Because of diminished healthy hormone levels, pelvic congestion, and decreased blood (and hormone) circulation, or because of sexually transmitted or urinary tract infections, a male may get prostatitis.

Prostatitis: Inflammation of the prostate. The causes may be varied, ranging from infection to portal congestion to cancer to increased adipose estradiol release in the middle-aged male to over-use.

Prosthesis: An artificial replacement of a missing or mal- functioning body part. Examples include false legs or arms fitted after amputation; artificial heart valves; artificial heart devices; cochlear implants to improve hearing; a bio-artificial pancreas (containing live pancreatic cells from pigs) now under development to treat diabetes.

Prosthetic Devices: A device that replaces all or a part of the human body because a part of the body is permanently damaged, is absent or is malfunctioning.

Protease: A digestive enzyme – also known as a proteolytic enzyme – that breaks down protein in food as part of the digestive process. The complex protein molecules are reduced to their con- stituent amino acids.

Proteasome: An enzyme complex found in all cells, responsible for breaking down proteins, including those involved in cell cycle regulation. Proteasome inhibition disrupts this process, leading to apoptosis or cell death.

Protected Health Information: Any information about health status, provision of health care or payment for health care that can be linked to an individual.

Protection of Pupil Rights Amendment: Department of education regulation that states the surveys, questionnaries and instructional materials for school children must be inspected by parents/guardians.

Protein Engineering: Technique for isolating and studying proteins and creating tailor-made proteins by altering the genes that direct their composition.

Proteinuria: Excretion of excessive amounts of protein (derived from blood plasma or kidney tubules) in the urine. The presence of protein in the urine, sometimes a symptom of kidney compromise.

Proteolysis: The mechanism by which complex protein molecules are broken down by digestive enzymes in the stomach and small intestine. The constituent amino acids are then absorbed into the bloodstream.

Proteolytic: An enzyme or agent that speeds up the breaking down or digestive hydrolysis of proteins into smaller proteins, peptides, polypeptides, oligopeptides, amino acids, and all that delicious nitrogenous slurry-stuff.

Proteome: The unique complement of proteins that reside in a cell.

Proteomics: The study of proteins.

Prothrombin: An inactive substance in the blood plasma that is the precursor of the enzyme, thrombin, which clots the blood. The conversion occurs when a blood vessel is damaged and the process of blood coagulation occurs.

Protocol Addendum: The written supplement or additions to the approved protocol.

Protocol Amendment: The written description of a change to or formal clarification of a protocol.

Protocol Approval: This marks completion of protocol development. The signature of the last reviewer on the protocol approval form has been obtained and signifies that all reviewer changes to the protocol have been incorporated.

Protocol Deviation: The non-compliance with the protocol schedule of events.

Protocol Modification: The written description of a change to a protocol.

Protocol Review Committee: A core group within an institution or organization that reviews trial protocols for completeness, accuracy, compliance and feasibility.

Protocol Signature Page: A document that acknowledge the investigator and/or sponsor

Protocol: A document and plan developed for each clinical trial. The plan is carefully designed to safeguard the health of the participants as well as answer specific research questions. A protocol describes what types of people may participate in the trial; the schedule of tests, procedures, medications, and dosages; and the length of the study. While in a clinical trial, participants following a protocol are seen regularly by the research staff to monitor their health and to determine the safety and effectiveness of their treatment.

Proton Pump: A key enzyme system in the parietal cells of the mucosal lining of the stomach: hydrogen ions are produced which acidify the stomach's secretions and convert pepsinogen to pepsin, an active participant in the digestion of food.

Protoplasm: The viscid, translucent, glue-like material containing fine granules and composed mainly of proteins, which makes up the essential material of plant and animal cells and has the properties of life.

Provider Directory: Provider directories are listings of providers who have contracted with a managed care network to provide care to its participants. Participants may refer to the directory to select in-network providers.

Provider Network: A panel of providers contracted by a health plan to deliver medical services to the enrollees.

Provider: A licensed health care facility, program, agency, physician or health professional that delivers health care services.

Proving: The most accurate method of ascertaining the action of medicines on human health. Medicines (usually potentized) are administered to healthy people to discover the symptoms they are capable of producing and thereby able to cure.

Proximal: A term of comparison applied to structures which are nearer the centre of the body or the median line as opposed to more distal, or distant, structures.

Proxy: A person authorized to act on

behalf of another person.

PRRTP (Psychosocial Residential Rehabilitation Treatment Programs): PRRTP is designed to provide a therapeutic residential environment for treatment for psychiatric conditions and psychosocial readjustment. VA defines PRRTP as inpatient care, and it is often viewed as an alternative to specialty psychiatry and substance abuse inpatient care.

Prurigo: An intensely itching form of eczema in which lichen takes a nodular form.

Pruritus: Another name for itching, it is a common symptom with many causes. It may accompany obvious skin disease such as urticaria, eczema or scabies.

Pseudoadaptation: Apparent adaptation of an organism to changing conditions of the environment (especially chemical) associated with stresses in biochemical systems that exceed the limits of normal (homeostatic) mechanisms: essentially there is a temporary concealed pathology that later on can be manifested in the form of explicit pathological changes sometimes referred to as "decompensation."

Pseudocyesis: Pseudocyesis means spurious or false pregnancy, a condition characterised by of the abdomen, and even enlargement of the breasts and early-morning sickness – the woman being quite convinced that she is pregnant.

Pseudohermaphrodite: A person in whom the gonads of only one sex are present in the body but in whom the external genitalia may not be obviously male or female. The condition is a result of a hormonal imbalance and can normally be treated by appropriate surgery and hormone drugs.

Pseudoxanthoma Elasticum: This is a hereditary disorder of elastic tissue. Degenerating elastic tissue in the skin produces lesions which look like soft yellow papules. Elastic tissue in the eye and blood vessels is also involved, giving rise to visual impairment, raised blood pressure and haemorrhages.

Psittacosis: Also called parrot disease. An infectious disease of parrots and other exotic birds which may be transmitted to humans and is caused by the micro-organism Chlamydia psittaci. It presents as pneumonia or a systemic illness in which the patient has an enlarged spleen and liver and pneumonitis. Tetracycline is an effective treatment, but relapses may occur.

Psoralen: A photosensitizing drug used in combination with UVA to treat psoriasis.

Psoriasis: A chronic skin condition with dull red lesions of the skin that come and go for many years. Usually painful or itchy, they tend to be worse in the winter and are often helped by increased exposure to the sun or moderate UV treatment. It is, at least to some degree, an inherited condition, auto-immune, and sometimes accompanied by joint pain. A chronic (long-lasting) skin disease of scaling and inflammation; understood to be a disorder of the immune system.

Psoriasis: This chronic, relapsing inflammatory skin disease is extremely common,

Psoriatic Arthritis: Joint inflammation that occurs in some people with psoriasis.

PSSF: Pharmacy Society of San Francisco

PSV: Pharmacy Software Vendor.

Psyche: The mind or soul of an individual and his or her mental – in contrast to the physical.

Psychiatry: That branch of medical science which treats mental disorder and disease and also helps with the management of people with learning dis- abilities.

Psychogeriatrics: The branch of psychiatry that investigates, diagnoses and treats the mental-health prob- lems of old people.

Psychometrics: The use of standardised psychological tests to measure differences in functions – for example, intelligence and personality – in individuals.

Psychoneurosis: A general term applied to various functional disorders

of the nervous system.

Psychopathic: Psychopathic disorder is defined by the Mental Health Act 1983 as a persistent disorder or disability of mind (whether or not including sig- nificant impairment of intelligence) which results in abnormally aggressive or seriously irresponsible conduct. The cardinal features are as follows: (1) Absence of normal feelings for other people such as love, affection, sympathy and condolence. (2) A tendency to antisocial impulsive acts with no forethought.

Psychosis: One of a group of mental disorders in which the affected person loses contact with reality. Thought processes are so disturbed that the person does not always realise that he or she is ill. Symptoms include delusions, hallucinations, loss of emotion, mania, depression, poverty of thought and seriously abnormal behaviour. Psychoses include schizophrenia, manic depression and organically based mental disorders.

Psychotropic: Exerting an effect upon the mind; capable of modifying mental activity.

Pterygium: A degenerative disorder of the conjunctiva which grows over the cornea medially and laterally. The overgrowths look like wings. They are commonly seen in people who live in areas of bright sunlight, particularly when reflected from deserts or snowfields. Treatment involves excision of the overgrowth.

PTF (Patient Treatment File): PTF is an automated system for recording and tracking events associated with VHA inpatient care. It is the principal source of inpatient VHA workload data.

Puberty: The change that takes place when childhood passes into manhood or womanhood. This change is generally a very definite one, occurring at about the age of 14 years, although it is modified by race, climate, and bodily health so that it may appear a year or two earlier or several years later.

Pubis: Pubis is the bone that forms the front part of the pelvis. The pubic bones of opposite sides

Public Health Impact Assessment: Applying risk assessment to a specific target population of known size, giving as the end product a quantitative statement about the number of people likely to be affected in a particular population.

Public Plan: The Public Prescription Drug Insurance Plan is administered by the Regie de 1'assurance maladie du Quebec and is intended for persons who don't have access to a private group insurance plan covering prescription drugs, for persons age 65 and over, and for employment assistance recipients and other holders of a claim slip (carnet de reclamation). Children of persons registered for the public plan are also covered by that plan.

Publication: Publishing the results for a

clinical trial in a peer-reviewed Journal.

Pubmed: The name of a computer system that enables people to read journals about clinical and basic sciences, nursing, dentistry, and health care planning and administration

Pudendum: The external genital organs. The term is usually used to describe those of the female.

Puerperal Depression: Also called postnatal depression, this is the state of depression that may affect women soon after they have given birth. The condition often occurs suddenly a day or so after the birth. Many women suffer from it and usually they can be managed with sympathetic support. If, however, he depression – sometimes called 'maternal blues' – persists for ten days or more, mild antidepressant drugs are usually effective. If not, psychiatric advice is recom- mended.

Puerperal Sepsis: An infection, once called puerperal fever, that starts in the genital tract within ten days after childbirth, miscarriage or abortion. Once a scourge of childbirth, with many women dying from the infection,

Puerperium: The period which elapses after the birth of a child until the mother is again restored to her ordinary health. It is generally regarded as last- ing for a month. One of the main changes to occur is the enormous decrease in size that takes place in the muscular wall of the womb.

Pulmonary Embolism: The condition in which an embolus or clot, is lodged in the lungs.

Pulmonary Fibrosis: A condition which may develop in both lungs (interstitial pulmonary fibrosis) or part of one lung. Scarring and thickening of lung tissues occur as a consequence of previous lung inflammation, which may have been caused by pneumonia or tuberculosis. Symptoms include cough and breathlessness and diagnosis is confirmed with a chest X-ray. The patient's underlying condition should be treated, but the damage already done to lung tissue is usually irreversible.

Pulmonary Function Tests: Tests to assess how the lungs are functioning. They range from simple spirometry (measuring breathing capacity) to sophisticated physiological assessments.

Pulmonary Hypertension: In this condition, increased resistance to the blood flow through the lungs occurs. This is usually the result of lung disease, and the con- sequence is an increase in pulmonary artery pressure and in the pressure in the right side of the heart and in the veins bringing blood to the heart.

Pulmonary Oedema: Collection of water in the lungs caused by left ventricular failure or mitral stenosis which produces back pressure in the lungs, thus forcing fluid into the tissues.

Pulmonary Stenosis: A disorder of the heart in which obstruction of the outflow of blood from the right ven- tricle occurs. Narrowing of the pulmonary valve at the exit of the right ventricle and narrowing of the pulmonary artery may cause obstruction. The condition is usually congenital, although it may be caused by rheumatic fever. In the congenital condition, pulmonary stenosis may occur with other heart defects and is then known as Fal- lot's tetralogy. Breathlessness and enlargement of the heart and eventual heart failure may be the consequence of pulmonary stenosis. Surgery is usually necessary to remove the obstruction.

Pulpitis: Inflammation, usually infectious, of the pulp of a tooth.

Pulsation: Also known as throbbing. An appearance seen or felt naturally below the fourth and fifth ribs on the left side, where the heart lies, and also at every point where an artery lies close beneath the surface. In other situations, it may be a sign of aneurysm. In thin people, pulsation can often be seen and felt in the upper part of the abdomen, due to the throbbing of the normal abdominal aorta.

Pulse Oximetry: Measurement of oxygen saturation of haemoglobin in a blood sample using a non- invasive device called a spectrophotometer.

Pulsus Paradoxus: A big fall in a person's systolic blood pressure when he or she breathes in. It may occur in conditions such as constrictive pericarditis and pericardial effusion, when the normal pumping action of the heart is hindered. Asthma may also cause pulsus paradoxus, as can chronic obstructive pulmonary disease (COPD).

Pumice Stone: A special foot care tool used to gently file calluses as instructed by your health care team.

Purchase Order: A written authorization issued by a buyer to a seller, indicating the type, quantities and agreed prices for products or services.

Purines: These are waste products or metabolites of nucleoproteins. They are not recyclable and are broken down further to the primary excretable form, uric acid. High purine presence in a tissue signifies a recent high turnover in nucleoproteins from injury or cell death, which is why some purines, such as allantoin, will stimulate cell regeneration. Many plants contain allantoin, most noticeably Comfrey. Some foods are heavy purine producers and can elevate serum uric acid levels. These include organ meats, seafood, legumes, and such politically correct foods as spirulina, chlorella, and bee pollen. Caffeine and theobromine are purine-based alkaloids and can mildly increase uric acid, but they pale beside algae, pollen, and glandular extracts from the chiropractor.

Purity: The absence of extraneous material in a drug product.

Purpura: A skin rash caused by bleeding into the skin from capillary blood vessels. The discrete purple spots of the rash are called purpuric spots or, if very small, petechiae. The disorder may be caused by capillary defects (non- thrombocytopenic purpura) or be due to a deficiency of platelets in the blood (thrombocytopenic purpura). Most worryingly, the rash may be due to a fulminant form of meningo- coccal septicaemia called purpura fulminans.

Pus: Pus is a thick, white, yellow or greenish fluid, found in abscesses, ulcers, and on inflamed and discharging surfaces generally. Its colour and consistency are due to the presence of white blood corpuscles, and superficial cells of granulation tissue or of a mucous membrane which die and are shed off in consequence of the inflammatory process. Bacteria that normally produce pus are streptococcus, pneumococcus and escherichia coli.

Pustular Psoriasis: Form of psoriasis in which blisters of noninfectious pus appear on the skin; attacks may be triggered by medications, infections, stress, or exposure to certain chemicals.

Pustule: A small collection of PUS. Malignant pustule is one of the forms of anthrax.

Putrefaction: The change that takes-place in the bodies of plants and animals after death, whereby they are ultimately reduced to carbonic acid gas, ammonia, and other simple substances. The change is almost entirely due to the action of bacteria, and, in the course of the process, various offensive and poisonous intermediate substances are formed. In the case of the human body, putrescine, cadaverine, and other alkaloids are among these intermediate products.

Putrid Fever: An old name for typhus fever.

PUVA: Treatment that combines oral or topical administration of a medicine called psoralen with exposure to ultraviolet A (UVA) light.

Pyaemia: A form of blood-poisoning in which abscesses appear in various parts of the body.

Pyelitis: An inflammation of the kidney pelvis, the interface between the urine- secreting inner surface of the kidney and the muscular ureter that drains into the bladder. It can be caused by kidney stones or an infection that has progressed up from the lower urinary tract.

Pyelography: The process whereby the kidneys are rendered radio-opaque, and therefore visible on an X- ray film. It constitutes a most important part of the examination of a patient with kidney disease.

Pyelolithotomy: Surgery to remove astone from the kidney via an incision in the pelvis of the kidney.

Pyelonephritis: Inflammation of the kidney, usually the result of bacterial infection. The inflammation may be acute or chronic. Acute pyelonephritis comes on suddenly, is com- moner in women, and tends to occur when they are pregnant. Infection usually spreads up the ureter from the urinary bladder which has become infected (cystitis). Fevers, chills and backache are the usual presenting symptoms. Antibiotics should be given, and in severe cases the intravenous route may be necessary. Septicaemia is an occasional complication. Chronic pyelonephritis may start in child- hood, and the usual cause is back flow of urine from the bladder into one of the ureters – per- haps because of a congenital deformity of the valve where the ureter drains into the bladder. Constant urine reflux results in recurrent infection of the kidney and damage to its tissue. Full investigation of the urinary tract is essential and, if an abnormality is detected, surgery may well be required to remedy it. Hypertension and renal failure may be serious complications of pyelonephritis.

Pylephlebitis: Inflammation of the portal vein. A rare but serious disorder, it usually results from the spread of infection within the abdomen – for example, appendicitis. The patient may develop liver abscesses and ascites. Treatment is by antibiotics and surgery.

Pyloromyotomy: Also called Ramstedt's operation, this is a surgical procedure to divide the muscle around the outlet of the stomach (pylorus). It is done – usually on babies – to relieve the obstruction caused at the outlet by congenital pyloric stenosis.

Pylorospasm: Spasm of the pyloric portion of the stomach. This interferes with the passage of food in a normal, gentle fashion into the intestine, causing the pain that comes on from half an hour to three hours after meals; it is associated with severe disorders of digestion. It is often produced by an ulcer of the stomach or duodenum.

Pylorus: The lower opening of the stomach, through which the softened and partially digested food passes into the small intestine.

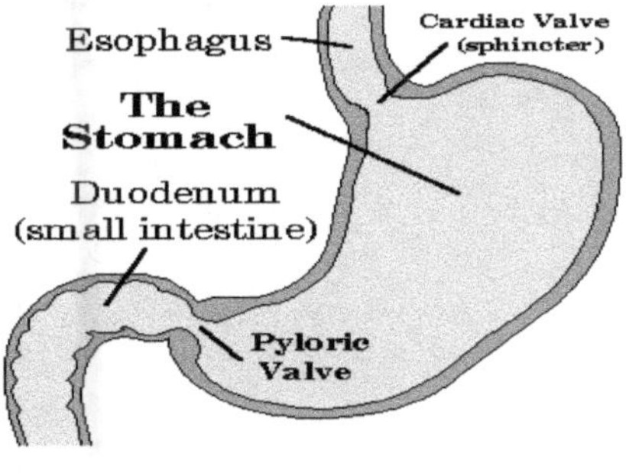

Pyo-: A prefix attached to the name of various diseases to indicate cases in which pus is formed, such as pyonephrosis.

Pyoderma Gangrenosum: This is a disorder in which large ulcerating lesions appear suddenly and dramatically in the skin. It is the result of underlying vasculitis. It is usually the result of inflammatory bowel dis- ease such as ulcerative colitis or crohn's disease but can be associated with rheumatoid arthritis.

Pyogenic Membrane: The membrane that lines and isolates abcesses.

Pyogenic: Pyogenic is a term applied to those bacteria which cause the formation of PUS and so lead to the formation of abscesses. Although many bacteria have this property, the most common cause of abscess is one of the rounded forms of bacterium (e.g. streptococcus).

Pyorrhea: Broadly, any discharge of pus, but usually referring to periodontitis or Pyorrhea alveolaris, with inflammatory and degenerative conditions in the gums, jaw bone and cementum. There may be alveolar bone resorption, teeth loss and receding gums...and hefty dental and oral surgery bills. These costs may be valid, but there is some thought in some radical dental circles that there is overdiagnosis of the condition.

Pyrexia: Condition in which the temperature of a human being or mammal is above normal.

Pyrogen: Any substance that produces fever.

Pyromania: A powerful urge in a person to set things on fire. Affected individuals, more commonly males, are called pyromaniacs. They usually have a history of fascination with fire since childhood and obtain pleasure or relief of tension from causing fires. Treatment is difficult and pyromaniacs commonly end up in the courts.

Pyrrolizidine Alkaloid: A type of alkaloid found in many plants of the Composite and Borage families, once termed a senecio alkaloid. Some of the pyrrolizidine group have been shown to cause several types of liver degeneration and blood vessel disorders. Several deaths have been attributed to improperly identified plant usage of a Senecio, and some of the desert boraginaceae annuals and Senecio annualsare overtly toxic. Young leaves and spring roots of comfrey hybrids should be avoided as well. Not all PAs are toxic, but those that are can produce spontaneous necrosis in the liver hepatocytes of a perfectly healthy person.

Pyuria: The presence of pus in the urine, in con-sequence of inflammation situated in the kidneys, urinary bladder or other part of the urinary tract.

Q Fever: A disease of worldwide distribution due to the organism Coxiella Burneti. It is characterised by fever, severe headache and often pneumonia. The disease was given the name 'q' fever, the q (as in question mark) referring to the unknown cause of the disease when first described. The principal reservoir of human infection in Britain is probably cattle and sheep in which the infection is usually sub-clinical.

QALY: This is an outcome measure of health care devised by health economists in the 1980s, and stands for quality adjusted life year. It takes a year of healthy life expectancy to be worth a grade of 1, and a year of unhealthy life expectancy to be worth less than 1. The worse the forecast of an unhealthy person's quality of life, the lower will be his or her rating. If someone is expected to live five years in a healthy state, the grading will be 5; ten years of life estimated to be only 25 per cent healthy will rate as 2.5.

QRS Complex: The section of an electrocardiogram (ECG) that precedes the s-t segment and registers contraction of the ventricle of the heart.

Q-T Interval: The interval in an electrocardiogram (ECG) that registers the electrical activity generated during ventricular contraction of the heart.

Quack: colloquial description of an unqualified person claiming to be a medical doctor.

Quadrantanopia: Inability to see in one quarter of the visual field. Homonymous quadrantanopia is loss of vision in the same quarter of the field in each eye.

Quadriceps: The large, four-headed muscle occupying the front and sides of the thigh, which straightens the leg at the knee-joint and maintains the body in an upright position.

Quadriplegia: Paralysis of the four limbs of the body.

Qualitative Variable: One that cannot be measured on a continuum and represented in quantitative relation to a scale (race or sex, for example). Data that fit into discrete categories according to their attributes.

Qualitative Variable: A variable that is based on the categorical data and cannot be measured numerically.

Qualitative: Relating to quality

Quality Assurance (QA): All those planned and systematic actions that are established to ensure that the trial is performed and the data are generated, documented (recorded), and reported in compliance with good clinical practice (GCP) and the applicable regulatory requirement(s).

Quality Assurance Report: A written document that captures the observation and findings of quality assurance audit.

Quality Assurance Unit: An independent group of personnel to ensure that the trial is performed and the data are generated, documented and reported in compliance with Good Clinical Practice (GCP) and the applicable regulatory requirements.

Quality Assurance: All those planned and systematic actions that are established to ensure that the trial is performed and the data are generated, documented and reported in compliance with Good Clinical Practice GCP) and the applicable regulatory requirement.

Quality assurance: All those planned and systematic actions necessary to provide adequate confidence that a product or service will satisfy given requirements for quality.

Quality Control (QC): The operational techniques and activities undertaken within the quality assurance system to verify that the requirements for quality of the trial- related activities have been fulfilled.

Quality Of Life Trials (or Supportive Care Trials): Refers to trials that explore ways to improve comfort and quality of life for individuals with a chronic illness, or an acute illness such as cancer.

Quality: Can be defined as a measure of the degree to which delivered health services meet established professional standards and judgments of value to the consumer. Quality may also be seen as the degree to which actions taken or not taken impact the likelihood of beneficial health outcomes and minimize risk and other undesired outcomes, given the existing state of medical science and art.

Quality: The pre-set standard for measuring the outcome.

Quantal Effect: Condition that can be expressed only as "occurring" or "not occurring," such as death or occurrence of a tumour.

Quantitative Digital: Radiography a radiological technique for detecting osteoporesis in which a beam of x-rays is directed at the bone-area under investigation – normally the spine and hip – and the calcium density measured. If the calcium content is low, preventive treatment can be started to reduce the likelihood of franctrues occurring.

Quantitative Structure-Activity Relationship (QSAR): Quantitative association between the physicochemical properties of a substance and/or the properties of its molecular substructures and its biological properties including its toxicity.

Quantitative Variable: One that can be measured (blood pressure, for example) and reported numerically, such as continuous data or counts.

Quantitative: Relating to quantity

Quarantine: The principle of preventing the spread of infectious disease by which people, baggage, mere- candies, and so forth likely to be infected or coming from an infected locality are isolated at frontiers or ports until their harmlessness has been proven to the satisfaction of the authorities. Originally quarantine, as its name implies, involved detention for 40 days; but the period now covers the incubation period of the disease, the presence of which is suspected.

Quartan Fever: description of intermittent fever with paroxysms developing every fourth day. Usually applied to malaria.

QUERI (Quality Enhancement Research Initiative): The Quality Enhancement Research Imitative (QUERI) is mandated to translate research discoveries and innovations into better patient care of our nation's veterans and VA systems improvement.

Query Management: Ongoing process of data review, discrepancy generation, and resolving errors and inconsistencies that arise in the entry and transcription of clinical trial data.

Query: Request from a sponsor or-sponsor's representative to an investigator to resolve an error or inconsistency discovered during data review. Request for clarification on a data item collected for a clinical trial.

Quickening: the first movements of a fetus in the womb as experienced by the mother, usually around the 16th week of pregnancy.

Quiescent: The description applied to a disease in an individual which is in an inactive phase and so likely to be undiagnosed.

Quinghaosu: A herbal drug used for two millennia in china to treat malaria. Its action derives from sesquiterpene lactone, a substance that cuts the number of blood-borne malarial parasites.

Quinidine: An alkaloid obtained from Cinchona bark and closely related in chemical composition and in action to quinine. It is commonly used in the form of quinidine sulphate to treat cardiac irregularities such as supraventricular tachycardia and ventricular arrhythmias.

Quinine: An alkaloid obtained from the bark of various species of cinchona trees. This bark is mainly derived from Peru and neighbouring parts of South America and the East Indies. Other alkaloids and acid substances are also derived from cinchona bark, such as Quinidine and Cinchonine. Action Quinine is a powerful antiseptic. Its best-known action is in checking the recurrence of attacks of malaria, as it destroys malarial parasites in the blood. In fevers it acts as an antipyretic. The most important use of quinine is its original one in malaria.

Quinolones: A group of chemically related synthetic antibiotics. Examples include Nalidixic acid, Cinoxacin and Norfloxocin which are effective in treating uncomplicated urinary-tract, respiratory-tract and gastrointestinal infections. They are usually effective against gram-negative and gram-positive bacteria.

Quinsy: A corruption of 'Cynanche', this is an old name for a peritonsillar abscess.

R

Rabies: An acute and potentially fatal disease, caused by a rhabdovirus called Lyssavirus, which affects the nervous system of animals, particularly car- nivora, and may be communicated from them to humans. Infection from person to person is very rare, but those in attendance on a case should take precautions to avoid being bitten or allowing themselves to be contaminated by the patient's saliva, as this contains the causative virus.

Racem: A flowering spike or cluster where the flowers are borne along the peduncle on pedicels of similar length.

RAD: A unit of ionising radiation absorbed by an individual. The acronym stands for radiation absorbed dose.

Radial Artery: This artery arises from the brachial artery at the level of the neck of the radius. It passes down the forearm to the wrist, where it is easily pal- pated laterally. It then winds around the wrist to the palm of the hand to supply the fingers.

Radiation Sickness: The term applied to the nausea, vomiting and loss of appetite which may follow exposure to radiation – for example, at work – or the use of radiotherapy in the treatment of cancer and other disease.

Radiation: Energy in the form of waves or particles. Radiation is mainly electromagnetic and is broadly classified as ionising and nonionising. Non-ionising radiation agitates the constituent atoms of nuclei but is insufficiently powerful to produce ions. Ionising radiation comprises x-rays, gamma rays and particle radiation. Non-ionising radiation includes ultraviolet light, radio waves, magnetic fields and ultra- sound.

Radioactivity: Breakdown of the nuclei of some elements resulting in the emission of energy in the form of alpha, beta and gamma rays. Because of this particle emission, the elements decay into other elements. Radium and uranium are naturally occurring radioactive elements.

Radiographer: An individual trained in the techniques of taking X-ray pictures of areas of the body is known as a diagnostic radiographer. One who is trained to treat patients with radiotherapy is a therapeutic radiographer.

Radiography: Diagnostic radiography is the technique of examining parts of the body by passing x-rays through them to produce images on fluorescent screens or photographic plates.

Radioimmunoassay: A technique introduced in 1960 which enables the minute quantities of natural substances in the blood such as hormones to be measured.

Radionuclide: Radionuclide is another word for a radioactive isotope. These isotopes are used in a scanning technique of body tissues. Different types of tissue – normal or abnormal tissues – absorb varying amounts of the isotopes; these difference are detected, recorded and displayed on a screen.

Radio-Opaque: Substances which absorb x-rays, rather than transmitting them, appear white on X-ray film and are described as radio-opaque. This is true of bones, teeth, certain types of gall-stones, renal stones and contrast media used to enhance the accuracy of radiographic imaging.

Radiotherapy: The treatment of disease (mainly cancer) with penetrating radiation. for many years radium and x-rays were the only sources available, but developments in knowledge led to the use of powerful x-rays, beta rays or gamma rays, either produced by linear accelerator machines or given off by radioactive isotopes.

Radium: The radiations of radium consist of: (1) alpha rays, which are positively charged helium nuclei; (2) beta rays – negatively charged electrons; (3) gamma rays, similar to x-rays but of shorter wavelength.

Rales: Abnormal sounds in the lungs, either from excess secretions or the narrowing of the bore by inflammation or congestion.

Random Allocation: Assignment of subjects to treatment (or control) groups in an unpredictable way. Assignment sequences are concealed, but available for disclosure in the event a subject has an adverse experience. The purpose of assigning trial subjects to treatment or control groups using an element of chance in order to reduce bias.

Random Assignment: The purpose of assigning trial subjects to treatment or control groups using an element of chance in order to reduce bias.

Random Number Table: Table of numbers with no apparent pattern used in the selection of random samples for clinical trials.

Random Sample: Members of a population selected by a method designed to ensure that each person in the target group has an equal chance of selection. A method of selection designed to ensure that each objects has an equal chance to getting selected.

Random: An element of chance or having no specific pattern.

Randomisation: A method based on chance by which study participants are assigned to a treatment group. Randomization minimizes the differences among groups by equally distributing people with particular characteristics among all the trial arms. The researchers do not know which treatment is better. From what is known at the time, any one of the treatments chosen could be of benefit to the participant.

Randomised Controlled Trial: A method of comparing the results between two or more groups of patients intentionally subjected to different methods of treatment – or sometimes of prevention.

Randomization Code: A list contains the randomization schedule to assign trial subjects to treatment or control groups.

Randomization into a Study: At this time, the subject has not yet met the inclusion/exclusion criteria for the trial.

Randomization List: A list contains the randomization schedule to assign trial subjects to treatment or control groups

Randomization: The process of assigning trial subjects to treatment or control groups using an element of chance to determine the assignments in order to reduce bias. The purpose of assigning trial subjects to treatment or control groups using an element of chance in order to reduce bias.

Randomized Control Trial: A trial that randomly assign participants to an intervention group or to a control group in order to measure the effects of the intervention.

Randomized Trial: Study in which participants are randomly (chance) assigned to one of two or more treatment arms of clinical trial.

Range: The difference between the

smallest and largest values of asset of measurements.

Ranitidine: An H_2-receptor antagonist drug used in the treatment of duodenal ulcer by reducing the hyperacidity of the gastric juice.

Ranula: A swelling which occasionally appears beneath the tongue, caused by a collection of saliva in the distended duct of a salivary gland.

Rape: A criminal offence in which sexual intercourse takes place with an unwilling partner, female or male, under threat of force or violence.

Raphe: A ridge or furrow between the halves of an organ.

Rarefaction: Diminution in the density of a bone as a result of withdrawal of calcium salts from it.

RAS: Remote Access Service, the name of a method provided by UCSF to connect to the Internet

Rat-Bite Fever: An infectious disease following the bite of a rat. There are two causative organisms – Spirillum minus and Actinobacillus muris. The disease is characterised by fever, a characteristic skin rash and often muscular or joint pains. It responds well to penicillin.

Rate Difference (RD): Absolute difference between two rates, for example, the difference in incidence rate between a population group exposed to a causal factor and a population group not exposed to the factor: in comparisons of exposed and unexposed groups, the term "excess rate" may be used as a synonym for rate difference.

Rate Ratio (RR): In epidemiology, the value obtained by dividing the rate in an exposed population by the rate in an unexposed population.

Rate: Measure of the frequency of a phenomenon: an expression of the frequency with which an event occurs in a defined population during a specified time interval.

Rational Use of Drugs: Method of drug use that considers the balance among drug cost, clinicar outcome, and the patient's quality of life

Rationale: The scientific basic or hypothesis of a clinical trial.

Ratticide: Substance intended to kill rats.

Raw Data: Researcher's records of subjects/patients, such as medical charts, hospital records, X-rays, and attending physician's notes. These records may or may not accompany an application to a Regulatory Authority, but must be kept in the researcher's file. All information in original records and certified copies of original records of clinical findings, observation or other activities in a clinical trial necessary for the reconstruction and evaluation of the trial, source data are contained in source documents.

Ray flowers: The margin flowers on a composite head, usually sterile, that resemble single petals. (Example: the white "petals" of a Daisy.)

Raynaud's Disease: Maurice Raynaud (1834–81), the Paris physician who published a thesis on the subject in 1862. This is a condition in which the circulation becomes suddenly obstructed in outlying parts of the body. It is supposed to be due to spasm of the smaller arteries in the affected part, as the result of them responding abnormally to impuilses from the sympathetic nervous system. Its effects are increased both by cold and by various diseases involving the blood vessels.

Raynauds: The first is less severe, characterized by blanching spasms of blood vessels leading to the hands and feet, initiated by cold, moisture, even emotional stress and low blood sugar. Sort of a finger migraine. After the spasm relaxes, the tissue distal becomes red, hot, even painful. R. Disease is more serious and perhaps deriving from different causes as well. The spasms may not subside, the effected tissues can become purplish, and in extreme cases, gangrenous.

RBC: Red blood cells or erythrocytes

RCT: Randomized clinical trial

RDIS: Research and Development Information System

RDP: Reference Drug Program.

Reactive Arthritis: An aseptic (that is, not involving infection) arthritis secondary to an episode of infection elsewhere in the body. it often occurs in associ- ation with enteritis caused by salmonella and certain shigella strains, and in both yersinea and campylo- bacter enteritis.

Read Codes: These form an agreed UK thesaurus of health- care terminology named after the general practitioner who devised them initially in the 1970s. The coding system provides a basis for computerised clinical records that can be shared across professional and administrative boundar- ies.

Readily Biodegradable: Arbitrary classification of substances that have passed certain specified screening tests for ultimate biodegradability; these tests are so stringent that such compounds will be rapidly and completely biodegraded in a wide variety of aerobic environments.

Real-Time Processing: Immediate electronic claim transaction allowing for an electronic pay or deny response within seconds of submitting the claim.

Reasonable Maximum Exposure (RME): Highest exposure that is reasonably expected to occur: typically the 95% upper confidence limit of the toxicant distribution is used: if only a few data points (6-10) are available, the maximum detected concentration is used.

Recalcitrance: Ability of a substance to remain in a particular environment in an unchanged form.

Receptors: Sites on the outside of cells where particular messenger molecules such as hormones can attach. This attachment to the receptor site causes corresponding changes inside the cell.

Recessive: Tending to recede. In genetic terms, a recessive gene is one whose expression remains dormant if paired with an unlike allele. The trait will only be manifest in an individual homozygous for the recessive gene.

Reclamation: Reclamation is the recovery of a material from a waste. Reclamation techniques differ from use and reuse techniques in that the recovered material is typically sold to another company.

Recombinant DNA: DNA or deoxyribonucleic acid containing genes from various sources that have been combined by genetic engineering.

Reconstructive (Plastic) Surgery: Reconstructive surgery on the skin and under-lying tissues that have been damaged or lost as a result of disease or injury. Surgeons graft healthy skin from another part of the body to repair skin dam- aged or destroyed by burns or injuries.

Record Retention: The length of time

for which trial record should be retained as per the stipulated guidelines.

Record: A collection of related data fields.

Recovered Material: Waste material and byproducts that have been recovered or diverted from solid waste; but such term does not include those materials and byproducts generated from, and commonly reused within, an original manufacturing process.

Recovery Factor: Fraction or percentage of the total quantity of a substance extracted under specified conditions.

Recovery Position: If an individual is unconscious – whether as a result of accident or illness or when in the post-operative recovery unit but is breathing and has a pulse, he or she should be placed in the recovery position.

Recovery: Process leading to partial or complete restoration of a cell, tissue, organ or organism following its damage from exposure to a harmful substance or agent. Term used in analytical and preparative chemistry to denote the fraction of the total quantity of a substance recoverable following a chemical procedure.

Recrudescence: The reappearance of a disease after a period without signs or symptoms of its presence.

Recruitment (Investigators): Process used by sponsors to select investigators for a clinical study.

Recruitment (Subjects): Process used by investigators to enroll appropriate subjects (those selected on the basis of the protocol's inclusion and exclusion criteria) into a clinical study.

Recruitment Period: Time period during which subjects are or are planned to be enrolled in a clinical trial.

Recruitment Target: Number of subjects that must be recruited as candidates for enrollment into a study to meet the requirements of the protocol. In multicenter studies, each investigator has a recruitment target.

Recruitments: The act of enrolling subjects with the proper inclusion criteria.

Rectification Analysis: The process to identifying the roots cause for a deviation/violation and the preventive measures taken to avoid it from happening again.

Rectum: The last part of the large intestine.

Recurrence: The reappearance of a disease after a period of remission.

Recurring Requirements: Requirements typically associated with on-going operations to maintain the programs, e.g., manpower, training, waste disposal, periodic sampling, permit fees, plans baseline surveys. These requirements will accrue from year to year.

Recyclability: The ability of a product or material to be recovered from, or otherwise diverted from, the solid waste stream for the purpose of recycling.

Recycle/Reuse: The process of minimizing waste generation by recovering (recycling, reusing) products that might otherwise become waste. Examples include the recycling of aluminum cans, waste paper, glass, and certain plastic containers.

Recycled Product: Items manufactured with recycled materials used in place of a raw or new material. These materials are derived from post-production and post-consumer waste that has served its intended use and been discarded. For example, industrial scrap, material derived from agricultural wastes, and other items, all of which can be used in the manufacture of new products.

Recycling: The series of activities, including collection, separation, and processing, by which products or other materials are recovered from the solid waste stream for use in the form of raw materials in the manufacture of new products other than fuel for producing heat or power by combustion.

Reference Concentration: Term used for an estimate of air exposure concentration to the human population that is likely to be without appreciable risk of deleterious effects during a lifetime.

Reference Distribution: Statistical distribution of reference values.

Reference Dose: Term used for an estimate (with uncertainty spanning perhaps an order of magnitude) of a daily exposure to the human population (including sensitive subgroups) that is likely to be without appreciable risk of deleterious effects during a lifetime.

Reference Drug Program: A pharma care program in which coverage is based on the cost of the reference drug (or drugs) in a therapeutic category.

Reference Drug/Drugs: The drug or drugs considered to be equally as efficacious and the most cost effective in a therapeutic category.

Reference Individual: Person selected with the use of defined criteria for comparative purposes in a clinical study.

Reference Interval: Area between and including two reference limits, for example the percentiles 2.5 and 97.5.

Reference Limit: Boundary value defined so that a stated fraction of the reference values is less than or exceeds that boundary value with a staled probability.

Reference Material: Substance for which one or more properties are sufficiently well established to be used for the calibration of an apparatus, the assessment of a measurement method, or for assigning values to other substances.

Reference Population: Group of all reference individuals used to establish criteria against which a population that is being studied can be compared.

Reference Sample Group: Selected reference individuals, statistically adequate numerically to represent the reference population.

Reference Value: According to IFCC, measured value of a property in a reference individual or sample from a reference individual.

Reference-Based Pricing: An established reimbursement level for a group of interchangeable drugs. Insurers reimburse suppliers up to this amount for any drug within this group. If the cost of a drug exceeds the reference price the consumer pays the difference.

References: A list of relevant published literature on a topic along with complete citation.

Referred Pain: Pain felt in one part of the body which is actually arising from a distant site (e.g. pain from the diaphragm is felt at the shoulder tip). This occurs because both sites develop from similar embryological tissue.

Refill Date (On The Pharmacy Receipt): The pharmacy receipt issued to persons covered by the public plan contains various types of information about their drug purchases during the month, including the refill date. The refill date is the date on which a person would theoretically have his or her prescription refilled, and is determined by the date of the previous drug purchase and the duration of the treatment. For example, the refill date for a 30-day prescription filled on September 1 would be October 1 (the 31st day after September 1).

Refill: The process of getting a valid, approved prescription filled one or more times after the initial prescription without consulting the prescribing physician again.

Reflex Action: One of the simplest forms of activity of the nervous system.

Reflux: Fluid flowing in the opposite direction to normal (i.e. back flow). Often refers to regurgitation of stomach contents into the oesophagus, or of urine from the urinary bladder back into the ureters.

Refraction: The deviation of rays of light on passing from one transparent medium into another of different density. The refractive surfaces of the eye are the anterior surface of the cornea and the lens

Refractory: Disease that does not respond to treatment.

Refractory: Unresponsive or resistant to treatment. Regimen a course of treatment – possibly combining drugs, exercise, diet, etc. – designed to bring about an improvement in health.

Regeneration: The regrowth of cells, tissues, organs or limbs.

Regimen: A strictly regulated course of medication designed to achieve optimal results.

Registered Nurse: A qualified nurse registered with there the applicable regulatory authority to practice nursing.

Registered Pharmacist: A qualified pharmacist registered with the applicable regulatory authority to prepare and dispense drug.

Registration and Premium Billing (R&PB): The branch of the Medical Services Plan (MSP) responsible for registering British Columbia residents for coverage, maintaining client demographic information, and billing clients for MSP premiums.

Registration Committee: A statutory committee of the college that guides council on all registration matters including entry-to-practice.

Registration Trial: As phase 3rd clinical trial intended to obtain marketing authorization for a drug product under investigations.

Regranulation: Granulation is the forming of connective tissue fibroblasts, epithelium and inflammatory cells around the nucleus of new capillaries in tissues that have been burned or scraped. This delicate tissue is often reinjured, and regranulation becomes a slower process, with more formation of scar tissue. Some plant resins will quickly stimulate the process, increase the complexity of healing, and lessen fibroblast scar formation.

Regulation: A government order with the force of law.

Regulatory Affairs: A job function that deals with regulatory authority.

Regulatory Agencies: The drug authorities or food and drug administration's that have authority to regulate.

Regulatory Approval: A permission of regulatory agencies to undertake a clinical trial or market a Drug.

Regulatory Authorities: Bodies having the power to regulate. In the ICH GCP guideline the term includes the authorities that review submitted clinical data and those that conduct inspections. These bodies are sometimes referred to as competent authorities.

Regulatory Compliance: The act of following regulations while under undertaking activities.

Regulatory Dose: Term used by the USEPA to describe the expected dose resulting from human exposure to a substance at the level at which it is regulated in the environment.

Regulatory Inspections: The act of reviewing documents, facilities, records and any other resources by a regulatory authority at a trial site or the sponsor and /or contract research organization's facilities or at other establishment deemed appropriate by regulatory authority.

Regulatory Letter: A letter issued by a Regulatory authority summarizing the observation of Regulatory inspections the actions required thereon.

Regurgitation: Regurgitation is a term used in various connec-tions in medicine. For instance, in diseases of the heart it is used to indicate a condition in which, as the result of valvular disease, the blood does not entirely pass on from the atria of the heart to the ventricles, or from the ventricles into the arteries. The defective valve is said to be incompetent, and a certain amount of blood leaks past it, or regurgitates back, into the cavity from which it has been driven. The term is also applied to the return to the mouth of food already swallowed and present in the gullet or stomach

Regurgitations, Tricuspid: Backflow of blood from the right ventricle (pumping deoxygenated thick venous blood into the lungs) into the right atrium (receiving used blood from the rest of the body) because of faulty closure of the tricuspid valve that guards between the two chambers.

Rehabilitation: The restoration to health and working capacity of a person incapacitated by disease – mental or physical by injury. Treatment designed to facilitate the process of recovery from injury, illness or disease as normal condition as possible.

Reinstated Benefits: When a policy has lapsed due to nonpayment of premiums, benefits may be reinstated at the company's option. It is common for the company to determine proof of insurability before it will do so.

Reinsurance: A mechanism to protect against part or all of the financial losses that may be incurred through insuring for risk. Reinsurance may be used for property and casualty losses as well as for life and health claims. It is a common "stop-loss" mechanism used by self-insured and insured entities throughout the economy, including business and industry, labor organizations, hospitals, HMOs, individual professionals, and even insurance companies.

Reiter's Syndrome: A condition probably caused by an immuno-logical response to a virus in which the patient has urethritis, arthritis and conjunctivitis.

Rejection: A term used in transplant medicine to describe the body's immunological response to foreign tissue. A process in which transplanted tissues or organs are degraded or destroyed

Relapse: The return of a disease that a patient has had and apparently recovered from. It may also be a worsening of a disease from which a person is still recovering.

Relapsing Fever: So-called because of the characteristic temperature chart showing recurring bouts of fever, this is an infectious disease caused by spirochaete.

Related Event: An adverse event that is related to the administration of investigational of drug.

Relative risk: Ratio of the risk of disease or death among the exposed to that among the unexposed. Ratio of the cumulative incidence rate in the exposed to the cumulative incidence rate in the unexposed; the cumulative incidence ratio.

Relative Value Scale (RVS): An approach to physician reimbursement based upon the time, effort, and skill each service or procedure requires from a physician when compared with other medical services or procedures.

Relaxation Therapy: This is a treatment in which patients are helped to reduce their levels of anxiety by reducing their muscle tone. It can be used on its own or in conjunction with a broader psychotherapy regime.

Relevant History: A clinical history that is relevant to a particular condition or occurrence of an adverse event.

Reliability: The degree to which the input-output relationship is reproducible if the relationship is studied repeatedly under comparable conditions.

REM Sleep: Rapid-eye-movement is a stage during sleep in which the eyes are seen to move rapidly beneath the lids and during which dreaming occurs. It occurs for several minutes at a time approximately every 100 minutes.

Remainder (on The Pharmacy Receipt): The pharmacy receipt issued to persons covered by the public plan contains various types of information about their drug purchases during the month, including the remainder. The remainder is the amount that persons covered by the public plan may be required to pay when purchasing insured drugs, until they reach their maximum monthly contribution.

Remedial action: A remedial action is a longer-term, more permanent and expensive solution to a more complex cleanup problem pursuant to cercla.

Remedy: Medicine, as in homeopathic remedy.

Remission: A period when a disease has responded to treatment and there are no signs or symptoms present.

Remission: When there is no evidence of cancer on examination. "Remission" is used instead of "cure" because doctors cannot be sure that the body is completely free of cancer.

Remittent Fever: The term applied to the form of fever in which, during remissions, the tempera- ture falls, but not to normal.

Remote Data Entry: Enter in the electronic case report form using a computer and modem through a distant location.

Removal Action: A removal action is a short-term, limited response to a more manageable cleanup problem pursuant to cercla.

Remuneration: Payment to the study subjects for participation in clinical trial.

Renal Clearance: Renal plasma (or blood) clearance C1R is the volume of plasma (or blood) freed of a substance by only renal mechanisms, per unit time.

Renal: Pertaining to the kidneys

Renewability: Guarantee that coverage cannot be discontinued as long as the insurer continues to do business in that particular market. However, the premium that can be charged is not necessarily protected.

Renewable At The Option of The Insurance Company: This refers to policy contract renewability. The insurance company can choose to cancel the policy on an individual basis.

Renewal: The term used when a prescription has exhausted all of its refills and requires the physician to authorize a new prescription. This is considered renewing a prescription.

Rennin: A milk-coagulating enzyme produced by the lining of the stomach. Rennin converts milk protein (caseinogen) into insoluble casein, thus ensuring that milk stays in the stomach for some time, during which it can be digested by various enzymes before passing into the small intestine.

Repellent: Substance used mainly to repel blood sucking insects in order to protect man and animals: also used to repel mammals, birds, rodents, mites, plant pests, etc.

Repertorize: To repertorize a case one looks up symptoms in a repertory.

Repertory: An index of the homeopathic materia medica by symptom. A list of remedies is indicated for each symptom. All modern day repertories use Kent's Repertory as their starting point.

Replacement: The act of enrolling a clinical trial subject to compensate for the withdrawal of another.

Replicate sampling: Act of taking multiple samples concurrently under comparable conditions; may be accomplished by taking samples adjacent in time or space.

Replication: Duplicated or repeated performance of an experiment under similar (controlled) conditions to reduce to a minimum the error, and to estimate the variations and thus obtain a more precise result: each determination, including the first is called a replicate. Process where by the genetic material is duplicated.

Report Cards: Information on health plan performance presented in a consumer friendly

Reporting Database: The database use to run the statistical reports/output.

Representative: An individual or juridical or other body authorized under applicable law to consent for the subjects participation in a clinical trial on behalf of a prospective subjects.

Reproducibility Conditions: Situation where test results are obtained with the same method on identical test material in different laboratories with different operators using different equipment.

Reproducibility: Closeness of agreement between test results obtained under reproducibility conditions.

Reproductive Studies: A type of toxicological testing (up to 9 month) in 2 species of animals to determine the effects of drug on fertility and reproduction and expose any teratogenic effects.

Reproductive Toxicant: Substance or preparation that produces non-heritable harmful effects on the progeny and/or an impairment of male and female reproductive function or capacity.

Reproductive Toxicology: Study of the adverse effects of substances on the embryo, fetus, neonate and prepubertal mammal and the adult reproductive and neuro-endocrine systems.

Research & Development (R&D): Basic or applied research for the purpose of creating new, or for improving existing materials, devices, products or processes.

Research Assistant: A person who assist the investigator in the conduct of research project.

Research Associate: A person who monitors the research projects in order to ensure compliance with the protocol and applicable regulatory guidelines.

Research Coordinator: A person who assists the investigator in the conduct of research project.

Research Ethics Board: An independent body (a review board or a committee) constituted of medical professionals and non medical members whose responsibility is to ensure the protection n of the rights ,safety and well being of human subjects involved in clinical trial and to provide public assurance of that protection by, among other things, reviewing and approving /providing favorable opinion on, the trial protocol the suitability of the investigator, facilities, trial to be used in obtaining and documenting informed consent of trial subjects.

Research Hypothesis: The research hypothesis is the conclusion a study sets out to support (or disprove); for example, "blood pressure will be lowered by [specific endpoint] in subjects who receive the test product." See also null hypothesis.

Research Misconduct: Willful misconduct (such as falsification of data) in the conduct of research project.

Research Nurse: Qualified nurse who assist the investigator in the conduct of research projects.

Research Projects: Systematic investigation designed to develop new/innovative products, process or services as well as improvisation of the existing product, processes or services new/innovative products, processed or services as well as improvisation of the existing product, processes or services.

Research Protocol: Documents that describes the objectives design, methodology, statistical consideration and organization of a trial it also gives the background rationale for the trial.

Research Record: All information in original records of clinical findings, observations or other activities in a clinical trial necessary for the reconstruction and evaluation of the trial.

Research Subjects: An individual who participates in a clinical trial or a BA/BE study.

Research: In medicine, the collation and assessment of existing facts and knowledge, and the critical systematic investigation of the normal and abnormal functioning of the body, along with the epidemiology of diseases and disorders affecting it – the aim being to increase the sum of knowledge in respect of the prevention, diagnosis and treatment of disease . Investigation and experimentation aimed at discovery, interpretations, and application of scientific data

Researcher: A person who carries out research; investigator.

Resection: The name of an operation in which a part of some organ is removed, as, for example, the resection of a fragment of dead bone or of a diseased section of intestine.

Reserpine: An alkaloid obtained from the root of rauwolfia that has been and continues to be used as an anti-hypertensive and a tranquillising agent.

Reserve Capacity: Physiological or biochemical capacity that may be available to maintain homeostasis when the body or an organism is exposed to an environmental change.

Reserve Volume: The additional amount of air that a person could breathe in or out if he or she were not using the full capacity of their lungs.

Residency Directory: An online publication, updated annually, is compiled by ASHP. It provides descriptions of each ASHP-accredited program, including location, contact person, type of residency, stipend, start date, and special features unique to that program.

Residency Showcase: An exhibit program conducted annually as part of the ASHP Midyear Clinical Meeting. The Showcase provides an opportunity for applicants to meet preceptors and current residents, and to get information on residency programs they may be interested in.

Residency, Accredited: Residency programs that meet the approved criteria and standards. The accreditation process begins with a self-study component and includes the submission of extensive documentation and an on-site visit by a trained team of pharmacy experts.

Residency, Pharmacy Practice: A program that focuses on the development of competence, skills, and application of drug therapy knowledge in providing the broad scope of pharmaceutical services to patients. Pharmacy practice residencies with emphasis in acute care, community care, home care, long-term care, and managed care are available/] These residencies concentrate on patient care, practice management, and other pertinent areas specific to the practice setting.

Residency, Specialized: A program that provides training in a focused area of pharmacy practice. Specialized residencies utilize and build upon the fundamental practice skills acquired during an ASHP-accredited pharmacy practice residency, and are typically undertaken following completion of a pharmacy practice residency.

Residency: A structured, directed, salaried, postgraduate training program in a defined area of pharmacy practice, typically lasting one year.

Resident Matching Program (RMP): Also called "The Match." The RMP is a formalized process conducted by the National Matching Service to match applicants with pharmacy practice residency programs. After the match is conducted, unmatched candidates receive a list of sites with positions still available. Applicants then communicate directly with the residency programs.

Resident: A long term care (permanent) patient in a licensed adult long term care (LTC) facility or a resident of the province of British Columbia.

Residential and Long-Term Care Centre: A subsidized private residential and long-term care centre receives government grants and complies with standards in force in the public sector. A non-subsidized private residential and long-term care centre is independent of the public sector and it does not receive government grants.

Residual Volume: The amount of air left in the lungs after an individual has breathed out as much as possible. It is a measure of lung function.

Resins: These are wax-containing plant oils, often secreted to fill in injured tissues, much like a blood clot, sometimes used to protect leaves from loss of water through evaporation or to render them unpalatable.

Resistance (in Toxicology): Ability to withstand the effect of various factors including potentially toxic substances.

Resolution: A term applied to infective processes, to indicate a natural subsidence of the inflammation.

Resonance: The lengthening and intensification of sound produced by striking the body over an air-containing structure such as the lung. Decrease of resonance is called dullness and increase of resonance is called hyper-resonance.

Resorptive Effect: Action of a substance after its resorption from the gut into the blood.

Respect For Person: An ethical principle having two separate moral requirement the requirement to acknowledge autonomy of an individual and the requirement to protect those individual with diminished autonomy.

Respiratory Arrest: Sudden stoppage of breathing which results from any process that strongly suppresses the function of the brain's respiratory centre. It leads to lack of oxygen in the tissues.

Respite Care: Temporary or periodic care provided in a nursing ho me or ahealthcare center so that the usual caregiver can take rest.

Respite: Short term; as in a respite resident/bed in a long term care facility.

Responder: A subject who demonstrate response to the treatmentunder investigation in a clinical trial.

Response: Proportion of an exposed population with a defined effect or the proportion of a group of individuals that demonstrate a defined effect in a given time at a given dose rate. Reaction of an organism or part of an organism (such as a muscle) to a stimulus.

Restless Legs Syndrome: A condition in which the patient experiences unpleasant sensations, and occasionally involuntary movements, in the legs when at rest, especially at night. No pathological changes have been identified. It is sometimes indicative of iron-deficiency anaemia, but in many cases the cause remains a mystery.

Restraints: A condition of being restrained.

Restricted listing (Close): A doctor must get special, written approval from the drug benefit plan before prescribing the drug. If approved, the drug benefit plan will pay for the drug cost (or part of it).

Restriction Enzyme: An endonuclease enzyme, extracted from bacteria, that is used to cut DNA into short segments – a process essential in genetic engineering.

Result Synopsis: The brief report prepared by biostatisticians summarizing primary (and secondary) efficacy results and key demographic information.

Retail Pharmacy: A chain or independent pharmacy in the retail business that dispenses drugs.

Retail Rate: The provider's charge for providing the same service to a private-pay patient.

Retardation: Slowing down; developmental delay. Psycho-motor retardation is a significant slowing down of speech and activity which eventually leads to a person being unable to cope with daily activities or to maintain personal hygiene. It is a symptom of severe depression.

Retching: Retching is an ineffectual form of vomiting.

Retention: Holding back within the body or within an organ, tissue or cell of matter that is normally eliminated. Holding in memory of what has been learned for later use as recall, recognition or relearning.

Reticulocytes: These are newly formed red blood corpuscles, in which a fine network can be demonstrated by special staining methods.

Reticulo-Endothelial System: This consists of highly specialised cells scattered throughout the body, but found mainly in the spleen, bone marrow, liver, and lymph nodes or glands. Their main function is the ingestion of red blood cells and the conversion of haemoglobin to bilirubin.

Retinoblastoma: A rare malignant growth of the retina which occurs in infants.

Retinoic Acid: A synthetic vitamin A derivative.

Retinoids: Any one of a collection of drugs that are derived from vitamin A.

Retinol: Retinol is the official chemical name of vitamin A.

Retractor: An instrument for pulling apart the edges of an incision to allow better surgical access to the organs and tissues being operated on.

Retrograde: Movement in a contrary or backward direction from normal.

Retrospect Scope: A mock-humorous term used by doctors to imply that one can always see things more.

Retrospective Study: Research design used to test aetiological hypotheses in which inferences about exposure to the putative causal factor(s) are derived from data relating to characteristics of the persons or organisms under study or to events or experiences in their past: the essential feature is that some of the persons under study have the disease or other outcome condition of interest, and their characteristics and past experiences are compared with those of other, unaffected persons. Persons who differ in the severity of the disease may also be compared.

Retrospective: Entry of clinical trial data is retrospective when it is recalled from memory rather than accomplished contemporaneously in real-time. Such retrospective entry is subject to "recall bias", which includes various differences that have been documented in psychological research between contemporaneous self-reported assessments and those that rely on memory.

Retroversion: an abnormal position of the uterus, occurring in about 20 per cent of women, in which its long axis is pivoted backwards in relation to the cervix uteri and vagina instead of forwards.

Retrovirus: A virus containing ribonucleic acid which is able to change its genetic material into deoxyribonucleic acid using an enzyme called reverse transcriptase.

Returned Effect of Poisons: Enhancement of the dose-effect relationship for a poison following repeated exposure to decreasing doses.

Revalidation: The periodic assessment of a doctor's professional competence.

Reversal: The submission of a transaction to cancel (reverse) a previous transaction. The reversal can be for a previous Pharma Net transaction or for a manual transaction.

Reverse Osmoses: Water treatment to remove impurities and aluminium from water before it is used to dilute dialysate concentrates.

Reverse Transcription: Process by which an RNA molecule is used as a template to make a single-stranded DNA copy.

Reversible Alteration: Change from normal structure or function, induced by a substance or other agent(s), that returns to normal status or within normal limits after cessation of exposure.

Review of Research: Ethical, scientific and medical review of a research proposal by an ethics committee.

Revision Summary: A point wise summary of additions and deletions made to an existing documents or process.

Revision: An amendment in the existing documents or processes.

Reye's Syndrome: A condition, now rare, which occurs predominantly in young children following a viral infection of the upper respiratory tract or a viral infection such as chickenpox or influenza.

Rhabdomyolysis: Acute, fulminating, potentially lethal disease of skeletal muscle that causes disintegration of striated muscle fibres as evidenced by myoglobin in the blood and urine.

Rheumatic Fever: An acute febrile illness, usually seen in children, which may include arthralgia, arthritis, chorea, carditis and rash.

Rheumatica: A form of rheumatism characterised by gross early-morning stiffness, which tends to ease off during the day, and pain in the shoulders and sometimes around the hips. It affects women more than men, and is rare under the age of 60. The cause is still obscure. It responds well to prednisolone, but treatment may need to be long continued. On the other hand the condi- tion is not progressive and does not lead to disability.

Rheumatism: Used broadly, rheumatism is a term meant to describe subjective sensations and not a specific disease, such as chronic joint inflammation, osteo or rheumatoid arthritis almost any chronic dull ache associated with the aging process. An obsolete medical term which no longer has a defined meaning. It remains a lay term covering any painful condition of the muscles and/or joints of the arms, legs or spine.

Rheumatoid Arthritis: A chronic disease involving the joint or joints of the body, leading to swelling, stiffness, pain, and possible loss of function.

Rheumatoid: Broadly, having dull aching in joints, muscles, eyes, and so forth. In a more literal sense, it is having an autoimmune response, usually between certain IgM and IgE antibodies, that may have started as a bacterial infection or as some autoimmune reaction. The severity is increased under emotional, physical, dietary, and allergic stress or any stress. Hans Selye showed a few years ago that once a chronic disease response occurs, any stress above metabolic tolerance will aggravate the chronic disease, which is why some people, stressed by cold, wet weather, must avoid it; but someone else is stressed by legumes, still another person gets upset (and stressed) by watching too much CNN.

Rheumatology: The medical speciality concerned with the study and management of diseases of the joints and connective tissue.

Rhinitis: Inflammation of the sinus membranes, sometimes extending to the eyes and ears. It may be caused by a head cold, hay fever, or a chemical irritant.

Rhinophyma: The condition characterised by swelling of the nose due to enormous enlargement of the sebaceous glands which may develop in the later stages of rosacea.

Rhinorrhoea: The persistent discharge of watery mucus from the nose. This is a usual symptom as a result of common cold or consequent upon allergy.

Rhizotomy: The surgical operation of cutting a nerve root, as, for example, to relieve the pain of TRI- geminal neuralgia.

Rhonchi: A description of the harsh cooing, hissing, or whistling sounds (wheezing) heard by auscultation over the bronchial tubes when they are the seat of infection.

Rhythm Method: A method of contraception which attempts to prevent conception by avoiding intercourse during the fertile part of the menstrual cycle.

Ribonucleic Acid: Linear, usually single stranded, polymer of ribonucleotides, each containing the sugar ribose in association with a phosphate group and one of 4 nitrogenous bases: adenine, guanine, cytosine, or uracil : it encodes the information for the sequence of amino-acids in proteins synthesized using it as a template.

Ribozyme: Sections of deoxyribo-nucleic acid (DNA) – The principal molecule in a cell carrying genetic information – that act as enzymes. The function of a ribozyme is to transform the messages encoded in DNA into proteins, using its property of catalysing chemical reactions in a cell.

Rickets: A disease of childhood characterised chiefly by a softened condition of the bones, and by other evidence of poor nutrition. Causes this disease is the result of deficiency of vitamin D in the diet.

Rickettsia: The general term given to a group of micro organisms which are intermediate between bacteria and viruses. They are the causal agents of typhus fever and a number of typhus-like diseases, such as rocky mountain spotted fever, Japanese river fever, and scrub typhus.

Rigidity: Stiffness, resistance to movement. The term is often used in neurology – for example, limb rigidity is a sign of parkinsonism.

Rigor: Shivering. If prolonged, it is generally accompanied by fever, and may be a sign of the onset of some acute disease such as influenza, pneumonia, or some internal inflammation. Rigor mortis is the name given to the stiffness that ensues soon after death.

Ringworm: Ringworm, or tinea, is the name given to inflammatory rashes caused by dermato- phytes of the genera microsporum, epidermophyton and trichophyton.

Rinnes Test: A hearing test in which a vibrating tuning fork is placed on the mastoid process. When the subject can no longer hear the ringing, it is placed beside the ear.

RIP (Rural Incentive Program): A program offered by Pharma care that assists those pharmacies in isolated communities by paying a subsidy per prescription to the pharmacy, subject to certain conditions.

Risk Acceptance: Decision that the risk associated with a given chemical exposure or an event leading to such exposure is low enough to be tolerated in order to gain associated benefits.

Risk Assessment Management Process: Global term for the whole process from hazard identification to risk management.

Risk Assessment: Identification and quantification of the risk resulting from a specific use or occurrence of a chemical or physical agent, taking into account possible harmful effects on individual people or society of using the chemical or physical agent in the amount and manner proposed and all the possible routes of exposure. Quantification ideally requires the establishment of dose effect and dose-response relationships in likely target individuals and populations.

Risk Associated With A Life Time Exposure: Probability of the occurrence of a specified undesirable event following exposure of an individual person from a given population to a specified substance at a defined level for the expected lifetime of the average member of that population.

Risk Aversion: Term used to describe the tendency of an individual person to avoid risk.

Risk Benefit Analyses: An analyses of potential benefits of a clinical trial along with the associated risk.

Risk Benefit Balance: Risk to individual subjects as potential benefits.

Risk Benefit Ratio: Risk to individual subjects as potential benefits.

Risk Characterization: Outcome of hazard identification and risk estimation applied to a specific use of a substance or occurrence of an environmental health hazard. The assessment requires quantitative data on the exposure of organisms or people at risk in the specific situation. The end product is a quantitative statement about the proportion of organisms or people affected in a target population.

Risk Communication: Interpretation and communication of risk assessments in terms that are comprehensible to the general public or to others without specialist knowledge.

Risk Estimation: Assessment, with or without mathematical modelling, of the probability and nature of effects of exposure to a substance based on quantification of dose- effect and dose-response relationships for that substance and the population(s) and environmental components likely to be exposed and on assessment of the levels of potential exposure of people, organisms and environment at risk.

Risk Evaluation: Establishment of a qualitative or quantitative relationship between risks and benefits, involving the complex process of determining the significance of the identified hazards and estimated risks to those organisms or people concerned with or affected by them.

Risk Factors: Traits that make it more likely that a person will get an illness. For example, a risk factor for getting type 2 diabetes is having a family history of diabetes.

Risk Identification: Recognition of a potential hazard and definition of the factors required to assess the probability of exposure of organisms or people to that hazard and of harm resulting from such exposure.

Risk Management: Decision-making process involving considerations of political, social, economic, and engineering factors with relevant risk assessments relating to a potential hazard so as to develop, analyse, and compare regulatory options and to select the optimal regulatory response for safety from that hazard. Essentially risk management is the combination of three steps : risk evaluation; emission and exposure control; risk monitoring.

Risk Marker: Attribute that is associated with an increased probability of occurrence of a disease or other specified outcome and that can be used as an indicator of this increased risk : not necessarily a causal or pathogenic factor.

Risk Monitoring: Process of following up the decisions and actions within risk management in order to check whether the aims of reduced exposure and risk are achieved.

Risk Perception: Subjective perception of the gravity or importance of the risk based on a person's knowledge of different risks and the moral, economic, and political judgment of their implications.

Risk Phrases: Word groups identifying potential health or environmental hazards required under CPL Directives (European Community); may be incorporated into Safety Data Sheets.

Risk Ratio: Value obtained by dividing the probability of occurrence of a specific effect in one group by the probability of occurrence of the same effect in another group, or the value obtained by dividing the probability of occurrence of one potentially hazardous event by the probability of occurrence of another. Calculation of such ratios is used in choosing between options in risk management.

Risk: In clinical trials, the probability of harm or discomfort for subjects. Acceptable risk differs depending on the condition for which a product is being tested. A product for sore throat, for example, will be expected to have a low incidence of side effects. However, unpleasant side effects may be an acceptable risk when testing a promising treatment for a life-threatening illness.

Risk-Benefit Ratio: The risk to individual participants versus the potential benefits. The risk/benefit ratio may differ depending on the condition being treated.

Risk-Specific Dose: Amount of exposure corresponding to a specified level of risk.

Risus Sardonicus: The term used for describing the facial appearance when the muscles of the forehead and the face go into spasm in tetanus.

RNA: Ribonucleic acid, which carries instructions from DNA in the nucleus to cell polyribosomes, where proteins are, made according to the RNA instructions.

Rocky Mountain Spotted Fever: A fever of the typhus group. It received its name from the fact that it was first reported in the Rocky Mountain States of the United States; these are still the most heavily infected areas, but the fever is now found in all parts of the US. The causative organism is Rickettsia rickettsi, which is transmitted to humans by tics.

Rodenticide: Substance intended to kill rodents.

Rombergism: A term applied to marked unsteadiness when a person stands with the eyes shut. It is found as a symptom in some nervous diseases, such as per- ipheral NEUROPATHY and tabes dorsalis (neurosyphilis).

Root Cause: The causative factor for a deviation or event.

Root Filling: Also called root-canal therapy, this is the treatment given when the nerve of a tooth has been exposed while the tooth is being prepared for a filling, or if it has died or become infected.

Rorschach Test: A psychological test for investigating personality and disorders of personality.

Rosacea: Common chronic inflammation of the facial skin, this condition is seen in middle and late life. Redness, obvious dilatation of venules and crops of acne-like papules and pustules affect mainly the central forehead, cheeks, nose and chin.

Roughage: Dietary fibre is that part of food which cannot be digested in the gastrointestinal tract, although it can be metabolised in the colon by the micro-organisms there. Roughage falls into four groups: cellulose, hemicelluloses, lignins and pectins, found in unrefined foods such as wholemeal cereals and flour, root vegetables, nuts and fruit.

Rouleau: A group of red blood cells arranged together like a roll of coins, usually only noticed on a slide under a microscope. Since red blood cells in a reasonably healthy person should have a mutually repelling membrane charge, this means that something like an inflammatory response or an elevation of liver-synthesized lipids (LDLs and VLDLs) is occurring. Inflammation makes the blood "sticky," and the lipids from the liver lower the charges. Remember, of course, that I am talking about subclinical imbalances, such things as rouleau can accompany some pretty gnarly diseases. Our kind of rouleau can give you a headache or make your hands and feet cold because it's hard to push rolls of coins through little bitty capillaries.

Rouleaux: The term applied to the heaps into which red blood corpuscles collect as seen under the microscope.

Rous Sarcoma: A malignant tumour of fowls which is caused by a virus. This tumour has been the subject of much experimental work on the nature of cancer.

Route of Exposure: Means by which a toxic agent gains access to an organism by administration through the gastrointestinal tract (ingestion), lungs (inhalation), skin (topical), or by other routes such as intravenous, subcutaneous, intramuscular or intraperitoneal routes.

Routes of Administration: The method by which a drug is administered into the body, such as by mouth (orally) or through an intravenous (IV) tube.

Routine Monitoring: The act of overseeing the progress of a clinical trial and of ensuring that it is conducted that it is ,recorded and reported in accordance with the protocol, standard operating procedure, good clinical practice and the applicable regulatory requirements.

Rubric: A symptom as written in a homeopathic repertory.

Rupture: A popular name for hernia.

Rural Pharmacy: A pharmacy is considered rural (or remote) if it is the sole pharmacy in a community and the next nearest pharmacy is at least 25 km away.

S

Saccharine: a sweetening agent that is 400 times as sweet as cane sugar, but with no energy content. Apart from its rather bitter aftertaste

Sacral Nerves: These are five pairs of CNS nerves that exit through the sacral foramen and sacral hiatus, and bring information in and out of the spinal cord. Much of their function relates to the sciatic nerve, and they bring information in from the skin sensory zones (dermatomes) of the heel, back of the legs, buttocks, and the pelvic floor. Sacs formed of muscular and fibrous tissue and lined by a mucous membrane, which is united loosely to the muscular coat so as freely to allow increase and decrease in the contained cavity. Bladders are designed to contain some secretion or excretion, and communicate with the exterior by a narrow opening through which their contents can be discharged. In humans there are two: the gall-bladder and the urinary bladder.

Sadism: the term applied to a form of sexual perversion, in which satisfaction is derived from the infliction of cruelty upon another person.

Safe Period: Period during the menstrual cycle when fertilization of the ovum is unlikely to occur. Ovulation usually occurs about 15 days before the onset of the menstrual period. A woman is commonly believed to be fertile for about 11 days in each menstrual cycle – in other words, on the day of ovulation and for five days before and five days after this; this would be the eighth to the 18th day of the usual 28-day menstrual cycle. Out- side this fertile period is the safe period:

Safety and Tolerability: The safety of a medical product concerns the medical risk to the subject, usually assessed in a clinical trial by laboratory tests (including clinical chemistry and hematology), vital signs, clinical adverse events (diseases, signs and symptoms), and other special safety tests (e.g., ECGs, ophthalmology). The tolerability of the medical product represents the degree to which overt adverse effects can be tolerated by the subject.

Safety Assessment: The assessment of adverse event and serious adverse event experienced by the participants in a clinical trial.

Safety Data Sheet: Single page giving toxicological and other safety advice, usually associated with a particular preparation, substance or process.

Safety of A Drug or other Substance in The Context of Human Health: The extent to which a substance may be used in the amount necessary for the intended purpose with a minimum risk of adverse health effects.

Safety Reports: Reports prepared from

a serious and unexpected adverse experience.

Safety: Relative freedom from harm. In clinical trials, this refers to an absence of harmful side effects resulting from use of the product and may be assessed by laboratory testing of biological samples, special tests and procedures, psychiatric evaluation, and/or physical examination of subjects. The high probability that injury will not result from exposure to a substance under defined conditions of quantity and manner of use, ideally controlled to minimize exposure.

Salicylates: Esters or salts of salicylic acid, such as aspirin, and including glycoside forms such as salicin. They are widely used as topical irritants and (especially) as anti-inflammatory and analgesic agents and prostaglandin inhibitors.

Saline: Normal saline is a solution containing 0·9 per cent of sodium chloride (common salt).

Saliva: The fluid secreted by the salivary glands into the mouth. The ingestion of food stimulates saliva production. saliva contains mucus and an enzyme known as ptyalin, which changes starch into dextrose and maltose

Salk Vaccine: a vaccine obtained by treating the poliomyelitis virus with formalin. This prevents the virus from causing the disease but allows it to stimulate the production of antibodies

Salmonella: A widespread genus of gram-negative motile-rod bacteria, some of them can cause moderate GI infections, while several can produce metabolites in food that cause serious toxic reaction when the food is eaten

Salpingitis: inflammation situated in the fallopian tubes; these run from the ovaries to the uterus and carry the ova or eggs. The disorder is commonly caused by infection spreading upwards from the vagina, cervix or uterus. It is one feature of pelvic inflammatory disease

Sample Size Adjustment: An interim check conducted on blinded data to validate the sample size calculations or re-evaluate the sample size.

Sample Size: The number of subjects in a clinical trial; number of subjects required for primary analysis. Subset of a larger population, selected for investigation to draw conclusions or make estimates about the larger population.

Sample: In statistics, a group of individuals often taken at random from a population for research purposes. One or more items taken from a population or a process and intended toprovide information on the population or process. Portion of material selected from a larger quantity in some manner chosen so that the portion is representative of the whole.

Sampling Error: Part of the total estimation error of a parameter (or alue of a property, such as concentration) caused by the random nature of the sample.

Sampling Plan: A plan that highlights the specific description of the data to be collected.

Sampling: Procedure used to obtain or constitute a sample.

Sanguineous: this term means containing blood.

SAP: Special Access Programme.

Saponin: Any plant glycoside with soapy action that can be digested to yield a sugar and a sapogenin aglycone. Many (but not all) saponins can be toxic and speed up hemoglobin degradation.

Sarcode: A tissue or glandular extract made into a homeopathic remedy.

Sarcoidosis: an uncommon chronic inflammatory disease of unknown origin which can affect many organs, particularly the skin, eyes and lungs

Sarcoma: Malignant tumour arising in a connective tissue and composed primarily of anaplastic cells resembling supportive tissue.

Sars: in 2003, an outbreak occurred of a previously unrecognized illness – termed sirs, or severe acute respiratory syndrome. it was caused by infection with a newly identified coronavirus, sars-cov. infection produced an illness with pneumonia as a prominent feature, but some patients developed other events such as loss of appetite, diarrhoea and bleeding from the stomach

Satellite Unit: A dialysis unit with a restricted number of staff who do not have a doctor on the premises.

Saturnism: Intoxication caused by lead.

Scabies: a common contagious itching disease caused by sarcoptes scabei hominis, which can live only on human skin.

Scanning Speech: a speech disorder in which articulated syllables are wrongly spaced and each is given the same vocal emphasis. The condition occurs as a result of disease in the cerebellum or its connecting nerves

SCAPE: A long flower-bearing stem or peduncle that arises from the ground. It is leafless, or the leaves are reduced to bracts.

Scar: the name applied to a healed wound, ulcer or breach of tissue

Scatter Plots: These displays relation between two numeric variable while distinguishing between levels of categorical variables.

Scheduled Continuation Review: The ongoing periodic review of trial progress by the ethics committee.

Schizogony: an asexual phase in the life-cycle of a sporozoan (sporozoa)

that occurs in red blood cells or liver cells.

Schizophrenia: an overall title for a group of psychiatric disorders typed by disturbances in thinking, behavior and emotional response.

Schwann cell: the cells that produce the myelin sheath of the axon of a modulated nerve.

Sciatica: This is neuralgia of the sciatic nerve. These are the two largest nerves in the body, composed of the tibial and common perineal nerves, bound together and containing elements of the lowest two lumbar and upper three sacral spinal cord nerves. Sciatica is felt as severe pain from the buttocks, down the back of the thighs, often radiating to the inside of the leg, even to the point of parasthesia or prickly numbness. Although tumors can cause the problem, far and away the most common causes are a lower back subluxation (responding to adjustment) or pelvic congestion and edema (responding to laxatives, exercise, and decreasing portal vein and lymphatic congestion).

Scientific Miss Conduct: Willful non compliance such as lack of supervision or falsification of data.

Scirrus: a hard form of cancer containing fibrous tissue

Sclerosis: Hardening of an organ or tissue, especially that due to excessive growth of fibrous tissue.

Scope of Practice: The custody, compounding and dispensing of drugs, the provision of nonprescription drugs, health care aids and devices, and the provision of information related to drug use.

Scotia: an area of blindness in the field of vision.

Scotoma: Area of depressed vision within the visual field, surrounded by an area of less depressed or normal vision.

Screen/Screening: Screening is the process by which substances are evaluated in a battery of tests or assays (screens) designed to detect a specific biological property or activity. It can be conducted on a random basis in which substances are tested without any pre selection criteria or on a targeted basis in which information on a substance with known activity and structure is used as a basis for selecting other similar substances on which to run the battery of tests.

Screening Level: Decision limit or cut-off point at which a screening test is regarded as positive.

Screening Log: A log that details of all the subjects of all the subjects screened for a clinical trial.

Screening Trials: Refers to trials that test the best way to detect certain diseases or health conditions which do not have signs or symptoms of the disease or condition being studied.

Screening: Carrying out of a test or tests, examination(s) or procedure(s) in order to expose undetected abnormalities, unrecognized (incipient) diseases, or defects : examples are mass X-rays and cervical smears.

Script: A program or a sequence of instructions that are interpreted or carried out by another program. A sequence of instruction that are interpreted or carried out by another programs.

Scrofula: this is a traditional term for tuberculosis of the lymph glands in the neck.

Search Engine: An online service that compares your search criteria with its database of information about the Internet and displays the results.

Seborrhea: A disorder of the sebaceous glands, with changes in the amount and quality of the oils secreted. Although it can occur in any part of the body, seborrhea of the scalp (dandruff) is most common.

Secondary Metabolite: Product of biochemical processes other than the normal metabolic pathways, mostly produced in micro-organisms or plants after the phase of active growth and under conditions of nutrient deficiency.

Secondary Sexual Characteristics: the physical characteristics that develop during puberty as the body matures sexually. Girls' breasts and genitals increase in size, and, like boys, they grow pubic hair. Boys also grow facial hair, their voice breaks and their genitals grow to adult size.

Secondary Variable: The primary outcome is the outcome of greatest importance. Data on secondary outcomes are used to evaluate additional effects of the intervention.

Secretion: Process by which a substance such as a hormone or enzyme produced in a cell is passed through a plasma membrane to the outside, for example the intestinal lumen or the blood (internal secretion) Solid, liquid or gaseous material passed from the inside of a cell through a plasma membrane to the outside as a result of cell activity.

Sedative: Substance that exerts a soothing or tranquillising effect.

Selenium Supplied: this is used as a shampoo in the treatment of dandruff and seborrhea dermatitis of the scalp.

Self: Cells or organs recognized as not foreign by the immune system

Self-Cleaning of Water: Water purification by natural biological and physico-chemical processes.

Self-Evident Change: A data discrepancy that can be easily and obviously resolved on the basis of existing information on the CRF, e.g., obvious spelling errors or the patient is male and a date of last pregnancy is provided.

Self-Monitoring Blood Glucose: A way for people with diabetes to find out how much glucose is in their blood. A drop of blood from the fingertip is placed on a special coated strip of paper that "reads" (often through an electronic meter) the amount of glucose in the blood.

Self-purification of the atmosphere: Purification of the atmosphere from contaminants by natural biological and physico-chemical processes.

Semen: fluid produced by the male on ejaculation from the penis at sexual orgasm. each ejaculate contrains up to 500 million spermatozoa (male germ cells) suspended in a fluid that is secreted by the prostate gland, seminal vesicles, and Cowper's glands – a pair of small glands (also called the bulb-urethral glands) that open into the urethra at the base of the penis.

Seminal Vesicles: These are a couple of spongy glands, l.5 to 2 inches long, that secrete high-sugar, acidic, and thick, ropy colloid into the ductus deferens (containing sperm from the testes) during ejaculation. The two fluids empty into the prostate, where they are mixed with alkaline prostatic fluids to form semen.

Senility: The aging related loss of mental faculties.

Seniorcare: SeniorCare is a prescription drug assistance program for Wisconsin residents who are 65 years of age or older who meet the eligibility criteria. The program is designed to help seniors with their prescription drug costs.

Sensation: The experience of a symptom, what it feels like. One of the parts of a complete symptom.

Senscequence: The reduction and decline of a gland's output over age.

Sensitivity: The lowest value of input that can be inferred with a given degree of validity and reliability from measurements of output. Analogous to the usage for the word "threshold" is the phrase "threshold dose". The engineer uses the word "threshold", however, to mean the smallest change in input that will result in change in output.

Sensitization: Immune process whereby individuals become hypersensitive to substances, pollen, dandruff, or other agents that make them develop a potentially harmful allergy when they are subsequently exposed to the sensitizing material (allergen).

Sensors: cells or tissues that monitor the internal and external environment, either neurologically or chemically, and can initiate compensatory action or communicate to other parts that can react.

Sensory Effect Level: Intensity, where the detection threshold level is defined as the lower limit of the perceived intensity rang.

Sepal: A leaf or segment of the calyx.

Sepsis: Like septicemia, an infection that has moved deeply into the body, involving the subcutaneous or submucosal layers, connective tissue, lymph system...or blood

Septic Shock: a dangerous disorder characterized by a severe fall in blood pressure and damage to the body tissues as a result of septicemia.

Septicemia: The presence of pathogenic bacteria or other microbes in the blood stream...a serious business, since most defenses are focused outside the bloodstream and the infection has bypassed them either due to its virulence, the depth and severity of the original focal infection or the weakened state of the body's immunity and life energy.

Septum: A membrane wall separating two or more cavities, such as the one between the nasal fossae and those separating the air sacs (alveoli) of the lungs.

Sequential Analysis: A type of statistical analysis where the sample size is not fix in advance . in sequential analysis data is evaluated as it is collected and further sampling is stopped in accordance to pre determined stopping rules as soon significant results are observed.

Sequential Design: A trial design that looks at the data at any dose results in death is life threading, requires inpatient hospitalization or prolongation of existing hospitalization result in persistent or significance disability/incapacity results in a anomaly/ birth defect.

Sequestrum: a fragment of dead bone cast off from the living bone in the process of necrosis

Serendipity: Discovery by chance.

Serious Adverse Drug Reaction: Any untoward medical occurrence that any dose result in death is life threaten requires impatient hospitalization or prolongation of existing hospitalization result in persistent or significant disability result in congenital anomaly/birth defect.

Serious Adverse Drug Reaction: Any untoward medical occurrence that any dose result in death is life threating requires impatient hospitalization or prolongation of existing hospitalization result in persistent or significant disability result in congenital anomaly/birth defect.

Serious Adverse Event (SAE) or Serious Adverse Drug Reaction (Serious ADR): Any untoward medical occurrence that at any dose: results in death, is life threatening, requires inpatient hospitalization or prolongation of existing hospitalization, results in persistent or significant disability/incapacity, or is a congenital anomaly/birth defect.

Serious Adverse Event (SAE): Any adverse event (AE) that is fatal, life-threatening, permanently disabling, or which results in hospitalization, initial or prolonged. Any untoward medical occurrence that any dose result in death is life threating requires impatient hospitalization or prolongation of existing hospitalization result in persistent or significant disability result in congenital anomaly/birth defect. Any experience that suggests a significant hazard, contra- indication, side effect or precaution.

Seriousness Criteria: Criteria that classifying adverse event as serious adverse event such as result in death is life threatening requires inpatient hospitalization or prolongation of existing hospitalization, results in persistent or significant disability result in congenital anomaly/birth effect.

Serotonin: An inhibitory neurotransmitter required for sleep.

Serous Membranes: Membranes that line many internal organs and cavities, secreting a thin, lymph-like fluid, that lubricates and slowly circulates.

Serum Sickness: a hypersensitivity reaction due to circulating antigen-antibody complexes. It was a not uncommon reaction to the administration of foreign serum which used to be given as a form of passive immunity before the days of anti-biotic.

Serum Sickness: Local and systemic immune response to foreign proteins

Serum: Watery proteinaceous portion of the blood that remains after clotting. Clear watery fluid especially that moistening the surface of serous membranes or that exuded through inflammation of any of these membranes.

Server: A computer that acts as a gateway for providing one or more services over a compute event :a severity of an adverse event that cause significant discomfort to a subject and requires treatment on an urgent basis.

Service Area: The geographical area covered by a network of health care providers.

Service Level Agreement: The agreement between HealthNet/ BC and each Pharmacy Software Vendor specifying roles, responsibilities, service standards, and escalation procedures.

Severe Employment Constraints: In addition to the basic benefits, employment assistance recipients whose physical or mental state has been proven to be significantly affected, either permanently or indefinitely, receive an allowance for severe employment constraints.

Severity: The intensity of an adverse event.

Sexual Deviation: any type of pleasurable sexual practice which society regards as abnormal

Sexual Dysfunction: inadequate sexual response may be due to a lack of sexual desire (libido) or to an inadequate performance; or it may be that there is a lack of satisfaction or orgasm. Lack of sexual desire may be due to any generalised illness or endocrine disorder, or to the taking of drugs that antagonise endocrine function

SGOT and SGPT: Liver enzymes that are normally only present in minute quantities in the blood, they become elevated under a variety of circumstances, particularly hepatitis.

Shigellosis: An acute, self-limiting intestinal infection, with diarrhea, fever, and abdominal pain, caused by one of the Shigella genus of gram-negative bacteria. The infection is contracted through food prepared by infected individuals or by direct contact with them. Raw sewage contamination can also be a source.

Shingles: Also called Herpes zoster. It is caused by the chickenpox virus, and usually occurs in middle-age, beginning as inflammation, sharp pain and finally vesicles, erupting at the edges of posterior ganglia of the trunk or face. Usually lasting two or three weeks, it is often triggered by stress or a concurrent viral infection, and can return again in some individuals.

Shipment/Inventory Forms: The documents that are sent along with a shipment and contains a complete description of the product being - shipped.

Short Term Exposure Limit (STEL): As used by US NIOSH, unless noted otherwise, the 15 minute time weighted average exposure that should not be exceeded at any time during a work day.

Shunt: A surgically created artificial external connection of a vein and an artery. The connection is usually made of non- irritant silastic tubes which are joined together when not in use.

Side Effects: Any unanticipated actions or effects of a drug or treatment not related to the primary purpose of the drug. Negative or adverse effects may include headache, nausea, hair loss, skin irritation, or other physical problems. Experimental drugs must be evaluated for both immediate and long-term side effects. Also known as “adverse drug reactions” (ADRs) or “adverse events” (AEs). In clinical trials, side effects would also include any injuries by overdosing, abuse/dependence, and unintended interactions with other medicinal products.

Siderosis: Pneumoconiosis resulting from the inhalation of iron dust.

Sidestream Smoke: Cloud of small particles and gases that is given off from the end of a burning tobacco product (cigarette, pipe, cigar) between puffs and is not directly inhaled by the smoker; the smoke that gives rise to passive inhalation on the -

part of bystanders.

Sign: Objective evidence of a disease, deformity or an effect induced by an agent, perceptible to an examining physician.

Significant or Statistically Significant: A statistical term indicating that the results of a study are stronger than would be expected from chance alone.

Signs: The physical abnormalities that are linked to an adverse event.

Silicosis: Pneumoconiosis resulting from inhalation of silica dust.

Simillimum: The most similar remedy corresponding to a case. As such, the remedy most likely to cure.

Simulation Test: Procedure designed to predict the rate of biodegradation of a compound under relevant environmental conditions.

Single Blind Study: A study in which th subjects is unaware of medication he/she is taking.

Single Center Study: A clinical trial which is conducted at one trial site only.

Single Masked: A study in which th subjects is unaware of medication he/she is taking.

Single-Blind Study: A study in which one party, either the investigator or the subject, does not know which medication or placebo is administered to the subject; also called single-masked study.

Sink: In environmental chemistry, an area or part of the environment in which, or a process by which, one or more pollutants is removed from the medium in which it is dispersed; for example - moist ground acts as a sink for sulfur dioxide in the air.

Sinusitis: Inflammation of the sinuses, with causes ranging from dust to hay fever. Obstinate cases can be caused by chronic sinus infections or the continued exposure to allergens from food, pets or environmental irritants.

Sister Chromatid Exchange (SCE): Reciprocal exchange of chromatin between two replicated chromosomes that remain attached to each other until anaphase of mitosis; used as a measure of mutagenicity of substances that produce this effect.

Site Assessment: A system review of investigator qualification facility infrastructure, patient load etc. by a sponsor's representative in order to select the sites for conducting a clinical trial.

Site Assessment Visit Report: A written report prepared by sponsor's representative that summarizes the observation and findings of site assessment visit.

Site Audit: A systematic and independent examination of trial related activities and documents at a trial site to determine whether the evaluate trial related activities were conducted and the data were recorded, analyzed and accurately reported according to the protocol, sponsor standard operating procedures and good clinical practice and the applicable regulatory requirements.

Site Close Out: The activity of the trail site at the and of the study or at study termination after all the requirements have been fulfilled.

Site Closure: The activity of the trail site at the and of the study or at study termination after all the requirements have been fulfilled.

Site Code: The unique number assigned individual trial site for its identification.

Site Contact List: A list that contact, name detail of all the participating sites in a trial.

Site Coordinator: A person employed at clinical investigator site to record the clinical trial data in compliance with protocol, GCP and applicable regulatory guideline .

Site Delegation Log: A document that in list specific trial related duties perform by individual study team members along with their signatures or initial .

Site Evaluation Report: The written report prepared by sponsors representative that summarizes the observation and finding of site visit.

Site Identification List: A list that contain name and contact of the participating sites in a trial.

Site Initiation: The activation of the site for imitating a clinical trial after the ethics committee and regulatory approval has been obtained and other trial specific requirements have been fulfilled.

Site Management Organization (SMO): Organization responsible for managing the investigator site in a clinical trial.

Site Management: The act of coordinating and managing a trail at a site.

Site Maser File: The file that contain all the essential trial document at a trial site.

Site Qualification Visit Report: Same as site evaluation report.

Site Selection Report: Same as site evaluation report.

Site Training: Imparting training to site personnel on study protocol, GCP and applicable regulatory processes.

Site Visit: The log that captures the details of visit made by sponsor/CRO's representative at a trail site.

Site: A facility where clinical trials are conducted.

Skeletal Fluorosis: Osteosclerosis due to fluoride.

Skilled Care: A type of a health care given when skilled nursing or rehabilitation staff is required to manage, observe, and evaluate the care.

Skilled Nursing Care: A level of healthcare that can only be performed safely and correctly by a licensed nurse.

Skilled Nursing Facility (SNF): A licensed facility that provides nursing care and related services for patients who do not require hospitalization in an acute care setting.

SLE: Systematic Lupus Erythematosus. A generalised connective tissue disorder.

Slimicide: Substance intended to kill slime-producing organisms (used on paper stock, water cooling systems, paving stones etc.).

Slope Facto: Value, in inverse concentration or dose units, derived from the slope of a dose-response curve; in practice, limited to carcinogenic effects with the curve assumed to be linear at low concentrations or doses

SNDA: Supplemental New Drug Application

SNP: Single nucleotide polymorphism

Social Worker: A person that provides social services.

Societal Risk: Total probability of harm to a human population including also the probability of adverse health effects to descendants and the probability of disruption resulting from loss of services such as industrial plant or loss of material goods and electricity.

Software: Computer programs, procedures, rules, and any associated documentation pertaining to the operation of a system.

Solid Waste: Solid waste is any garbage, refuse, sludge from a waste treatment plant, water supply treatment plant, or air pollution control facility and other discarded material, including solid, liquid, semisolid, or contained gaseous materials resulting from industrial, commercial, mining, and agricultural activities and from community activities.

Solvent Abuse: Deliberate inhalation (or drinking) of volatile solvents, in order to become intoxicated.

Solvent Recovery System: Equipment associated with capture, transportation, collection, concentration, and purification of organic solvents. It may include enclosures, hoods, ducting, piping, scrubbers, condensers, carbon absorbers, distillation equipment, and associated storage vessels.

Solvent: A liquid substance that is used

to dissolve or dilute another substance.

Soporific: Substance producing sleep.

Sorption: Noncommittal term used

instead of adsorption or absorption when it is difficult to discriminate experimentally between these two processes.

Source Data/Document Verification: The verification of source documents and other trial records for accuracy, compliance with protocol, GCP and applicable regulatory guidelines.

Source Data: All information in original records (and certified copies of original records) of clinical findings, observations or other activities in a clinical trial necessary for the reconstruction and evaluation of the trial. Source data are contained in source documents.

Source Documents: Original documents ,data and records (e.g. hospital records, clinical and office charts laboratory notes, memoranda, subjects diaries or evaluation checklist, pharmacy dispensing records, recorded data from automated instrument, copies or transcription certified after verification as being accurate copies, microfiches photographic negatives ,microfilms or magnetic media, x-rays, subjects files ,and record kept at the pharmacy, at laboratories and at medico :technical departments involved in the clinical trial.)

Source Reduction: The practice that reduces the amount of any hazardous substance, pollutant, or contaminant entering any waste stream or otherwise released into the environment (including fugitive emissions) prior to recycling, treatment, or disposal; and reduces the hazard to public health and the environment associated with the release of such substances, pollutants, or contaminants

Source: Any building, structure, facility, or installation from which there is or may be a discharge of pollutants.

Special Access Programme (SAP): This programme, administered by the Therapeutic Products Directorate of Health Canada, is responsible for authorizing the sale of pharmaceutical, biological, and radio-pharmaceutical products that are not for sale in Canada. The programme may occasionally allow physicians to treat patients with drugs not approved for sale in Canada in cases of serious or life-threatening illness when conventional therapies have failed, are unsuitable, are unavailable or offer limited options.

Special Authority: An exemption that makes a particular drug eligible for full or partial Pharma Care benefit.

Specialist: A person who specializes in a particular field (e.g. cardiologist, oncologist, auditor etc.).

Specialists: Providers whose practices are limited to treating a specific disease (e.g., oncologists), specific parts of the body (e.g., ear, nose and throat), a specific age group (e.g., pediatrician), or specific procedures (e.g., oral surgery).

Specialty Pharmacy Services: Pharmacy services associated with the delivery of high-cost medications (example: specialty injectables) for the treatment of specific conditions that often require special handling such as refrigeration. These services may be provided by a vendor that contracts with the-PBM.

Speciation: Determination of the exact chemical form or compound in which an element occurs in a sample, for instance - determination of whether arsenic occurs in the form of trivalent or pentavalent ions or as part of an organic molecule, and the quantitative distribution of the different chemical forms that may coexist.

Species Differences in Sensitivity: Quantitative or qualitative differences of response to the action(s) of a potentially toxic substance on various species of living organisms.

Species: In biological systematics, group of organisms of common ancestry that are able to reproduce only among themselves and that are usually geographically distinct.

Species-Specific Sensitivity: Quantitative and qualitative features of response to the action(s) of a potentially toxic substance that are characteristic for particular species of living organism.

Specific Death Rate : Death rate computed for a subpopulation of individual organisms or people having a specified characteristic or attribute, and named accordingly (for example, age-specific death rate, the number of deaths of persons of a specified age during a given period of time, divided by the total number of persons of that age in the population during that time).

Specific Pathogen Free (SPF): Describing an animal removed from its mother under sterile conditions just prior to term and subsequently reared and kept under sterile conditions.

Specificity (of a screening test): Proportion of truly non- diseased persons who are identified by the screening test.

Specimen: Specifically selected portion of any substance, material, organism (specifically tissue, blood, urine or faeces) or environmental medium assumed to be representative of the parent substance etc. at the time it is taken for the purpose of diagnosis, identification, study or demonstration.

Speech Therapy: Treatment to correct a speech impairment that resulted from birth or from disease, injury or prior medical treatment.

Spenddown: A Senior Care level of participation during which the participant is required to pay for his or her prescription drugs an amount equal to the amount the income exceeds 240% of the federal poverty level.

Spleen: The large organ lying to the left of, below, and behind the stomach. This organ is partially responsible for white blood cell formation (red blood cells in childhood), and it is lined with resident macrophages that help it filter the blood, remove and recycle old and dead red blood cells, and send this all up to the liver in the portal blood. The liver, in fact, does most of the recycling of splenic hemoglobin derivatives. The spleen initiates much resistance and immunologic response, being made mostly of lymph pulp, and it stores and concentrates a large number of red blood cells. These can be injected into the bloodstream for immediate use under flight or fight stress, since the spleen is covered with capsule and vascular muscles that constrict in the presence of adrenalin or sympathetic adrenergic nerve stimulus.

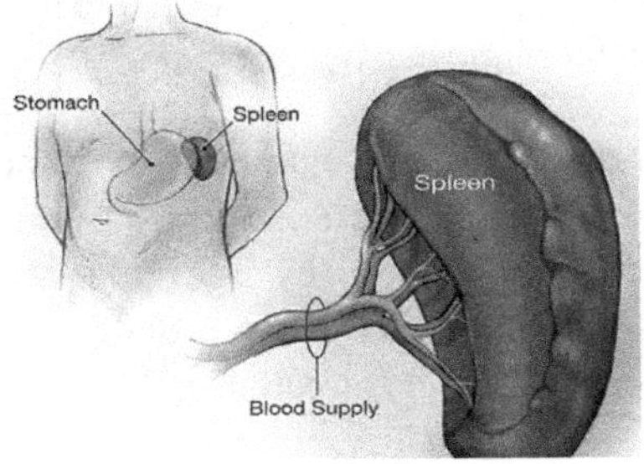

Splenitis: Inflammation of the spleen, caused by a variety of conditions ranging from exposure to hemolytic chemicals, systemic infections lodged in the spleen, even cancer.

Splenomegaly: For practical purposes a term interchangeable with splenitis, since neither will have the usual symptoms associated with inflammation. Splenomegaly is often associated with viral hepatitis, mononucleosis, typhoid fever and abnormally high levels of red blood cells or platelets.

Sponsor Investigator: An individual who both initiate and conduct alone or with others a clinical trial and under whose immediate direction the investigational product is administered to, dispensed to, or used by a subject. The obligations of a sponsor and those of an investigator.

Sponsor Master File: The time that contains all the essential trial documents at sponsor's facility.

Sponsor: An individual, company, institution, or organization that takes responsibility for the initiation, management, and/or financing of a clinical trial. A corporation or agency whose employees conduct the investigation is considered a sponsor and the employees are considered investigators. The entity responsible for conducting and financing a clinical trial. Sponsors can be governmental agencies, companies, medical or educational institutions, or even private individuals.

Sponsor-Investigator: An individual-who both initiates and conducts, alone or with others, a clinical trial, and under whose immediate direction the investigational product is administered to, dispensed to, or used by a subject. The term does not include any person other than an individual (e.g., it does not include a corporation or an agency). The obligations of a sponsor-investigator include both those of a sponsor and those of an investigator [ICH]. Under FDA regulations, the term does not include any person other than an individual, e.g., corporation or agency.

Spouse for Pharma Care Purposes: A spouse is a person who is either married to or living and cohabitating in a marriage-like relationship with the registrant and may be of the same gender as the registrant.

Spouse: Two persons (of the opposite sex or the same sex) are considered spouses if they : are married or have entered into a civil union; have been living together for 12 months (separations of less than 90 days do not interrupt the 12- month period); or are living together (regardless of for how long) and together have had or have adopted a child.

Spreader: Agent used in some pesticide formulations to extend the even disposition of the active ingredient.

SSN (Social Security Number): The nine-digit Social Security number is used by the U.S. federal government to track workers' contributions to the Social Security retirement system. Because it is unique to each individual, it is widely used as a personal identification number. Most VA healthcare databases identify patients by their true SSNs or by scrambled versions of them.

SSO (Software Support Organization): A person or organization providing a product to a HealthNet/BC participant for the purpose of accessing Ministry applications. Also referred to a Pharmacy Software Vendor.

Stability Data: The data on the stability of a drug product under routine and accelerated stability condition.

Stability Half-Life (Half-Time): Time required for the amount of a substance in a formulation to decrease, for any reason, by one-half (50 %).

Stability: The time period and storage condition under which a drug is stable.

Stable Disease (SD): Where patients neither improve nor get worse.

Staff Signature Log: The document that enlisted the specific trial related duties performed by individual study team members along with their signatures and /or initials.

Staging: Criteria used to determine the extent or stage of cancer.

Stamens: The male, pollen-producing organs in flowering plants. A staminate flower is only male, with pistillate (female) flowers on the same or different plants. Most flowering plants have both parts on the same flower, although they may mature at different times to avoid self-pollination.

Standard Deviation: Indicator of the relative variability of a variable around its mean; the square root of the variance.

Standard Drugs: Viewed as offering little or no therapeutic advantage over existing therapies and receive lower review status by the FDA

Standard Error: The estimated standard deviation or error of a series of measurements.

Standard Material: Reference material (or calibration material) for which, for specified element concentrations, values are recommended by some official body. Substance sufficiently well defined to be used for calibration and quality control of measurement techniques.

Standard of Care: Medical management of a patient based on established regimens or guidelines. Treatment regimen or medical management based on state-of-the art patient care.

Standard Operating Procedures (Sops): Detailed, written instruction to achieve uniformity in the performance of a specific task/activity.

Standard Treatment: A treatment currently in wide use and approved by regulatory agencies, such as the United States Food & Drug Administration or the Committee for Medicinal Products for Human Use (CHMP) in Europe, considered to be effective in the treatment of a specific disease or condition.

Standard: That which is established as a measure or model to which others of a similar nature should conform. Technical specification, usually in the form of a document available to the public, drawn up with the consensus or general approval of all interests affected by it, based on the consolidated results of science, technology and experience, aimed at the promotion of optimum community benefits and approved by a body recognized on the national, regional or international level.

Standardization: Making any substance, drug or other preparation conform to type or precisely defined characteristics. Establishment of precisely defined characteristics, or precisely defined methods, for future reference. Definition of precise procedures for administering, scoring and evaluating the results of a new method that is under development.

Standardized) Mortality (Morbidity) Ratio (SMR): Ratio of the number of events observed in the study group or population to the number of deaths expected if the study population had the same specific rates as the standard population, multiplied by 100.

Stanguria: Slow and painful discharge of the urine, due to spasm of the urethra and bladder.

Staph: This is short for Staphylococcus, a genus of micrococci bacteria with many members that can cause disease. They are gram-positive, nonmotile bacteria that are aerobic-(unless they need to be anaerobic). Staph of various types are responsible for boils and carbuncles; they may be involved in impetigo, toxic shock syndrome, endocarditis, osteomyelitis, and urinary tract infections, as well as some food poisoning.

Startup Meeting: The investigator meeting conducted before initiating a clinical trial and attended by all the participating investigator and respective site personnel.

Stasis: Static, atonic, unable to resolve or initiate change, resulting in lympatic and venous stasis, congestion or stagnation...such as an intestinal blockage.

Statement of Investigator: FDA form signed and dated by the investigator before initiating a phase 1:3 trial that outlines the obligation of investigator.

Statement of Work: Description of services to be provided under the scope of an agreement.

Static Lung Volumes And Capacities Can Be Measured: these include vital capacity – the maximum volume of air that can be exhaled slowly and completely after a maximum deep breath; forced vital capacity is a similar man- oeuvre using maximal forceful exhalation and can be measured along with expiratory flow rates using simple spirometry; total lung cap- acity is the total volume of air in the chest after

Statistical Analysis Plan: A statistical analysis plan is a document that contains a more technical and detailed elaboration of the principal features of the analysis described in the protocol, and includes detailed procedures for executing the statistical analysis of the primary and secondary variables and other data.

Statistical Analysis Plan: A written document that contains detailed information and rationale on the statistical analysis to be undertaken for a clinical trial.

Statistical Analysis: Providing meaning to a set of data (which otherwise would be a collection of numbers and/or values) by applying an appropriate statistical method.

Statistical Error: Errors that can occur in a clinical trial such as a situation where the effect of therapy is confounded with another factor or when the patients on trial are not sufficiently representative of the targeted patient population .the common statistical errors are Type :1

error (alpha error)and Type: 2 error (beta error).

Statistical Method: A method of analyzing or representing the statistical data.

Statistical Significance: State that applies when a hypothesis is rejected. Whether or not a given result is significant depends on the significance level adopted. For example, one may say "significant at the 5% level". This implies that a level of significance has been applied such that when the null hypothesis is true there is only a 1 in 20 chance of rejecting it or that the observed result has led to rejection of the null hypothesis.

Statistician: A person having knowledge and competence of statistics.

Statistics: The process of collecting, recording and summarizing data that are collected from experiments, records and surveys .

Status change: A lifestyle event that may cause a person to modify their health benefits coverage category. Examples include, but are not limited to, the birth of a child, divorce or marriage.

Steatorrhea: The presence of undigested fat in the feces. This may be the result of failing to inoculate fatty foods with enough surfactant (biliary "soap") to digest them, the failure of the lower small intestine to absorb them, or simply too much fat for even normal digestion to handle. Sometimes this can indicate liver, gall bladder or lipid metabolism diseases. Usually the causes are subclinical and treatable with less invasive approaches...like herbs.

Stem Cells: (Used in this context for blood stem cells.) Immature cells that give rise to all the normal components of blood. Stem cells are normally found in the bone marrow, but some will circulate in, and can be collected from the blood.

Stenosis: Narrowing of a stricture of a duct or a canal.

Step Therapy: When a new, expensive drug comes out, and there are less costly alternatives available, PBM's (formularies) may require step therapy or documented proof that the lower-cost alternatives were tried first, and failed before approving the more expensive drugs.

Sterility: The absence of viable contaminating microorganism.

Steroid Hormone: These are fats similar to, and usually synthesized from, cholesterol, starting with Acetyl-CoA, moving through squalene, past lanosterol, into cholesterol, and, in the gonads and adrenal cortex, back to a number of steroid hormones. Nearly all of the classic hormones are proteins or smaller peptides; they don't get inside a cell (the membrane keeps them out); instead, they bind to, and initiate, cell changes from the outside. The exceptions are the thyroxines (from the thyroid) and the steroid hormones. They move into the cell, bind with receptors, and initiate changes in the way a cell regenerates itself or synthesizes new compounds. Because the steroid hormones stimulate cell growth, either by changing the internal structure or increasing the rate of proliferation, they are often called anabolic steroids. Estrogen, an ovarian steroid, when secreted into the bloodstream, will be bound within a short time by internal receptors inside those cells that need estrogen for their growth; the unused portion is partially broken down, mostly in the liver, and partially stored in a less active form by adipose tissue. Since luteinizing hormone from the pituitary is surged in pulses an hour apart, the estrogen is also surged from the reacting ovaries, and by the time more estrogen is available, the binding cells need more; their program of synthesis has run out and needs to be started again.

Steroids, Plant: The previous subject is obviously an endless one, but as this is the glossary of an herbal nature, let me assure you, virtually no plants have a direct steroid hormone-mimicking effect. There are a few notable exceptions with limited application, like Cimicifuga and Licorice. Plant steroids are usually called phytosterols, and, when they have any hormonal effect at all, it is usually to interfere with human hormone functions. Beta sitosterol, found in lots of food, interferes with the ability to absorb cholesterol from the diet. Corn oil and legumes are two well-endowed sources that can help lower cholesterol absorption. This is of only limited value, however, since cholesterol is readily manufactured in the body, and elevated cholesterol in the blood is often the result of internal hormone and neurologic stimulus, not the diet. Cannabis can act to interfere with androgenic hormones, and Taraxacum phytosterols can both block the synthesis of some new cholesterol by the liver and increase the excretion of cholesterol as bile acids; but other than that, plants offer little direct hormonal implication. The first method discovered for synthesizing pharmaceutical hormones used a saponin, diosgenin, and a five-step chemical degradation, to get to progesterone, and another, using stigmasterol and bacterial culturing, to get to cortisol. These were chemical procedures that have nothing to do with human synthesis of such hormones, and the plants used for the starting materials- Mexican Wild Yam, Agave, and Soy were nothing more than commercially feasible sources of compounds widely distributed in the plant kingdom. A clever biochemist could obtain testosterone from potato sterols, but no one would be likely to make the leap of faith that eating potatoes makes you manly (or less womanly), and there is no reason to presume that Wild Yam (Dioscorea) has any progesterone effects in humans.

Steroids: Chemical substances that are related in structure and contain the same chemical skeleton. Steroids closely resemble a hormone called Cortisol that is produced by your adrenal glands. Used to treat inflammatory diseases and conditions, steroids reduce the immune system's activity and decrease inflammation.

Sthenic: Strong of body or function, even to an excess.

Stimulatory Neurotransmitter: A neurotransmitter that increases electro-chemical activity in the nerve cells. Nor epinephrine is a stimulatory neurotransmitter.

Stipules: A leafy appendage formed at the juncture of leaf and main stem.

Stochastic Effect: Consequence for which the probability of occurrence depends on the absorbed dose: hereditary effects and cancer induced by radiation are considered to be stochastic effects. The term "stochastic" indicates that the occurrence of effects so named, would be random. This means that, even for an individual, there is no threshold of dose below which the effect will not appear, and the chance of periencing the effect increases with increasing dose.

Stochastic: An analysis involving a random variable or arising from chance and hence involving probability and obeying the laws of probability.

Stock Transfers: Pharmacy transactions that include the sale of drug inventories to other pharmacies, the transfer of inventories to long-term care facilities, and the return of drugs to wholesalers. Stock transfers are not PharmaNet transactions.

Stoloniferous: A plant that tends to form lateral roots, sometimes green and potentially stemming, sometimes blanched and tending to root from the nodes...or both.

Stopping Rules: A statistical criterion that, when met by the accumulating data, indicates that the trial can or should be stopped early to avoid putting participants at risk unnecessarily or because the intervention effect is so great that further data collection is unnecessary.

Strangury: Painful, sporadic and drop-by-drop urination, caused by the presence of kidney stones, chronic inflammation such as interstitial cystitis, or urethral scar tissue. This is not a specific disease, but a symptom, like nausea or a sore joint.

Stratification (in Epidemiology): Process of or result of separating a sample into several subsamples according to specified criteria such as age groups, socio-economic status, etc.

Stratification : A tool used during the randomization process to ensure an exact balance between the treatment arms with respect to key patient factors that are strongly related to the outcome variable.

Stratified Sample: Subset of a population selected according to some important characteristic.

Strep: A genus of gram-staining chain-forming cocci bacteria. Some are responsible for common and potentially serious human infections, ranging from scarlet fever and strep throat to bacterial endocarditis and pus pockets. Most of the disease-potential streps are also a normal part of the skin, mouth and upper respiratory flora.

Stringent: Strict adherence to a set standard

Stroke: Damage to a part of the brain that happens when the blood vessels supplying that part are blocked, such as when the blood vessels are clogged with fats (a condition sometimes called hardening of the arteries).

Structural Biology: Structure of molecules within cells and how these molecules interact and function

Structure-Activity Relationship (SAR): Association between the physicochemical properties of a substance and/or the properties of its molecular substructures and its biological properties including its toxicity.

Structured Practical Training (SPT): A structured, competency- based training period with scheduled assessments.

Student: A person registered with the College to complete his/ her initial training period. Able to perform the controlled acts under direct supervision of a pharmacist

Students t-Test: A comparison test to determine whether the mean value of a numeric variable (or response variable) in a given population is same as that in another population. For e.g. a study that investigate whether the mean hemoglobin level after a fixed treatment period of drug a is same as that of drug B.

Study Arm: The treatment group of a clinical trial.

Study Closeout Visit Report: A written report to document that all activities required for study closeout have been completed and all the essential documents are achieved in a appropriate files

Study Closeout: The closing a clinical study after the same has been completed or prematurely terminated/suspended.

Study Closure: Closing a clinical study after the same has been completed or prematurely terminated/suspended.

Study Coordinator: A person employed at clinical investigator's site to record the clinical trial data in compliance with protocol, GCP and applicable regulatory guidelines. The investigator delegates specific duties to a study coordinator.

Study Design: Plan for the precise procedure to be followed in a clinical trial, including planned and actual timing of events, choice of control group, method of allocating treatments, blinding methods; it assigns a subject to pass through one or more epochs in the course of a trial. Specific design elements, e.g., crossover, parallel; dose- escalation.

Study Drug: An investigational product or marketed product (i.e. active control) place be used in a clinical trial.

Study Endpoint: Primary or secondary outcome (s) used to judge the effectiveness of a treatment.

Study Initiation Visit Report: A report that documents the activation of a clinical study after the ethics committee and regulatory approval has been obtained and other trial specific requirements have been fulfilled.

Study Materials: A complete set of supplies (e.g. documents, investigational product. Laboratory kits etc.) to be used in a clinical trial .

Study Monitor: A person employed by the sponsor or CRO. Who reviews study records at trial site (s) to determine that a study is being conducted in accordance with the protocol and applicable regulatory guidelines.

Study Number: A unique identifier assigned to a clinical study for its easy identification.

Study Nurse: A qualified nurse who assist the investigator in the conduct of research project.

Study Subject: An individual who participates in a clinical trial either as a recipient of the investigational product or as a control.

Study Supplies: A complete set of supplies (e.g. documents, investigational product , laboratory kits etc.) to be used in a clinical trial.

Study Team: A group of individuals that constitutes the research team.

Study Timeline: The sequence of study related activities arranged on a time scale.

Study Treatment: The modality of medical management of a trial patient.

Study Type: The primary investigative techniques used in an observational protocol; types are Purpose, Duration, Selection, and Timing.

SU: The building letter code for the Surge Research Building, 90 Medical Center Way, San Francisco

Sub Acute Toxicity: A type of toxicological testing of a drug in 2 species that usually lasts for 6 months.

Sub Contract: A contract to perform apart of all of the obligation of another contract.

Sub Contracting: The act of executing a sub contrast.

Sub Investigator: Any individual member of the clinical trial team designated and supervised by the investigator to perform trial related procedure and to make important trial designs at a site.

Sub-acute (Sometimes called Subchronic) Effect: Biological change resulting from multiple or continuous exposures usually occurring over about 21 days. Sometimes the term is used synonymously with subchronic effect and care should be taken to check the usage any particular case.

Sub-acute Rehabilitation: Post-acute phase of rehabilitation. Capacity to treat for 6-24 months. Need not be hospital based. Stay based on demonstrated improvement.

Sub-acute: Having characteristics of both acute and chronic. This is the state in a disease when most of the aches and pains have subsided and you are likely to overdo things and not completely recover. The chest cold that lingers for weeks as a stubborn cough is a subacute condition, as is the tendonitis that lingers because you won't stop playing tennis long enough to completely heal.

Subchronic (Subacute) Effect: Biological change resulting from an environmental alteration lasting about 10 % of the lifetime of the test organism. In practice with experimental animals, such an effect is usually identified as resulting from multiple or continuous exposures occurring over 3 months (90 days). Sometimes a subchronic effect is distinguished from a subacute effect on the basis of its lasting for a much longer time.

Subchronic Toxicity: Adverse effects resulting from repeated dosage or exposure to a substance over a short period, usually about 10 % of the life span. The capacity to produce adverse effects following subchronic exposure.

Subchronic: Related to repeated dose exposure over a short period, usually about 10 % of the life span; an imprecise term used to describe exposures of intermediate duration.

Subclinical Effect: Biological change following exposure to an agent known to cause disease either before symptoms of the disease occur or when they are absent.

Subclinical: This is our turf, the period of time when a potential disease is still potential, and a functional imbalance or tendency has not caused any organic disruption. Years of poor digestion, heartburn, and systematic suppression of upper GI function by adrenalin stress have not become overt gastritis, ulcers, or IBS. You have symptoms of distress (subclinical) but no real, ripened clinical disease.

Subcutaneous: Below the surface of the skin, but probably above the following term well anyway, definitely lower than the top of the skin

Subdermal: Below the surface of the skin, and probably below the previous term, which should really be suprasubdermal higher up than the muscles.

Subgroup Analysis: Exploratory secondary analysis performed to gain more insight into various aspects of the trial.

Sub-Investigator: Any individual member of the clinical trial team designated and supervised by the investigator at a trial site to perform critical trial-related procedures and/or to make important trial-related decisions (e.g., associates, residents, research fellows).

Subject Enrollment Log: A log that captures the date of enrolment and other protocol required visits of a clinical trial subjects.

Subject Identification Code: A unique identifier assigned by the investigator to each trial subject to protect the subject's identity and used in lieu of the subject's name when the investigator reports adverse events and/or other trial-related data.

Subject Identification List: A confidential list maintained by a trial site that captures names of all the trial subjects vis : a : vis. Their allocated subjects number /code.

Subject Number: A unique identifier assigned to trial subjects for protecting their identity

Subject/Trial Subject: An individual who participates in a clinical trial, either as recipient of the investigational product(s) or as a control.

Subject: An individual who participate in a clinical trial either as recipient of investigation product or as a control.

Subjective Environment: Surrounding conditions as perceived by persons living in these conditions.

Subjects Identification Code: A unique identifier assigned to trial subjects for protecting their identity.

Submission date: The date the claim was submitted and/or received by CIGNA HealthCare.

Subscriber: An individual who is a member of a benefits plan. For example, in the case of family coverage, one adult is ordinarily the subscriber. A spouse and children would ordinarily be dependents.

Succus Entericus: Intestinal Juice. These are enzyme-rich secretions produced by the lining of the upper small intestines. Apparently the enzymes produced compensate for any pancreatic enzymes that are deficient for that particular meal.

Succussion: The process of forcefully striking a homeopathic remedy against a firm surface.

Sudorific: Substance that causes sweating.

Sufficient Evidence: According to the USEPA's Guidelines for Carcinogen Risk Assessment, sufficient evidence is a collection of facts and scientific references that is definite enough to establish that an adverse effect is caused by the agent in question.

Suggested No Adverse Response Level: Maximum dose or concentration that on current understanding is likely to be tolerated by an exposed organism without producing any harm.

Summary Sheet: Two-to-four page summary of a risk assessment.

Summation: Process of addition of separate postsynaptic responses caused by stimuli that are adjacent in time and space. Excitation of a synapse evokes a graded potential change in the postsynaptic membrane that may be below the threshold required to trigger an impulse.

Superiority Trial: A trial with the primary objective of showing that the response to the investigational product is superior to a comparative agent (active or placebo control).

Superoxide Dismutase (SOD): A zinc and copper or manganese containing enzyme which reacts with superoxide radicals to convert them to less dangerous chemical entities.

Superoxide Radical: A free radical thought to play a central role in arthritis and cataract formation.

Supersede: Replacing a document with the new one after the same has been revised.

Supervisor: A person with the authority to oversee the work of a person or group.

Supplemental Coverage: Supplemental coverage is offered by certain private plans to persons age 65 and over. This coverage does not replace the basic coverage, but may add to such coverage, for example, by paying for uninsured drugs or by paying a portion of costs not reimbursed to persons who have only the basic coverage.

Supplier: Supplier of products and services recognized by PharmaCare, i.e., prosthetics. Also called provider.

Support Group: A group of people who share a similar problem or concern. The people in the group help one another by sharing experiences, knowledge, and information.

Surface Layer: Region of space comprising and adjoining the phase boundary between a solid and liquid phase, between a solid and gas phase, or between a liquid and gas phase within which properties of matter are significantly different from the values in the adjoining bulk phases.

Surrogate Endpoints: Physiological or biological markers that can be measured quickly and easily and that are taken as being predictive of important clinical outcomes.

Surrogate Marker: A measurement of a drug's biological activity that substitutes for a clinical endpoint such as death or pain relief.

Surrogate Variable: A variable that provides an indirect measurement of effect in situations where direct measurement of clinical effect is not feasible or practical.

Surrogate: Relatively well studied toxicant whose properties are assumed to apply to an entire chemically and toxicologically related class; for example, benzo(a)pyrene data may be used as toxicologically equivalent to that for all carcinogenic polynuclear aromatic hydrocarbons.

Surveillance: Ongoing scrutiny, generally using methods distinguished by their practicability and uniformity, and Fequently by their rapidity, rather than by complete accuracy. Its main purpose is to detect changes in trend or distribution in order to initiate investigative or control measures.

Survival Data: Any time to event in clinical trial common example of survival data includes time to response time to relapse time in remission time to discontinuation.

Survivor Function: The probability that an individual will survive up to at last a certain time the estimate for the survivor function is called as Kaplan-meier estimate.

Susceptibility: Condition of lacking the power to resist a particular disease or infection; thus in susceptible people "normal expected" results occur but with a lower exposure (or dose) than in the rest of the population.

Sympathetic: A division of the autonomic or involuntary nervous system that works in general opposition to the parasympathetic division (q.v.). Many of the sympathetic functions are local, specific, and involve secretion of acetylcholine, like any other of your normal nerves stimulating or suppressing a specific muscle, gland, or whatever. A certain number of these nerves, however, unlike any others in the body, secrete epinephrine (adrenalin) and norepinephrine (noradrenalin). These are called adrenergic. Since the adrenal medulla also secretes the same substances into the bloodstream as hormones, all the muscles or glands that are affected by the adrenergic sympathetic nerves also react in to the epinephrine secreted into the blood. This forms the basis for a potentially lifesaving emergency fight or flight response and is meant for short, drastic activities. A chronic excess of the adrenergic response, however, is a major cause of stress- and a major contributor to many types of chronic disease. The more you use a particular nerve pathway or induce a particular group of functions, the more blood, fuel storage, and mitochondria are produced to strengthen that group of actions. Using adrenergic energy excessively gives literal dominance to those things that are stimulated or suppressed, and the effects of adrenalin stress linger in the body after the adrenalin is long gone. Since one of the first subjective symptoms of subclinical malnutrition, metabolic imbalances, and environmental pollution is irritability of the central nervous system, hypersympathetic function acts as an intermediate between poor diet, pollution, and disease.

Sympatholytic: Blocking transmission of impulses from the adrenergic (sympathetic) postganglionic fibres to effector organs or tissues. Agent that blocks transmission of impulses from the adrenergic (sympathetic) postganglionic fibres to effector organs or tissues.

Sympathomimetic: A substance that mimics at least part of adrenalin or catecholamine responses. The term is a little biased towards the minority of sympathetic functions that are adrenergic. A better name might be adrenalomimetic, epinephromimetic, catecholamimetic. or speedomimetic. Examples: coffee, ephedrine, amphetamines. Producing effects resembling those of impulses transmitted by the postganglionic fibres of the sympathetic nervous system. Agent that produces effects resembling those of impulses transmitted by the postganglionic fibres of the sympathetic nervous system.

Symptom: Any subjective evidence of a disease or an effect induced by a substance as perceived by the affected subject.

Symptomatology: General description of all of the signs and symptoms of exposure to a toxicant: signs are the overt (observable) responses associated with exposure (such as convulsions, death, etc.) whereas symptoms are covert (subjective) responses (such as nausea, headache, etc.).

Synapse: Functional junction between two neurones, where a nerve impulse is transmitted from one neurone to another.

Syndrome: Set of signs and symptoms occurring together and often characterizing a particular disease-like state.

Synergies: When compounds are combined and their effects are more than the sum of their individual effects; the compounds are said to have positive synergy. Many of the nootropic compounds have positive synergy effects with each other, they become synergistic.

Synergism: Pharmacological or toxicological interaction in which the combined biological effect of two or more substances is greater than expected on the basis of the simple summation of the toxicity of each of the individual substances.

Synergistic Effect: Biological effect following exposure simultaneously to two or more substances that is greater than the simple sum of the effects that occur following exposure to the substances separately.

Synthetic Chemistry: Deliberate manufacture of pure substances of defined structure.

System: People, machines, software, applications and/or methods organized to accomplish a set of specific functions or objectives.

Systematic Sample: Subset selected according to some simple rule such as specified date or alphabetic classification.

Systemic Effect: Consequence that is of either a generalized nature or that occurs at a site distant from the point of entry of a substance: a systemic effect requires absorption and distribution of the substance in the body.

Systemic Treatment: A treatment, such as a tablet or capsule, taken internally.

Systemic: Relating to the body as a whole, throughout the entire body.

Systolic: The measurement of arterial blood pressure at the point of heart contraction (greatest pressure); the higher of the two BP numbers, with diastolic (q.v.) being the lower.

T

T CELL: A type of white blood cell that is part of the immune system and normally helps protect the body against infection and disease. In psoriasis, it also can trigger inflammation and excessive skin cell reproduction.

T Lymphocyte: Animal cell which possesses specific cell surface receptors through which it binds to foreign substances or organisms, or those which it identifies as foreign, and initiates immune system responses.

T_4: Also termed tetraiodothyronine, nicknamed is thyroxine. Secreted by the thyroid along with T3 (tri iodothyronine also called thyroxine), this thyroxine is mostly conjugated in the blood by TBG (thyroxine-binding globulin), whereas the more active T3 tends to float free. T4 is broken down to T3 and forms a stable feeder reserve, preventing rapid shifts in its more labile relative

Tabes: This means, literally, a wasting disease, and is a traditional name applied to various diseases such as tabes dorsalis (tertiary syphilis) and tuberculosis accompanied by enlargement of glands.

Tachy: Prefix meaning rapid as in tachycardia and tachypnoea.

Tachycardia: Abnormally fast heartbeat.

Tachyphylaxis: Rapidly developing tolerance to a drug.

Tachypnoea: Abnormally fast breathing.

Tacrolimus: An immunosuppressant drug used for primary immunosuppression in recipients of kidney or liver transplants

Taeniacide: Substance intended to kill tapeworms.

Talipes: Also known colloquially as club-foot, this is a deformity apparent at birth, affecting the ankle and foot: the foot is twisted at the ankle-joint so that the sole does not rest on the ground when standing.

Talus: The square-shaped bone which forms the lower part of the ankle-joint and unites the leg bones to the foot.

Tampon: A plug of compressed gauze or cotton wool inserted into a wound or orifice to arrest bleeding.

Tamponade: A potentially life-threatening compression of the heart by the accumulation of fluid in the pericardial sac.

Tannins: A group of simple and complex phenol, polyphenol, and flavonoid compounds, bound with starches, and often so amorphous that they are classified as tannins simply because at some point in degradation they are astringent and contain variations on gallic acid. Produced by plants, tannins are generally protective substances found in the outer and inner tissues, often breaking down in time to phlebotannins and, finally, humin. All of the tannins are relatively resistant to digestion or fermentation, and either decrease the ability of animals to easily consume the living plant, or, as in deciduous trees, cause shed parts of the plant to decay so slowly that there is little likelihood of infection to the living tree from rotting dead material around its base. All tannins act as astringents, shrinking tissues and contracting structural proteins in the skin and mucosa. Tannin-containing plants can vary a great deal in their physiological effects and should be approached individually.

Tantulum: A heavy metal that is used in surgery because it is easy to mould and does not corrode. It is particularly suitable for repairing defects in the skull bones.

Tapotement: A massage technique in which a part of the body is hit repeatedly and quickly with the hands. The technique is useful in helping patients with bronchitis to loosen the mucus in the air passages of their lungs, thus helping them to cough it up.

Tapping: The popular name for the withdrawal of edema fluid from the cavities or the subcutaneous tissues of the body.

Tardive Dyskinesia: Also known as orofacial dyskinesia, this is characterised by involuntary chewing and grimacing, usually the result of years of taking antipsychotic drugs, particularly in the eld- erly when these drugs are sometimes used to sedate troublesome patients.

Target Enrollment: The number of subjects in a class or group (including the total for the entire trial) intended to be enrolled in a trial to reach the planned sample size. Target enrollments are set so that statistical and scientific objectives of a trial will have a likelihood of being met as determined by agreement, algorithm or other specified process.

Target population (epidemiology): Collection of individuals, items, measurements, etc. about which we want to make inferences: the term is sometimes used to indicate the population from which a sample is drawn and sometimes to denote any reference population about which inreferences are required. Group of persons for whom an intervention is planned.

Target Selection : A drug discovery approach that involves choosing a disease or biological target such as enzyme receptor. To treat and then developing a model for that disease.

Target: Human being or any organisms, organ tissue, cell, resource, or any constituent of the environment, living or not, that is subject to the activity of a pollutant or other chemical or physical activity or other agent.

Tartar: A concretion that forms on the teeth near the margin of the gum, consisting chiefly of phosphate of lime deposited from the saliva. Mixed with this are food particles and this is an ideal medium for bacteria to flourish in.

Taxanes: A group of cytotoxic drugs administered intravenously for the treatment of advanced ovarian cancer

Taxicodynamics: Process of interaction of a potentially toxic substance with target sites, and the biochemical and physiological consequences leading to adverse effects.

Taxicogenetics: Study of the influence of hereditary factors on the effects of potentially toxic substance on individual organisms.

Taxis: The method of pushing back, into the abdominal cavity, a loop of bowel which has passed through the wall in consequence of a rupture.

Tay Sachs Disease: An inherited recessive condition in which there is abnormal accumulation of lipids in the brain. The result is blindness, mental retardation and death in early childhood.

Tazarotene (Tazorac): A tropical retioned cream or gel used in the treatment of psoriasis.

Technology Change: Technology changes are oriented toward process and equipement modifications to reduce waste, primarily in a production setting. Technology changes can range from minor changes that can be implemented in a matter of days at low cost, to the replacement of processes involving large capitat costs.

Technology Provider: A health care technology company that develops applications to support physicians with practice management, electronic prescribing, electronic medical records, or similar office automation products. Person, company, or other entity who develops, produces, and sells software applications and/or hardware for use in conducting clinical trials and/or in analyzing clinical trial data and or submitting clinical trial information for regulatory approval.

Teichopsia: This refers to zigzag lines that patients with migraine often experience as a visual aura preceding an attack.

Telangiectasis: Abnormal dilatation of arterioles and venules. In the skin it is seen in spider Naevus and rosacea particularly.

Telemedicine: A broad term used to describe medicine at a distance through a communications link.

Telephone Log: A list / log that capture the detail of telephonic discussion in clinical trial.

Telnet: A method of connecting to a computer to access information, such as e-mail.

Template: A predesigned form document that include standard fill in the blanks spaces for capturing the standard information.

Temporary Acceptable Daily Intake: Value for the acceptable daily intake proposed for guidance when data are sufficient to conclude that use of the substance is safe over the relatively short period of time required to generate and evalulate further safety data, but are insufficient to conclude that use of substance is safe over a lifetime. A higher-than-normal safety factor is used when establishing a temporary ADI and an expiration date is established by which time appropriate data to resolve the safety issue should be available.

Temporary Maximum Residue Limit: Temporary maximum residue limit is established for a specificified, limited period when, only a temporary acceptable daily intake has been established, the residue data are inadequate for firm maximum residue recommendations.

Temporary Premium Assistance: A program of the Medical Services Plan (MSP) which provides qualified (low-income) persons / families with reduced MSP premiums on a short-term basis.

Tendon Transfer: Reconstructive surgery in which the tendon from an unimportant muscle is removed and used to repair or replace a damaged tendon of a major muscle.

Tenesmus: A symptom of disease affecting the lower part of the large intestine, such as dysentery, piles (Haemorrhoids) or tumour. The painful expelling cramps of the tubular smooth muscles and ducts. Normal peristalsis of various types produce no pain or sensation (except for the dreaded borborygmies); only the energetic expulsion contraction can induce referred pain. Examples: Nausea, gas pain, uterine cramps, gall bladder pain.

Tentorium: A wide flap of dura mater forming a partition between the cerebrum and cerebellum and supporting the former.

Teratogen: Agent that, when administered prenatally (to the mother), induces permanent structural malformations or defects in the offspring.

Teratogenic :The substance or agent that can interfere with normal embryonic development.

Teratogenicity: Potential to cause or the production of structural malformations or detects in offspring.

Teratoma: A tumour that consists of partially developed embryonic tissues.

The most common sites of this tumour are the ovary

Termination (of Subject): Now considered nonstandard.

Termination of Study: Pre mature suspension of a clinical study due to lack of efficacy or safety or funding.

Terpenes: Any of a group of hydrocarbons that are made up of building blocks of isoprene (C_5H_8) or similar five-carbon units, with a monoterpene made up of two units (example: limonene and pinene), a sesquiterpene made up of three units (example: humulene, a Hops aromatic), and a diterpene made up of four units. The terpenes, in our context, are the primary constituents in the aromatic fractions of our scented plants.

Testamentary Trust: A trust that is created upon death by the terms of a persons's will.

Testator: An individual who dies leaving a will or testament in force.

Testing of Chemicals: In toxicology, evaluation of the therapeutic and potentially toxic effects of substances by their application through relevant routes of exposure with appropriate organisms or biological system so as to relate effects to dose following application. In chemistry, qualitative or quantitative analysis by the application of one or more fixed methods and comparisons of the results with established standards.

Testosterone: The principal reproductive androgen of males, largely responsiblefor sexual maturation, some libido, and a range of metabolic reactions that, while supplying short-term strengths, creates a long-term fragility and brittleness if not in balance with less garish but more sustainable metabolic buffers. It is secreted by the Leydig cells of the testes, as well as smaller amounts in the adrenal cortices of both sexes. It is made under the direction of LH from the pituitary, and, if oversecreted, can be inhibited by sperm-producing cells, diminished pituitary support, and a rise in blood levels of its waste-product, stored in adipose tissues.

Tetanic: Pertaining to tetanus, characterized by tonic muscle spasm.

Tetanus: Also called lockjaw, this is a bacterial infection of the nervous system. Increased excitability of the spinal cord results in painful and prolonged spasms of the voluntary muscles throughout the body, rapidly leading to death unless treated. Causes the disease is caused by the bacillus Clostridium tetani,

Tetany: A condition characterised by spasm of muscle, usually caused by a fall in blood calcium levels. This results in hyperexcitability of muscles which may go into spasm at the slightest stimulus. This is well demonstrated in two of the classical signs of the disease: Chvostek's sign, in which the muscles of the face contract when the cheek is tapped over the facial nerve as it emerges on the cheek; and Erb's sign, in which muscles go into spasm in response to an electrical stimulus which normally causes only a contraction of the muscle

Tetralogy of Fallot: The most common form of cyanotic con-genital heart disease. The tetralogy consists of stenosis of the pulmonary valve; a defect in the septum separating the two ventricles; the aorta over-riding both ventricles; marked hypertrophy of the right ventricle. Surgery is required to remedy the defects.

Tetraplegia: Paralysis of the body's four limbs, also called quadriplegia.

Thalassaemia: Also known as Cooley's anaemia, this is a condition characterised by severe anaemia, due to an abnormal form of haemoglobin in the blood. It is an inherited disease.

Therapeutic Index: Ratio between toxic and therapeutic doses (the higher the ratio, the greater the safety of the therapeutic dose.

Therapeutic Substitution: The prescribing physician authorizes the pharmacist to dispense an alternate chemical entity from the same therapeutic class instead of the drug originally prescribed.

Therapeutic Window: The concentration range over which a drug has a therapeutic effect without having unacceptable toxicity.

Therapeutic: Referring to the cure management of a disease.

Therapeutics: The general name applied to different methods of treatment and healing.

Therapy: The treatment of disease.

Thermography: A method of detecting the amount of heat produced by different parts of the body. This is done with an infra-red sensitive photographic film. High blood flow in an area shows up as a heat zone and thus tumours such as breast can- cer can be identified

Thermoluminescent Dosimeter: A commonly used device for measuring people's exposure to radiation. It contains activated sodium fluoride which luminesces in proportion to the radiation dose to which it is exposed

Therpaeutic Interchange: Used in both managed and closed formularies to substitute a preferred drug for a chemically different non- preferred prescription. This is accomplished by contacting the physician to obtain permission for an alternate medicine.

Thiersch's Graft: The term given to a method of skin-grafting in which strips of skin are shaved from a normal area and placed on a burned, injured or scarred area to be grafted

Third Party Out Sourcing : The act of sub contracting a part of all of the obligation of a contract to a third party not include in the initial contract.

Third- Party Payment: Payment or reimbursement amounts that are established by third- party drug programs for prescriptions and services dispensed to recipients.

Thomsonian Medicine: That school of medical philosophy and therapy founded by the American messianic nature therapist Samuel Thomson (1769). Thomson's great axiom was, "Heat is life, and cold is death." He lived in New England, which explains some of this. Thomsonians made great use of vomiting, sweating, and purging to achieve these ends crude by present standards, but saner than standard medicine of the times (mercury, lead, bleeding, etc.).

Thoracic Duct: This is the bodies' main lymph collecting vessel. It starts in the little collecting bladder in the abdomen (the cisterna chyli), moves up the center of the body in front of the spinal chord, alongside the esophagus and aorta to the neck, where it drains into the left subclavian vein. It drains the lymph from the entire body, except the head, right thorax and arm, which collects lymph separately and drains into the right subclavian vein. Lacking the ability to contract and expand, the thoracic duct relies on its valves and the kinetic energy of breathing and nearby arterial pumping to drain lymph upwards.

Threshold Limit Value (TLV): Concentration in air of a substance to which it is believed that most workers can be exposed daily without adverse effect (the threshold between safe and dangerous concentrations.)These values are established (and revised annually) by the (American Conferences of Governmental Industrial Hygienists) and are time-weighted concentrations for a 7 or 8 hour workday and a 40 hour workweek. For most substances the value may be exceeded, to a certain extent, provided there are compensatory periods of exposure below the value during the workday (or in some cases the week).for a few substances (mainly those that produce a rapid response) the limit is given as a ceiling concentration (maximum permissible concentration – designated by "C") that should never be exceeded.

Threshold: Dose or exposure concentration below which an effect is not expected.

Thrombocytopenia: Decrease in the number of blood platelets (thrombocytes).

Thromboembolism: The formation of a thrombus (blood clot) in one part of the circulatory system from which a T portion becomes detached and lodges in another blood vessel, partially or completely obstructing the blood flow (an embolism). Most commonly a thrombus is formed in the veins of the leg deep vein thrombosis

(DVT) and the embolism lodges in the pulmonary (lung) circulation.

Thrombolysis: The breakdown of a blood clot by enzymic activity. Naturally occurring enzymes limit the enlargement of clots, and drugs.

Thrombophlebitis: Inflammation of the veins combined with clot formation.

Thrombosis: The formation of a blood clot within the circulatory system. It may form in the roughened vein wall in a varicosity, form around arteriosclerotic plaques, or result from trauma and surgery. The tendency rises with thick blood, age, obesity and in those once physically active and now sedentary.

Thumb-Sucking: Also called finger-sucking, this is a universal and harmless habit in infancy

Thymoma: A tumour of the thymus gland. Such tumours are rare and are classified according to the variety of thymus tissue from which they develop.

Thymus: The master gland of the immune system located behind the breastone.

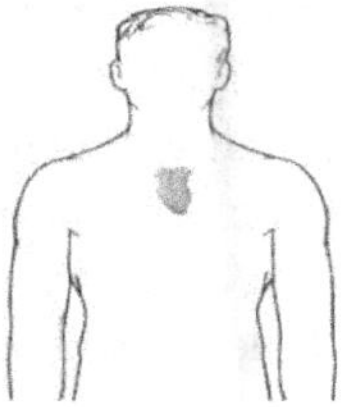

Thymus gland

Thyroglobulin: The iodine- containing protein that is stored in the thyroid gland. It is converted into circulating thyroxines when the thyroid is stimulated by TSH (Thyroid Stimulating Hormone).

Thyroid: The gland located in the centre of the brain responsible responsible (amongst other things) for temperature regulation.

Thyrotoxicosis: A pathologic thyroid

hyperfunction. It is sometimes referred to as exophthalmic goiter. An overt disease, sometimes life-threatening, it is very different from the moderately elevated basal metabolism some constitutional types manifest under stress.

TIBC: Total Iron binding Capicity. A tool for measuring plasma transferin.

Tic Douloureux: Another name for trigeminal neuralgia due to some affection of the fifth cranial nerve, and characterised by pain – situated somewhere about the temple, forehead, face, or jaw – and sometimes by spasm in the muscles of the affected region.

Tidal Volume: Quantity of air or test gas that is inhaled and exhaled during one respiratory cycle.

Tier: Level of co payment an insurer asks its insured to bear for services and drugs.

Tiered Copayment: A plan that reimburse the insured with one price for generic drugs and a lower price for brand name drugs.

Time Cost: Earnings that investors forego on funds used to support R&D activities before any return can be earned.

Time To Progression (TTP): The length of time after a disease is diagnosed (or treated) until the disease starts to get worse.

Time –Weighted Average Exposure (TWAE) or Concentration (TWAC): Concentration in the exposure medium at each measured time interval multiplied by that time interval and divided by the total time of observation: for occupational exposure a working shift of eight hours is commonly used as the averaging time.

Timeout: A network interruption that occurs when the pharma net host computer cannot process transactions within a predetermined length of time. Timeouts can occur for various reasons.

Tincture: An extract, usually herbal, and usually made with a mixture of water and alcohol, although there were official tinctures that also used acetic acid, chloroform and glycerin. Only a few tinctures are still official in the U.S., including Tincture of Arnica and Compound Tincture of Benzoin. In herb commerce, the term should really only be appropriate when the extract at least resembles the formerly official methods for making plant extracts. The strength should be listed, usually as a ratio (1:5 being the most common) or a percentage (20%...the same strength as 1:5). Green Tinctures of fresh plants, are usually appropriate when defined as 1:2 or 50%. The alcohol percentage should be given, and, if below 45%, is made incorrectly. Dry plant tinctures, the norm, are official when percolated (usually), although maceration was and is allowed as an alternative method. The term Tincture is still pharmaceutical in implication, so the FDA periodically objects to its use in the herb industry.

Tinea Versicolor: A chronic skin fungus, often without symptoms-except the light skin splotches of infected surfaces that don't tan. It seems easily transmitted from one part of the body to another or one person to another. It is also called Pityriasis Versicolor.

Tineas: A dermatomycosis; various skin fungus infections, such as ringworm, athlete's foot, and so forth. It is generally slow to acquire and hard to get rid of.

Tinnitis: A ringing in the ears. It may be caused by viral infections of the middle and inner ear, allergies, stress, even drugs or environmental agents. Tenacious for some people, it often seems to occur when you have lots of things to do and little tolerance anyway.

Tinnitus: Continual noise in the ears, such as ringing, buzzing, roaring, or clicking.

Tipping Fee: A fee for the unloading or dumping of waste at a landfill, transfer, recycling center, or waste -to-energy facility, usually stated in dollrs per ton; also called a disposal or service fee.

Tissue Dose: Amount of a substance or physical agent (radiation) absorbed by a tissue.

Tissue Typing: Procedure that determi-

nes the HLA type of the donor and recipient.

Titubation: A regular nodding movement of the head that sometimes involves the trunk. The term can also refer to a staggering or reeling condition, especially due to disease of the spinal cord or cerebellum.

TMJ: The temporomandibular joint. These are the two joints that connect the jawbone to the skull under the zygomatic arch. TMJ syndrome involves pain in the joint, clicking in the joint from degradation of the sinovial fluids, and sharp, shooting pain when chewing. The two main causes are malocclusion (improper tooth alignment) and tension. Some people grind their teeth, others clench their jaws, perhaps from the inability to say what is felt. Chiropractors and osteopaths love helping these folks, some even specializing in TMJ work.

Tocolytic Therapy: This therapy works to prevent premature labor by halting contractions and slowing down the birth process.

Tolerable Daily Intake: (TDI): Regulatory value equivalent to the acceptable daily intake established by the European Commission Scientific Committe on food. Unlike the ADI, the TDI is expressed in mg/ person, assuming a body weight of 60 kg.

Tolerable Risk : Probability of suffering disease or injury that can, for the time being, be tolerated, taking into account the associated benefits, and assuming thtat the risk is minimized by appropriate control procedures.

Tolerance: Adaptive state characterized by diminished effects of a particular dose of a substance: the process leading to tolerance is called "adaption". In food toxicology, dose that an individual can tolerate can without showing an effect. Ability to experience exposure to potentially harmful amounts of a substance without showing an effect. In immunology, state of specific immunological unresponsiveness.

Tomentose: Having woolly hairs.

Tomography: A technique using X-rays or ultrasound to build up a focused image of a 'slice' through the body at a given level. By producing a series of such slices at different depths, a three- dimensional image of the body structures can be built up.

Tonic: A substance taken to strengthen and prevent disease, especially chronic disease. Formerly, tonics were widely available both as over-the-counter and prescription formulas. Unfortunately, the increased sophistication of medicine has led to the abandonment of preventative or strengthening approaches that utilize the innate abilities of an organism (like ourselves) to right itself with a little prodding in the correct direction. The last several decades have seen increased focus on disease-at-a-time medicine, with more and more patients receiving treatment at acute care facilities like hospitals and clinics, circumstances that delegate against preventative or tonic approaches. Tonics tend to stimulate deficient functions, therefore are best suited for functional disorders, not organic ones.

Tonics: These are placebos and may be used in conditions for which there is no known remedy, to strengthen and support the patient.

Tophus: The name given to urate based deposits which form in connection with joints or tendon sheaths as the result of attacks of gout

Topical Agent: A treatment, such as an ointment or cream that is applied to the surfaces of the skin rather than taken internally.

Topical: Pertaining to a particular area , as in topical effect that involves only the area to which the causative substance has been applied.

Torpor: A condition of bodily and mental inactivity, not amounting to sleep, but interfering greatly with the ordinary habits and pursuits. It is often found in people suffering from fever, and is a common symptom in aged people whose arteries are diseased.

Total Diet Study: Study designed to established the pattern of pesticide residue intake by a person consuming defined diet undertaken to show the range and amount of various foodstuffs in the typical diet or to estimate the total amount of a specific substance in a typical diet.

Total Parentral Nutritional (TPN): Administration of supplemental nutrition intravenously (through the veins) for patients who are unable to digest or absorb food.

Tourniquet: A device – usually a rubber cord or tight bandage – that is bound around a limb to stop temporarily the arterial blood supply and so control severe bleeding. Tourniquets should be applied with caution; unless untied within 15 minutes

Toxic Dose: Amount of a substance which produce intoxication without lethal outcome.

Toxic Substance: Material causing injury to living organisms as a result of a physiochemical interactions.

Toxic: Poisonous, everything, including water and oxygen is toxic in sufficiently high doses.

Toxicity Equivalency Factor (TEF): Factor used in risk assessment to estimate the toxicity of a complex mixture, most commonly a mixture of a chlorinated dibenzo-p-dioxins, furans and biphenyls: in the case, TEF is based on relative toxicity to 2,3,7,8 – tetrachlorodibenzo-p-dioxin (TEF = 1).

Toxicity Test: Experimental study of the adverse effects of exposure of a living organism to a substance for a defined duration under defined conditions.

Toxicity: An adverse effect produced by a drug that is detrimental to the participant's health. The level of toxicity associated with a drug will vary depending on the condition that the drug is used to treat. If toxicity prevents people from taking more of an experimental drug, the toxicity is called dose limiting. Capacity to cause injury to a living organism defined with reference to the quantity of substance administered or absorbed, the way in which the substance is administered (inhalation, ingestion, topical application, injection) and disturbed in time (single or repeated doses), the type and severity of injury, the time needed to produce the injury, the nature of organism(s) affected and other relevant conditions. Adverse effects of a substance on a living organism defined with reference to the quantity of a substance administered or absorbed, the way in which substance is administered (inhalation, ingestion, topical application, injection) and distributed in time (single or repeated doses), the type and severity of injury, the nature of organism(s) affected, and other relevant conditions.

Toxicofication: metabolic conversion of a potentially toxic substance to a product that is more toxic.

Toxicokinetics: Process of the uptake of potentially toxic substance by the body, the biotransformation they undergo, the distribution of the substances and their metabolites in the tissues, and the elimination of the substances and their metabolites of the body. Both the amounts and the concentrations of the substances and their metabolites are studied. The term has essentially the same meaning as pharmacokinetics, but the later term should be restricted to the study of pharmaceutical substances.

Toxicological Data Sheet: Document that gives in a uniform manner data relating to toxicology of a substance, its production and application, properties and methods of identification; the data sheet may also include recommendations on protective measures.

Toxicological Testing: Studies to determine the potential risk of a compound man and enviourment

Toxicology: Scientific discipline involving the study of the actual or potential danger presented by the harmful effects of substances (poisons) on a living organism ecosystem, of the relationship of a such a harmful effects to exposure, and the mechanisms of action, diagnosis, prevention and treatment of intoxications.

Toxicometry: Term sometimes used to indicate a combination of investigative methods and techniques for a making a quantitative assessment of toxicity and the hazards of potentially toxic substance.

Toxicophoric (Toxophoric) Group: structural moety that upon metabolic activation exerts toxic effects: the presence of a toxicophoric group indicates only potential and not necessarily actual toxicity of a drug or other or other substances.

Toxicovigilance: active process of identification, investigation and evaluation of various toxic effects in the community with a view to taking measures to reduce or control exposures involving the substances which produces these effects.

Toxigenomics: Application of genetic and genomics methods to the study of toxicology.

Toxin: poisonous substance produced by a biological organism such as microbe, animal or plant.

Toxinology: scientific discipline involving the study of the chemistry, biochemistry, pharmacology and toxicology of toxins.

Toxoid: A toxin which has been rendered non-toxic by certain chemicals, or by heat, or by being partly neutralised by antitoxin.

Tracer: A compound introduced into the body, the progress of which can subsequently be followed and information obtained about the body's metabolic activities. Radioactive tracers are compounds labelled with radionuclide which give off radiation, for example a radioactive isotope may reolace a stable chemical element in atoxic compound enabling the toxicokinetics to be followed. Labelled member of a population used to measure certain properties of that population.

Trachea: The cartilage tube that brings air from the larynx to the two bronchi of the lungs. It is lined with mucus membranes and ciliated epithelia.

Trachoma: Trachoma is a severe type of conjunctivitis. This chronic contagious condition is caused by Chlamydia trachomatis, a bacterium with virus-like characteristics.

Trademark: A name ,word symbol or phase used to identify a particular product.

Training Log: A documented trail of all the training undertaking by clinical research personnel. It generally include the topic of the training modality completion date and signature of the personnel.

Training Records: A documented trail of all the training undertaking by clinical research personnel. It generally include the topic of the training modality completion date and signature of the personnel.

Trance: A profound sleep from which a person cannot for a time be aroused, but which is not due to organic disease. The power of voluntary movement is lost, although sensibility and even consciousness may remain

Transaction: TIP this prescriber identification transaction on pharmanet is used to obtain information on a practitioner name or by the unique identification number assigned to the practitioner by the appropriate regulatory body.

Transcription: Process of transforming dictated or otherwise documented information into an electronic format.

Transfusion: The administration of any fluid into a person's vein using a drip.

Transgeric: Adjective used to describe animals carrying a gene introduced bt micro organisms.

Transitional Benefits/ Plans: When an employer changes, transition plans enable participants already in treatment to transition to an network heath provide a specific number of days to contact CIGNA Health care in order to discuss the patient's treatment plan and obtain authorization to continue treatment at all in-network benefit level for a specified period of time , or to transition to a contracted professional.

Transitional Rehabilitation: To prepare for community re- entry. Non-medically based. Emphasis on functional skills for maximum independence. May be in group home or part of a continuum of rehabilitation centre. Typical stay 4-8 months. Greater focus on compensation vs. restoration.

Translation: The act of transforming document text or phrase or one phase from one language

Translocation Science Research: Studies involving the translation of basic science research finding into applications that benefit people.

Transporter: A transporter is a person engaged in the off-site transportation of hazardous waste by air, rail, highway, or water. Anyone who moves hazardous waste off the waste the site where it is generated or the site where it is being treated, stored, or disposed of, will be subject to transporter standards.

Trans-Sexualism: The psycho-sexual condition characterised by feelings of belonging to the gender opposite to that of the genitalia and the secondary sex characteristics.

Transudation: The passage of liquid – called the transudate –through a membrane: for example, the passage of blood through the wall of a capillary vessel.

Trauma: A term used in a physical sense as a wound or injury.

Treatability: In relation to waste water, the amenability of substance to removal without adversely affecting the normal operation of biological treatment processes (such as sewage treatment plant).

Treatment Effect: An effect attributed to a treatment in a clinical trial. In most clinical trials the treatment effect of interest is a comparison (or contrast) of two or more treatments.

Treatment Ind : A type of non commercial investigational new drug application where by food and drug administration can permit an investigational drug to be used while the final clinical work and review is still ongoing ,, if there is preliminary evidence drug efficacy and the drug is intended to treat a serious or life threading debases, or if there is no comparable alternative drug or thereby available to treat a serious or life life treating disease , or if there is no comparable alternative drug or there by available to treat that stage of disease in the intended patient population.

Treatment Trials: Refers to trials which test new treatments, new combinations of drugs, or new approaches to surgery or radiation therapy.

Treatment, Storage and Disposal: Facilities subject to permit requirements and to a comprehensive set of EPA regulations governing location, design, facility operation, and closure of TSD facilities.

Treatment: Actions to change the physical, chemical, or biological character or composition of waste to neutralize, recover energy or resources, reduce toxicity, reduce volume, or make it safer to store or discard.

Treatment-Emergent Adverse Event: An event that emerges during treatment, having been absent pretreatment, or worsens relative to the pretreatment state.

Trellis Plots: Graphical display of relationship between variable. These are useful for investigation the need of subgroup analysis.

Tremor: A fine involuntary movement. Tremors may be seen in projecting parts like the hands, head and tongue, or they may involve muscles

Triage: Assessment of sick, wounded and injured persons following a disaster to determine priority needs for efficient use of available medical facilities.

Trial Arm: The treatment group of a clinical trial.

Trial Audit: A systematic and independent examination of trial related activities and documents to determine whether the evaluated trial related activities were conducted and the data were recorded , analyzed and accurately reported according to the protocol standard operating procedure, GCP and the applicable regulatory requirements.

Trial Drug: A drug that is initially dispensed in a trial quantity (14 day supply) under the Trial Prescription Program.

Trial Master File: A central file in which all the essential clinical trial documents are filed.

Trial Number: A unique identifier assigned to a clinical study for its easy identification.

Trial Prescription Program: A Pharma Care program that pays the dispensing fee associated with providing a small initial quantity of a drug to a consumer who has not used the drug before, in order to determine whether use of the drug is effective or subject to adverse side effects.

Trial Site: The location(s) where trial-related activities are actually conducted.

Trial Site: The location or facility where a clinical trial is conducted.

Trial Statistician: A statistician who has a combination of education/training and experience sufficient to implement the principles in the ICH E9 guidance and who is responsible for the statistical aspects of the trial.

Trial Subject: An individual who participates in a clinical trial either as a recipient of the investigational product or as a control.

Trial Suspension: Pre mature termination of a clinical trial due to lack of efficacy or safety or funding.

Trichiasis: A condition in which the eyelashes become ingrown

Trifoliate: Having three leaflets in a compound leaf, like a clover.

Trigeminal Neuralgia: Facial neuralgia or tic doulourex. This is pain of the gasserian ganglion or one or more branches of the trigeminal nerves. It is felt as pain along the side or top of the head, the scalp and around the eyes, a "skin headache" and sometimes accompanied by facial muscle cramps. It is usually initiated by trigger points, with blood sugar irregularities and substance sensitivities often lowering their threshold of irritation.

Trigone: This is the triangular basement muscle of the urinary bladder. It differs in structure and nerves from the top of the bladder, the detrusor muscle, which expands as the bladder fills, and contracts during urination under parasympathetic nerve stimulus. The trigone does not expand, is under sympathetic nerve stimulus, and supplies the rigidity and sphincter support for the urethra in front and the ureters in back.

Trimester: The three three-month sections of a pregnancy.

Tripinnate: Thrice pinnately compound leaf.

Triple Blind: A study design in which the knowledge of treatment assignment is concealed from the personnel who organize and analyze the data as well as from the investigators and study subjects.

Triple-Blind Study: A study in which knowledge of the treatment assignment(s) is concealed from the people who organize and analyze the data of a study as well as from subjects and investigators.

Triplicate prescription program: A College of physician & Surgeons program that monitors the prescribing and use of certain drugs with a potential for a misuse.

Trophic level: Amount of energy in terms of food that an organism needs : organism not needing organic food, such as plants, are said to be on a low trophic level, whereas predator species needing food of high energy content are said to be on a high tropic level. The tropic level indicates the level of the organism in the food chain.

Trophic: A term applied to the influence that nerves exert with regard to the healthiness and nourishment of the parts to which they run. When the nerves become diseased or injured, this influence is lost and the muscles waste, while the skin loses its healthy appearance and is liable to break down into ulcers

Trust: A written legal instrument created by a grantor during his or her lifetime or at death for the benefit of another.

Trustee: The individual or institution enstrusted with the duty of managing property placed in the trust. A "co-trustee" serves as trustee with another. A "contingent trustee" becomes trustee upon the occurrence of a specified future event.

T-Test: A statistical test used to compare the means of two groups of test data.

Tubal Pregnancy: Also known as ectopic pregnancy. Implantation of the embryo in one of the fallopian tubes, rather than in the lining of the uterus. The patient usually complains of pain between six and ten weeks' gestation and, if the Fallopian tube is not removed, there may be rupture with potentially life-threatening haemorrhage.

Tuber: A fleshy, underground part of a stem or root. Example: potato.

Tumor Necrosis Factor: A chemical substance, which deliver messages between cells in the body, which is involved in inflammation.

Turbinates: The three nasal conchae, bone ridges that help spiral and flutter inhaled air, increasing the efficiency of heating, moistening and cleansing

Two Stage Design: A study in which response rate is assessed after a predetermined number of patients have been observed and the study is either terminated or allowed to continue depending on the outcome.

Type 1 (or Type I) Error: Error made when a null hypothesis is rejected but is actually true. Synonym: false positive.

Type 1 diabetes: A Condition in which the pancreas make so little insulin that the body can't use blood glucose as energy. Type 1 diabetes most often occurs in people younger than age 30 and must be controlled with daily insulin injections.

Type 2 (or Type II) Error: Error made when an alternative hypothesis is rejected when it is actually true. Synonym: false negative.

Type 2 diabetes: A Condition in which the body either makes too little insulin or can't use the insulin it makes to use blood glucose as energy. Type 2 diabetes most often occurs in people older than age 40 can often be controlled through meal plans and physical activity plans. Some people with type 2 diabetes have to take diabetes pills or insulin.

Type 3 (or Type III) Error: Some statisticians use this designation for an error made when calling the less effective treatment the more effective one.

U

U.S.P.-N.F: ***USP and National Formulary. The U.S.P. was first published in 1820 and ever ten years thereafter until the Second World War, after which it has been revised every five years. It has always been meant to define the physical, chemical, and pharmaceutical characteristics of the most accepted and widely used drugs of the time, and to set the standards for purity. The N.E was first published in 1888, and, up until 1980, in the same year as the United States Pharmacopoeia. Since 1980, both have been issued in the same volume. The National Formulary was originally intended as a list of the official recipes for pharmaceutical formulas; characteristics of those drugs or plants used in the formulas or that were still recognized as secondary drugs; and the substances needed for the manufacturing of drugs but that were not active, like gelatin or pill binders.***

UCSF link: The name of the VPN service provider by UCSF that enables you to connect to UCSF-restricted serves from a standard Internet connection.

UD: Unit Dose Dispensing Fee. Reimbursement to providers when a qualified unit dose dispensing system used. The drugs may be packaged into unit doses by the labeller or the provider.

UF: Ultrafiltration. The removal of fluid down to pressure gradient.

Ulcer: Defect, often associated with inflammation, occurring locally or at the surface of an organ or tissue owing to sloughing of necrotic tissue.

Ulcerative Colitis: Chronic inflammation of the lining of the colon and rectum. Symptoms the onset may be sudden or insidious. In the acute form there is severe diarrhoea and the patient may pass up to 20 stools a day.

Ulna: The inner of the two bones in the forearm.

Ultra Filtration: Filtration carried out under pressure. Blood undergoes ultrafiltration in the kidneys to remove the waste products, urea and surplus water that constitute urine.

Ultra Sonography: The use of ultrasound to produce images of structures in the body that can be viewed on a television screen and transferred to photo- graphic film.

Ultrasound: Ultrasound, or ultrasonic, waves comprise very high frequency sound waves above 20,000 Hz that the human ear cannot hear. Ultrasound is widely used for diagnosis and also for some treatments. In obstetrics, ultrasound can assess the stage of pregnancy and detect abnormalities in the fetus it is a valuable adjunct in the investigation of diseases in the bladder, kidneys, liver, ovaries, pancreas and brain

Ultraviolet Rays (UVR): Invisible light rays of very short wavelength beyond the violet end of the sun's spectrum. Ultraviolet-C wavelength < 290 nm

Umbel: A flowering head where the pedicels (individual flower stems) all spring from one point, usually the end of the peduncle. Compound umbels, found in some Umbelliferae, have umbels branching from peduncle umbels that themselves are branching from the main stem.

Umbilical Cord: The fleshy tube containing two arteries and a vein through which the mother supplies the fetus with oxygen and nutrients. The cord, which is up to 60 cm long, ceases to function after birth and is clamped and cut about 2·5 cm from the infant's abdominal wall.

Umbilicus: The scientific name for the navel, a circular depression in the abdomen that marks the areas where the umbilical cord was attached when the fetus was in the uterus.

Unblinding: Identification of the treatment code of a subject or grouped results in studies where the treatment assignment is unknown to the subject and investigators. The act of breaking the blinding codes of a clinical trial.

Uncertainly Factor: 1. In essay methodology, confidence interval or fiducial limit used to assess the probable precision of an estimate. 2. In toxicology, value used in extrapolation from experimental animals to man (assuming that man may be more sensitive) or from selected individuals to the general populations: for example, a value applied to the no-observed effect level (NOEL); or no-observed adverse effect level (NOAEL) to derive an acceptable daily intake or reference dose (the NOEL or NOAEL is divided by the value to calculate the acceptable daily intake). The value depends on the nature of the toxic effect, the size and the type of population to be protected, and the quality of the toxicological information available.

Uncinate Fit: A type of temporal lobe epilepsy in which a patient has a hallucination of smell or of taste; it may also be the result of a tumour pressing on that part of the brain concerned with the appreciation of smell and taste.

Unconscious: A state of unconsciousness or a description of mental activities of which an individual is unaware.

Unconsciousness: The brain is the organ of the mind. Normal conscious alertness depends upon its continuous adequate supply with oxygen and glucose, both of which are essential for the brain cells to function normally. If either or both of these are interrupted, altered consciousness results.

Uncontrolled Clinical Trial: A trial design that does not contain a control arm.

Unequal Randomization: A technique used to allocate subjects into groups at a differential rate; for example, three subjects may be assigned to a treatment group for every one assigned to the control group.

Unexpected Adverse Drug Reaction:

An adverse reaction, the nature or severity of which is not consistent with the applicable product information (e.g., investigator's brochure for an unapproved investigational product or package insert/summary of product characteristics for an approved product).

Unexpected Adverse Drug Reaction: An adverse reaction the nature or severity of which is not consistent with the applicable product information.

Unguentum: The Latin name for ointment.

Unified Credit: A federal tax credit that offsets gift tax and estate tax liability. For gift tax purposes, the unified credit remains at $345,800 through 2009, which is equivalent to an applicable exclusion amount of $1 million. For estate tax purposes, the unified credit is being gradually increased from $345,800 in 2003 to $1,455,800 in 2009, which is equivalent to an applicable exclusion amount of $1 million in 2003 to $3.5 million in 2009.

Uniform Hazardous Waste Manifest: A control and transport document that - accompanies hazardous waste at all times.

Unipolar: Having only one polarity; primarily in reference to individuals who only manifest a manic or depressive phase in personality or thyroid bipolarity.

Unit Dose Drug Dispensing: A technology that distributes all medications to the clinical unit in individual packages that identify its name and dose. The packaging is safeguard that ensures identification of the drug up to the moment it is administered to the patient. Unit-dose packaging is a major safety measure that helps to reduce the risk of drug distribution errors.

Unit Risk: Incremental upper-boundary lifetime risk estimated to result from lifetime exposure to an agent if it in air at a concentration of 1mg/L.

Unit: The term applied to a quantity assumed as a standard for measurement

UNOS (United Network for Organ Sharing): A government funded organization overseeing organ transplant statistics and allocation in the United State.

Unplanned Analyses: A situation where the trial data is analyzed at a time point which is not specified in the protocol or statistical analysis plan.

Upper boundary: Estimate of the plausible upper limit to the true value of a quantity: this is usually not a statistical confidence limit.

Upper Limb Disorders: A group of injuries resulting from overuse of a part of the limb. One example is Tennis Elbow (epicondylitis) caused by inflammation of the tendon attaching the extensor muscles of the forearm to the humerus because of overuse of the muscles. Overuse of the shoulder muscles may cause inflammation andpain around the joint.

Uptake: Entry of a substance into the body, into an organ, into a tissue, into a cell, or into the body fluids by passage through a membrane or by other means.

Urachus: A corded structure which extends from the bladder up to the navel, and represents the remains of the canal which in the fetus joins the bladder with the allantois.

Uraemia: Retention of excessive by-products of protein metabolism in the blood due to nephrons being unable to excrete urea and other products of protein metabolism. (a) Pre-Renal – Uraemia due to inhibition of renal function by disorders not related to urinary tract.

Urate: The salts of uric acid, found in the urine, some kidney stones, and (unfortunately) in gouty joints.

Urea: One of the chief waste products of the body. When the body breaks down food, it uses what it needs and throws the rest away as waste. The kidneys flush the waste from the body in the form of urea, which is the urine.

Ureteralgia: Spasm or pain of the ureters, the ducts that milk urine from the kidneys to the bladder.

Urethral Syndrome: A group of symptoms of unknown cause. It mainly affects women, and occasionally men, with pain and discomfort in the lower abdomen, a frequent urge to urinate and, in women, pain in the area of the vulva

Urethritis: Any inflammation of the urethra, whether from external irritation, overly acidic or scalding urine, passage of stones, or an active infection of the canal.

Urethroplasty: Surgical repair of the urethra, usually to relieve a stricture

Urgent Care: When prompt medical attention is needed in a non-emergency situation, that's called "urgent" care. Examples of urgent care needs include ear infections, sprains, high fevers, vomiting and urinary tract infections. Urgent situations are not considered to be emergencies.

Uric Acid: The final end product of certain native or dietary proteins, especially the nucleoproteins found in the nucleus of cells. Unlike the much smaller nitrogenous waste product urea, which is mostly recycled to form many amino acids, uric acid is an unrecyclable metabolite. It is a bent nail that won't restraighten, and it must be excreted: nucleoprotein to purine to uric acid to the outside in the urine or the sweat.

Uricosuric Drug: A drug that increases the amount of uric acid excreted in the urine. Among the drugs used are probenecids or a sulfa derivative. Uricosurics are used to treat gout and other disorders which cause raised blood-uric-acid concentrations.

Urinary Tract (UT): The kidneys and the lower urinary tract, which includes the ureters, bladder, and urethra.

Urinometer: A simple instrument designed for estimating the specific gravity of urine – a test that can be helpful in diagnosing disorders of the urinary tract.

Urobilinogen: A chemical compound formed when bacteria in the intestine act on bilirubin. Some is reabsorbed and returns to the liver and some is eliminated in the faeces.

Urocele: A cystic swelling that develops in the scrotum when urine escapes from the urethra usually after injury.

Urodynamics: The measurement of the pressures within the urinary bladder as well as the pressures of the urethral sphincter.

Urokinase: Urokinase is an enzyme obtained from urine which dissolves blood clots. It is used to treat thrombolysis in the eye, in arteriovenous shunts and deep-vein thrombosis.

Ursodeoxycholic Acid: A preparation used in the treatment of cholesterol gallstones when laparoscopic cholecystectomy and endoscopic biliary procedures cannot be used

Urticaria: Vascular reaction of the skin marked by the transient appearance of smooth, slightly elevated patches that are redder or paler than the surrounding skin and often attended by severe itching.

Use and Reuse: The return of a waste material either to the originating process as a substitute for an input material, or to other another process as an input material.

User Id: The name or alphanumeric code by which the user is identified and grains acess to a computer system or network.

USP: United States Pharmacopaeia.

Usual and Customary Charge: the amount charged by the provider for the same service when provided to private pay patients.

UTI: Urinary Tract Infection.

UVA Phototherapy: Artificial light treatment with a long wavelength that penetrates deeper into the skin than UVB for the treatment of psoriasis.

UVB Phototherapy: The more common of the treatment used for mild psoriasis to slow the abnormal growth of skin cells.

Uvea: Uvea is a term applied to the middle coat of the eye, including the iris, ciliary body and choroid.

Uvula: The small mass of muscle covered by mucous membrane that hangs down from the middle of the soft palate on its posterior aspect. Its function is not certain and it seldom causes problems.

V

Vaccination: A shot given to protect against a disease

Vaccine: A preparation, often made of weakened viruses of a disease, used to inoculate a person in order to protect him or her from that disease by causing the formation of antibodies against it.

Vacuole: Membrane-bound cavity within a cell.

Vagina: A Canal in females from external genitalia (vulva) to the cervix of the uterus.

Vaginitis: An inflammation of the vagina, either from simple tissue irritation or from an infection

Vaginosisl: A vaginal infection characterized by a smelly discharge and the presence of Gardnerella, Mycoplasma, and other anaerobic bacteria, with the lack of Lactobacillus species.

Vagus Nerve: Also called the pneumogastric nerve, this is the tenth cranial nerve, with many fibers leading to parasympathetic ganglia in internal organs, and can be considered the presynapse starter for the upper parts of the parasympathetic functions.

Validation Certificate: A document to demonstrate that a procedure process and activity will consistently lead to the expected results.

Validation of Data: Procedure carried out to ensure that the data contained in the final clinical trial report match original observations.

Validity of a Study: Degree to which the inferences drawn from study, especially generalizations extending beyond the study sample are warranted when account is taken of the study methods, the representativeness of the study sample, and the nature of the population from which it is drawn.

Validity: the degree to which output reflects, i.e. input , the degree to which output is a function of known input and it alone. The accuracy of the relationship between two or more variables.

Valium: The proprietary name for diazepam, a widely used anxiolytic drug.

Valsalva's Manoeuvre: This is carried out by closing the mouth, holding the nose and attempting to blow hard.

Valves: These cup-like structures are found in the heart, veins, and lymphatic vessels.

Valvotomy: An operation that opens a stenosed heart valve and allows it to function properly again.

Valvuloplasty: An operation to repair or reconstruct a defective heart valve.

Van Den Bergh Test: A test done on serum from patients with jaundice to discover whether the excess bilirubin in the blood which causes the jaundice is conjugated or unconjugated. If conjugated, this indicates that haemolysis is causing the jaundice; if unconjugated, disease of the liver or bile duct.

Vancomycin: An antibiotic derived from streptomyces, which is active against a wide range of gram-positive organisms.

Vaporiser: A device that via a narrow nozzle turns water or a drug into a fine spray, thus enabling medicine to be taken by inhalation.

Variance: A measure of the variability in a sample or population. It is calculated as the mean squared deviation (MSD) of the individual values from their common mean. In calculating MSD, the divisor n is commonly used for a population variance and the divisor n-1 for a sample variance.

Varicosities: Enlarged veins or an engorged complex of smaller vessels. Various bismuth chelates and complexes, such as sucralfate, effective in healing gastric and duodenal ulcers are available. They may act by a direct toxic effect on gastric helicobacter pylori, or by stimulating mucosal prostaglandin or bicarbonate secretion.

Vasa Efferentia: Efferent seminal ducts of the testis; these carry semen from the testis to the head of the epididymis.

Vascular: Pertaining to blood vessels or relating to the blood vessels.

Vasculitis: Inflammation of the blood vessels. This may damage the lining of the vessels and cause nar- rowing or blockage, thus restricting blood flow.

Vasectomy: The surgical operation performed to render men sterile, or infertile. It consists of ligating, or tying, and then cutting the ductus, or vas, deferens.

Vasocholinergic: An agent that stimulates blood flow to the viscera, and more closely mimicking the balance of circulation induced by parasympathetic states. This is one way to oppose excessive adrenergic circulatory states.

Vasoconstriction: Decrease of the calibre of the blood vessels, leading to a dereased blood flow. A nerve, agent or substance that narrows blood vessels.

Vasodialation: Increase in the calibre of the blood vesseks leading to an increased blood flow.

Vasodilation, peripheral: The increase of blood into the skin, resulting from the relaxation of the small arterioles that lead into the capillary beads at the edges of the body. This is a gentle way to lessen early high blood pressure, decreasing the difficulty of pushing columns of arterial blood through miles of capillaries.

Vasodilator: Nerves, hormones or substances (like herbs) that induce the relaxation of blood vessels.

Vasoneurosis: Spasms and cramps of blood vessels that are caused by neurologic factors. Also called angioneurosis

Vasopressin: The fraction isolated from extract of the posterior pituitary gland lobe which stimulates intestinal activity, constricts blood vessels, and inhibits the secretion of urine.

Vasovagal Attack: The temporary loss of consciousness caused by an abrupt slowing of the heartbeat. This may happen following shock, acute pain, fear, or stress.

Vector: An animal that is the carrier of a particular infectious disease.

Vegetations: Roughenings, comprising fibrin and blood cells, that appear upon the valves of the heart, usually as the result of acute rheumatism.

Vehicle: Substance(s) used to formulate active ingredients for administration or use(general term for solvents, suspending agents, etc).

Veins: The vessels which return the blood to the heart after it has circulated through the tissues.

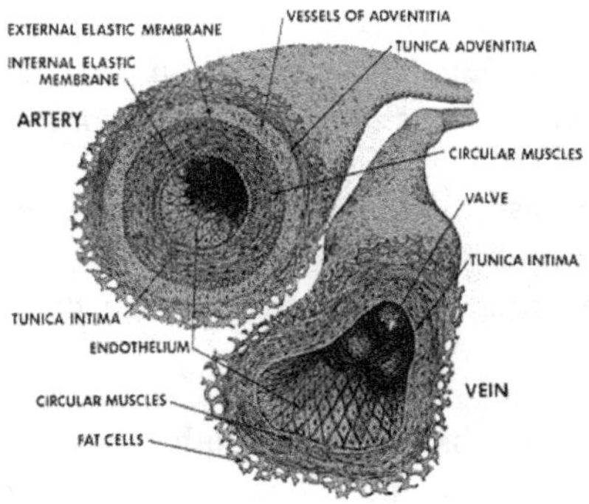

Vendor: Supplier of goods or services.

Venepuncture: The insertion of a needle into a vein, usually for the purpose of injecting a drug or withdrawing blood for hematological or biochemical analysis.

Venereal Diseases: Sexually transmitted diseases (STDS).

Venereal Warts: Caused by human papillomavirus (HPV) and also known as condylomata acuminata, anal warts, and genital warts. It is nearly always transmitted from person to person by sexual contact, can increase the risk for women of cervical cancer, and occurs in near epidemic proportions in sexually active teenage women.

Venom: Animal toxin generally used for self defence or predation and usually delivered by a bite or sting.

Venosities: Enlarged varicose veins on the cervix of the uterus, often accompanying ulcerations or long-term pelvic congestion. A symptom only of congestion or impaired circulation, they can occur in both semi-trivial and serious conditions.

Venosity: An area where there is a buildup of excess venous blood, with enlarged veins and tissue congestion.

Venous Stasis: Having congested venous blood and lymph. Usually a larger condition effecting tissue or organ function, as opposed to the more vascular implications of venosities and varicosities.

Venous: Pertaining to the veins, or more broadly to include both venous AND lymphatic circulation.

Ventilation, Artificial: The procedure usually carried out in an operating theatre or intensive-care unit, in which a device called a ventilator, takes over a person's breathing. This is done for someone who is unable to breathe normally.

Ventilation: Process of supplying a building or room with fresh air. Process of exchange of air between the ambient atmosphere and the lungs, in physiology, the amount of air inhaled per day. Oxygenation of blood.

Ventricular Fibrillation: Irregular heartbeat characterized by uncoordinated contractions of the ventricle.

Ventricular Hypertrophy: Enlargement of the ventricular chambers of the heart, a common complication of hypertension and coronary artery disease

Ventricular Septal Defect: An inherited defect of the heart. The septum (partition) separating the two ventricles is pierced by a hole which, if large, results in blood being diverted to the lungs at a greater pressure than normal.

Ventrosuspension: This is a surgical procedure to fixate a displaced uterus to the front wall of the abdomen.

Verapamil: Verapamil is a drug used in the treatment of hypertension, disordered rhythms of the heart, and angina pectoris. The drug is one of the calcium channel blockers

Vermicide: Substance intended to kill worms.

Vermifuge: Substance that causes the explusion of intestinal wrms.

Vernacular Language: The native language of a country or a place.

Vernix: The white, cheese-like substance that covers the skin of a newborn infant. It consists of dead cells in a fatty secretion, protects the infant's skin, and helps lubricate its passage through the cervix and vagina during delivery.

Version Control : The act that captures

the details of personnel who visits a clinical trial site.

Version: The number assigned to an essential document in use. Version number is important to provide an audit trial.

Version: The name given to an operation in obstetrics which consists in turning the fetus in the uterus where the fetus is lying in an abnormal position which may make eventual delivery difficult.

Vertebra: One of the irregularly shaped bones that together form the vertebral column. Vertebrobasilar Insuffiency Intermittent incidents of double vision, dizziness, weakness and speaking difficulties caused by a reduced blood supply to parts of the brain

Vertigo: Dizziness; the illusion of movement as if the external world were revolving around an individual or as if the individual were revolving in space.

Vesical Irritation: In my context, irritation of the bladder and urethra.

Vesicle: Small sac or bladder containing fluid. Blisterlike elevation on the skin containing serous fluid.

Vesico-Ureteric Reflux: The back flow of urine from the urinary bladder into the ureters. The cause is defects in the valves which normally prevent this reflux from occurring.

Vesicovaginal Fistula: A false communication between the urinary bladder and the vagina. The result is urinary incontinence. Surgical damage to the bladder during operations for gynaecological dis- orders is one possible cause.

Vesicular Breathing: Normal breath sounds heard in the lung by means of a stethoscope. These are soft regular sounds which become altered by disease.

VHA (Veterans Health Administration): VHA is a division of VA responsible for providing medical care to eligible veterans.

Vincent's Angina: An ulcerative inflammation of the throat, often foul-smelling, and caused by large, spindle- shaped bacilli (fusobacterium) and spirilla.

Vincents Infection: Trench mouth or NUGS. It is usually a symptom of extreme physical stress, nutritional deficiencies and heavy metal poisoning.

Viraemia: A condition occurring at various times in some viral infections in which the infecting virus is present in large amounts in the blood.

Virilisation: The masculinisation of women suffering from excessive production of the male hormone androgen. The person develops temporal balding, a male body shape, increased muscular bulk, deepening of the voice, an enlarged clitoris and hirsutism. Virilisation may also occur in women who take synthetic androgens,

Virilism: The condition in which masculine characteristics develop in the female; it is commonly the result of an overactive suprarenal gland, or of a tumour of its cortex. It may also result from an androgen-secreting ovarian tumour and also from the polycystic ovary syndrome.

Viscera: The general name given to the larger organs lying within the cavities of the chest and abdomen.

Visit: A clinical encounter for a subject in a trial. Visits are frequently referred to as occurring on Day X or during Week Y; there may be gaps between visits, which can take place within an epoch or span an epoch.

Visiting professor: Senior scientist, who is often from a foreign country, conducting research in a specific field of study.

Visitor's Log: A log that captures the details of personnel who visits a clinical trial site.

VISN (Veterans Integrated Service Network): The VA medical system is organized into 22 geographic networks known as VISNS. Each VISN contains two more VAHCSs (VA health care systems) as well as outpatient clinics.

VISTA (Veterans Health Information Systems And Technology Architecture): VISTA is the primary responsibility of clinical, administrative/financial and infrastructure data in VA. It consists of computer systems at each VA medical centre and the national network that links them.

Visual Evoked Response: Stimulation of the retina of the eye with light causes changes in the electrical activity of the cerebral cortex. These changes can be measured from outside the skull and can give valuable information about the state of the visual pathway from the retinal ganglion cells to the occipital cortex.

Vital Capacity: The amount of air that can be forcibly exhaled from the lungs after a deep inspiration

Vital force: the energy that maintains life in the individual.

Vitamin: A term applied to a group of substances which exist in minute quantities in natural foods, and which are necessary to normal nutrition, especially in connection with growth and development. Some A, D, E and K are fat-soluble and can be stored in the body. The remainder C, B_{12} and other members of the B complex are water-soluble and are quickly excreted

Vitiligo: A disease in which small or large areas of skin lose their pigment and become white because of a reduction in the body's production of melanin. The hair may be similarly affected.

Vitrectomy: An operation to remove the blood that sometimes collects at the back of the eyes when a person has eye disease.

Vitreous Body: A semi-fluid, transparent substance which fills most of the globe of the eye behind the lens.

Viviparous: A term used to describe animal groups including most mammals in which the embryos develop inside the mother's body with the young being born alive

VLDL: Very Low Density Lipids. These are blood transport fats, consisting mainly of triglycerides (made from sugar by the liver) and loosely covered in specialized proteins

Vocal Resonance: The air carrying the voice produced in the larynx passes through the throat, mouth and nose. The shape and size of these structures will influence the timbre of the voice, or vocal resonance.

Volar: A term relating to the palm or sole.

Volatile Organic Compound: Constituents that evaporates at their temperature of use and which, by a photochemical reaction, will potentially produce smog under favourable climatic conditions.

Volkmann's Contracture: A rare condition in which, as a result of too great a pressure from splint or bandage in the treatment of a broken arm, the flexor muscles of the forearm contract and thus obstruct free flow of blood in the veins; the muscles then swell and ultimately become fibrosed.

Volume of Distribution: Apparent (hypothetical) volume of fluid required to certain the total amount of a substance in the body at the same concentration as that present in the plasma assuming equilibrium has been attained.

Voluntary Action Indicated: A classification of regulatory inspection outcome that requires voluntary action.

Voluntary: The act of giving ones own free will without any correction or undue inducement.

Volunteer: An individual who voluntary participates in a clinical trial either as a recipient of the product or as a control.

Von Recklinghausen's Disease: An inherited disease, now called neurofibromatosis. About one case occurs every 3,000 live births. The disease is characterised by tumours along the course of nerves which can be felt beneath the skin. Soft tumours may also develop beneath the skin.

Von Willebrand's Disease: A genetically determined blood disorder in which the affected person suffers episodes of spontaneous bleeding similar to that occurring in people with haemophilia.

Voting Members: The ethics committee member who participant in a meeting and provide their openion on a clinical trial proposal.

Voyeurism: The regular viewing of people who are naked or part-naked or who are taking part in intimate sexual activities. The voyeur's subjects are unaware that they are being watched. The voyeur, nearly always a man, usually becomes sexually excited and may induce orgasm by masturbation.

Vulnerable Population: Individual whose willingness to volunteer in a clinical trial may be unduly influnced by the expection, whether justified or not of benefits associated with participation or a retalitory response from senior member of a hieerchy in case of refusal to participates.

Vulnerable Subjects: Individuals whose willingness to volunteer in a clinical trial may be unduly influenced by the expectation, whether justified or not, of benefits associated with participation, or of a retaliatory response from senior members of a hierarchy in case of refusal to participate. Examples are members of a group with a hierarchical structure, such as medical, pharmacy, dental and nursing students, subordinate hospital and laboratory personnel, employees of the pharmaceutical industry, members of the armed forces, and persons kept in detention. Other vulnerable subjects include patients with incurable diseases, persons in nursing homes, unemployed or impoverished persons, and patients in emergency situations, ethnic minority groups, homeless persons, nomads, refugees, minors, and those incapable of giving consent.

W

WAC (Wholesale Acquisition Cost): Wholesale acquisition cost is a value meant to approximate a wholesaler's cost for a particular pharmaceutical.

Warfarin: An anticoagulant, It is indicated for the prophylaxis of embolisation in rheumatic heart disease and atrial fibrillation after prosthetic heart valve; prophylaxis and treatment of venous thrombosis and pulmonary embolism; and transient ischaemic attacks or episodes

Warning Letter: A written communication from FDA notifying an individual or firm that the agency considers one or more products, practices, processes, or other activities to be in violation of the Federal FD&C Act, or other acts, and that failure of created pages that draw on a database associated with the site.

Warts: Warts are small, solid outgrowths from the skin arising from the epidermis and caused by various subtypes of 'human papilloma virus'.

Wash Back: Method to minimise blood loss to the patient after haemodialysis. Saline is normally used to wash blood back from the dialysis.

Washout Period: A period in a clinical study during which subjects receive no treatment for the indication under study and the effects of a previous treatment are eliminated (or assumed to be eliminated).

Wassermann Reaction: A test introduced for the diagnosis of syphilis by examination of the blood. It has now been largely supplanted by other, more specific tests.

Waste Minimization: The reduction, to extent feasible, of hazardous waste that is generated or subsequently treated, stored, or disposed of, it includes any source reduction or recycling undertaken by a generator.

Waste Prevention: Any change in the design, manufacturing, purchase or use of materials or products (including packaging) to reduce their amount or toxicity before they become municipal solid waste. waste prevention also refers to the reuse of products or materials.

Waste Reduction: Preventing or decreasing the amount of waste being generated through waste prevention, and recycling or by purchasing recycled and environment preferable products.

Waste Stream: The by product produced during the life cycle of any given product or material. This material may be a contaminated that may or may not exceed maximum acceptable contaminated levels.

Waste: Anything that us discarded deliberately or otherwise disposed of on the assumption that it is no further use to the primary user.

Wasting Syndrome: Disease marked by weight loss and atrophy of mascular and other connective tissues that is not directly related to a decrease in fod and water consumption.

Water Bed: A bed with a water filled mattress can help prevent bed sores in patients confined to bed for more than a few days.

Water Intoxication: A disorder resulting from excessive retention of water in the brain. Main symptoms are dizziness, headache, confusion and nausea. In severe cases the patient may have fits or lose consciousness.

Water Brash: A symptom of indigestion; during the course of digestion, the mouth fills with tasteless or sour fluid, which is generally saliva, but is sometimes brought up from the stomach. This is accompanied by a burning pain often felt at the pit of the stomach or in the chest.

Wax: This is used in medicine as an ingredient of ointments, plasters, and suppositories. It is used either as yellow wax derived directly from honeycomb, or as white wax, which is the same substance bleached. It is also used in the form of paraffin wax to apply heat in the relief of rheumatic pains.

WBC: White Blood Cells, including those of innate immunity, including basophils, neutrophils, eosinophils, monocytes, macrophages (and others) and those of acquired immunity, the various types of lymphocytes. Also called leukocytes.

WCB (workers compensation board): A British Columbia provincial government agency.

Weals: Raised white areas of oedema in the skin with reddened margins, which may result from sharp blows, or accompanied by itching, may be a symptom of urticaria.

Weaning: The process by which a baby is introduced to solid foods after having only had breast milk or artificial milk to drink.

Web Browser: A computer program that interprets HTML and other Internet languages and protocols and displays Web pages on your computer monitor.

Web Page: A single page on a Web site, such as a home page.

Web Server: A computer program that delivers HTML pages or files. Sometimes the computer on which a server program runs is also referred to as a server.

Web Site: A collection of Web pages and other files. A site can consist of a single Web page, thousands of pages, or custom endpoints might be facts about assessments. When a "variable" is captured or measured, there is no necessary sense that any evaluation or judgment is involved. However, when a variable is to be measured that obviously or actively pertains to subject status, which is always the concern of the physician, that variable becomes or will always be an assessment. The term assessment is intended to invoke some degree of evaluation or judgment concerning subject status. See also primary variable, secondary variable, supporting variable.

Webbed Fingers: Or toes, A deformity sometimes present at birth, and which tends to run in families. The web may be quite a thin structure, or the fingers may be closely united by solid tissue.

Weber's Test: A test with a tuning fork that is used to assess a person's deafness.

Weighting: An adjustment in a value based on scientific observations within the data.

Weight-of-Evidence for Toxicity: Extent to which the available biomedical data support the hypothesis that a substance causes a defined toxic effect such as cancer in humans.

Well Being: A state of physical and mental soundness. The physical and mental integrity of the subjects participating in a clinical trial.

Werner's Syndrome: A rare inherited condition in which the affected person suffers from premature ageing from adolescence onwards.

Wernicke's Encephalopathy: Also - called the Wernicke- Korsakoff syndrome, this uncommon disorder is characterized by mental confusion or delirium It is caused by a deficiency of vitamin B1 (thiamine) which affects the brain and nervous system.

Werthheim's Hysterectomy: A major operation done to remove cancer of the uterus or ovary

West Nile Virus: A mosquito-borne viral infection that is normally harmless to healthy people who, if infected, develop a mild flu-like illness.

Wheal: An inflammatory response to mild skin irritation, with a well-defined, raised redness, lasting for perhaps an hour and then disappearing. The cause is usually atopic allergies in an IgE-excess person, although mild, subclinical adrenocortical deficiency can be another factor.

Wheezing: A popular name applied to the various sounds produced in the chest when the bronchial tubes are narrowed. It is applied particularly to the long-drawn breathing of asthma, and to the whistling or purring noises that accompany breathing in cases of bronchitis when there is a small amount of fluid in the pleural cavity in the chest.

Whiplash Injury: An injury to the neck region caused by the neck being forcibly bent backwards and forwards.

White Blood Cells: A general term for different types of blood cells that form part of the body's immune system.

White Finger: Spasm of the blood vessels in the finger, resulting in a white appearance. The condition is common in Raynaud's disease but it can be caused by the long-term use of percussion implements such as mechanical road drills or hammers.

White Goods: Large household appliances such as referigerators, stoves, air conditions and washing machines.

White Hair: The greying or whitening of hair which takes place with age is due to a loss of its pigment, melanin, and the collection of air bubbles in the shaft of the hair.

WHO: World Health Organization.

Wholesale Acquisition Cost (WAC): The price paid by a wholesaler's supplier typically the manufacture of the drug. On financial statements, the total these amounts equals the wholesaler's cost of goods sold.

Wholesaler: A company that serves a bridge a drug manufacture and a covered entity. This means any entity to which the labler sells the covered outpatient drug, but that does not relable or repackage the covered outpatient drug.

Whooping-Cough: Whooping-cough, or pertussis, is a respiratory- tract infection caused by Bordetella pertussis and spread by droplets.

Wide Area Network: A computer communications network that connects geographically separated areas.

Wilcoxin Rank Sum Test: A statistical test to determine whether two population groups have the same value of a response variable. This test requires the presence of two comparison groups.

Wilcoxon Signed Rank Test: A non parametric equivalent of the paired t test. This test requires poulation group with, 2 measurement of numeric variable per patient for example: a study that investigate with in group change from baseline to endpoint of a given numeric variables, where the change variable does not follow a normal distribution.

Will: A legally executed document that directs how and to whom a person's property is to be disturbed after death.

Wilms' Tumour: Also called nephroblastoma. This is the commonest kidney tumour in infancy.

Wilson's Disease: Wilson's disease, or hepatolenticular degeneration, is a familial disease in which there is an increased accumulation of copper in the liver, brain, and other tissues including the kidneys. Its main manifestation is the development of tremor and rigidity, with difficulty in speech.

Windpipe: The popular name for the trachea, which extends from the larynx above to the point in the upper part of the chest where it divides into the two large bronchial tubes, one to each lung. It is about 10 cm (4 inches) in length

Winter Vomiting Disease: Winter vomiting disease, or epidemic nausea and vomiting, is a condition caused by sub- types of the genus Norwalk-like virus and is characterized by nausea, vomiting, diarrhoea and giddiness, which occurs during the winter.

Wisdom Tooth: A popular name for the last molar tooth on either side of each jaw. These teeth are the last to appear and should develop in early adult life

Withdrawal Bleeding: Loss of blood from the uterus via the vagina occurring when the women's level of oestrogen hormones, progesterone hormone or progestogen drugs falls quickly.

Withdrawal Effect: adverse event following withdrawal from aperson or animal of a drug to which they have been chronically exposed or on which they have become dependent.

Withdrawal: The act of reducing the degree of participation by a subject in a clinical trial. Subjects may withdraw permission for Sponsor use of data derived from study participation, privacy waivers, informed consent, or withdraw from active treatment component of a clinical trial but continue to be observed. Full withdrawal from participation in a study is called discontinuation.

Within-Subject Differences: In a crossover trial, variability in each subject is used to assess treatment differences.

Work Information Management System: Environmental subsystem (WIMS-ES) – an authorized management tool for program areas within the environmental arena. These programs areas are identified as "modules" of WIMS-ES. As the name indicates, WIMS-es is a subsystem the air force- wide civil engineering Wang-based WIMS.

Working Zone: Space measuring up to 2m over the level of floor or platform that contains a worker's permanent or temporary station.

World Wide Web: All the resources and users on the Internet that are using HTTP protocols.

Wrinkle: A natural furrow in the skin commonly associated with ageing. Wrinkles are most prominent on the face and other exposed body parts.

Writer's Cramp: A spasm which affects certain muscles when a person is writing, and which may not occur when the same muscles are employed in other acts.

Wry-Neck: A condition in which the head is twisted to one side. It may be caused by the contraction of a scar, such as that resulting from a burn or by paralysis of some of the muscles; as a result of injury at birth; or trauma to the area later in life.

X

X Chromosome: One of two sex chromosomes. Every normal female body cell has a pair of X chromosomes. Men have only one X chromosome and this is paired with a Y chromosome.

X Disease: Hyperkeratotic disease in cattle following exposure to chlorinated dibenzo-p-dioxins, naphthalenes and related compounds.

Xanthelasma Palpebrarum: These yellow smooth nodules of lipid-laden cells occur in and around the eyelids

Xanthoma: (Plural: xanthomata) A deposit of fatty material in the skin, subcutaneous fat and tendons. The presence of a xanthoma may be the first sign that a person has primary or secondary hyperlipidaemia- a raised con- centration of lipids in the blood. When fatty deposits occur in various parts of the body – skin, brain, cornea, internal organs and tendons – the condition is called xanthomatosis.

Xenobiotic: Strictly, any substance interacting with an organism that is not a natural component of that organism.

Xenograft: A transplant from one animal to another of a different species. It is also known as a.

Xenotransplantation: Transplantation of organs from one species to another – for example, from pigs to humans. It seems likely that stem cell research will provide a more realistic source of tissues for transplanta- tion to replace diseased organs in humans.

Xeroderma Pigmentosum: A rare disease in which DNA repair mechanisms fail, rendering the skin especially vulnerable to damage from ultraviolet light

Xeroderma: The term means dry skin. Normal skin may become dry when exposed to very low ambient humidity and is then vulnerable to irritation by soaps, detergents and other chemicals which cause 'chapping'.

Xerophyte: A plant that is adapted to, and needs, dry desert climate or is particularly hardy in periodic droughts.

Xerosis: Abnormal dryness, especially of the eye.

Xerostomia: Dryness of the mouth due to lack of saliva.

Xiphoid Process: Also known as the xiphisternum or xiphoid cartilage, this is the small oval-shaped projection forming the lowest of the three parts of the sternum or breastbone.

X-Rays: Also known as Röntgen rays, these were discovered in 1895 by Wilhelm Conrad Röntgen. Their use for diagnostic imaging (radiology) and for cancer therapy is now an integral part of medicine. Many other.

Xylose: A sugar containing five carbon

atoms involved in the metabolic conversion of carbohydrates inside cells. Xylose is used in diagnostic tests on the functioning of the intestine.

Y Chromosome: One of two sex chromosomes that is present in every male body cell where it is paired with an X chromosome. The sex or germ cells in women as well as men contain one X and one Y chromosome

Yawning: An involuntary opening of the mouth, which is accompanied by marked dilatation of the pharynx, a characteristic distortion of the face and usually stretching of the limbs. It is classically regarded as a sign of drowsiness or boredom, but it it may be the result of raised concentrations of carbon dioxide (CO_2) in the blood the physiological aim being to cut the amount of CO_2 and raise the level of oxygen in the blood.

Yaws: A non-venereal spirochaetal infection caused by Treponema pertenue; it was formerly widespread in most tropical and subtropical regions amongst the indigenous population, florid disease being more common in children than adults. Clinically, the primary stage is characterised by a granulomatous lesion, or papule (framboesioma or 'mother yaw') at the site of infection – usually the lower leg or foot; this enlarges, crusts, and heals spontaneously.

Yeast Infection: A vaginal infection that is usually caused by a fungus. Women who have feel itching burning when urinating, and pain, and some women a vaginal discharge. Yeast infections occur more frequently in women with diabetes.

Yeast: This consists of the cells and spores of unicellular fungi belonging to the family of Sac- charomycetaceae. The main species of yeast used in medicine is Saccharomyces cerevisiae, which is used in the fermentation industries, such as brewing. It is a rich source of the vitamin B complex.

Yellow Fever: An acute arbovirus infection caused by a flavivirus of the togavirus family, transmitted from animals to humans by various species of forest mosquito (jungle/sylvan yellow fever), and from human to human by Aëdes aegypti (urban yellow fever). Clinically, yellow fever is characterized by jaundice, fever, chills, head ache, gastrointestinal haemorrh-ages, and albuminuria.

Yersinia: A genus of bacteria which plague, yersinia pestis.

Yoga: ***A system of Hindu philosophy and physical discipline involving special breathing techniques and a series of prescribed physical poses. These are intended to relax the body and teach the individual mental and physical control.***

Yoghurt: Sour milk curdled with one of the lactic acid producing bacilli, such as Lactobacillus acidophilus or Lactob - acillus bulgaricus. It contains all the protein, fat, calcium, and vitamins of the original milk, and is therefore a -

...od, but there is no ... that it has any unique ...ncial properties of its own.

Yohimbine: This substance is derived from the bark of Pausinystalia yohimbie, a West African tree. Once widely used as an aphrodisiac an action for which there is no good evidence, it is now being used in the treatment of certain cases of postural hypotension and for the treatment of impotence.

Zalcitabine: A nucleoside Reverse Transcriptase Inhibitor used in the treatment of HIV, in combination with other antiretroviral drugs.

Zanamivir: An antiviral drug which ameliorates the symptoms of influenza.

Zidovudine: An antiviral drug used to treat AIDS and its related conditions, such as pneumocystis pneumonia. The drug slows down the growth of human immunodeficiency virus (HIV) but does not cure the disease.

Zinc: A metal, several salts of which are used in medicine for external application. It is essential for growth and development in animals and plants. The average human body contains a total of 1–2 grams, and most human diets contain 10–15 mg. In human beings, deficiency of zinc results in lack of growth, slow sexual development and anaemia. Deficiency is also associated with a skin disorder known as acrodermatitis enteropathica.

Zollinger-Ellison Syndrome: ***A rare disorder in which severe peptic ulcers recur in the stomach and duodenum. It is caused by a tumor in the pancreas that produces a hormone, gastrin, which stimulates the stomach and duodenum to produce excess acid: this causes ulceration.***

Zona and Zoster: Two names for the eruption popularly known as shingles

Zonulolysis: The process whereby the zonule is dissolved by an enzyme (chymotrypsin) as part of intracapsular cataract surgery. Once the zonule has been dissolved the cataract can be lifted out of the eye

Zoocide: Substance intended to kill animals.

Zoonoses: Animal diseases which can be transmitted to humans.

Zygoma: The name given to a bridge of bone formed by the union of a process from the temporal bone in the skull with one from the malar (cheek) bone. It lies in the region of the temple, gives attachment to the powerful masseter muscle which moves the lower jaw, and forms a protection to the side of the head.

Zygote: Cell such as fertilized egg resulting from the fusion of two gametes 2. Cell obtained as a result of complete or partial fusion of cells produced by meiosis.

Zymolysis: The mechanism of ferment-ation initiated by an enzyme during digestion.